AF780215

Geomechanics and Geotechnical Engineering Problems in the Design and Construction of Underground Buildings

Geomechanics and Geotechnical Engineering Problems in the Design and Construction of Underground Buildings

Guest Editors

Shaobo Chai
Yongqiang Zhou
Erdi Abi
Longlong Lv

MDPI

Basel • Beijing • Wuhan • Barcelona • Belgrade • Novi Sad • Cluj • Manchester

Guest Editors
Shaobo Chai
School of Civil Engineering
Chang'an University
Xi'an
China

Yongqiang Zhou
Institution of Rock and Soil
Mechanics
Chinese Academy of Sciences
Wuhan
China

Erdi Abi
National Inland Waterway
Regulation Engineering
Research Center
Chongqing Jiaotong
University
Chongqing
China

Longlong Lv
School of Civil and Hydraulic
Engineering
Ningxia University
Yinchuan
China

Editorial Office
MDPI AG
Grosspeteranlage 5
4052 Basel, Switzerland

This is a reprint of the Special Issue, published open access by the journal *Buildings* (ISSN 2075-5309), freely accessible at: https://www.mdpi.com/journal/buildings/special_issues/OO8DGP0W81.

For citation purposes, cite each article independently as indicated on the article page online and as indicated below:

Lastname, A.A.; Lastname, B.B. Article Title. *Journal Name* **Year**, *Volume Number*, Page Range.

ISBN 978-3-7258-4667-2 (Hbk)
ISBN 978-3-7258-4668-9 (PDF)
https://doi.org/10.3390/books978-3-7258-4668-9

© 2025 by the authors. Articles in this book are Open Access and distributed under the Creative Commons Attribution (CC BY) license. The book as a whole is distributed by MDPI under the terms and conditions of the Creative Commons Attribution-NonCommercial-NoDerivs (CC BY-NC-ND) license (https://creativecommons.org/licenses/by-nc-nd/4.0/).

Contents

About the Editors

Shaobo Chai

Shaobo Chai, PhD, is mainly engaged research work in rock joint mechanical characteristics and the propagation of explosion vibration waves in discontinuous rock mass. He has presided over more than 10 scientific research projects with the National Natural Science Foundation of China. He has published 40 papers as the first or corresponding author. His other research interests include the seismic resistance of underground structures, stability of high-voltage tower foundations, and improvement of mechanical properties of collapsible loess.

Yongqiang Zhou

Yongqiang Zhou, Associate Research Fellow at the Institute of Rock and Soil Mechanics, Chinese Academy of Sciences, specializes in static and dynamic constitutive modeling of rock materials, seismic dynamic response analysis in geotechnical engineering, and advanced numerical simulation methods. Dr. Zhou has undertaken multiple research projects funded by the National Natural Science Foundation of China, sub-projects of the National Key R&D Program, the General Program of Hubei Province, and Wuhan Municipal Innovation Projects. As a core member, he has contributed significantly to several major national-level research initiatives, including the National Basic Research Program (973 Program), NSFC Major Research Plans, and Joint Funding Projects. His academic achievements include nearly 50 SCI- and EI-indexed papers, 2 academic monographs as Chief Editor, and approximately 20 authorized invention patents and software copyrights. Dr. Zhou also serves as a peer reviewer for several leading journals in the field of geotechnical engineering.

Erdi Abi

Erdi Abi, Associate Professor of Chongqing Jiaotong University is mainly engaged in research on unloading rock mechanics, rock impact damage mechanics, disaster-causing mechanisms, and protective measures in water conservancy projects, etc. He is currently in charge of over 10 scientific research projects, including those funded by the National Natural Science Foundation of China, and has published more than 40 academic papers. He was awarded the Special Prize of the Science and Technology Innovation Award of the Yangtze River Navigation Administration and Second Prize for the Science and Technology Award of the China Institute of Navigation.

Longlong Lv

Longlong Lv, is a researcher at Ningxia University, College of Civil and Hydraulic Engineering. His main interests lie in the mechanics and engineering of loess and the structural properties, water sensitivity, and strength characteristics of rock and soil materials. He received his PhD from Xi'an Jiaotong University. He has been responsible for two provincial and ministerial-level longitudinal projects and five enterprise-organized transverse projects.

Preface

Due to the nature of underground engineering being located or partially located below the surface, the geotechnical problems of underground engineering remain throughout the entire process, including site selection, planning, investigation, design, construction, use, maintenance, transformation, reinforcement, demolition, and restoration. For underground engineering, geomechanics and geotechnical problems affect the whole process, both each particular element and extensively throughout.

For instance, for high-rise buildings, there are no mature theoretical methods for the design of deep foundations and settlement issues. For foundation pits formed by large-scale excavation, the stability of the foundation pit is closely related to the properties of the soil and the surrounding loads. Foundation pit support is a very complex project. Furthermore, with the development of transportation engineering, tunnel structures with large cross-sections, long distances, and high burial depths have gradually emerged, which are affected by higher ground stresses and more complex surrounding environments. Whether the analysis of the deformation of the tunnel structure and the design of the tunnel support can be performed well is of vital importance to tunnel projects. In addition, in recent years, many buildings have had to be constructed in special rock and soil mass areas, such as those of soil–rock mixtures, saturated clay, collapsible soil, and soil with disintegration characteristics. For cases of these special soils being used as building foundations, there are no relatively complete theoretical methods and reports of their mechanical properties and deformation characteristics, which poses obstacles to the implementation of engineering design and construction.

In conclusion, there are still a great many geotechnical engineering problems in the area of underground projects such as tunnel engineering, foundation pit engineering, and subway construction. Researchers and practitioners have yet to overcome scientific or technical issues such as special soil mechanical properties, the interaction laws between structures and rock and soil, and the stability of underground structures in complex environments. It is necessary to continuously conduct research on these problems so as to accumulate more experience.

This Reprint aims to introduce and disseminate the latest research progress in geotechnical engineering issues related to underground structures, including the seismic response of the induced joints in subway stations; the shock absorption and energy dissipation effects of improved foundations; the stability, support effects, and seismic response laws of tunnels in different rock and soil masses; the stability and support methods of deep foundation pits in different environments; and the mechanical characteristics of different types of rock and soil foundations, etc. The research results presented in this Reprint provide data for the design and construction of underground building structures.

Shaobo Chai, Yongqiang Zhou, Erdi Abi, and Longlong Lv
Guest Editors

buildings

MDPI

Article

3-D Numerical Simulation of Seismic Response of the Induced Joint of a Subway Station

Dengzhou Quan, Shaobo Chai *, Yuling Wang, Zhishuang Fan and Yonghong Bu

School of Civil Engineering, Chang'an University, Xi'an 710064, China; qdz0809@chd.edu.cn (D.Q.); 2019228011@chd.edu.cn (Y.W.); 2022128048@chd.edu.cn (Z.F.); byh956@chd.edu.cn (Y.B.)
* Correspondence: shbchai@chd.edu.cn

Abstract: In recent times, induced joints have been set along the length of subway stations in order to avoid disordered cracking of the main structures occurring due to temperature stress, concrete shrinkage, creep, or uneven foundation settlement. At present, the use of induced joints in subway station structures is mainly based on engineering experience. The seismic response of induced joints has not yet been well explained, much less mastered. In this study, a 3-D numerical model of a subway station incorporating certain sorts of induced joints is established systematically. Then, the seismic response of those induced joints applied in different positions and various forms has been studied under different seismic waves by varying the spectral characteristics and peak acceleration values of the waves. The results show that the horizontal relative sliding displacement of the structures on both sides of an induced joint increases gradually from bottom to top along the structure of the subway station. While the vertical sliding displacements that occur along the section width are larger at both ends of the induced joints than in the middle. What is more, with an increase in seismic intensity, the horizontal relative sliding displacement becomes larger, while the vertical displacement becomes even smaller. In addition, the relative sliding displacement can be reduced by increasing the residual longitudinal reinforcement ratio of the induced joint. Furthermore, it is discovered that the setting of key grooves at the bottom plate of the induced joint section has a certain effect on controlling the horizontal relative sliding displacement, as well as a significant effect on preventing the vertical relative dislocation of the structures on both sides of the induced joint.

Keywords: induced joint; subway station; numerical simulation; seismic response; underground structure dislocation

1. Introduction

At present, the longitudinal length of subway stations constructed in China is relatively large, usually reaching 120–300 m. In order to avoid disordered cracking of the subway station structure caused by temperature stress, concrete shrinkage, creep, or uneven settlement of the foundation, induced joints are usually set along the longitudinal direction of the subway station structure, as shown in Figure 1. The amount of longitudinal reinforcement used at the induced joint of the subway station structure is 30–40% of the normal section, and concrete is poured in stages, forming a weak link in crack concentration distribution. Under the action of an earthquake, the induced joints of the subway station structure in loess stratum may be damaged initially, which then leads to the overall damage of the structure. In addition, the presence of the induced joint can easily cause excessive horizontal relative sliding, vertical relative movement, and bending of the track structure, which endangers the safety of running trains. When considering the spatial effects of seismic damage in transfer subway stations with complex shapes, the seismic damage to the induced joints and structures can be more severe, resulting in significant economic, life, and property losses, which are difficult to repair after an earthquake.

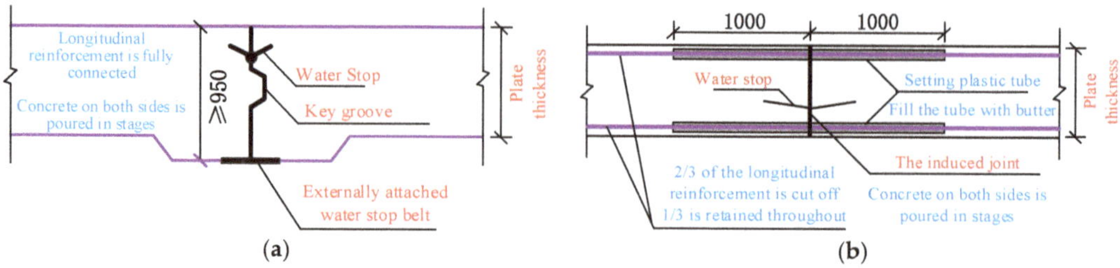

Figure 1. Schematic diagram of induced joint structure: (**a**) Induced joint structure of the floor; (**b**) Induced seam structure of the roof.

In recent years, a lot of research has been conducted on the seismic performance of underground structures and the response of different foundations and structures under earthquake action through theoretical analysis [1–3], experiment [4–6], and numerical simulation [7–9]. Some studies took the subway structure as the main research object and obtained results that provided a rich theoretical basis and practical experience for the improvement of earthquake resistance in subway construction. In terms of theoretical studies, Jeng and Lu [10] proposed a theoretical method to analyze the vibration of an underground circular tunnel with jointed linings subjected to seismic waves. This method treats the surrounding medium as a linear elastic medium, but the results obtained by this method can better reflect the vibration distribution law and provide a reference basis for related problem experiments and numerical simulation research. Based on the theory of dynamic interaction between soil and structure, Lin et al. [11] proposed a new model for the seismic response of underground structures for scattering and diffraction analysis and verified the accuracy and efficiency of this method through numerical examples. Carlos et al. [12] proposed an improved simple calculation method for seismic internal forces of shallow-buried rectangular underground structures. The results indicated that this method could predict structural internal forces more accurately than the existing simplified methods. Mehdi et al. [13] carried out a theoretical study on the seismic response of an elastic homogeneous ground surface in the presence of unlined horseshoe-shaped underground cavities subjected to obliquely propagating incident SH-waves by using the time-domain half-plane boundary element method. Zhao et al. [14] developed the response spectrum method (RSM) for the seismic analysis of underground structures, including seismic soil-structure interaction (SSI), and established an SSI analysis model composed of the underground structures and their adjacent soil. The research shows that this method has high accuracy and can meet engineering requirements. However, there are few theoretical studies on the seismic response of subway station structures, and the abovementioned analysis methods cannot accurately predict the seismic response and failure characteristics of subway underground structures, let alone structures with induced joints.

In terms of experimentation, the main research method is shaking table tests. For instance, Iwatate et al. [15] used a shaking table test to study the destruction process of subway underground structures in the Kobe earthquake. The results showed that the collapse of the Dakai Station structure was caused by the inability of the horizontal lateral stiffness of the central column to resist seismic shear loads. By taking the buried open-cut subway tunnel in San Francisco Bay as the research object, Chou et al. [16] carried out a centrifugal shaking table test to study the influence of actual site conditions of loose sand and gravel backfill in the offshore bay on structural seismic response. Cilingir et al. [17] used aluminum alloy and loose dry sand to make model structures and foundations, respectively, conducted centrifugal shaking table tests of circular and square tunnels, and revealed the dynamic internal force response of tunnel structures. Baziar et al. [18] studied rectangular underground structures in sand through centrifugal shaking table tests and analyzed the influence of underground structures on the ground acceleration response. Ulgen et al. [19] conducted a series of dynamic centrifuge tests on a box-shaped flexible underground structure and obtained the responses of soil and buried structure models under harmonic motions with different accelerations and frequencies. Dashti et al. [20] evaluated the seismic effects of high-rise buildings on adjacent shallow-buried underground structures through a series of centrifugal shaking table tests. Masoud et al. [21] conducted a series of shaking table tests to study the influence of a circular subway tunnel on ground motion amplification patterns. Quan et al. [22] carried out a large-scale shaking table test of the subway station in loess regions, revealed the seismic response law of the subway station under the action of earthquakes, and listed the weak parts of subway station structures. Then, through numerical simulation analysis of the seismic response of subway stations in loess regions, the horizontal displacement, peak acceleration, contact soil pressure, and structural damage distribution of the system were studied. In addition, shaking table test studies of subway underground structures in soft clay, slightly inclined sites, and liquefiable sites have fully revealed the interaction mode of soil and underground structures under seismic load [23–27]. Shaking table tests and centrifugal shaking table tests show great advantages in studying seismic response characteristics of underground structures, and can intuitively obtain characteristics such as failure modes, weak nodes, and displacement changes. However, these tests cost too much, and a single experiment cannot comprehensively conduct a full variable parameter analysis. In addition, the existing experimental studies are all based on the overall model, and the experimental studies on the seismic response of induced joints in subway underground structures are rarely reported.

Due to the disadvantages of the high cost of shaking table tests and the inability to carry out multiple working condition analyses at the same time, numerical simulation has become an effective and economical method to study the dynamic response of the structure. The main numerical methods applied to underground structural seismic response are the finite element method and the improved finite element method. The main difficulties in simulation are the input of seismic motion, boundary processing, and the accurate establishment of the model. Through nonlinear finite element simulation, Parra et al. [28] analyzed the seismic response of the Dakai subway station under the action of an earthquake by numerical simulation and compared it with actual earthquake damage. In view of the important influence of soil–structure interaction on the seismic resistance of structural systems, some scholars have tried to use numerical methods to study the dynamic interaction of subway station–soil–adjacent high-rise building systems and analyze the influence of subway stations on the seismic response characteristics of adjacent structures [29,30]. As the site characteristics and soil stratum also greatly affect the seismic response of underground structures, relevant studies pay more and more attention to the influence of soil properties. For example, Conti et al. [31] and Abate et al. [32] carried out a numerical analysis based on a centrifugal shaking table test of a tunnel structure in a sandy soil foundation. The study showed that the acceleration response of numerical simulation was consistent with that of a model test, but there were certain differences in the dynamic internal forces of the structure. Keykhosropour et al. [33] conducted a numerical simulation study on the seismic response characteristics of deep-buried underground flexible structures in sandy soil and analyzed the effects of internal friction Angle, cohesion, and structural stiffness on the development of seismic earth pressure and wall deformation. Yoo et al. [34] developed a dynamic numerical analysis model based on PLAXIS2D and conducted a series of dynamic numerical analyses for deep underground structures under various earthquake conditions. Sun et al. [35] carried out numerical simulations to study the critical difference in the seismic performances of three- and four-sided box culverts. In this study, a variety of burial depths, flexibility ratios, and foundation widths of the culverts were considered. The findings of this work shed light on the critical role of the bottom slab in the seismic responses of box culverts. Wang et al. [36] established a numerical model of the composite structure of a subway station combined with a flyover under earthquake action through the finite element method and analyzed the dynamic response of the subway station and bridge pier and pile under earthquake action. Alejandro et al. [37] proposed a numerical method based on the indirect boundary element method (IBEM) to calculate the seismic amplification in the weak parts of floating and underground tunnels. The above extensive research results fully demonstrate that the finite element and improved finite element methods can accurately and effectively simulate the seismic response of underground structures in complex geological environments. Therefore, it is feasible to study the seismic response of induced joints in subway stations through finite element numerical simulation.

Induced joints have been widely used in subway stations in many large and medium-sized cities in China, but the seismic design method of the inducted joints has not been given in the current specifications. There is still a lack of research on the seismic resistance of subway station structures with induced joints, especially on the horizontal relative sliding effect of the structure at the induced joints, and the force transfer effect of remaining longitudinal reinforcement has not been reported. In view of this, in the present study, based on the overall dynamic time-history analysis method, a 3-D non-linear finite element model of the seismic response of a typical subway station in loess soil with induced joints is established, and its reliability is verified. The seismic response characteristics of induced joints in subway stations under the action of seismic waves with different spectral contents and different peak accelerations are studied. In addition, the influence of different factors on the seismic response of induced joints of subway stations is analyzed by variable parameters. This research can reveal the basic law of seismic response of induced joints in subway stations and the influencing factors, which will be of great significance for controlling the deformation of induced joints in subway stations under earthquake action and can provide an important reference for the seismic design of induced joints in subway stations.

2. Shaking Table Test and Numerical Simulation Method

In the preliminary research [22], large-scale shaking table tests were carried out to obtain the seismic response law of subway stations in loess regions under earthquake action. In this paper, the numerical simulation method is used to carry out a follow-up study in order to obtain the seismic response of induced joints. The prerequisite of an accurate numerical simulation is to verify the accuracy of numerical methods. Therefore, in this section, the accuracy of the numerical method is verified by comparing the results of model tests under the same working conditions as the shaking table tests.

2.1. Information Related to Previous Shaking Table Test

In the shaking table test, the under-artificial quality model was adopted for the similarity model based on the structure of the subway station. Length, elastic modulus, and acceleration were selected as the basic physical quantities for the similarity design. The size of the table was 3.36 m × 4.86 m (vibration direction). The size of the model box was 3.7 m (vibration direction) × 2.2 m (longitudinal direction) × 1.7 m (vertical direction). The length similarity ratio was $\lambda_l = 1/30$, and the similarity ratio of the modulus of elasticity was $\lambda_E = 1/5$.

In the preliminary shaking table test, particulate concrete with similar mechanical properties to ordinary concrete was used to simulate ordinary concrete. The compressive strength of the particulate concrete cube was 8.11 MPa, and the elastic modulus was 6602 MPa, which met the requirements of the similarity relationship between the model materials and prototype structural materials. The steel reinforcement of the prototype concrete structure was simulated by galvanized steel wire. The model structure is shown in Figure 2. In the shaking table test, the model foundation soil was taken from the foundation pit of Feitian Road Station of Xi'an Metro Line 4, located 6–8 m below the ground surface.

(a)

(b)

Figure 2. Model structure and model soil box: (**a**) Model structure; (**b**) laminated shear model soil box.

The model soil box used for the shaking table test was a laminated shear model soil box, as shown in Figure 2b. When making the model foundation, layers of soil samples were loaded into the model soil box, and the water content and density were controlled according to the natural water content and density of the soil in the prototype site. The surface of the lower soil body was roughened before each layer of loess was added.

2.2. Numerical Simulation of the Shaking Table Test

Based on the shaking table test in the previous study, a three-dimensional numerical model of the seismic response of a subway station was established using ABAQUS (version number 6.14) finite element software. The model structure size was determined according to the geometric similarity ratio and was 3.5 m (vibration direction) × 2.0 m (longitudinal direction) × 1.4 m (vertical direction). In order to ensure the calculation accuracy as well as reduce the calculation time cost, the numerical model was calculated along the longitudinal half structure under symmetric boundary conditions. The numerical calculation model is shown in Figure 3. The number and location of monitoring points in the model remain the same as in the shaking table test. The equivalent linear model was used to simulate the non-linear dynamic characteristics of loess. The stratified and transitive parameters of the loess foundation are shown in Table 1, where the values of each parameter are measured in the laboratory, consistent with the previous study [22]. In the preliminary shaking table test, the structure of a subway station was made of particulate concrete model material after repeated debugging. Through the mechanical properties test, it was found that the CDP model is reasonable to use in simulating the dynamic damage evolution process of structural particulate concrete, and its model parameters are shown in Table 2. The ideal elastoplastic model was used to simulate the mechanical deformation characteristics of galvanized steel wire in the structure. The full integration unit C3D8 was used for the model structure, the reduced integration unit C3D8R was used for the model foundation, and the 3-D truss unit T3D2 was used for the model reinforcement. The bond between steel bars and concrete was simulated by embedding steel bars into the concrete using the Embedded command. The soil grid was divided according to the principle of gradually sparse from near to far, and the induced cracks and the adjacent grids were properly encrypted. Songpan, Taft, and Xi'an artificial waves were selected as input seismic waves.

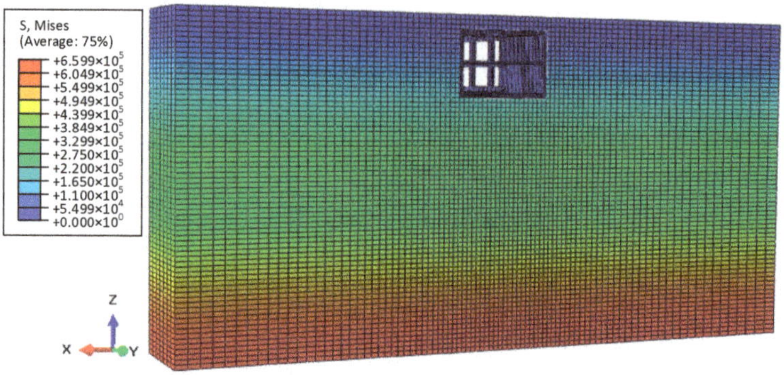

Figure 3. Numerical model of shaking table test for a subway station in Loess soil.

Table 1. Soil stratification and basic physical parameters.

Layer	Thickness h (m)	Density ρ (g/cm^3)	Poisson's Ratio	G_{max} (MPa)
1	0.15	1.67	0.31	6.8
2~6	0.25	1.67	0.31	6.8

Table 2. Parameters of concrete damage plastic model of the finite element model.

Model Parameter	Values	Model Parameter	Values
Density (kg/m^3)	2500	Dilatancy angle Ψ (°)	30
Elastic modulus E (MPa)	0.66×10^4	Viscosity coefficient μ	0.0005
Poisson's ratio υ	0.2	Invariant stress ratio K_c	0.6667
Ultimate compressive stress	5.39	Damping ratio ξ	0.1

2.3. Acceleration Response Comparison

Under the action of three kinds of seismic waves, the acceleration time history curves and corresponding Fourier spectra of two measuring points in the shaking table test and numerical simulation were compared as shown in Figure 4. The two measuring points were A12 on the model foundation and A21 on the model structure, respectively.

It can be seen from the figure that the acceleration response time history curve waveform, peak acceleration, Fourier transform spectra composition characteristics, basic frequency, and corresponding Fourier amplitude at each measuring point in the shaking table test and numerical simulation are largely the same. A slight difference exists between the Fourier transform spectra of the acceleration response at each measuring point in the shaking table test and numerical simulation in the range of 5–20 Hz, but the difference gradually disappears with the increase in burial depth of the measuring point. When Songpan waves, Taft waves, and Xi'an artificial waves with a peak acceleration of 0.2 g are respectively input, the horizontal relative displacement of the foundation at different depths is calculated through the shaking table test and numerical simulation, as shown in Figure 5. As can be seen, the horizontal relative displacement distributions of the model foundation in the shaking table test and numerical simulation are consistent, and both increase with the decrease in soil depth. Under the three seismic wave conditions, except for the surface measurement point, the calculated results of other measurement points are very close, and the maximum error of the two methods is about 16%. Hence, the numerical analysis method and the three-dimensional numerical model established in this paper can well simulate the seismic response characteristics of the whole system during the dynamic interaction between a loess soil site and a subway station, and have good reliability. Thus, it can be inferred that the numerical method is also accurate and feasible for use in carrying out a study of the three-dimensional seismic response of the induced joints of subway stations under loess site conditions.

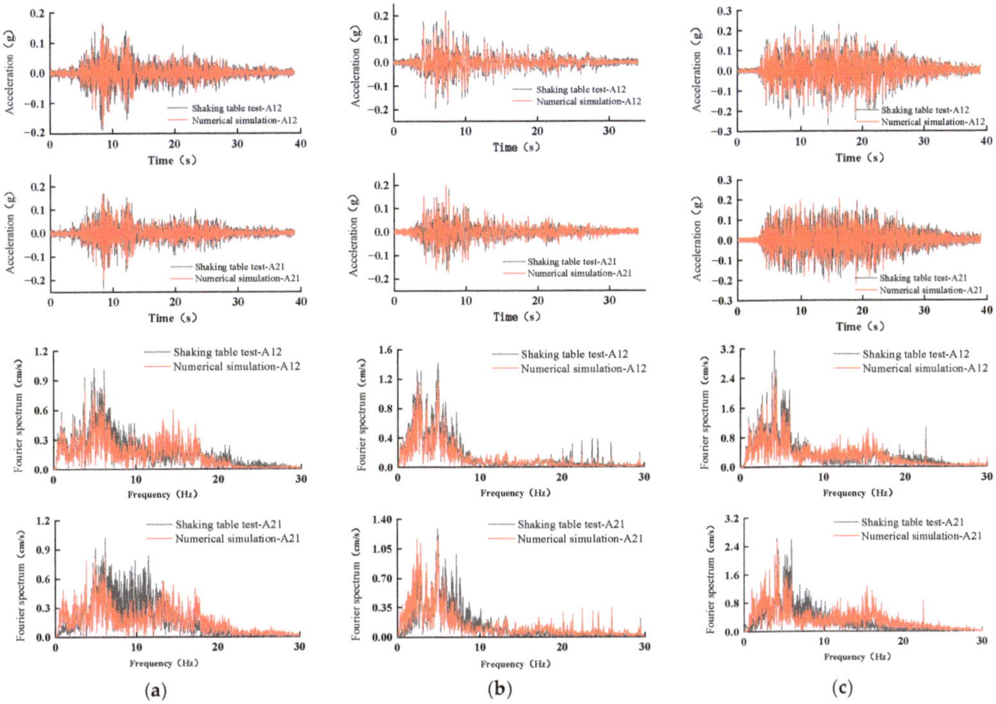

Figure 4. Comparison diagram of model foundation acceleration and Fourier spectrum: (**a**) Songpan wave; (**b**) Taft wave; (**c**) Xi'an artificial wave.

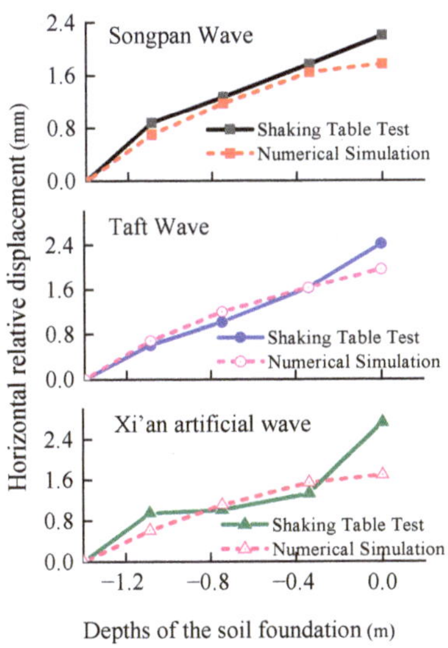

Figure 5. Comparison of horizontal relative displacement of model foundation.

3. Numerical Model of the Induced Joint of the Subway Station Structure

3.1. Model Design and Boundary Setting

This paper takes Feitian Road station of Xi'an Metro Line 4 as the research object. The station is a two-story double-span box section structure, as shown in Figure 6. The model structure section is designed to match the size of the actual structure, with a longitudinal length of 21 m between two adjacent induced joints of the station. The induced joint of the model structure is located at one-third of the distance between columns, where two-thirds of the longitudinal reinforcement is removed and one-third is retained throughout the entire length of the subway station. Key grooves are set at the induced joint of the bottom plate to enhance the engagement of the structures on both sides. According to the Code for Seismic Design of Urban Rail Transit Structures [38], during dynamic time history analysis, the distance between the artificial boundary on the side of the model foundation and the underground structure during dynamic time history analysis shall not be less than three times the horizontal effective width of the underground structure. In addition, the artificial boundary on the bottom should be taken to the design seismic action datum plane and the distance from the structure shall not be less than three times the vertical effective height of the underground structure. Accordingly, considering the requirements of the specification and the calculation time cost, the foundation width of this model is set as 134.4 m and the depth is 70 m, while the longitudinal length is consistent with the model structure.

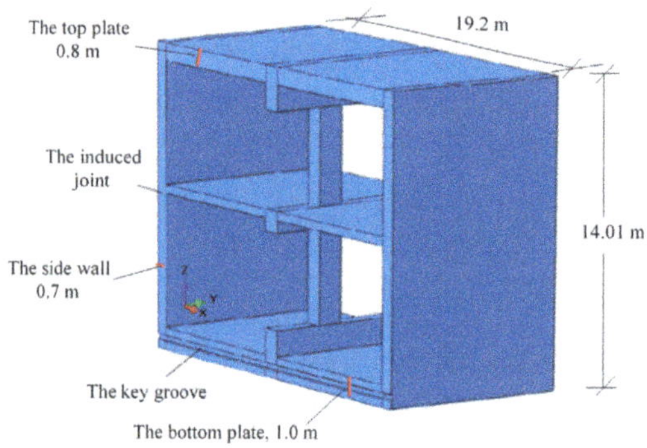

Figure 6. Schematic diagram of the numerical model of the subway station structure.

Du and Zhao. [39] established the stress type viscoelastic artificial boundary based on the linear elastic constitution of an infinite medium, which has high accuracy for finite element analysis. Therefore, in this paper, the stress type viscoelastic artificial boundary (viscoelastic artificial boundary for short) is selected in the numerical simulation. The spring stiffness parameters K_i and damper parameters C_i (i = 1, 2, 3) on viscoelastic artificial boundary nodes are:

$$K_1 = K_2 = \frac{1}{1+a} \cdot \frac{G}{R}\sum_{i=1}^{I} A_i \quad (1)$$

$$K_3 = \frac{1}{1+a} \cdot \frac{\lambda + 2G}{R}\sum_{i=1}^{I} A_i \quad (2)$$

$$C_1 = C_2 = b\rho C_s \sum_{i=1}^{I} A_i, C_3 = b\rho C_p \sum_{i=1}^{I} A_i \quad (3)$$

where R is the distance from the geometric center of the near-field area to the soil boundary where the artificial boundary point is located; G is the shear modulus of the medium; λ is the medium lame constant; ρ is the medium density; $C_p = \sqrt{\frac{\lambda+2G}{\rho}}$ and $C_s = \sqrt{\frac{G}{P}}$ are P wave velocity and S wave velocity, respectively. The dimensionless parameter a represents the ratio of plane waves to scattered waves, and dimensionless parameter b represents the relationship between physical wave velocity and apparent wave velocity. The values of parameters a and b can be obtained through numerical experiments, and will be taken as $a = 0.8$, $b = 1.1$ in the present work based on experience [40]. $\sum A_i$ is the area represented by the nodes on the artificial boundary. The equivalent seismic load is input through the bottom boundary node of the model.

3.2. Material Parameters and Seismic Wave Setting

In this paper, the nonlinear dynamic characteristics of loess are simulated using the equivalent linear model proposed by Du and Zhao [40]. The equivalent linear model is applicable to a wide range and is more accurate in the seismic response analysis of horizontally layered sites. The loess site composition and mechanical parameters of Feitian Road station are shown in Table 3. The CDP model is used to simulate the dynamic damage evolution process of the concrete of the station structure. The model parameters are shown in Table 4. Considering the differences in the structural quality on both sides of the induced joint in the subway station, the density ratio of the structures on both sides of the induced joint is set as 1:3. The ideal elastic–plastic model is used to simulate the stress and deformation characteristics of the reinforcement in the station structure. The elastic modulus of the reinforcement is taken as 2×10^{11} Pa, Poisson's ratio is taken as 0.3, and the yield limit is taken as 2.1×10^8 Pa. In order to ensure the calculation accuracy and reduce the calculation time cost, the model structure is set as the full integral element C3D8, the model foundation is set as the reduced integral element C3D8R, and the model reinforcement is set as the three-dimensional truss element T3D2.

Table 3. Soil layer and mechanical parameters of the loess region.

Layer	Thickness (m)	Density (g/cm^3)	Poisson's Ratio	Shear Wave Velocity (m/s)	G_{max} (MPa)	γ_Y (10^3)	λ_{max}	β
New loess	0~6	1960	0.26	205	43.6	2.060	0.100	0.2720
New loess	6~13	2010	0.26	241	48.7	3.772	0.058	0.2420
ancient soil	13~17	1540	0.26	271	82.3	2.258	0.137	0.3963
ancient loess	17~24	1670	0.29	298	97.5	2.364	0.156	0.5951
ancient soil	24~36	1760	0.29	317	123.5	2.872	0.109	0.4829
ancient loess	36~44	2060	0.30	339	63.9	3.764	0.130	0.2428
ancient soil	44~56	2000	0.29	383	95.0	1.531	0.175	0.4751
ancient loess	56~64	1970	0.31	434	95.0	2.309	0.135	0.4318
ancient loess	64~70	1980	0.31	466	92.6	2.000	0.150	0.5401

Table 4. Parameters of concrete damage plastic model.

Model Parameter	Values	Model Parameter	Values
Density (kg/m^3)	2500	Dilatancy angle Ψ (°)	36
Elastic modulus E (MPa)	3.25×10^4	Viscosity coefficient μ	0.0005
Poisson's ratio υ	0.2	Invariant stress ratio K_c	0.6667
Ultimate compressive stress	26.8	Damping ratio ξ	0.1

In order to study the seismic response of subway stations with induced joints on a loess region under the action of seismic waves with different spectral characteristics, Songpan waves, Taft waves, and Xi'an artificial waves are selected as input seismic waves in this paper. The peak accelerations of the original seismic waves are adjusted to 0.1 g, 0.2 g, 0.4 g, and 0.6 g, respectively, and then input into the numerical model to study the seismic response law of the induced joints of the subway station under the action of different peak acceleration seismic waves. The design of the numerical simulation conditions in this paper is shown in Table 5.

Table 5. The working condition of the input seismic waves.

Working Condition	Seismic Wave	Peak Accelerations	Working Condition	Seismic Wave	Peak Accelerations
B1	Songpan wave	0.2 g	B4	Taft Wave	0.2 g
B2	Xi'an artificial wave	0.2 g	B5	Taft Wave	0.4 g
B3	Taft Wave	0.1 g	B6	Taft Wave	0.6 g

3.3. *Location of Observation Section and Test Point*

In the model structure, the X and Y axes are the width direction and longitudinal direction of the model, respectively, while the Z axis is the vertical direction. The origin of the coordinate system is located in the section where the induced joint is located, as shown in Figure 7a. Taking the section of the induced joint as the boundary, the structure located in the positive direction of the Y-axis is called the +Y side structure, and the negative direction of the Y-axis is called the −Y side structure. The section of the induced joint of the +Y side structure is set as observation section No. 1 and the −Y side structure is set as observation section No. 2. The purpose of setting these two observation sections is to obtain the vertical dislocation values on both sides of the induced joint. The section of the middle column of the −Y side structure is set as observation section No. 3. Monitoring points PE1–PE4 for plastic strain are set in the remaining longitudinal reinforcements at the top plate, middle plate, bottom plate, and side wall of the induced joint section of the structure, respectively, as shown in Figure 7b. Horizontal displacement measuring points are arranged at different heights in the side walls of observation sections No. 1, 2, and 3, as shown in Figure 7c. Vertical displacement measuring points are arranged at different widths in the bottom plates of sections No. 1 and 2, as shown in Figure 7d.

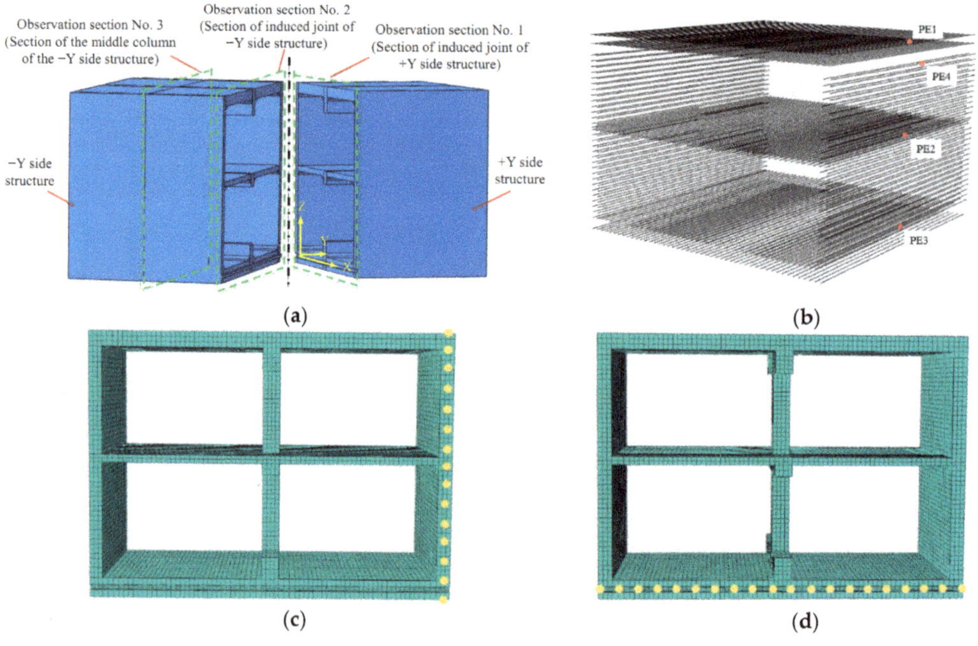

Figure 7. Seismic response measurement points of model structure: (**a**) Model structure; (**b**) laminated shear model soil box; (**c**) Position of horizontal displacement measuring point; (**d**) position of vertical displacement measuring point.

3.4. Dynamic Damage of the Subway Station Structure

Under different seismic wave types with the same peak acceleration, namely B1, B2, and B4 conditions, the compression damage cloud diagrams of the final state of the subway station structure are shown in Figure 8a,b,d. It can be seen from the figures that under the action of seismic waves with different frequency spectrum characteristics, the distribution positions of structural compression damage are largely the same, and obvious compression damage occurs at the connection of the top plate, middle plate, bottom plate, and side wall as well as at the end of the column. However, there are some differences in the degree of structural compression damage in the three working conditions. Under the action of a Xi'an artificial wave, the degree of structural compression damage is the largest, followed by a Songpan wave, with a Taft wave being the smallest.

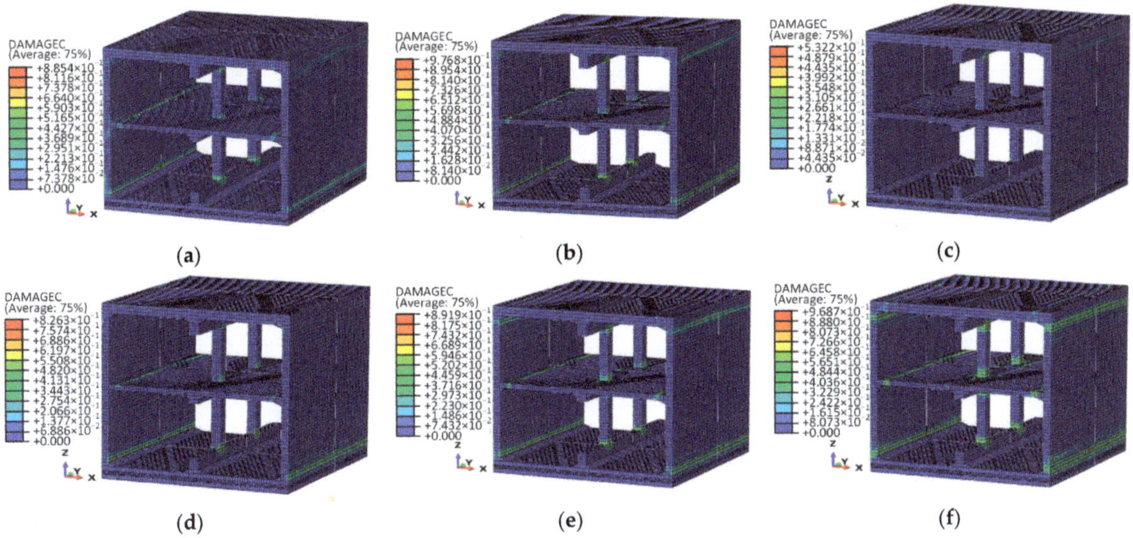

Figure 8. Pressure damage cloud diagram of the structure under different working conditions: (**a**) Working condition B1; (**b**) Working condition B2; (**c**) Working condition B3; (**d**) Working conditio B4; (**e**) Working condition B5; (**f**) Working condition B6.

Under the conditions of B3, B4, B5, and B6, the compressive damage of the subway station structure in the final state is shown in Figure 8c–f. As can be seen from the figure, the distribution locations of compression damage of subway station structures under Taft waves with different peak accelerations are largely the same, and obvious compression damage areas appear at the end of the middle column and the joints of the roof, middle plate, bottom plate, and side wall. When the peak acceleration is small, compression failure appears at the top of the middle column appears first. With an increase in peak acceleration, compression failure also appears at the bottom of the middle column, and the failure at the top is more serious. When the peak acceleration is large, the failure area at the top and bottom of the middle column expands rapidly and the column gradually loses its bearing capacity. The redistribution of the internal forces of the structure leads to significant failure at the top and bottom of the side wall, which bears a large bending moment. The above analysis shows that the failure mode of the model structure is manifested as the redistribution of internal forces caused by the failure of the middle column and the collapse of the bending moment at the connection between the side wall and the plate member, which leads to the failure of the structure.

Cloud diagrams of structural compression damage at different observation sections under the B2 working condition are shown in Figure 9a–c. From the figure, it can be seen that the structures at different observation sections all suffer compression damage at the connection between the side wall and plate components. However, compared with the central column section, the damage area at the connection between the side wall and the plate component of the induced joint section is larger and the damage is more serious. In addition, the middle part of the roof and the side wall also have local damage, indicating that the induced joint section is more prone to local damage. Similar characteristics can be seen under the B1 and B4 working conditions.

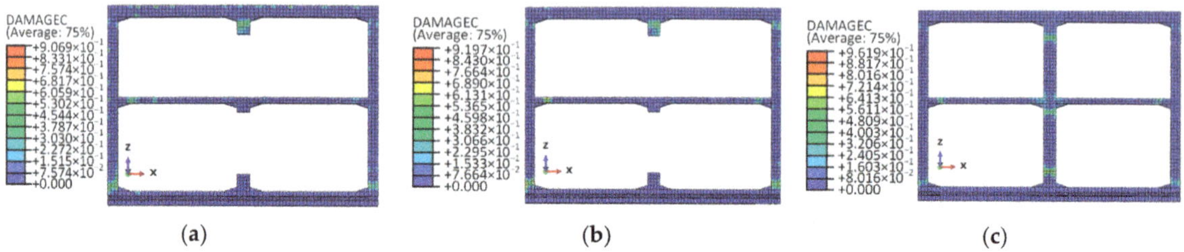

Figure 9. Pressure damage cloud diagram of different observation sections under the B2 working condition: (**a**) Observation section No. 1; (**b**) Observation section No. 2; (**c**) Observation section No. 3.

4. Seismic Response of the Induced Joint in the Subway Station Structure

4.1. Influence of Seismic Waves with Different Spectral Characteristics

4.1.1. Strain of Remaining Longitudinal Reinforcement at Induced Joint

The plastic strain time history curve of the remaining longitudinal reinforcement at measuring points PE1, PE2, PE3, and PE4 at the induced joint under working conditions, B1, B2, and B4 is shown in Figure 10.

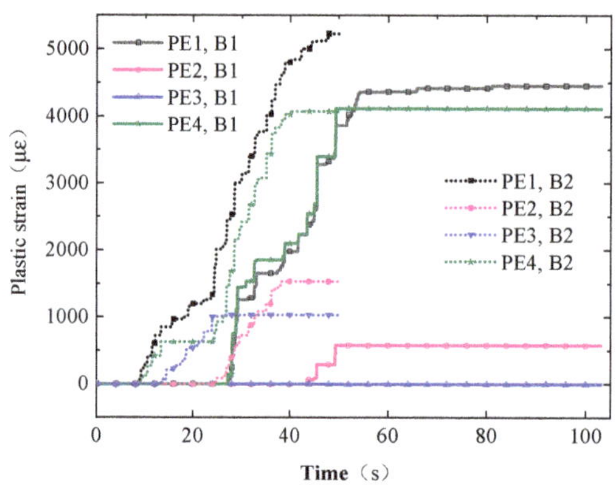

Figure 10. Plastic strain time-history curve of residual reinforcement in the induced joint under different working conditions.

It can be seen that, under the action of seismic waves with different spectral characteristics, the plastic deformation accumulation process of the remaining longitudinal reinforcement at each measuring point of the induced joint section is different. Under the action of Songpan waves, the remaining longitudinal reinforcements at the top plate and side wall of the induced joint section first undergo plastic deformation, followed by the remaining longitudinal reinforcements at the middle plate, while the remaining longitudinal reinforcements at the bottom plate never enter the plastic stage. The plastic deformation development of the remaining longitudinal reinforcement at each measuring point of the induced joint section is mainly concentrated in the duration of strong earthquake action, and the time history curve is steep. Under the action of Xi'an artificial waves, plastic deformation occurs sequentially in the remaining longitudinal reinforcements at the top plate, side wall, bottom plate, and middle plate of the induced joint section. The plastic deformation develops throughout the whole process of seismic action, and the time-history curve shows a continuously increasing feature, with a significant cumulative effect of plastic strain. However, under the action of Taft waves, no plastic deformation occurs in the remaining longitudinal reinforcements at all members of the induced joint section.

The maximum plastic strain of the remaining longitudinal reinforcement at each measuring point of the induced joint under the three working conditions is shown in Table 6. It can be seen from the table that the maximum plastic strain of the remaining longitudinal reinforcement at the top plate of the induced joint section is the highest, followed by the side wall, and is relatively small at the middle plate and the bottom plate. Under the action of seismic waves with different spectral characteristics, except for individual measuring points, the maximum plastic strain of the remaining longitudinal reinforcement in the induced joint section shows obvious differences—that is, its value is the highest under the action of Xi'an artificial waves, followed by the action of Songpan waves, and is a minimum under the action of Taft waves.

Table 6. The maximum plastic strain of longitudinal reinforcement remaining in the induced joint under different working conditions.

Working Conditions	Maximum Plastic Strain (με)			
	PE1	PE2	PE3	PE4
B1	4450	578	0	4116
B2	5226	1536	1030	4068
B4	0	0	0	0

4.1.2. Horizontal Relative Sliding of Structure on Both Sides of Induced Joint

Horizontal displacement measuring points are arranged at different heights on the side walls of observation sections No. 1, No. 2, and No. 3, and the horizontal relative displacements of different measuring points of each observation section are calculated based on the bottom measuring point. The horizontal relative displacement of measuring points at each observation section under B1, B2, and B4 conditions is shown in Figure 11. It can be seen from the figure that the horizontal relative displacement of the side wall of each observation section gradually increases from bottom to top. At the same height, the horizontal relative displacement of the side wall of the induced joint section on the +Y side is larger than that on the −Y side, and that of the side wall of the column section on the −Y side is the smallest. This indicates that the section of the induced joint has a larger seismic response and is more likely to be damaged by large horizontal relative displacement, and is the weak surface of the subway station structure. By comparing the curves at observation Section No. 1 under the three working conditions, it can be seen that the horizontal relative displacement of the induced joint section is larger under the action of Songpan waves and Xi'an artificial waves, and the smallest under the action of Taft waves. Under the action of Songpan waves, there is a sudden change in the slope of the horizontal relative displacement curve at the height of the middle plate, which indicates that the displacement angles between the bottom floor and the top floor of the subway station structure are larger. However, under the action of Xi'an artificial waves and Taft waves, there is little difference in displacement angle between the bottom floor and the top floor of the subway station structure.

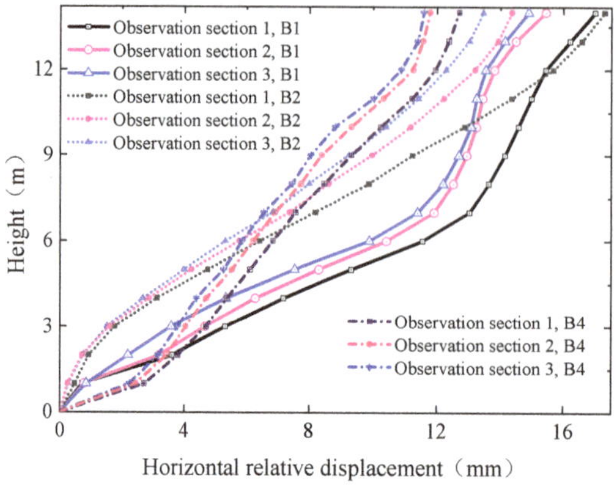

Figure 11. Horizontal relative displacement of structural side wall under B1, B2 and B4 working conditions.

In order to ensure the safety of subway operation, it is necessary to focus on the horizontal relative sliding of the structures on both sides of the induced joint under earthquake action, which can be represented by the peak value of the time history curve of the horizontal displacement difference of the measuring point at the same height in the side wall of observation sections No. 1 and No. 2. The horizontal relative sliding of the structures on both sides of the induced joint under the working conditions B1, B2, and B4 are shown in Figure 12. It can be seen that the horizontal relative sliding gradually increases along the subway station structure from bottom to top. Due to a higher number of low-frequency components in Songpan and Xi'an artificial waves, the seismic response of the loess–subway station structure system is more intense under the corresponding conditions. Thus, the horizontal relative sliding of structures on both sides of the induced joint is relatively large under the action of Songpan and Xi'an artificial waves, while that under the action of Taft waves is relatively small.

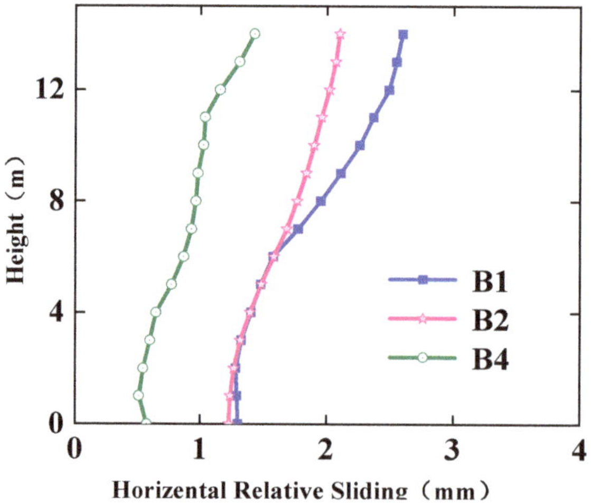

Figure 12. Horizontal relative sliding of structures on both sides of the induced joint under B1, B2 and B4 working conditions.

4.1.3. Vertical Relative Dislocation of Structures on Both Sides of Induced Joint

The vertical relative dislocation of the structures on both sides of the induced joint under earthquake action can be represented by the peak value of the vertical displacement difference time history curve of the measuring points at the same width on the bottom plates of observation sections No. 1 and No. 2. The calculated results under working conditions B1, B2, and B4 are shown in Figure 13. It can be seen that the vertical relative dislocations of the structures on both sides of the induced joint along the width direction of the section of the subway station show a general rule that the dislocations at the two ends are larger and that at the middle is smaller. This is mainly due to the concentrated action of the mass of the walls on both sides of the structure on both ends of the floor. However, in general, the value of vertical dislocation is very small. In particular, the calculation results under Songpan waves and Xi'an artificial waves are less than 1 mm.

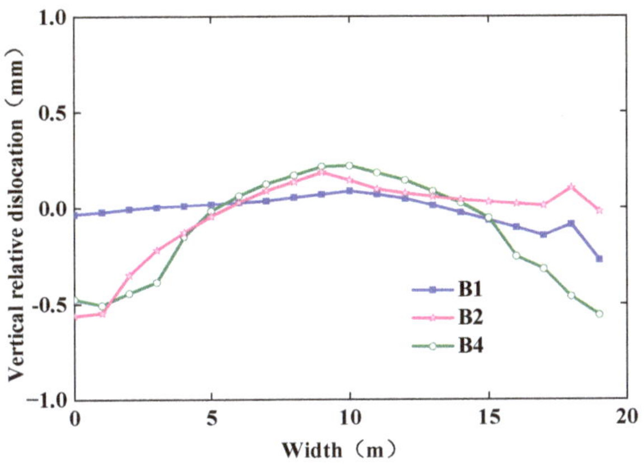

Figure 13. Vertical relative dislocation of structures on both sides of the induced joint under B1, B2 and B4 working conditions.

4.2. Influence of Seismic Waves with Different Peak Acceleration

4.2.1. Strain of Remaining Longitudinal Reinforcement at Induced Joint

The plastic strain time-history curves of the remaining longitudinal reinforcement at measuring points PE1–PE4 at the induced joint under working conditions, B5 and B6, are shown in Figure 14. When the peak acceleration of the input seismic wave is greater than 0.2 g, the plastic deformation of the remaining longitudinal reinforcement at the top plate and side wall of the induced joint section occurs first, followed by that at the middle plate, while the remaining longitudinal reinforcement at the bottom plate has very little plastic deformation. Comparing the results of PE1 and PE2 under two working conditions, it is found that under the action of seismic waves with different peak accelerations, the plastic deformation accumulation process of longitudinal reinforcement at each measuring point of the induced joint section is the same, and the plastic deformation is mainly concentrated in the duration of strong earthquake action, and the time-history curve is steep. According to the maximum plastic strain of remaining longitudinal reinforcement at each measuring point of the induced joint under different working conditions, when the peak acceleration of the input seismic wave is less than or equal to 0.2 g, no plastic deformation occurs in the remaining longitudinal reinforcement. When the peak acceleration of the input seismic wave is greater than 0.2 g, the value of maximum plastic strain increases rapidly with an increase in input seismic peak acceleration.

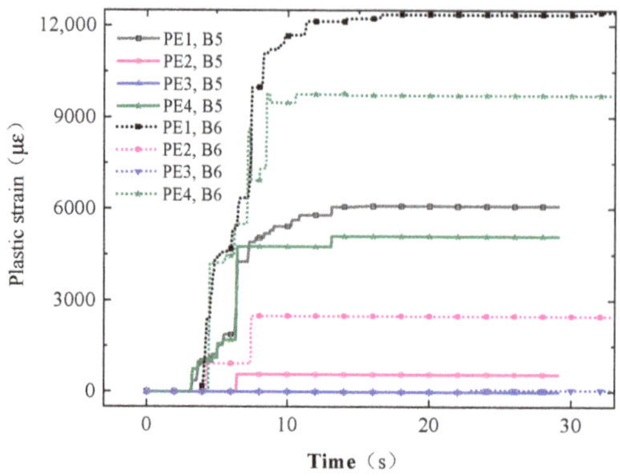

Figure 14. Plastic strain time history curve of residual reinforcement in the induced joint under B5 and B6 working conditions.

4.2.2. Horizontal Relative Sliding of Structures on Both Sides of Induced Joint

The horizontal relative displacements of different measuring points in each observation section are calculated based on the bottom measuring point. The horizontal relative displacements of each observation point under the B3, B4, B5, and B6 working conditions are shown in Figure 15.

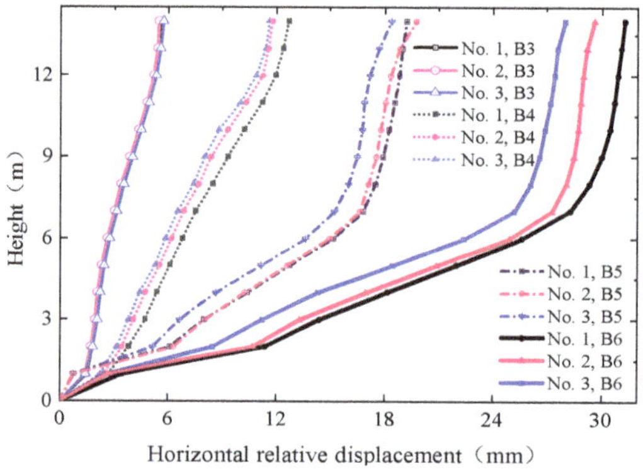

Figure 15. Horizontal relative displacement of structural side wall under B3, B4, B5 and B6 working conditions.

With an increase in peak acceleration of the input seismic wave, the horizontal relative displacement differences at different observation sections at the same height gradually increase. When the peak acceleration of the input seismic wave is less than or equal to 0.2 g, the horizontal relative displacement of the side wall changes largely linearly along the height, indicating that there is little difference in the displacement angle between the bottom floor and the top floor of the subway station structure. When the peak acceleration is greater than 0.2 g, the slope of the horizontal relative displacement curve of the side wall has a sudden change at the height of the mid-plate, indicating that the displacement angles between the bottom and the top floors of the subway station structure are significantly different, and the displacement angles between the bottom floors are significantly larger than that of the top floors.

4.3. Influence of Remaining Longitudinal Reinforcement

At present, the empirical practice of induced joint is to make one-third of the longitudinal reinforcement pass through and the rest of the reinforcement cut-off, but there is no corresponding theoretical basis for this practice. In order to study the influence of the remaining longitudinal reinforcement ratio on the seismic response of the induced joint of the subway station structure, the remaining longitudinal reinforcement ratio is taken as the parameter for design conditions G1, G2, G3, and G4, and the corresponding remaining longitudinal reinforcement ratio for four conditions is 0, 1/3, 2/3, and 1, respectively. The seismic waves are all set as Songpan waves with 0.4 g peak acceleration.

4.3.1. Strain Analysis of Remaining Longitudinal Reinforcement at Induced Joint

The plastic strain time-history curves of the remaining longitudinal reinforcement at PE1–PE4 measuring points at the induced joint under the G2, G3, and G4 working conditions are shown in Figure 16. With an increase in the reinforcement ratio of the remaining longitudinal reinforcement at the induced joints, the plastic deformation accumulation process of the remaining longitudinal reinforcement at each measuring point during strong earthquake action gradually slows. The value of maximum plastic strain decreases gradually with an increase in the reinforcement ratio at the induced joint. When the reinforcement ratio increases from 1/3 to 2/3, the maximum plastic strain of the remaining longitudinal reinforcement at the induced joint section decreases the most, except for at individual measuring points.

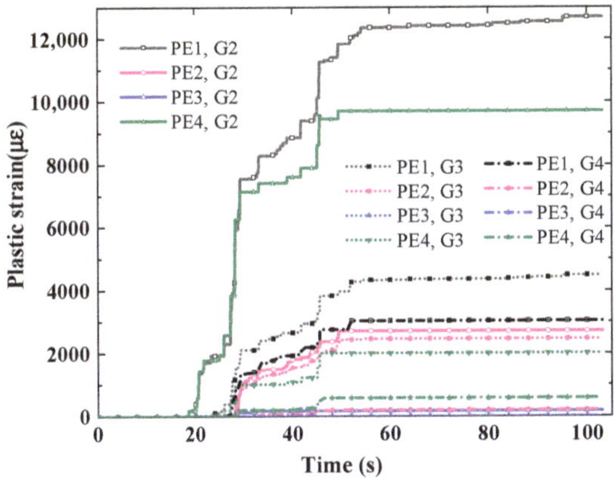

Figure 16. Plastic strain time history curve of residual reinforcement in the induced joint under G2, G3 and G4 working conditions.

4.3.2. Horizontal Relative Sliding of Structures on Both Sides of Induced Joint

The horizontal relative sliding of structures on both sides of the induced joint under the G1, G2, G3, and G4 working conditions is shown in Figure 17.

It can be seen from the figure that under G1 working condition, the horizontal relative sliding of the structures on both sides of the induced joint at the bottom layer gradually decreases from bottom to top along the subway station structure, while the change law of the horizontal relative sliding of the top layer is opposite to that of the bottom layer. The horizontal relative sliding is visibly larger at places with a large concentrated mass of the bottom plate and top plate. Under G2, G3, and G4 working conditions, the bottom layer shows an overall swing. The top layer of the structure on both sides of the induced joint is subject to relative torsional deformation along the longitudinal direction. With an increase in the remaining longitudinal reinforcement ratio at the induced joint, the horizontal relative sliding of the side wall gradually decreases. The overall horizontal relative sliding under the G1 working condition is the largest and reaches 13.33 mm at the bottom plate. This is because all the longitudinal reinforcement at the induced joint is cut off, and the effect of resisting the horizontal relative sliding of the structures on both sides only by the friction between concrete interfaces is limited.

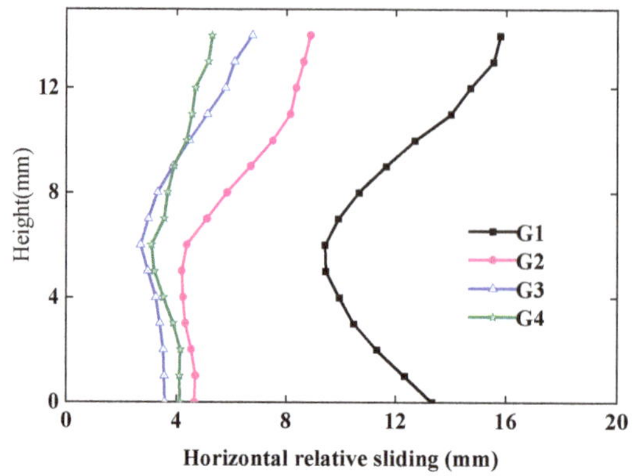

Figure 17. Horizontal relative sliding of structures on both sides of the induced joint under G1, G2, G3, and G4 working conditions.

4.3.3. Vertical Relative Dislocation of Structures on Both Sides of Induced Joint

The vertical relative dislocation of structures on both sides of the induced joint under the G1, G2, G3, and G4 working conditions is shown in Figure 18. It can be seen that the calculated values under the G2, G3, and G4 working conditions are very close, and the maximum value is less than 1 mm. However, under the G1 working condition, the vertical relative dislocation is visibly greater than that under other conditions, and gradually increases from left to right along the width direction of the section of the subway station, with a maximum value of 4.2 mm. This is due to the absence of through longitudinal reinforcement at the induced joint and poor structural connectivity on both sides. Although key grooves are set at the induced joint of the bottom plate, the structure on both sides may have longitudinal relative displacement or rotation under earthquake action, which may cause the key grooves to detach or damage, causing serious vertical relative dislocation of the structures on both sides of the induced joint section.

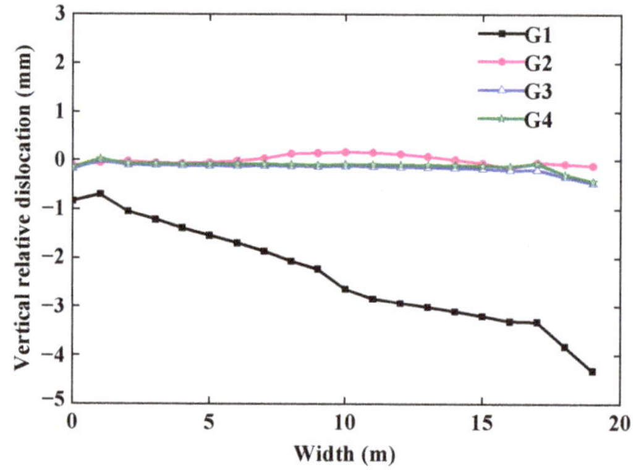

Figure 18. Vertical relative dislocation of structures on both sides of the induced joint under G1, G2, G3, and G4 working conditions.

4.4. Influence of the Key Groove

In order to study the influence of the bottom plate key groove on the seismic response of the induced joint, the G2 working condition is selected for comparison. The bottom plate key groove of the induced joint section under the G2 working condition is removed and marked as the J2 working condition. The plastic strain time-history curves of the remaining longitudinal reinforcement at the PE1–PE4 measuring points at the induced joint under the J2 working condition are shown in Figure 19.

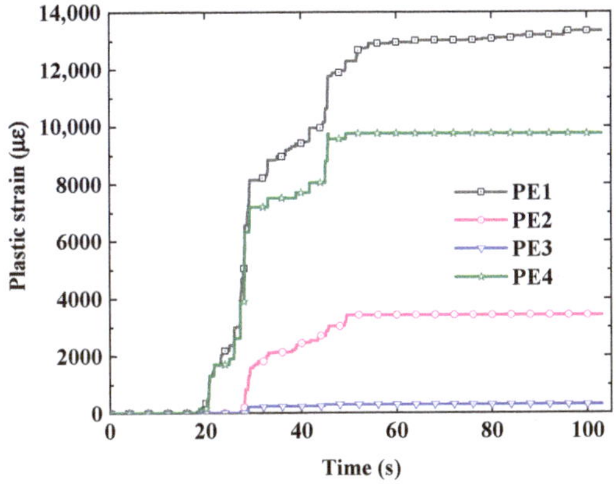

Figure 19. Plastic strain time history curve of residual reinforcement.

Compared with Figure 16, it is found that the plastic deformation of the longitudinal reinforcement remaining at each measuring point is larger during the whole vibration time when no key groove is set at the bottom plate of the induced joint section. Under the J2 working condition, the maximum plastic strain of remaining longitudinal reinforcement at each measuring point of induced joint section points PE1–PE4 is 13,341, 3416, 294, and 9751, respectively. Obviously, the maximum plastic strain under the working condition of not setting the bottom plate key groove is greater than that of the corresponding measuring point under the working condition of setting the bottom plate key groove. The reason is that the lack of a key groove at the bottom plate of the induced joint section will reduce the connectivity of the structures on both sides, and the remaining longitudinal reinforcement will bear greater shear force under the coupling effect of gravity and seismic force.

To better analyze the change in horizontal relative displacement, the horizontal relative sliding of structures on both sides of the induced joint under the G2 and J2 working conditions is compared, as shown in Figure 20a. The horizontal relative sliding under the J2 working condition is only a little greater than that under the J1 working condition, and the maximum difference between the two is only 1.05 mm. This indicates that the key groove set on the bottom plate has a certain effect on controlling the horizontal relative sliding of the structures on both sides of the induced joint on the bottom layer. The vertical relative dislocation of the structures on both sides of the induced joint under the two working conditions is shown in Figure 20b. It is found that the vertical relative dislocation under the J1 working condition is significantly smaller than that under the J2 working condition. The vertical relative dislocation of each measuring point under the J1 working condition is less than 1 mm, while the vertical relative dislocation of each measuring point under the J2 working condition is more than 2 mm, with a maximum value of 2.7 mm. This shows that the key groove on the bottom plate has a significant effect on controlling the vertical relative dislocation of the structures on both sides of the induced joint.

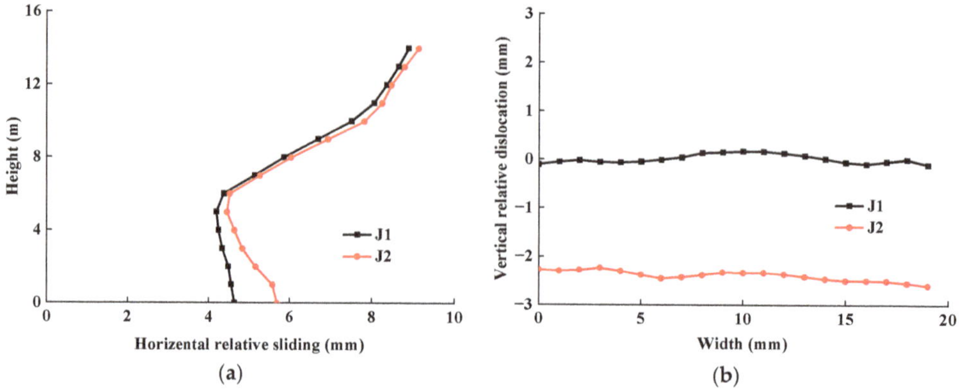

Figure 20. Displacement on both sides of the induced joint under different working conditions: (**a**) Horizontal relative sliding; (**b**) Vertical relative dislocation.

5. Conclusions and Suggestions

In this paper, the seismic response of induced joints in subway stations on loess sites under the action of seismic waves is studied by numerical simulation. The horizontal relative sliding and vertical relative dislocation of the structures on both sides of the induced joints of the subway station structure under different working conditions, and the strain response of the remaining longitudinal reinforcement at the induced joint, are mainly analyzed. The specific conclusions are as follows:

(1) The structure at the induced joint section is more likely to be damaged due to large horizontal relative displacement and is the weak section of the subway station structure. The horizontal relative sliding of the structures on both sides of the induced joint gradually increases from bottom to top along the subway station structure. With an increase in earthquake intensity, the horizontal relative sliding of the structures on both sides of the induced joint becomes larger, and the horizontal relative sliding of the top structure at the section of the induced joint of the subway station is more obvious.

(2) The horizontal relative sliding and vertical relative misalignment of the structures on both sides of the induced joint can be reduced by increasing the reinforcement ratio of the remaining longitudinal reinforcement in the induced joint.

(3) The vertical relative dislocation of structures on both sides of the induced joint along the width direction of the subway station section is generally larger at both ends and smaller in the middle. With an increase in earthquake intensity, the vertical relative dislocation is smaller.

(4) Under the action of Songpan and Taft waves, the plastic deformation of the remaining longitudinal reinforcement at the induced joint is mainly concentrated in the duration of strong earthquake action. However, under the action of Xi'an artificial waves, the plastic deformation of the remaining longitudinal reinforcement at the induced joint runs through the whole process of earthquake action, and the cumulative effect of plastic strain is significant. With an increase in earthquake intensity, the plastic deformation of the remaining longitudinal reinforcement at the induced joint increases.

(5) The key groove on the bottom plate has little effect on the horizontal relative sliding of the structures on both sides of the induced joint at the bottom layer but has a significant effect on controlling the vertical relative dislocation of the structures on both sides of the induced joint.

In this present study, the seismic response of induced joints in subway station structures is revealed and some meaningful conclusions are obtained. However, most of the research results are qualitative and lack a more accurate quantitative analysis. It is suggested to further use incremental dynamic analysis (IDA) for quantitative analysis to draw more accurate conclusions to provide a scientific basis for controlling and reducing excessive deformation and the failure of induced joints.

Author Contributions: D.Q.: Methodology, Conceptualization, Data curation, Validation; S.C.: Reviewing, Funding acquisition, Supervision, Writing; Y.W.: Methodology, Writing, Data analysis; Z.F.: Data analysis; Y.B.; Supervision, Reviewing. All authors have read and agreed to the published version of the manuscript.

Funding: This research was funded by the National Natural Science Foundation of China (grant number 41902277), the Natural Science Basic Research Plan in Shaanxi Province, (grant number 2023-YBGY-085), and Fundamental Research Funds for the Central Universities of China (grant number 300102282201).

Data Availability Statement: The data used to support the findings of this study are included in the article.

Conflicts of Interest: The authors declare no conflict of interest.

References

1. Hashash, Y.M.A.; Hook, J.J.; Schmidt, B.; Yao, J.I. Seismic design and analysis of underground structures. *Tunn. Undergr. Space Technol. Inc. Trenchless Technol. Res.* **2001**, *16*, 247–293. [CrossRef]
2. Smerzini, C.; Avilés, J.; Sánchezlogesma, F.J.; Paolucci, R. Analytical solutions for the seismic response of underground structures under SH wave propagation. *Aip Conf. Am. Inst. Phys.* **2008**, *1020*, 674–683.
3. Naggar, H.E.; Hinchberger, S.D.; Naggar, M. Simplified analysis of seismic in-plane stresses in composite and jointed tunnel linings. *Soil Dyn. Earthq. Eng.* **2008**, *28*, 1063–1077. [CrossRef]
4. Kazemeini, M.J.; Haghshenas, E.; Kamalian, M. Experimental evaluation of seismic site response over and nearby underground cavities (study of subway tunnel in city of Karaj, Iran). *Iran. J. Sci. Technol. Trans. Civ. Eng.* **2015**, *39*, 319–322.
5. Dadfar, B.; Naggar, M.E.; Nastev, M. Experimental and analytical study of seismic site response of discontinuous permafrost. *Can. Geotech. J.* **2016**, *53*, 1363–1375. [CrossRef]
6. Hushmand, A.; Dashti, S.; Davis, C.; McCartney, J.S.; Hushmand, B. A centrifuge study of the influence of site response, relative stiffness, and kinematic constraints on the seismic performance of buried reservoir structures. *Soil Dyn. Earthq. Eng.* **2016**, *88*, 427–438. [CrossRef]
7. Amorosi, A.; Boldini, R.; Elia, R. Parametric study on seismic ground response by finite element modelling. *Comput. Geotech.* **2010**, *37*, 515–528. [CrossRef]
8. Ichimura, T.; Tanaka, S.; Hori, M.; Yamamoto, Y.; Dobashi, H.; Osada, M.; Ohbo, N.; Yamada, T. Full three-dimensional seismic response analysis of underground structures with large complex cross sections and two-step analysis method for reducing the computational costs. *J. Earthq. Tsunami* **2016**, *10*, 1640016. [CrossRef]
9. Schanze, E.; Schanze, E.; Gomez, M.; Lopez, A. Numerical study of the seismic response of an instrumented building with underground stories. *Appl. Sci.* **2021**, *11*, 3190. [CrossRef]
10. Jeng, D.S.; Lu, J.F. Vibration analysis of a circular tunnel with jointed liners embedded in an elastic medium subjected to seismic waves. *J. Vib. Acoust. Trans. ASME* **2009**, *131*, 021005. [CrossRef]
11. Lin, G.; Lu, S.; Liu, J. Transmitting boundary for transient analysis of wave propagation in layered media formulated based on acceleration unit-impulse response. *Soil Dyn. Earthq. Eng.* **2016**, *90*, 494–509. [CrossRef]
12. Carlos, G.M.; Jesus, G.G.; Claudio, O.M. Enhanced simplified method for pseudostatic analysis of seismic internal forces in rectangular underground structures. *J. Geotech. Geoenviron. Eng.* **2021**, *147*, 04021090.
13. Mehdi, P.; Saeed, M.H.; Amirabbas, F. Seismic response of the ground surface including underground horseshoe-shaped cavity. *Transp. Infrastruct. Geotechnol.* **2021**, *9*, 338–355.
14. Zhao, M.; Gao, Z.D.; Du, X.L.; Wang, J.J.; Zhong, Z.L. Response spectrum method for seismic soil-structure interaction analysis of underground structure. *Bull. Earthq. Eng.* **2019**, *17*, 5339–5363. [CrossRef]
15. Iwatate, T.; Kobayashi, Y.; Kusu, H.; Rin, K. Investigation and shaking table tests of subway structures of the Hyogoken-Nanbu earthquake. In Proceedings of the 12th World Conference on Earthquake Engineering, Auckland, New Zealand, 30 January–4 February 2000; New Zealand Society for Earthquake Engineering: Wellington, New Zealand, 2000.
16. Chou, J.C.; Kutter, B.L.; Travasarou, T.; Chacko, J.M. Centrifuge modeling of seismically induced uplift for the BART Transbay Tube. *J. Geotech. Geoenviron. Eng. ASCE* **2011**, *137*, 754–765. [CrossRef]
17. Cilingir, U.; Madabhushi, S.P.G. A model study on the effects of input motion on the seismic behaviour of tunnels. *Soil Dyn. Earthq. Eng.* **2011**, *31*, 452–462. [CrossRef]
18. Baziar, M.H.; Moghadam, M.R.; Kim, D.S.; Choo, Y.W. Effect of underground tunnel on the ground surface accelerationl. *Tunn. Undergr. Space Technol.* **2014**, *44*, 10–22. [CrossRef]

19. Ulgen, D.; Saglam, S.; Ozkan, M.Y. Dynamic response of a flexible rectangular underground structure in sand: Centrifuge modeling. *Bull. Earthq. Eng.* **2015**, *13*, 2547–2566. [CrossRef]
20. Dashti, S.; Hashash, Y.M.A.; Gillis, K.; Musgrove, M.; Walker, M. Development of dynamic centrifuge models of underground structures near tall buildings. *Soil Dyn. Earthq. Eng.* **2016**, *86*, 89–105. [CrossRef]
21. Masoud, R.M.; Mohammad, H.B. Seismic ground motion amplification pattern induced by a subway tunnel: Shaking table testing and numerical simulation. *Soil Dyn. Earthq. Eng.* **2016**, *83*, 81–97.
22. Quan, D.Z.; Chai, S.B.; Chen, S.; Wang, Y.H. Experimental study on seismic response of subway station built in loess. *Lat. Am. J. Solids Struct.* **2020**, *17*, e282. [CrossRef]
23. Moss, R.; Crosariol, V.A. Scale model shake table testing of an underground tunnel cross section in soft clay. *Earthq. Spectra* **2013**, *29*, 1413–1440. [CrossRef]
24. Tabatabaiefar, H.R. Development of synthetic soil mixture for experimental shaking table tests on building frames resting on soft soils. *Geomech. Geoengin.* **2016**, *12*, 28–35. [CrossRef]
25. Zhuang, H.Y.; Wang, X.; Miao, Y.; Yao, E.; Chen, S.; Ruan, B.; Chen, G.X. Seismic responses of a subway station and tunnel in a slightly inclined liquefiable ground through shaking table test. *Soil Dyn. Earthq. Eng.* **2019**, *116*, 371–385.
26. Adampira, M.; Derakhshandi, M.; Ghalandarzadeh, A. Experimental study on seismic response characteristics of liquefiable soil layers. *J. Earthq. Eng.* **2021**, *25*, 1287–1315. [CrossRef]
27. Wang, J.N.; Yang, J.; Zhuang, H.Y.; Ma, G.W.; Sun, Y.J. Seismic responses of a large unequal-span underground subway station in liquefiable soil using shaking table test. *J. Earthq. Eng.* **2022**, *26*, 8446–8467. [CrossRef]
28. Parra, M.G.J.; Bobet, A.; Ramirez, J.A. Evaluation of soil-structure interaction and structural collapse in Daikai subway station during Kobe earthquake. *ACI Struct. J. Am. Concr. Inst.* **2006**, *103*, 113–122.
29. Abate, G.; Massimino, M.R.; Maugeri, M.; Muir Wood, D. Numerical modelling of a shaking table test for soil-foundation-superstructure interaction by means of a soil constitutive model implemented in a FEM code. *Geotech. Geol. Eng.* **2010**, *28*, 37–59. [CrossRef]
30. Long, H.; Chen, G.X.; Zhuang, H.Y. Analysis of seismic response influence of high rise building nearby subway station in soft site. *Appl. Mech. Mater.* **2012**, *166–169*, 2105–2111. [CrossRef]
31. Conti, R.; Viggiani, G.M.; Perugini, F. Numerical modelling of centrifuge dynamic tests of circular tunnels in dry sand. *Acta Geotech.* **2014**, *9*, 597–612. [CrossRef]
32. Abate, G.; Massimino, M.; Maugeri, M. Numerical modelling of centrifuge tests on tunnel–soil systems. *Bull. Earthq. Eng.* **2015**, *13*, 1927–1951. [CrossRef]
33. Keykhosropour, L.; Lemnitzer, A. Seismic response behavior of deep flexible underground structures in sand-insights from an experimental-numerical investigation. *Bull. Earthq. Eng.* **2022**, *20*, 5205–5231. [CrossRef]
34. Yoo, M.; Kwon, S.Y.; Hong, S. Dynamic response evaluation of deep underground structures based on numerical simulation. *Geomech. Eng.* **2022**, *29*, 269–279.
35. Sun, Q.Q.; Peng, D.; Dias, D. Seismic performances of three- and four-sided box culverts: A comparative study. *Geomech. Eng.* **2020**, *22*, 49–63.
36. Wang, Q.; Geng, P.; Guo, X.Y.; Wang, X.; Li, P.; Zhao, B. Case study on the seismic response of a subway station combined with a flyover. *Undergr. Space* **2021**, *6*, 665–677. [CrossRef]
37. Alejandro, R.C.; Victor, M.C.; Fabian, I.P. Seismic response of underground and floating tunnels. *Pure Appl. Geophsics* **2022**, *179*, 973–992.
38. *GB 50909-2014*; Code for Seismic Design of Urban Rail Transit Structures. China Planning Publishing House: Beijing, China, 2014.
39. Du, X.L.; Zhao, M.; Wang, J.T. Artificial stress boundary conditions for near field wave simulation. *J. Mech.* **2006**, *1*, 49–56.
40. Du, X.L.; Zhao, M. Seismic response analysis method of arch dam based on viscoelastic boundary. *J. Water Conserv.* **2006**, *9*, 1063–1069.

Disclaimer/Publisher's Note: The statements, opinions and data contained in all publications are solely those of the individual author(s) and contributor(s) and not of MDPI and/or the editor(s). MDPI and/or the editor(s) disclaim responsibility for any injury to people or property resulting from any ideas, methods, instructions or products referred to in the content.

buildings

MDPI

Article

Numerical Simulation of Impact Effect on Stability of Transmission Tower Foundation

Lang Song [1,2], Shaobo Chai [2,*], Lianzeng Chai [2], Xianpeng Li [2] and Jinhao Liu [2]

1 Department of Civil and Architectural Engineering, Xi'an Jiaotong University City College, Xi'an 710018, China; songlang0712@163.com
2 School of Civil Engineering, Chang'an University, Xi'an 710064, China; 2021128033@chd.edu.cn (L.C.); 2021228066@chd.edu.cn (X.L.); 2022228041@chd.edu.cn (J.L.)
* Correspondence: shbchai@chd.edu.cn

Abstract: The impact effect can cause structural damage to a transmission tower's foundation and affect its overall stability. In order to study the influence of the impact effect on the stability of transmission tower foundations, a three-dimensional finite element numerical simulation method was used to investigate the variations in the extent of damage, displacement, and inclination degree of a transmission tower foundation under different impact velocities, impact durations, impactor shapes, and impact locations. The results show that as the impact velocity increases, the damage value of the transmission tower foundation continuously increases, and the damaged area expands. The lateral displacement value increases continuously with the duration of the impact effect, and the variation in lateral displacement follows a linear function distribution. The inclination degree of the transmission tower foundation increases continuously with increased impact duration and can lead to overturning failure. A smaller impact contact area results in a larger compressive damage value for the transmission tower foundation, and different impact contact areas lead to different modes of failure for the transmission tower foundation. The damage value and damaged area of the transmission tower foundation vary with the location of the impact. By comparing the deformation of the transmission tower foundation before and after reinforcement, it is evident that the reinforcement design can significantly improve the deformation resistance and anti-overturning capacity of the transmission tower foundation.

Keywords: impact effect; transmission tower foundation; stability; deformation; numerical simulation

1. Introduction

In order to keep up with the rapid development of the national economy and the ever-increasing demand for electricity and achieve wider implementation of optimized resource allocation and the construction of modern intelligent power grid, the construction of transmission engineering infrastructure needs to continue at a fast pace [1]. As a carrier for high-capacity power transmission and a crucial lifeline project which is part of the lifeblood of national economic construction and development [2,3], the safe and stable operation of transmission engineering infrastructure is vital for the country's economic growth and the normal needs of people's lives. Once failure or damage occurs, it will cause significant economic losses and trigger various secondary disasters [4–10]. Transmission tower foundations, which play the role of pillar in the transmission engineering infrastructure, have an irreplaceable role in maintaining the stability of transmission towers and transferring load. Ensuring the safety and stability of transmission tower foundations under adverse external factors is the key focus of engineering design and operational maintenance. Therefore, it is of great engineering significance to study the deformation and stability of transmission tower foundations under adverse effects.

With the increasing number of transmission tower foundations installed on the Earth's surface, the safety and stability of transmission tower foundations have drawn significant attention from scholars. Relevant research on the stability of transmission tower foundations has been carried out by some scholars [11–13]. For example, Gao et al. [14] studied the failure location of transmission towers and the stress characteristics of key components through a full-scale experiment. Taking a transmission tower in a coal-mining subsidence area as an example, Shu et al. [15] adopted finite element numerical simulation to analyze the deformation resistance of the transmission tower under the action of foundation displacement, wind load, and ice cover load. Zhou et al. [16] studied the influence of the number, distribution characteristics, and development model of soil caves on the stability of transmission tower foundations using a numerical simulation method. Based on the construction project of a double-line loess tunnel near a transmission tower, Yao et al. [17] established a finite element numerical model with the ABAQUS software (https://www.hindawi.com/journals/ace/2023/2533212/, accessed on 16 October 2023) to study the influence of different tunnel construction methods on the deformation and stress of transmission tower foundations. Wang et al. [18] studied the dynamic response of tunnels in the surrounding rock and a high-voltage transmission tower under an explosion load by numerical analysis in a FLAC finite difference program. In order to analyze the frost heave characteristics of a prefabricated cone column foundation, Xin et al. [19] carried out physical experiments on the frost heave characteristics of a prefabricated cone column foundation under different temperature conditions and studied the temperature field distribution characteristics and the frost heave displacement mode. Yuan et al. [20] studied the failure mode of transmission tower foundations under large spans and then proposed improvement measures to enhance the stability of transmission tower foundations. In summary, the above studies have investigated the stability of transmission tower foundations under conditions such as soil cave development, construction disturbance, and ultimate loads and revealed the variations in the stability of transmission tower foundations under adverse effects to some extent. However, transmission tower foundations often suffer various impact effects in practical engineering, yet there is a lack of research on the influence of impact loading on the deformation and stability of transmission tower foundations. In particular, for transmission tower foundations installed in farmland areas, due to the popularization of automation of farming operations such as sowing and harvesting, agricultural equipment and agricultural vehicles frequently appear in such farmland areas. Due to limited visibility and improper handling of personnel, incidents of agricultural equipment and agricultural vehicles hitting transmission tower foundations occur frequently. The impact effect in such a case will cause certain structural damage to and displacement of a transmission tower foundation, which will seriously affect the overall stability of the foundation. There is even a risk of instability and collapse of the upper transmission tower, which would result in the paralysis of the power grid system and serious economic losses.

Based on the above problems, in this paper, a three-dimensional finite element numerical simulation method was used to investigate the influence of impact loading on the stability of transmission tower foundations, and then the variations of damage degree, displacement, and inclination degree of transmission tower foundations under different impact speeds, different impact durations, different impactor shapes, and different impact positions are analyzed. Further comparative analysis of the deformation characteristics and damage values of transmission tower foundations before and after reinforcement and the influence law of the impact effect on the deformation and stability of transmission tower foundations are revealed. Based on this paper, this research enriches the study of the stability of transmission tower foundations under adverse conditions and provides references for the assessment and safety analysis of the stability of transmission tower foundations under impact effects in practical engineering.

2. Numerical Modeling

The ABAQUS software is used in this study to conduct three-dimensional finite element numerical simulation, and the model is established according to the specifications of a specific transmission tower in practical engineering. The transmission tower foundation has a length of 6.8 m, a height of 3.5 m, and a column width of 0.8 m. The concrete grade used for the transmission tower foundation is C30, and the load borne by the transmission tower foundation is determined based on the actual conditions. The numerical model includes the soil mass, impactor, transmission tower, and transmission tower foundation, in which a tie constraint between the transmission tower foundation and soil mass is adopted. Constraints are applied to the surrounding boundaries of the soil mass in the form of normal displacement, while three-directional displacement constraints are imposed at the bottom of the soil mass. The top surface of the soil mass is a free boundary. A vertical displacement constraint is applied to the bottom of the impactor, and the horizontal velocity is applied to the impactor to realize the impact to the transmission tower foundation. A dynamic implicit step is used for impact analysis in the numerical simulation process. In order to eliminate the influence of boundary constraints and enhance the computational efficiency, the horizontal distance between the boundary of the numerical model and the edge of the transmission tower foundation is 3 times the width of the transmission tower foundation, and the vertical distance is 7 times the height of the transmission tower foundation. A solid element is adopted for the model of the impactor, transmission tower foundation, and soil mass. The grid adopts the solid reduction integration C3D8R element, including the Mohr–Coulomb model of soil mass, and the concrete damaged plasticity (CDP) model is used for the transmission tower foundation. It has been proven that the CDP model can effectively simulate the stress state, damage degree, and failure form of concrete materials [21,22], and its uniaxial compression and tensile stress–strain relationships are determined by referencing existing literature [23]. The numerical model is divided into 78,252 grid cells, as shown in Figure 1. The detailed parameters of the numerical model are shown in Table 1, and the parameters of the CDP model are shown in Table 2.

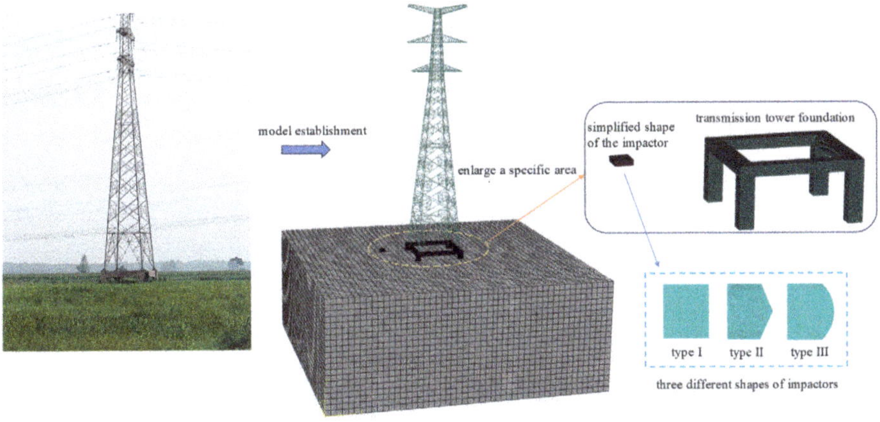

Figure 1. Schematic diagram of the numerical model of the transmission tower foundation.

Table 1. Material parameters of individual components of the numerical model.

Component	Density (kg/m^3)	Elastic Modulus (MPa)	Poisson's Ratio	Friction Angle (°)	Cohesive Force (kPa)
Clay	1700	28	0.35	20	20
Limestone	2100	56	0.32	25	17
Transmission tower structure	7850	206,000	0.30	-	-
Transmission tower foundation	2500	30,000	0.15	-	-
Impactor	7850	206,000	0.30	-	-

Table 2. Parameters of the concrete damage plastic model.

Parameter	Dilation Angle	Eccentricity	fb0/fc0	Invariant Stress Ratio K	Viscosity Parameter
Value	30	0.1	1.16	0.6667	0.001

In the actual farmland areas, it is often agricultural equipment and agricultural vehicles that impact transmission tower foundations. However, due to the diverse and complex specifications of agricultural equipment and agricultural vehicles and the difficulty of modeling, simplified impactors are adopted to simulate actual agricultural equipment and agricultural vehicles; three different shapes of impactors are used to simulate the different impact surfaces of agricultural equipment and agricultural vehicles. The authors of [24] conducted numerical simulation studies on the direct impact of impactors on a transmission tower structure. The focus of this study is to analyze the deformation and stability of a transmission tower foundation under the impact effect, which means the impactor does not directly impact the transmission tower structure; that is, the influence of the impact effect on the upper transmission tower structure cannot be considered, so the transmission tower structure is not considered in the calculation and analysis process. In order to further improve the calculation efficiency of the numerical model, an equivalent force is applied to the top of the transmission tower foundation to replace the self-weight of the transmission tower structure and the load transmitted by the transmission lines. The vertical force and horizontal force applied to the transmission tower foundation are 1000 kN and 150 kN, respectively.

3. Result Analysis

3.1. Different Impact Velocities

In practical engineering, agricultural vehicles will impact a transmission tower foundation at different speeds. Considering that these vehicles do not travel at high speeds in farmland areas, impact velocities of 10, 20, and 30 km/h are selected to analyze the deformation and damage to the transmission tower foundation under different impact velocities. The simulation results indicate that the stress and lateral displacement of the transmission tower foundation at the initial impact time are different under different impact velocities, and the values increase with increasing impact velocity. Taking the type I impactor as an example, Figure 2 shows the compression damage contour of the transmission tower foundation at the initial impact time under different impact velocities. From Figure 2, it can be observed that the transmission tower foundation exhibits a certain degree of damage at the initial impact time, and the damage values are different for different impact velocities. When the impact velocity is taken as 10 km/h, the maximum damage value is 0.265; when the impact velocity is taken as 20 km/h, the maximum damage value is 0.318, which is an increase of 20% compared with the impact velocity of 10 km/h; when the impact velocity is taken as 30 km/h, the maximum damage value is 0.601, which is an increase of 131% compared with the impact velocity of 10 km/h. It is evident that with the increase of the impact velocity, the damage value of the transmission tower foundation increases significantly. It can also be found from Figure 2 that with the increase of the impact velocity, the damage area of the transmission tower foundation expands continuously, which indicates that the damage degree of the transmission tower foundation is more serious under high-velocity impacts.

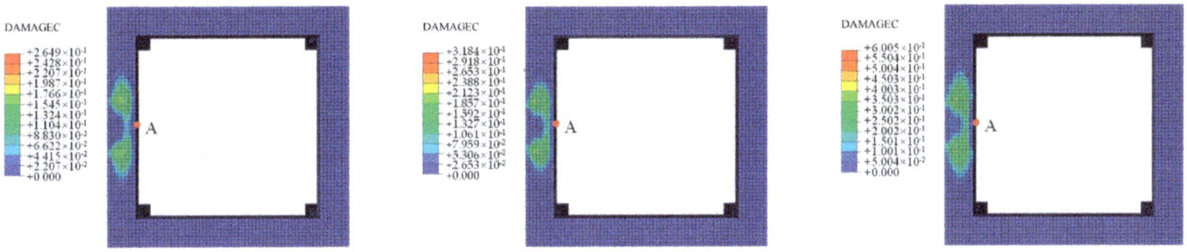

(**a**) (**b**) (**c**)

Figure 2. Compression damage contours of the transmission tower foundation at initial impact time under different impact velocities: (**a**) v = 10 km/h; (**b**) v = 20 km/h; (**c**) v = 30 km/h.

In order to further investigate the variations in the lateral displacement of the impacted area under different impact velocities, a unit node of the transmission tower foundation numerical model was selected for analysis (point A in Figure 2), and the lateral displacement time history curves of unit node A under different impact velocities within the time range of 20 ms after the initial impact time were extracted, as shown in Figure 3. Under different impact velocity conditions, the lateral distance between the impactor and the transmission tower foundation remains unchanged, and it takes a certain time for the impactor to impact the transmission tower foundation after the velocity is given to the impactor. The difference in the given velocity causes the moment of impact on the transmission tower foundation to be different. It can be seen from Figure 3 that within the time range of 20 ms after the initial impact time, when the impact velocity is taken as 10, 20, and 30 km/h, the lateral displacement increases from 0 mm to 64 mm, 122 mm, and 186 mm respectively. The results show that with the increase of the impact velocity, the incremental value of the lateral displacement within the same impact duration increases continuously. At the same impact velocity, the lateral displacement value of unit node A increases with the duration of the impact effect. It can be found through function fitting that the growth trend of the lateral displacement value conforms to linear function distribution, and the slopes of the fitting curves are 2.91, 2.96, and 8.90 when the impact velocity is taken as 10, 20, and 30 km/h, respectively. The different slopes of the lateral displacement numerical fitting curves at different impact velocities indicate that the varying rate of the lateral displacement is different, which indicates that the varying rate of the lateral displacement of the transmission tower foundation increases with increasing impact velocity.

3.2. *Different Impact Durations*

Agricultural equipment and agricultural vehicles often stop when impacting a transmission tower foundation in practical engineering; that is, the impact duration is an instant. In some cases, agricultural equipment and agricultural vehicles still maintain a certain speed after impacting a transmission tower foundation, causing sustained damage to the transmission tower foundation. Therefore, it is necessary to study the deformation responses of a transmission tower foundation under different impact durations. Figure 4 illustrates the lateral displacement contours of the transmission tower foundation under different impact durations of the type I impactor.

As can be seen from Figure 4, a 4.4 mm lateral displacement occurred at the central point of the transmission tower foundation at the initial impact time. With the increase of the impact duration, the lateral displacement value increased continuously, and the area where the lateral displacement occurred gradually expanded. When the impact duration is taken as 10 ms, 20 ms, 30 ms, and 40 ms, the lateral displacement values of the impact center are 96.8 mm, 186.2 mm, 278.8 mm, and 368.8 mm, respectively. The lateral displacement values of individual points along the longitudinal center line of the connected beam under the impact effect were extracted (as shown in Figure 5). It can be found that the lateral displacement value increases with a smaller distance from the impact center point, and the lateral displacement values exhibit a parabolic shape along the longitudinal center line of the connected beam. In addition, it can be seen that with the increase of the impact duration, the lateral displacement curves of individual points along the longitudinal center line of the impacted connected beam move to the right continuously, and the bending degree of the impacted connected beam increases with the increase of the impact duration.

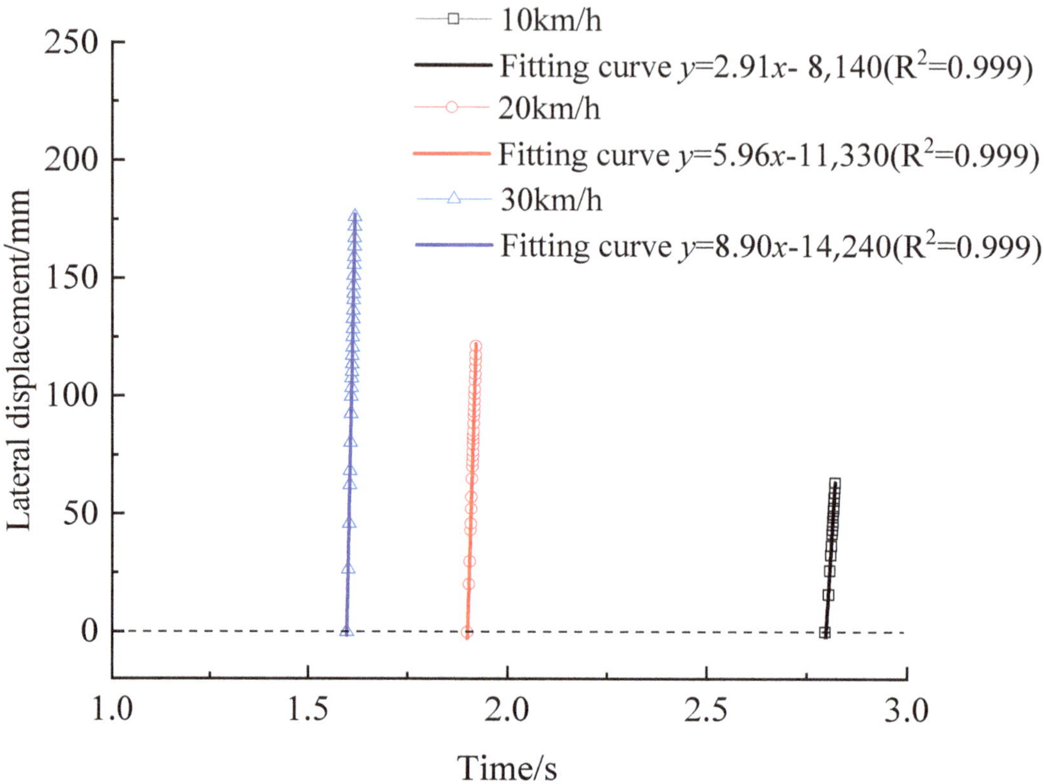

Figure 3. Lateral displacement time history curves of unit node A under different impact velocities.

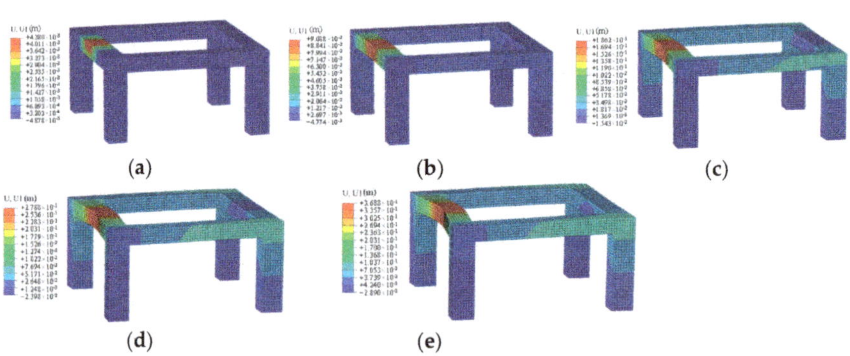

Figure 4. Contours of the lateral displacement of the transmission tower foundation under different impact durations: (**a**) initial impact time; (**b**) 10 ms; (**c**) 20 ms; (**d**) 30 ms; (**e**) 40 ms.

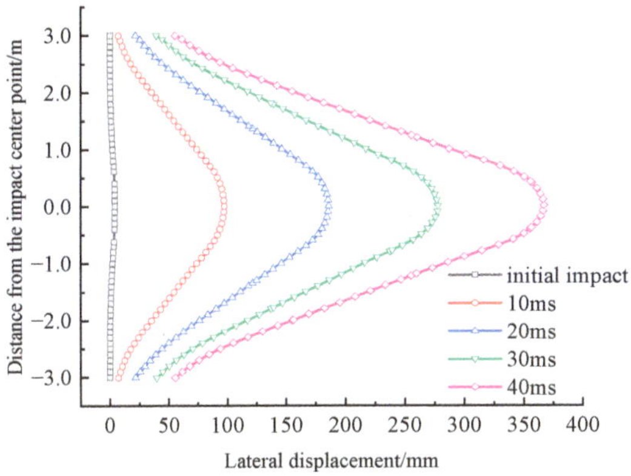

Figure 5. Lateral displacement values of individual points along the longitudinal center line of the impact beam.

Figure 6 shows the vertical displacement contours of the transmission tower foundation under different impact durations with the type I impactor. It can be observed that with increasing impact duration, the left side of the transmission tower foundation will uplift to a certain extent, while the right side will sink to a certain extent, indicating that the transmission tower foundation has an obvious tendency to overturn in the rightward direction. Overturning failure is a common mode of instability failure in transmission tower foundations. The inclination degree index is defined as the ratio of the vertical maximum displacement difference between the near and far ends of the transmission tower foundation to the length of the transmission tower foundation. According to the maximum inclination degree rate of 6‰ stipulated in the Technical Code for Design of Overhead Transmission Line Foundation (DL/T519-2014) [25], the inclination degrees of the transmission tower foundation under different impact durations with the type I impactor are calculated. When the impact duration is taken as initial impact time plus 10, 20, 30, and 40 ms, the inclination degrees of the transmission tower foundation are 0.2‰, 1.7‰, 4.0‰, 5.9‰, and 7.8‰, respectively, indicating that the continuous impact effect will have a serious influence on the inclination degree of the transmission tower foundation. It is also found that when the impact duration exceeds 30 ms, the inclination degree of the transmission tower foundation surpasses the prescribed limit value, which indicates that the transmission tower foundation will sustain an obvious overturning failure under the continuous impact of the type I impactor.

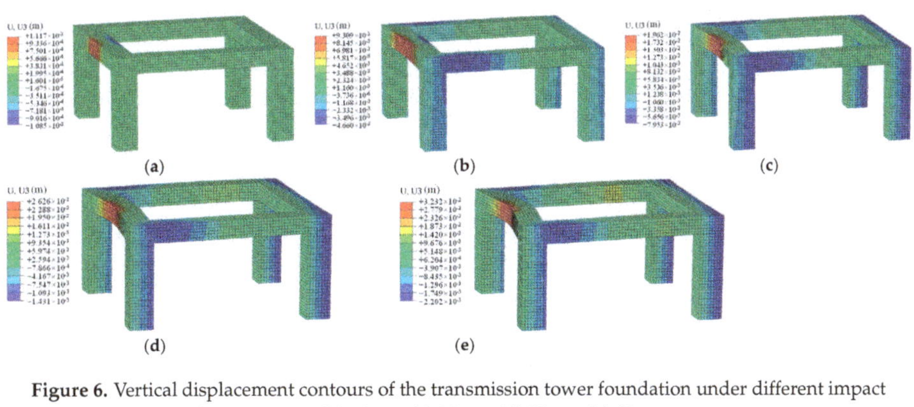

Figure 6. Vertical displacement contours of the transmission tower foundation under different impact durations: (**a**) initial impact time; (**b**) 10 ms; (**c**) 20 ms; (**d**) 30 ms; (**e**) 40 ms.

3.3. Different Shapes of Impactor

In practical engineering, there are various shapes and specifications of agricultural equipment and agricultural vehicles, resulting in a variety of different shapes of impact surfaces. In this study, these are simplified into a square impact surface, triangular impact surface, and semi-circular impact surface, corresponding to the type I, type II, and type III impactors, respectively. Figure 7 shows the pressure damage contours of the type I, type II, and type III impact objects on the transmission tower foundation at an impact speed of 30 km/h and an impact time of 5 ms.

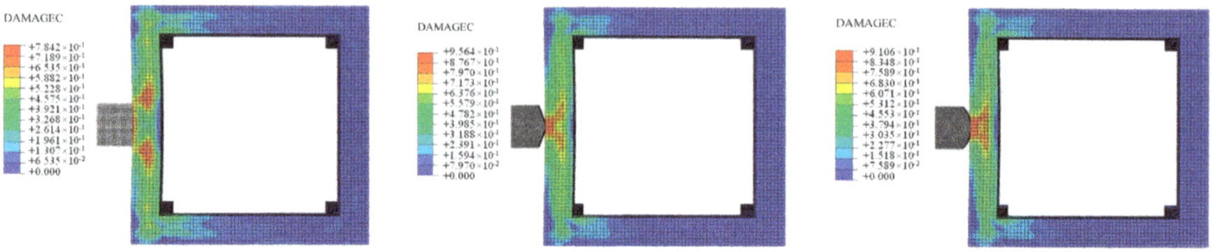

(a) (b) (c)

Figure 7. Contours of the compression damages of the transmission tower foundation under different impactor shapes: (**a**) type I; (**b**) type II; (**c**) type III.

It can be seen from Figure 7 that the damage position distributions and damage degrees of the transmission tower foundation are different under the impact effects of differently shaped impactors. The damage degree of the type I impactor is small in the front position, and there are two obvious damage areas at the corner position of the type I impactor. The obvious damage areas of the transmission tower foundations using the type II and type III impactors appear at the front positions of the impactors; that is, the obvious damage areas are mainly concentrated at the end positions of the impactors. It can also be seen from Figure 7 that under the impact effects of the type I, type II, and type III impactors, the compression damage values of the transmission tower foundation are 0.784, 0.956, and 0.911, respectively, which indicates that at the same impact time, the sharper the impactor is—that is, the smaller the impact contact area—the greater the compression damage value of the transmission tower foundation. With the continuation of the impact, the compression damage value of the transmission tower foundation under the Impact of the type II impactor reaches 0.99 first, which indicates that the sharper the impact object, the more likely the transmission tower foundation is to experience structural failure at an earlier stage.

In order to further compare the inclination degrees of the transmission tower foundation under the impact effects of the differently shaped impactors, the inclination degrees of the transmission tower foundation under different impact durations of the type I, II, and III impactors are shown in Figure 8. It can be seen that at the initial impact time, the inclination degrees of the transmission tower foundation for the different shapes are, from largest to smallest: type II, type III, and type I. This indicates that the smaller the impact contact area, the more susceptible the transmission tower foundation is to deformation under the impact effect. The inclination degree of the transmission tower foundation increases with increasing impact duration, and the variation law approximately follows a linear relationship. With the continuation of the impact, the inclination degree of the transmission tower foundation with the type I impactor exceeds the prescribed critical value of instability. Although the inclination degrees of the transmission tower foundation with the type II and type III impactors do not reach the prescribed critical values of instability, the small impact contact area causes the phenomenon of stress concentration, and the affected beam will take the lead in a structural failure during the impact process. The above results show that different impact contact areas will lead to different damage forms of the transmission tower foundation. When the impact contact area is small, structural failure is more likely to occur, while when the impact contact area is large, the transmission tower foundation is prone to overturning failure.

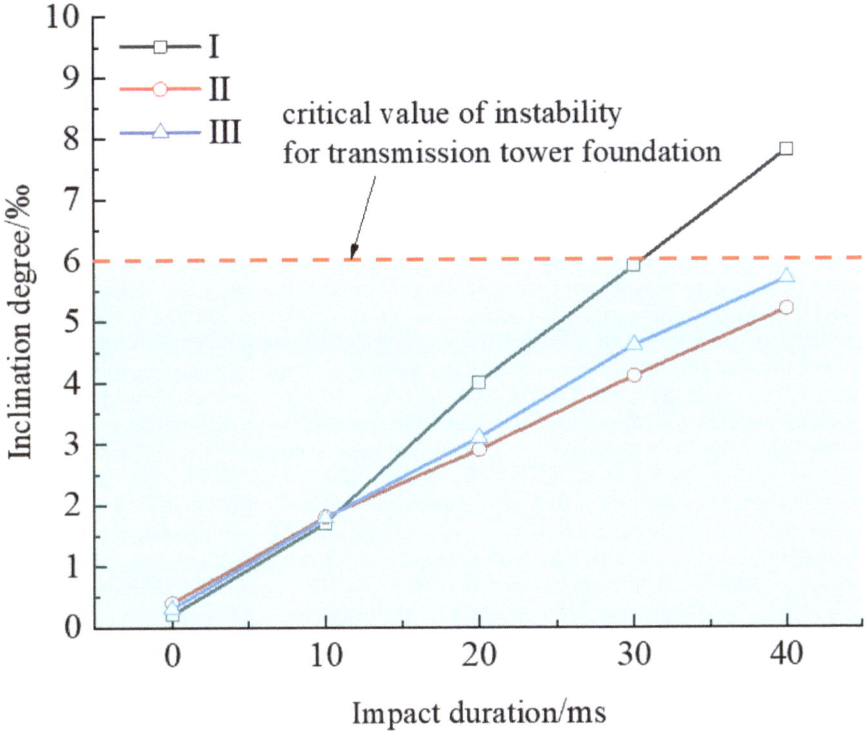

Figure 8. Inclination degrees of the transmission tower foundation under different impactor shapes.

3.4. Different Impact Positions

When the position of the impact is different, the damage to the transmission tower foundation is also different. The center points of impacts are selected as the mid-span, quarter-span, and edge positions of the connecting beams of the transmission tower foundation. Figure 9 shows the compression damage contours of the transmission tower foundation when the type I impactor impacts different positions at the initial impact time.

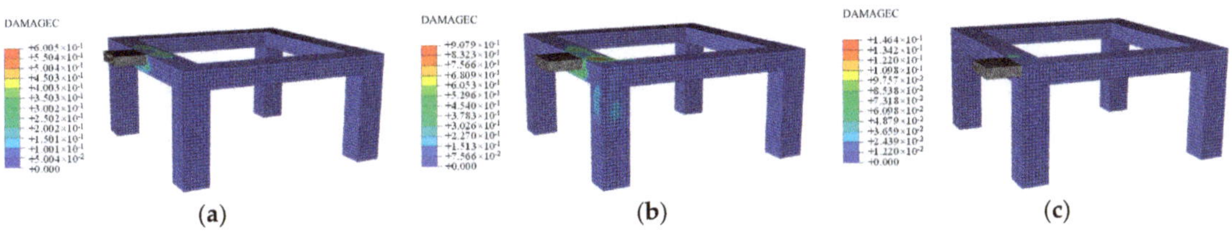

Figure 9. Compression damage contours of the type I impactor at different locations at the initial impact time: (**a**) impact mid-span position; (**b**) impact quarter-span position; (**c**) impact edge position.

As can be seen from Figure 9, there are significant differences in the damage to the transmission tower foundation when it is impacted at different locations at the initial impact time, and the maximum damage value is 0.60 when the impact position is the mid-span point; the maximum damage value is 0.91 when it is impacted at the quarter-span position; the maximum damage value is 0.15 when it is impacted at the edge position.

A comparison reveals that when impacted at the edge position, the damage area of the transmission tower foundation is the smallest; that is because the impact load is shared by the connecting beam and the column foundation, and the impact load can be transmitted to the adjacent connecting beam. However, when impacted at the quarter-span position, the damage value and damage area of the transmission tower foundation are the largest, and certain damage areas also appear at the column position, among which the edge of the connecting beam connected to the column foundation is the most seriously damaged. With the continuation of the impact, the damage value of this position first reaches 0.99, indicating that this position is the weakest position of the transmission tower foundation and should be paid special attention in engineering design and protection.

During the impact process, the deformation and failure of the soil mass are observed at the same time. Because the deformation and failure of the soil mass can easily lead to rainwater infiltration and decrease the bearing capacity of the foundation, the deformation and failure of the soil mass are related to the stability of the foundation and can further affect the long-term stability of the transmission tower foundation. The variation curves of the uplift values of unit node B at the left front corner of the transmission tower foundation during the impact process were extracted, as shown in Figure 10. It can be observed that the uplift values of unit node B increase with continuous impacting, and the increase rates of the uplift values of unit node B are different when different positions are impacted. At the same impact time, the uplift value of unit node B is the largest when the impact occurs at the edge position; the uplift value of unit node B at the quarter-span position is the second largest; the uplift value of unit node B is the smallest when the impact occurs at the mid-span position, and the uplift value of unit node B when the impact occurs at the edge position is significantly greater than the uplift values of element node B in the other two impact cases. It can be seen that an impact at the edge position has a significant influence on the deformation of the surrounding soil of the transmission tower foundation.

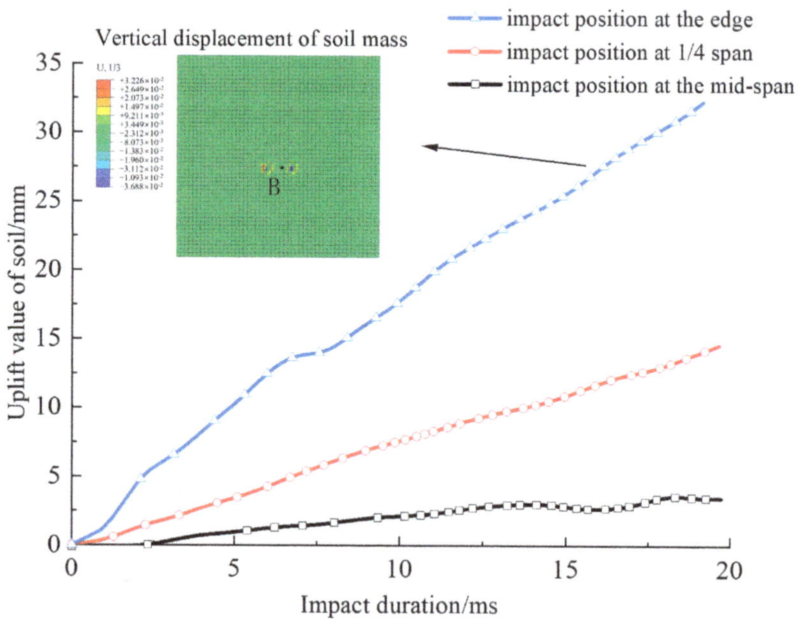

Figure 10. Variation curves of the uplift value of unit node B at the left front corner of the transmission tower foundation.

4. Comparative Analysis before and after Reinforcement Design

Longitudinal reinforcement bars and stirrups installed inside the transmission tower foundation can improve the bearing capacity and resistance to external loads of the transmission tower foundation and increase its overall stability, as shown in Figure 11. The density of the rebars in the model is 7800 kg/m^3, the elastic modulus is 2.06×10^5 MPa, and Poisson's ratio is 0.3. The element was the three-dimensional truss element T3D2, and an ideal elastic–plastic model was used to simulate the stress and deformation characteristics of the rebars. Then, the rebar model was built into the transmission tower foundation model by embedding command [26,27].

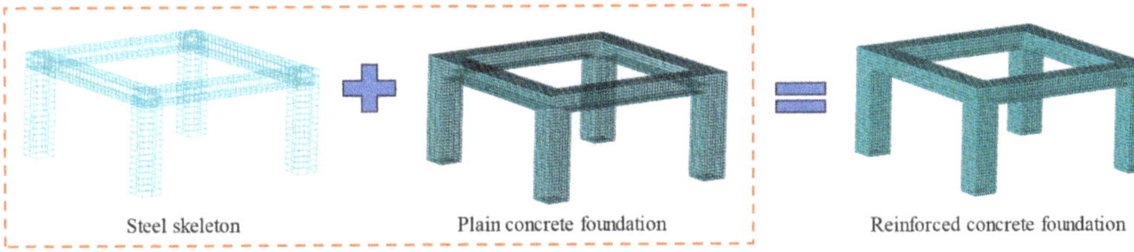

Figure 11. Schematic diagram of the numerical model of the reinforced transmission tower foundation.

Rebars of different types, that is, of different strengths, are used to reinforce the transmission tower foundation. Table 3 shows the deformation and damage values of the transmission tower foundation strengthened with rebars of different strengths when the impact velocity is 30 km/h and the impact duration is 20 ms. It can be seen that with the increase of the reinforcement strength, the maximum lateral displacement, maximum vertical displacement, inclination degree, and damage value of the transmission tower foundation decrease continuously. The reinforcement design using rebars of type HRB400 is taken as an example. Figure 12 shows the lateral displacement and vertical displacement contours of the reinforced transmission tower foundation when the impact velocity is 30 km/h and the impact duration is 20 ms. It can be seen that only the impacted beam of the reinforced transmission tower foundation deforms to a certain extent under the impact effect, while the impact has minimal effect on the other parts of the transmission tower foundation. In order to compare and analyze the deformation characteristics of the transmission tower foundation before and after reinforcement, Figure 12a,b are compared with Figures 4c and 6c. It can be found that the maximum lateral displacement and vertical displacement of the transmission tower foundation before reinforcement are 186.2 mm and 19.6 mm, respectively, at the same impact time, while the maximum lateral displacement and vertical displacement of the transmission tower foundation after reinforcement are 138.7 mm and 12.7 mm, respectively, and the lateral displacement and vertical displacement of the transmission tower foundation after reinforcement have been reduced by 25.5% and 35.2%, respectively. This indicates that the lateral displacement and vertical displacement of the transmission tower foundation are significantly reduced, and the deformation area is significantly reduced after reinforcement. It is further calculated that the inclination degree of the transmission tower foundation after reinforcement is reduced by 47%, indicating that the reinforcement design can significantly improve the anti-deformation and anti-overturning ability of the transmission tower foundation. The main reason for the above phenomenon is that the steel skeleton participates in resisting the external impact load during the impact process. It can also be seen from the stress contours of the steel skeleton (Figure 12c) that the rebars built into the connecting beam generate a large resistance to stress and share the stress of the original concrete structure. In addition, during the impact, the steel skeleton dissipates a portion of the impact energy through plastic deformation, thereby reducing the overall deformation of the structure, which can better prevent the expansion of cracks and the occurrence of fractures.

Table 3. Deformation and damage values of the transmission tower foundation after reinforcement with rebars of different strengths.

Rebar Type	Maximum Lateral Displacement (mm)	Maximum Vertical Displacement (mm)	Inclination Degree (‰)	Damage Value
HRB335	152.3	15.8	2.6	0.836
HRB400	138.7	12.7	2.1	0.818
HRB500	127.9	11.2	1.9	0.792

Figure 12. Contours of the lateral displacement, vertical displacement, and rebar cage stress of reinforced transmission tower foundation: (**a**) lateral displacement; (**b**) vertical displacement; (**c**) steel skeleton stress contour.

5. Conclusions

In response to engineering accidents involving impacts on transmission tower foundations by agricultural equipment and agricultural vehicles in practice, a finite element numerical simulation method is used to study and analyze the variations in the damage degree, displacement, and inclination degree of a transmission tower foundation under different impact velocities, different impact durations, different shapes of impactor, and different impact positions. An exploratory study was conducted to investigate the influence of impacts on the stability of the transmission tower foundation. This study provides guidance for assessments of the stability and safety analyses of transmission tower foundations under impact loads in actual engineering applications. The main conclusions can be drawn as follows:

(1). With increasing impact velocity, the damage value of the transmission tower foundation increases, and the damage area expands. The lateral displacement value of the transmission tower foundation increases with continuous impacting, the variation trend of the lateral displacement accords with a linear function distribution, and the lateral displacement of the transmission tower foundation increases with increasing impact velocity.

(2). With the increase of the impact duration, the lateral displacement curves of individual points along the longitudinal center line of the impacted connected beam move to the right continuously, and the lateral displacement values show a parabolic shape along the longitudinal center line of the connected beam. The inclination degree of the transmission tower foundation increases with increasing impact duration, which indicates that continuous impacting will seriously affect the inclination degree of the transmission tower foundation and even cause overturning failure.

(3). Different shapes of impactors have different damage position distributions and damage degrees for the transmission tower foundation under the impact effect. The sharper the impactor is—that is, the smaller impact contact area—the greater the compression damage value of the transmission tower foundation, and the more likely the transmission tower foundation is to experience structural failure at an earlier stage. The study also indicates that different impact contact areas lead to different modes of failure in transmission tower foundations. When the impact contact area is small, structural failure is more likely to occur, while when the impact contact area is large, a transmission tower foundation is prone to overturning failure.

(4). The damage values and damage areas of the transmission tower foundation are different when the positions of the impactors are different. The damage value and damage area of the transmission tower foundation are the largest when the impact occurs at the quarter-span position, and the edge of the beam connected with the column foundation is the most seriously damaged, which means that this position is the weakest position of the transmission tower foundation. At the same impact time, the uplift value of unit node B is the largest when the impact occurs at the edge position, which means that an impact at the edge position has a significant influence on the deformation of the surrounding soil of the transmission tower foundation.
(5). By using reinforcement bars of different strengths, the maximum lateral displacement, maximum vertical displacement, inclination degree, and damage values of the transmission tower foundation are continuously reduced. During the impact process, the steel skeleton participates in resisting the external impact load and shares the stress of the original concrete structure. Additionally, the steel skeleton dissipates a portion of the impact energy through plastic deformation. The reinforcement design significantly improves the deformation resistance and overturning resistance of the transmission tower foundation.

Author Contributions: L.S.: writing, methodology, conceptualization, data curation, and validation; S.C.: reviewing and funding acquisition; L.C.: supervision and writing; X.L.: data analysis and reviewing; J.L.: methodology and data analysis. All authors have read and agreed to the published version of the manuscript.

Funding: This research was funded by the National Natural Science Foundation of China (Grant No. 41902277 and 42172302) and Fundamental Research Funds for the Central Universities, CHD (Grant No. 300102282201), for which the authors are very thankful.

Institutional Review Board Statement: The study did not require ethical approval, and can exclude this statement.

Informed Consent Statement: The study did not involve humans, and can exclude this statement.

Data Availability Statement: The data used to support the findings of this study are included in the article.

Conflicts of Interest: The authors declare no conflict of interest.

References

1. Liang, H.B.; Xie, Q.; Bu, X.H.; Cao, Y.X. Shaking table test on 1000 kV UHV transmission tower-line coupling system. *Structures* **2020**, *27*, 650–663. [CrossRef]
2. Tian, L.; Pan, H.Y.; Ma, R.S. Probabilistic seismic demand model and fragility analysis of transmission tower subjected to near-field ground motions. *J. Constr. Steel Res.* **2019**, *156*, 266–275. [CrossRef]
3. Fu, Z.Y.; Tian, L.; Liu, J.C. Seismic response and collapse analysis of a transmission tower-line system considering uncertainty factors. *J. Constr. Steel Res.* **2022**, *189*, 107094. [CrossRef]
4. Park, H.; Choi, B.H.; Kim, J.J.; Lee, T.H. Seismic performance evaluation of high voltage transmission towers in South Korea. *KSCE J. Civ. Eng.* **2015**, *20*, 2499–2505. [CrossRef]
5. Albayrak, U.; Morshid, L. Evaluation of seismic performance of steel lattice transmission towers. *Civ. Eng. J.* **2020**, *6*, 2024–2044. [CrossRef]
6. Nazemi, M.; Dehghanian, P. Seismic-resilient bulk power grids: Hazard characterization, modeling, and mitigation. *IEEE Trans. Eng. Manag.* **2020**, *67*, 614–630. [CrossRef]
7. Pan, H.Y.; Tian, L.; Fu, X.; Li, H.N. Sensitivities of the seismic response and fragility estimate of a transmission tower to structural and ground motion uncertainties. *J. Constr. Steel Res.* **2020**, *167*, 105941. [CrossRef]
8. Li, J.K.; Gao, F.; Wang, L.H.; Ren, Y.N.; Liu, C.C.; Yang, A.Q.; Yan, Z.; Tao, J.; Li, C.B. Collapse mechanism of transmission tower subjected to strong wind load and dynamic response of Tower-Line system. *Energies* **2022**, *15*, 3925. [CrossRef]
9. Ma, L.Y.; Khazaali, M.; Bocchini, P. Component-based fragility analysis of transmission towers subjected to hurricane wind load. *Eng. Struct.* **2021**, *242*, 112586. [CrossRef]
10. Bernuzzi, C.; Crespi, P.; Montuori, R.; Nastri, E.; Simoncelli, M.; Stochino, F.; Zucca, M. Resonance of steel wind turbines: Problems and solutions. *Structures* **2021**, *32*, 65–75. [CrossRef]

11. Mozakka, I.; Zeynalian, M.; Hashemi, M. A feasibility study on construction methods of high voltage transmission towers' foundations. *Arch. Civ. Mech. Eng.* **2021**, *21*, 41. [CrossRef]
12. Liu, H.Y.; Du, M.R.; Zhang, B.Y.; Lin, Z.B.; Liu, C.W.; Wang, F. Study on the combined mining scheme for coal resources under high-voltage pylons and the reinforcement for pylons. *Energies* **2022**, *15*, 3978. [CrossRef]
13. Yang, F.L.; Yang, J.B.; Han, J.K.; Zhang, Z.F. Study on the limited values of foundation deformation for a typical UHV transmission tower. *IEEE Trans. Power Deliv.* **2010**, *25*, 2752–2758. [CrossRef]
14. Gao, X.Y.; Yi, R.; Zhang, L.Q.; Jiang, X.; Li, J.X. Failure analysis of transmission tower in full-scale tests. *Buildings* **2022**, *12*, 389. [CrossRef]
15. Shu, Q.J.; Yuan, G.L.; Guo, G.L.; Zhang, Y.F. Limits to foundation displacement of an extra high voltage transmission tower in a mining subsidence area. *Int. J. Min. Sci. Technol.* **2012**, *22*, 13–18. [CrossRef]
16. Zhou, Y.B.; Zhou, L.; Duan, Z.Q.; Ke, F.C.; Liu, H. Analysis of the influence of the distribution and development of soil caves on the stability of High-Voltage Transmission Tower foundations. *Adv. Civ. Eng.* **2022**, *2022*, 2856947. [CrossRef]
17. Yao, A.J.; Lin, J.B.; Ren, B. The influence of small clear-distance tunnel construction on adjacent high-voltage transmission tower foundation. *Adv. Civ. Eng.* **2023**, *2023*, 2533212. [CrossRef]
18. Wang, F.; Zhang, G.H.; Li, W.W.; Nie, H.W. Numerical investigation into the effects of controlled tunnel blast on dynamic responses of the transmission tower. *Arch. Civ. Mech. Eng.* **2023**, *2023*, 6021465. [CrossRef]
19. Xin, W.S.; Liu, J.K.; Li, X.; Hu, T.F.; Kolos, A. Frost jacking characteristics of prefabricated cone cylindrical tower foundation in cold regions. *Cold Reg. Sci. Technol.* **2023**, *211*, 103847. [CrossRef]
20. Yuan, H.P.; Zhao, P.; Wang, Y.X.; Zhou, H.L.; Luo, Y.H.; Guo, P.P. Mechanism of deformation compatibility and pile foundation optimum for Long-Span tower foundation in Flood-Plain deposit zone. *Int. J. Civ. Eng.* **2016**, *15*, 887–894. [CrossRef]
21. Chi, Y.; Yu, M.; Huang, L.; Xu, L.H. Finite element modeling of steel-polypropylene hybrid fiber reinforced concrete using modified concrete damaged plasticity. *Eng. Struct.* **2017**, *148*, 23–35. [CrossRef]
22. Van Nguyen, D.; Kim, D.; Nguyen, D. Nonlinear seismic soil-structure interaction analysis of nuclear reactor building considering the effect of earthquake frequency content. *Structures* **2020**, *26*, 901–914. [CrossRef]
23. Xiang, S.; Zeng, L.; Liu, Y.H.; Mo, J.X.; Ma, L.L.; Zhang, J.C.; Chen, J. Experimental study on the dynamic behavior of T-shaped steel reinforced concrete columns under impact loading. *Eng. Struct.* **2020**, *208*, 110307. [CrossRef]
24. Bian, M.H.; Peng, J.N.; Qin, S.L.; Zhang, X.S.; Li, J.H. A simplified analytical method for lateral dynamic responses of a transmission tower due to rockfall impact. *Front. Mater.* **2023**, *10*, 1229327. [CrossRef]
25. *DL/T5219-2014*; Power Industry Standard of the People's Republic of China, Technical Code for Design of Overhead Transmission Line Foundation. China Planning Press: Beijing, China, 2014.
26. Lee, S.; Abolmaali, A.; Shin, K.; Lee, H. ABAQUS modeling for post-tensioned reinforced concrete beams. *J. Build. Eng.* **2020**, *30*, 101273. [CrossRef]
27. Quan, D.Z.; Chai, S.B.; Wang, Y.L.; Fan, Z.S.; Bu, Y.H. 3-D numerical simulation of seismic response of the induced joint of a sub-way station. *Buildings* **2023**, *13*, 1244. [CrossRef]

Disclaimer/Publisher's Note: The statements, opinions and data contained in all publications are solely those of the individual author(s) and contributor(s) and not of MDPI and/or the editor(s). MDPI and/or the editor(s) disclaim responsibility for any injury to people or property resulting from any ideas, methods, instructions or products referred to in the content.

buildings

MDPI

Article

Study on the Influence of a Rubber-Modified Soil Isolation Layer on the Isolation Performance of Frame Structures with Different Foundation Forms

Shaoqiang Chai [1], Yong Chen [1], Dongbo Cai [1], Wei Wang [1], Qihao Chen [1] and Jinhao Liu [2,*]

1 The Seventh Engineering Co., Ltd. of CFHEC, Zhengzhou 451450, China; chaishaoqi@chaishaoqi.ntesmail.com (S.C.); photonman810@126.com (Y.C.); caidongbo@caidongbo.ntesmail.com (D.C.); zhongjiao_wang@126.com (W.W.); 18595416600@163.com (Q.C.)
2 School of Civil Engineering, Chang'an University, Xi'an 710064, China
* Correspondence: sdlyliu1119@163.com

Abstract: In order to investigate the seismic performance of a rubber-modified soil isolation layer, a three-dimensional finite element model was constructed using finite element analysis software, utilizing a two-story frame structure as the engineering background. Nonlinear dynamic time history analysis and comparisons were performed against the seismic performance of the structure. The evaluation was based on several parameters, including the contact area of the base, the thickness of the rubber-particle-modified soil isolation layer, ground motion records with varying amplitudes, and seismic frequency spectrum characteristics. The research findings indicate that the implementation of a rubber-modified soil isolation layer effectively mitigates the peak acceleration, horizontal displacement, and shear stress of the frame structure. This not only enhances the seismic performance of the structure but also enlarges the contact area of the base. Increasing the thickness of the rubber-modified soil isolation layer will effectively decrease the peak acceleration, horizontal displacement, and shear stress of the structure during seismic events. The effectiveness of the isolation provided by the rubber-modified soil layer improves as the intensity of the ground motion record increases.

Keywords: frame structure; rubber-modified soil; seismic performance; dynamic time history analysis; seismic isolation

1. Introduction

A significant quantity of waste rubber tires is generated every year in China, resulting in considerable environmental pollution. The rubber-modified soil, produced by recycling and crushing waste tires and blending them with loess, exhibits favorable physical and mechanical properties, and effectively addresses the issue of environmental pollution. Rubber materials have found extensive applications in the field of civil engineering. An example of this application is the amalgamation of waste rubber tires with sand to create a cost-effective and practical structural foundation, which demonstrates excellent performance. The rubber-modified soil offers several advantages, including a straightforward structure, positive environmental influence, dependable load-bearing capacity, and cost-effectiveness. This technique has the potential to significantly enhance the safety factor of rural building structures. Another approach involves mechanically breaking down waste rubber tires into fragments or particles. These fragments or particles are then combined with sand and soil in specific proportions to create a new material known as rubber sand or rubber soil mixture, which is practical for engineering applications. The blend of rubber sand and rubber soil is characterized by strong performance in terms of elastic deformation, excellent elastic recovery, a lightweight composition, a low shear modulus, and high damping properties. It has been proven that rubber material is a promising new material with commendable performance and environmental benefits.

Currently, several scholars have conducted experimental research on the mechanical properties of rubber particles when integrated into soil samples, and yielded meaningful results. For instance, Soltani et al. [1] conducted a study on the influence of fine and coarse recycled tire rubber on the expansion and shrinkage of high expansive soil mixture. Two types of rubber were blended into the soil at four distinct content levels (i.e., the mass ratio of rubber to dry soil) of 5, 10, 20, and 30%, respectively. The results demonstrated that rubber with a maximum content of 10% was deemed the most suitable option. In cases where environmental concerns take precedence and strength and stiffness are not significant issues, up to 20% of rubber content can also be deemed acceptable. Raeesi et al. [2] investigated the mechanical properties of a substantial permeable pavement test site situated in South Australia. This pavement was composed of tires and rock-derived aggregates (TDA and RDA) bonded together using a polyurethane (PUR)-based adhesive. Various TDA hybrid designs were employed, involving different RDA content, size, shape, and diverse PUR content. In total, an area spanning approximately 400 square meters was laid as part of the study. Increasing the amount of RDA aimed to enhance the interaction and friction between particles, promoting a more pliable rigid body. This in turn elevated the strength and stiffness of the pavement while concurrently limiting its susceptibility to strain during development. Akbarimehr et al. [3,4] employed three distinct forms of rubber (granular, fibrous, and flake) to assess the shear strength of waste clay infused with rubber. The results indicated that an increase in rubber particle size led to enhancements in both the strength and shear strain of the mixture. At varying ultimate stress levels, the mixture incorporating rubber dust exhibited a strength that was 10–25% higher compared to the mixture containing rubber powder. Saparudin et al. [5] incorporated 5%, 10%, and 15% of tire rubber fragments into the soil samples, and conducted tests to assess the physical properties of clay sand, including particle size distribution and plasticity index. The test results demonstrated that the tire rubber powder met the requirements for effectively stabilizing the clay sand soil in highway construction roadbeds. Sadek and El-Attar [6] incorporated tire fragments of various sizes into the mixture for producing cement bricks. A novel type of wall was created, and its mechanical properties were thoroughly investigated. The study revealed that both the size and content of rubber tire fragments significantly influenced the strength of the blocks. Chen Yong et al. [7] studied the shear mechanical properties of modified loess improved by incorporating rubber particles and EICP technology. The results demonstrated that the addition of rubber powder had a positive influence on enhancing the shear strength of the loess to a certain extent. Furthermore, when combined with EICP technology, the shear strength of the modified loess increased by nearly 50%. These studies collectively indicate that appropriate rubber particles can significantly enhance the mechanical properties of soil. The se conclusions provide valuable guidance for the treatment and utilization of rubber in various applications.

Regarding structural shock absorption, researchers have conducted various forms of research and foundation, and have successfully demonstrated that enhancing the energy absorption capacity of foundations can effectively mitigate structural vibrations. For instance, Zhou et al. [8] based on dynamic numerical analysis of nonlinear time history analysis of different types of local damping effect of shock absorption layer. Select six actual ground motions with different main frequencies and set their peak ground acceleration (PGA) to 0.2 g, 0.4 g, 0.6 g, and 0.8 g respectively. The damage index of secondary lining, total dissipated energy and equivalent plastic strain of surrounding rock are used as indexes to reflect the damping performance of local damping layer. The numerical results show that the damping performance of the double local damping layer is the best, and the damping performance of the damping layer under large earthquake is better than that of the small earthquake damping layer. Bandyopadhyay et al. [9] performed a comparative analysis between pure sand cushion and rubber sand cushion using a small-scale shaking table test. The results demonstrated that the pure sand mat was only able to isolate strong vibration inputs exceeding 0.65 g. In contrast, the rubber sand mat with 50% rubber content proved to be a cost-effective and efficient foundation isolator. Hazarika H. et al. [10] conducted structural shaking table isolation tests on rubber particles and sand isolation pads in submerged conditions. The analysis and conclusions highlighted that these isolation pads effectively reduced acceleration during earthquake events and played a significant role in shock absorption. Alhan and Gavin [11] employed frequency domain analysis and seismic time history analysis to investigate the influence of isolation damping on higher-mode effect and the interlayer displacement ratio. Due to the significance of higher-mode effect and bidirectional ground motion in the dynamic performance of these structural systems, a straightforward comparison of the isolation damping mechanism using a single degree or two degrees of freedom was not feasible. To integrate these critical aspects into the investigation of the dynamic behavior of these structures, a series of tests was conducted on eight-story prototype building models. The results demonstrated that the isolator displacement decreased with an increase in the damping. Conversely, suitable levels of isolation stiffness and damping can effectively restrict basement drift without causing a notable impact on floor acceleration and the interlayer drift ratio. D'Amato et al. [12] demonstrated isolation technology on Italian heritage sites, particularly focusing on reinforced concrete frame constructions with historical significance. A three-dimensional finite element model was employed to simulate the seismic responses of both the existing building and an enhanced isolation system, comprising an elastomer and a sliding isolator. Additionally, this paper introduces a novel method for estimating structural seismic resistance and its application.

The research method, involving structural vibration tests, has proven effective. However, it is important to note that this approach is time-consuming and expensive and offers limited working conditions for analysis. Moreover, analytical methods encounter challenges in accurately analyzing complex structural responses. Hence, numerical simulation offers convenience and efficiency for the effective analysis of the structural vibration response under complex working conditions [13–15]. Various numerical methods, including the boundary element method, finite difference method, and finite element method, have been employed to investigate the dynamic interaction between the soil and structure at a specific site. For instance, Cui et al. [16] studied the influence of the thickness of the damping layer on the damping effect. Taking a shallow double-arch rectangular tunnel project in a city as an example, the numerical simulation software ABAQUS 2022 was used to compare the damping effect of 50 kinds of damping layers with thickness of 50, 100 and 150 mm respectively. When the shock absorbing layer with a thickness of 4mm is applied, the convergence of the side wall decreases by 65.14%, the maximum and minimum principal stresses decrease significantly, among which the maximum and minimum principal stresses decrease by 03.6%, the maximum shear stresses decrease by 17.42%, and the minimum safety factor increases by at least 72.100%. The shock absorbing layer of 50 mm thickness is better than the shock absorbing layer of 150 mm and 100mm thickness. Li et al. [17] proposed a numerical stability analysis method based on the subsystem of a three-dimensional viscoelastic artificial boundary element explicit algorithm. This method identified the subsystem that governs the stability of the overall value system of artificial boundaries. It combined the three-dimensional viscoelastic artificial boundary element explicit integral algorithm with an analytical solution for stability conditions. Building on this foundation, the method can effectively address issues related to determining the maximum time increment for the viscoelastic artificial boundary element. Brunet et al. [18] utilized a concentrated mass model to simulate the subsoil structure system located 1–5 m behind the rubber sand cushion for nonlinear analysis. The results demonstrated that the isolation effect was more pronounced with peak values of seismic input. The recommended cushion thickness was 2–3 m and the optimal rubber sand content was found to be 35%. Quan et al. [19] established a three-dimensional numerical model of a subway station with multiple sensing seams. The research explored the seismic response of various induced joints in different positions and forms using different ground motion records by altering the frequency spectrum characteristics and acceleration peak values.

Zheng et al. [20] selected a two-story brick and concrete unstructured column building in a typical village as a prototype, and designed a vibration table comparison test with a 1:2 ratio between the isolated and non-isolated layers. Firstly, the response spectrum analysis method was utilized to select three ground motion records of both isolated and non-isolated structures. Subsequently, various input scenarios were investigated using numerical analysis to record the vibration of the structure and understand the seismic response of the structure. Based on the results obtained from the numerical simulations, the sensor layout and loading scheme for the structural shaking table test were designed. Cao et al. [21] conducted simulated seismic shaking table tests on four structural models of rural buildings. Among these models, one was a sliding foundation masonry building, and another was a seismic-resistant building. The structural models tested included a special-shaped insulation block masonry house with a structural column and an ordinary insulation block masonry house. The findings indicate that the sliding foundation isolation structure is characterized by its simplicity, cost-effectiveness, and reliable performance.

The above research shows that rubber-particle-modified soil offers various advantages, including a high elastic modulus, a significant damping ratio, and good resilience. Utilizing rubber-particle-modified soil as a building foundation can enhance the energy absorption capabilities without compromising its strength of foundation. Furthermore, the cost and construction complexity associated with rubber-particle-modified loess are lower compared to traditional isolation technologies. Therefore, in this study, the modified loess containing rubber particles laid on the bottom of the foundation of the frame structure is regarded as a seismic isolation layer used to explore the influence of its energy absorption on the vibration response of the frame structure. In this paper, a numerical simulation method is employed to comprehensively study the mechanical properties of the enhanced loess, the influence of the thickness of the improved foundation, and various types of ground motion records.

2. ABAQUS Finite Element Calculation Model

Model Construction

The model in this paper is a two-layer frame structure designed following the specifications outlined in the General Code for Concrete Structures [22]. For the columns in this model, the section size is 300 mm × 300 mm, the reinforcement comprises 6Φ12 bars, and the spacing of the stirrups is Φ8@200 mm. The beam section size in this model is 300 mm × 600 mm, the reinforcement comprises 6Φ12 bars, and the spacing of the stirrups is Φ8@200 mm. The concrete grade for both the column and beam is C40, while the steel bar grade is Q345, and the stirrup grade is HRB400. The fundamental period of the structure is set within the range of 0.016–0.2 s. The finite element model in this paper comprises four main components: the two-layer frame structure, the rubber-modified soil damping layer, and the foundation soil. The foundation forms are categorized into raft foundation and strip foundation, with a layer of rubber-modified soil placed at the base of the foundation. The analysis models for both the non-isolated and isolated structures were constructed in the finite element analysis software ABAQUS 2022. The upper structure of this structural model comprises a two-story frame structure with dimensions of 9.6 m in length, 5.6 m in width, and 3 m in height. This structural model is derived from a real engineering case, which provides the basis for the analysis and simulations conducted in this study. Given that a significant portion of rural houses in China employs this straightforward frame structure, it has been chosen as the focus of this study. By doing so, the research outcomes can be effectively utilized and applied in the subsequent promotion and application research for this common two-story structure. The dimensions of the soil area are specified as 14 m in length, 10 m in width, and 8 m in depth. Equivalent linearization treatment is implemented for the analysis. In the finite element model, an elastic–plastic constitutive model (CDP model) is used to simulate the dynamic damage evolution process of structural granular concrete. The reinforcement employs a double broken line constitutive model to simulate mechanical deformation characteristics, while the soil is modeled using a linear elastic model. The contact between the rubber-particle-modified soil isolation layer, the foundation, and the underlying soil is defined as surface-to-surface contact. Artificial viscoelastic boundaries are utilized as the boundary conditions. Reinforcements are embedded within the frame structure. The model structure employs full integral element C3D8, the model foundation utilizes a simplified integral element C3D8R, and the reinforcement is represented using a three-dimensional truss element T3D2.

In this paper, a constructed finite element model is utilized to simulate the seismic response of the two-layer frame structure, considering varying foundation forms and different thicknesses of the rubber-modified soil damping layer. Firstly, in order to investigate the influence of the base contact area on the seismic isolation effect of the structure, two foundation forms are employed in establishing the finite element model: the raft foundation and strip foundation. In order to facilitate the investigation of how the rubber-modified soil damping layer affects the displacement of the frame structure, and considering the symmetry of frame structure, one side of the column is chosen for observation. The model is illustrated in Figure 1, and the material parameters of each material are provided in Table 1.

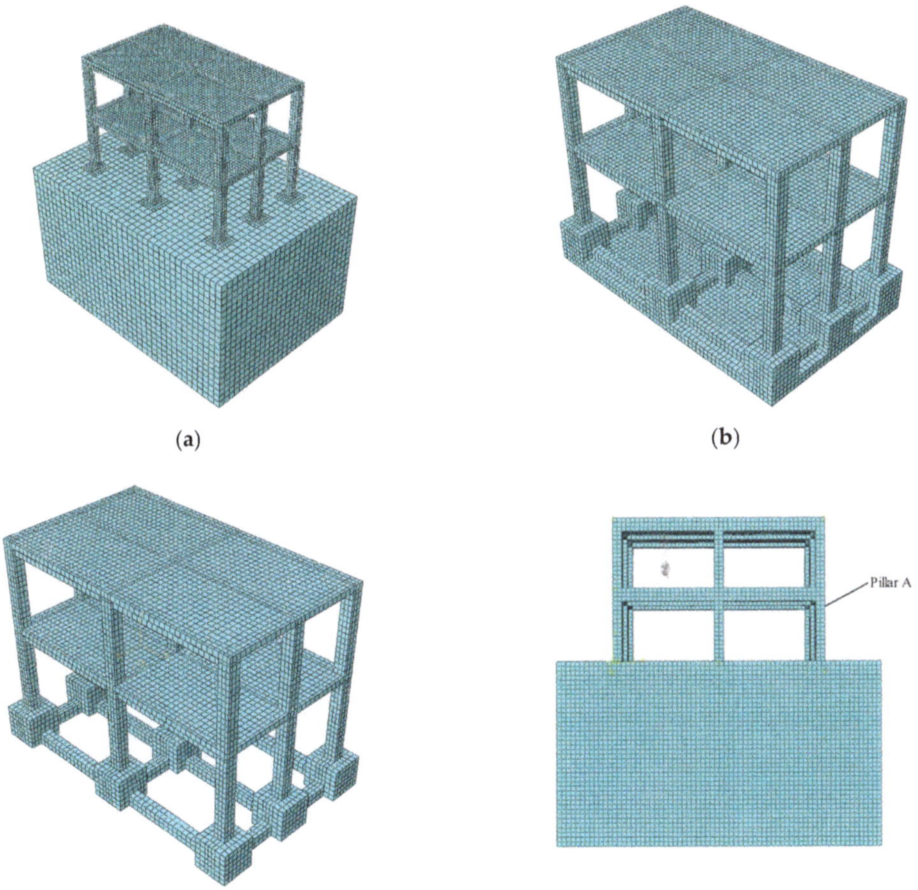

(**a**) (**b**) (**c**) (**d**)

Figure 1. Model diagram: (**a**) whole model figure; (**b**) raft foundation; (**c**) strip foundation; (**d**) position diagram of observation pillar A.

Table 1. Material parameters.

Material	Density (kg/m^3)	Modulus of Elasticity (MPa)	Poisson's Ratio	Damping Ratio
Concrete	2500	32,500	0.2	0.05
Foundation soil	1960	109.87	0.26	0.05
Rubber-modified soil	1460	50	0.49	0.3
Steel reinforcement bar	7850	206,000	0.3	0.04

The viscoelastic boundary condition [23] takes into account the radiation damping of the foundation and effectively simulates the absorption and dispersion of ground motion records by the modified soil layer in the semi-infinite body of the original soil layer. This contributes to more reasonable seismic results for the structure. Hence, to ensure precision and reliability in the Abaqus finite element simulation software, the viscoelastic artificial boundary [24] is adopted as the soil boundary.

3. Ground Motion Record Selection and Input

3.1. Ground Motion Record Selection

When using the time history analysis method, the acceleration–time history curve from actual strong earthquake records and artificial simulations should be carefully chosen based on the building site type and the designated earthquake design group [25]. In this paper, according to the "General Code for Seismic Design of Buildings and Municipal Engineering", the ground motion records include natural and artificial seismic ground motion records, with natural ground motion records accounting for no less than 2/3. Given that this study is fundamental research and not engineering design, three seismic motions with different spectral compositions were selected as input motions to explore the influence of various types of seismic ground motion records on the seismic response of the model. This included two real ground motion records and one artificial ground motion record. All three ground motion records were not scaled. The two historic earthquakes included were as follows: the Friuli (Italy) earthquake of 6 May, 1976 (Mag: 6.5; Rjb: 14.97 km; Rrup: 15.82 km; Duration: 36.32 s) and the San Fernando earthquake at Santa Anita Dam, 1971, (Mag: 6.61; Rjb: 30.7 km; Rrup: 30.7 km). In order to investigate the influence of various types of ground motion on the seismic response of the model, three ground motions characterized by distinct spectra were chosen as the input ground motions. One example conformed with the standard inclusion of artificial ground motion records as part of the response spectrum. To investigate a ground motion record with the same peak acceleration, the peak acceleration of the original ground motion record was adjusted to 0.2 g and the modified input numerical model was utilized for calculation. The time history curve and Fourier spectrum of the ground motion record acceleration are presented in Figure 2.

3.2. Ground Motion Record Input

As the finite element model in this paper incorporates a viscoelastic artificial boundary, an equivalent load input method is employed for introducing ground motion. This ground motion record is in the form of a vertically propagated shear wave. The seismic input mode is the equivalent nodal force input mode. The velocity–time history and displacement–time history are determined by integrating and doubly integrating the acceleration–time history, respectively.

3.3. Calculation Scheme

To investigate the influence of the contact area between the rubber-modified soil damping layer and various foundation forms, as well as the thickness of the isolation layer, on the damping effect of the frame structure, two foundation forms—the raft foundation and strip foundation—are individually defined. The isolation layer's thickness is set at 0 m, 0.4 m, 0.6 m, and 0.8 m. Additionally, three types of ground motion record acceleration time history curves are utilized. The seismic performance simulation was conducted for a total of 24 project scenarios, as outlined in Table 2.

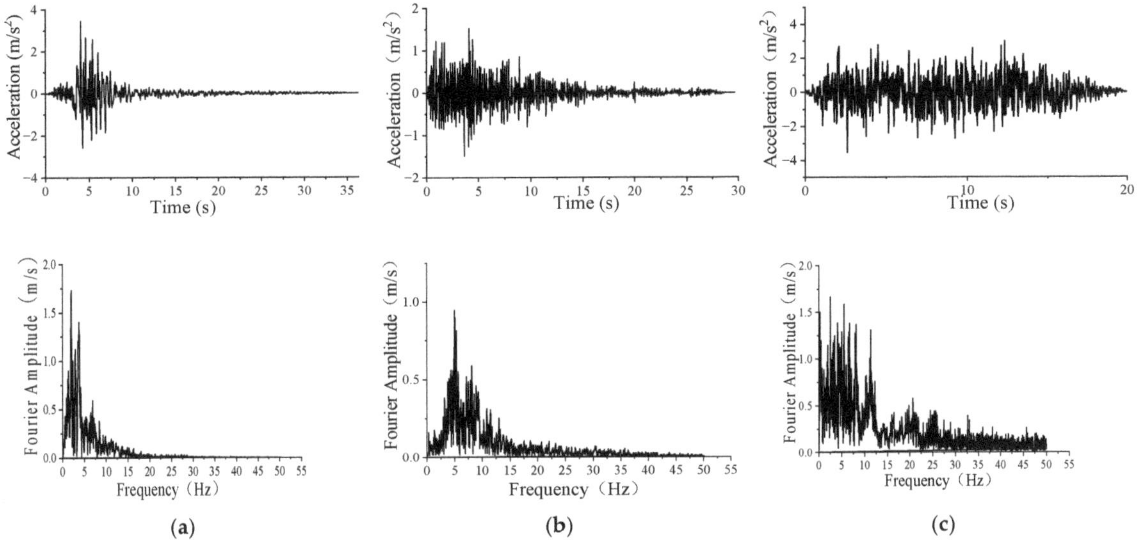

Figure 2. Time history curve of ground motion record acceleration: (**a**) Friuli; (**b**) San Fernando; (**c**) artificial ground motion record.

Table 2. Project status table.

Project Status	Type of Foundation	Ground Motion Record	Seismic Isolation Layer Thickness	Project Status	Type of Foundation	Ground Motion Record	Seismic Isolation Layer Thickness
1	Raft foundation	Friuli	0 m	13	Strip foundation	Friuli	0 m
2	Raft foundation	Friuli	0.4 m	14	Strip foundation	Friuli	0.4 m
3	Raft foundation	Friuli	0.6 m	15	Strip foundation	Friuli	0.6 m
4	Raft foundation	Friuli	0.8 m	16	Strip foundation	Friuli	0.8 m
5	Raft foundation	San Fernando	0 m	17	Strip foundation	San Fernando	0 m
6	Raft foundation	San Fernando	0.4 m	18	Strip foundation	San Fernando	0.4 m
7	Raft foundation	San Fernando	0.6 m	19	Strip foundation	San Fernando	0.6 m
8	Raft foundation	San Fernando	0.8 m	20	Strip foundation	San Fernando	0.8 m
9	Raft foundation	Artificial ground motion record	0 m	21	Strip foundation	Artificial ground motion record	0 m
10	Raft foundation	Artificial ground motion record	0.4 m	22	Strip foundation	Artificial ground motion record	0.4 m
11	Raft foundation	Artificial ground motion record	0.6 m	23	Strip foundation	Artificial ground motion record	0.6 m
12	Raft foundation	Artificial ground motion record	0.8 m	24	Strip foundation	Artificial ground motion record	0.8 m

The analysis explores the influence of the base contact area, rubber-particle-modified soil isolation layer, ground motion spectrum, and ground motion intensity on the structure's isolation effect under various working conditions.

4. Calculation Results and Analysis

4.1. Influence of Base Contact Area on Isolation Effect

4.1.1. Acceleration Response Analysis of the Top Layer of the Structure

In this section, the finite element model is utilized to simulate the seismic response of a two-story frame structure considering varying foundation forms and different thicknesses of the rubber-modified soil damping layer. Firstly, the influence of the base contact area on the seismic isolation effect of the structure is explored. Two foundation forms, namely the raft foundation and strip foundation, are employed to establish the finite element model. An artificial ground motion record with a peak acceleration of 0.2 g is chosen as the dynamic load for the finite element model. The artificial ground motion record is a shear ground motion record propagating vertically upward. The time history curves of the vertex acceleration of the structure are illustrated in Figure 3.

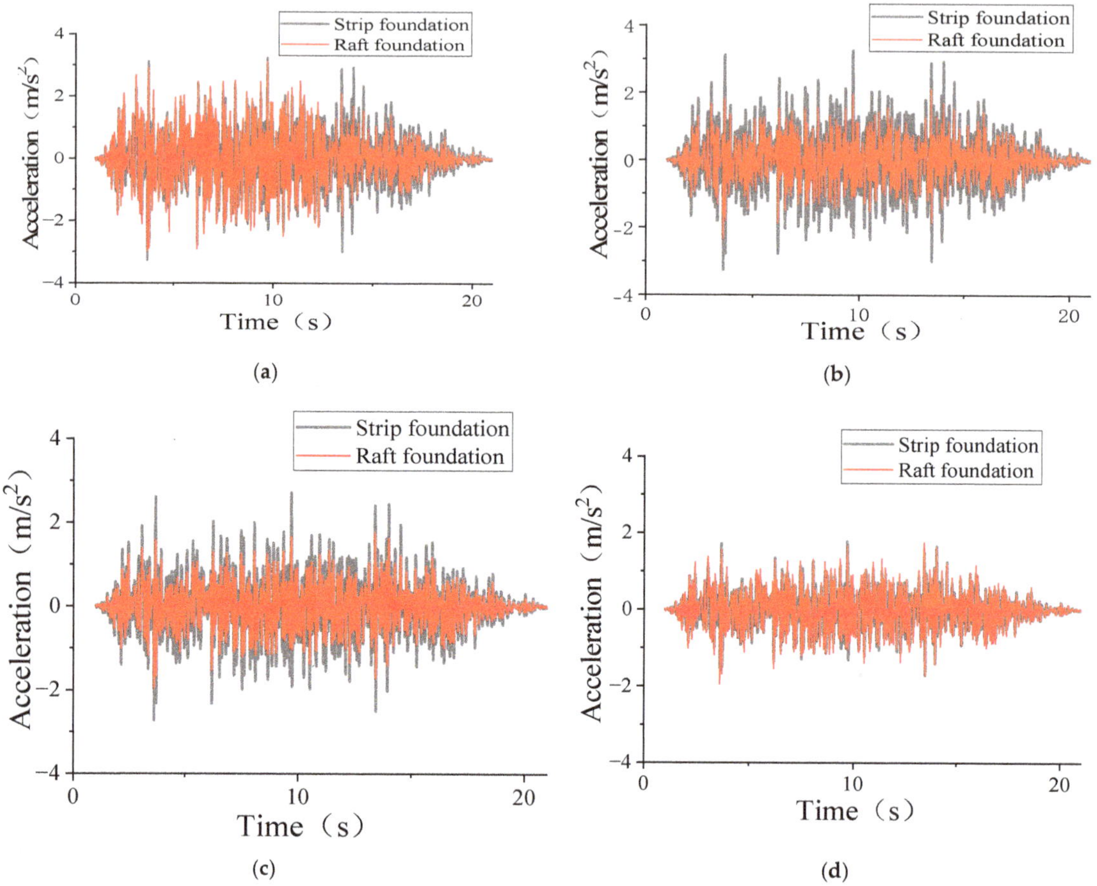

Figure 3. Improved soil thickness under the different foundation forms with the same acceleration–time history curve of contrast: (**a**) 0 m seismic isolation layer thickness; (**b**) 0.4 m seismic isolation layer thickness; (**c**) 0.6 m seismic isolation layer thickness; (**d**) 0.8 m seismic isolation layer thickness.

The following two points can be derived from the qualitative analysis presented in Figure 3:

(1) Common points: the shape, duration, and time of the peak acceleration curve of vertex acceleration for the two foundation forms are similar. This indicates that different foundation forms have no influence on the shape, duration, and peak time of the acceleration curve of the frame structure when the ground motion record used as input is the same.

(2) Different points: there are notable differences in the peak value of the peak acceleration–time history curve of the structure under the varying foundation form working

conditions. The peak value of the acceleration–time history curve for the structure with the raft foundation is lower compared to that with the strip foundation, implying that the structure with the raft foundation exhibits a more effective seismic mitigation effect. The findings demonstrate that increasing the contact area between the base and the rubber-particle-modified soil isolation layer enhances the damping performance of the frame structure.

Further quantitative analysis is carried out based on the data provided in Table 3:

Table 3. Peak acceleration difference under the same isolation layer thickness and different foundation forms.

Seismic Isolation Layer Thickness	Peak Acceleration of Strip Foundation (m/s^2)	Peak Acceleration of Raft Foundation (m/s^2)	Reduction of Peak Acceleration (%)
0 m	3.26	3.24	0.6
0.4 m	2.87	2.69	6.2
0.6 m	2.54	2.32	8.6
0.8 m	2.26	2.03	10.1

As indicated in Table 3, with an increase in the thickness of the rubber-particle-modified soil isolation layer, the peak acceleration of the structure with the raft foundation gradually decreases compared to that with the strip foundation. Specifically, when the thickness of the rubber-particle-modified loess isolation layer is 0 m, 0.4 m, 0.6 m, and 0.8 m, the reduction in the peak acceleration for the raft foundation is 0.6%, 6.2%, 8.6%, and 10.1%, respectively, compared to that of the strip foundation. Hence, as the thickness of rubber-particle-modified soil isolation layer increases, the isolation effect on the frame structure is more pronounced when utilizing a raft foundation.

4.1.2. Comparison of the Column Height–Maximum Displacement Curve

In order to facilitate the investigation of the influence of the rubber-modified soil damping layer on the displacement of the frame structure, an observation point is designated every meter along the previously mentioned observation column. The maximum displacement of each observation point is then extracted to construct the column height–maximum displacement curve. The peak curves of column height–absolute displacement under different foundation forms are presented in Figure 4.

The following two points can be obtained from the qualitative analysis in Figure 4a–d:

(1) Common point: the change trend of the column height–maximum displacement curve is consistent under both foundation forms, where displacement increases with a rise in column height. Additionally, the maximum displacement of the observation column decreases with an increase in the thickness of the rubber-particle-modified soil isolation layer for both foundation forms. This indicates that increasing the thickness of the rubber-particle-modified soil isolation layer is effective in reducing the maximum displacement of the frame structure.

(2) Different points: there is a noticeable gap between the curves of column height and maximum displacement for the two foundation forms with the same thickness of the rubber-particle-improved soil layer. The maximum displacement of the frame structure with the raft foundation at the same height is smaller than that of the frame structure with the strip foundation. This observation suggests that augmenting the contact area between the base and the rubber-particle-improved soil isolation layer effectively diminishes the maximum displacement of the frame structure. Furthermore, as the thickness of the rubber-particle-modified soil isolation layer increases, the gap between the column height–maximum displacement curves for the two foundation forms widens. This highlights that the influence of the contact area between the different foundations and the rubber-particle-modified soil isolation layer on

the height–maximum displacement of the frame structure columns becomes more pronounced.

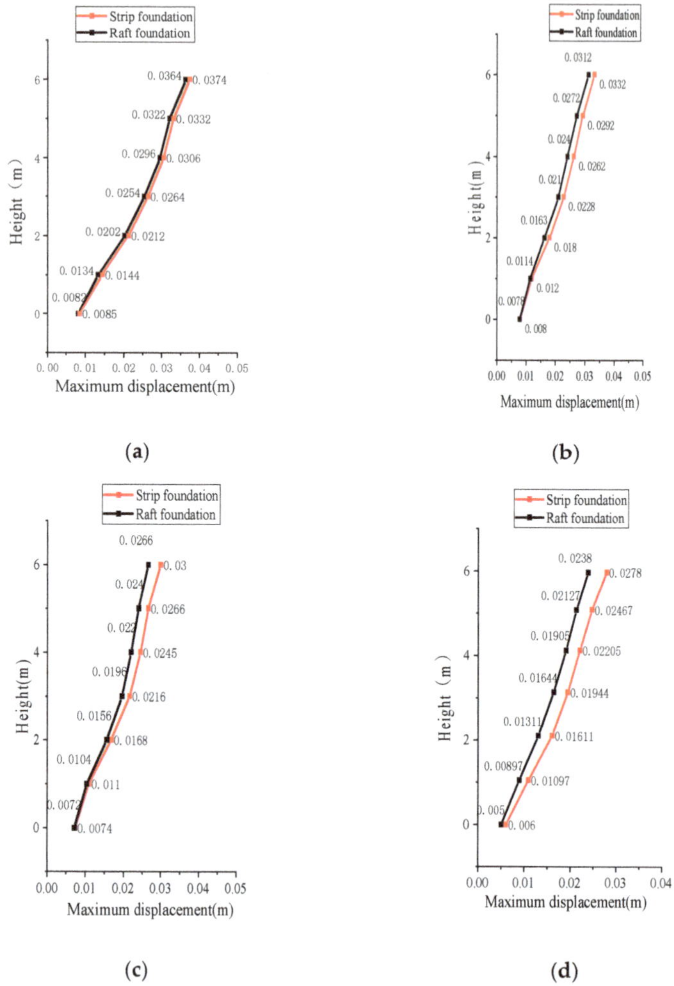

Figure 4. The same thickness of the damping layer between different foundation forms under the column height—maximum displacement curve: (**a**) 0 m seismic isolation layer thickness; (**b**) 0.4 m seismic isolation layer thickness; (**c**) 0.6 m seismic isolation layer thickness; (**d**) 0.8 m seismic isolation layer thickness.

Further quantitative analysis is conducted based on the data presented in Table 4:

Table 4. Maximum displacement difference between columns of different foundation forms with an isolation layer of the same thickness.

Seismic Isolation Layer Thickness	Maximum Displacement of Strip Foundation (m)	Maximum Displacement of Raft Foundation (m)	Maximum Displacement Difference (m)
0 m	0.0374	0.0364	0.0010
0.4 m	0.0332	0.0312	0.0020
0.6 m	0.03	0.0266	0.0034
0.8 m	0.0278	0.0238	0.0040

Under the condition of an identically thick rubber-particle-modified soil isolation layer, the maximum displacement of the frame structure with the raft foundation is smaller compared to that with the strip foundation. When the thickness of the rubber-particle-modified soil isolation layer is 0 m, 0.4 m, 0.6 m, and 0.8 m, the maximum displacement difference between the two foundation forms is 0.0012 m, 0.0022 m, 0.0024 m, and 0.0026 m, respectively. As the thickness of the rubber-particle-modified soil isolation layer increases, the maximum displacement difference between the two foundation forms also increases. This indicates that increasing the thickness of the rubber-particle-modified soil isolation layer can amplify the influence of the base contact area on the isolation effect of the frame structure.

4.1.3. Column Height—Maximum Shear Stress Curve

The qualitative analysis based on Figure 5a–d reveals the following two points:

(1) Common point: the trend in the interlayer shear curve is consistent for both foundation forms, where the maximum shear stress decreases with increasing height. Furthermore, the interlayer shear stress decreases with an increase in the thickness of the rubber-particle-modified soil isolation layer, suggesting that augmenting the thickness of the rubber-particle-modified soil layer effectively reduces the maximum shear stress of the frame structure.

(2) Different point: in the rubber-modified soil isolation layer of the same thickness, the interlayer shear stress of the raft foundation frame structure is smaller than that of the strip foundation frame structure. This indicates that increasing the contact area between the base and the rubber-modified soil isolation layer can effectively reduce the interlayer shear stress of the frame structure. With an increase in the thickness of the rubber-modified soil isolation layer, the distance between the interlayer shear stress curves for the two foundation forms also increases. This indicates that increasing the thickness of the rubber-modified soil isolation layer can amplify the difference in contact areas between different substrates.

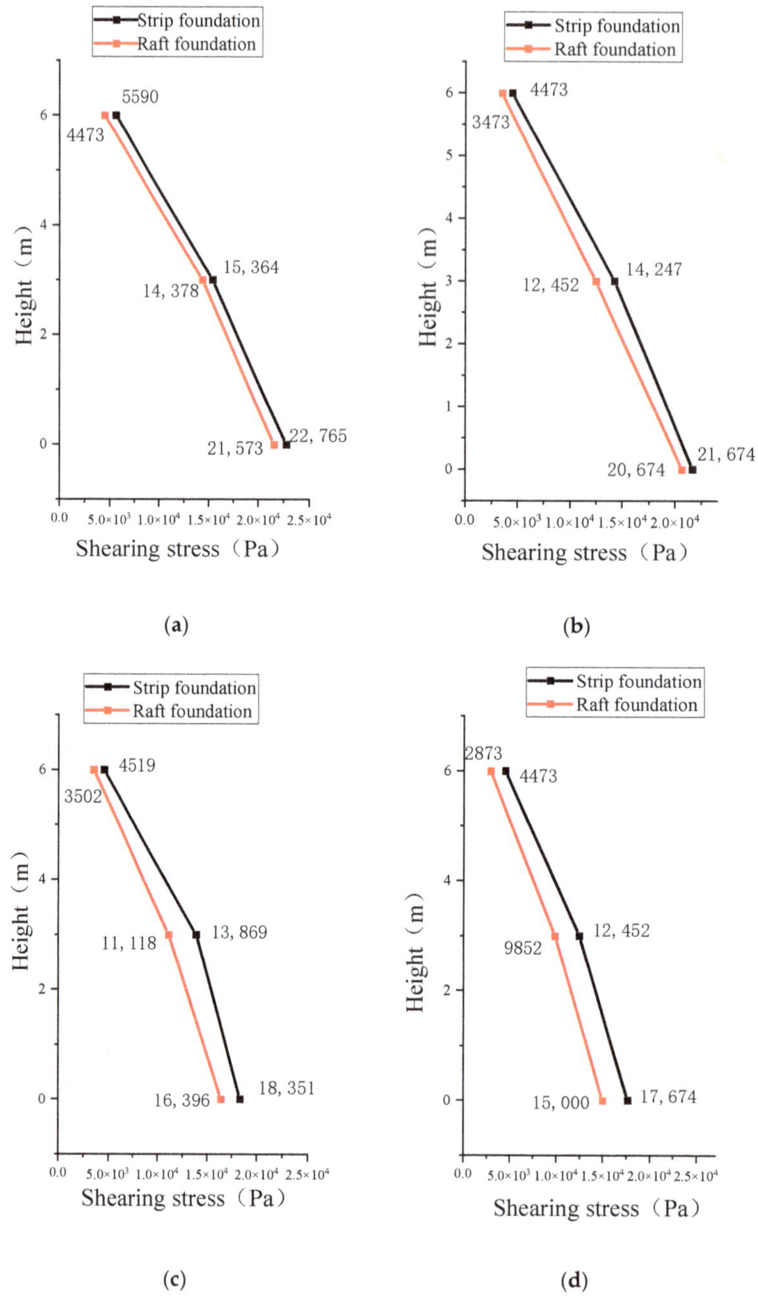

Figure 5. Comparison diagram of the maximum shear stress curves between columns of different foundation forms under an isolation layer of the same thickness: (**a**) 0 m seismic isolation layer thickness; (**b**) 0.4 m seismic isolation layer thickness; (**c**) 0.6 m seismic isolation layer thickness; (**d**) 0.8 m seismic isolation layer thickness.

4.2. Influence of Isolation Layer Thickness on Isolation Effect

4.2.1. Acceleration Response Analysis of the Top Layer of the Structure

The comparison of the peak acceleration–time history curves for a finite element structure with an input peak value of a 0.2 g artificial ground motion record under the conditions of different thickness isolation layers is illustrated in Figure 6.

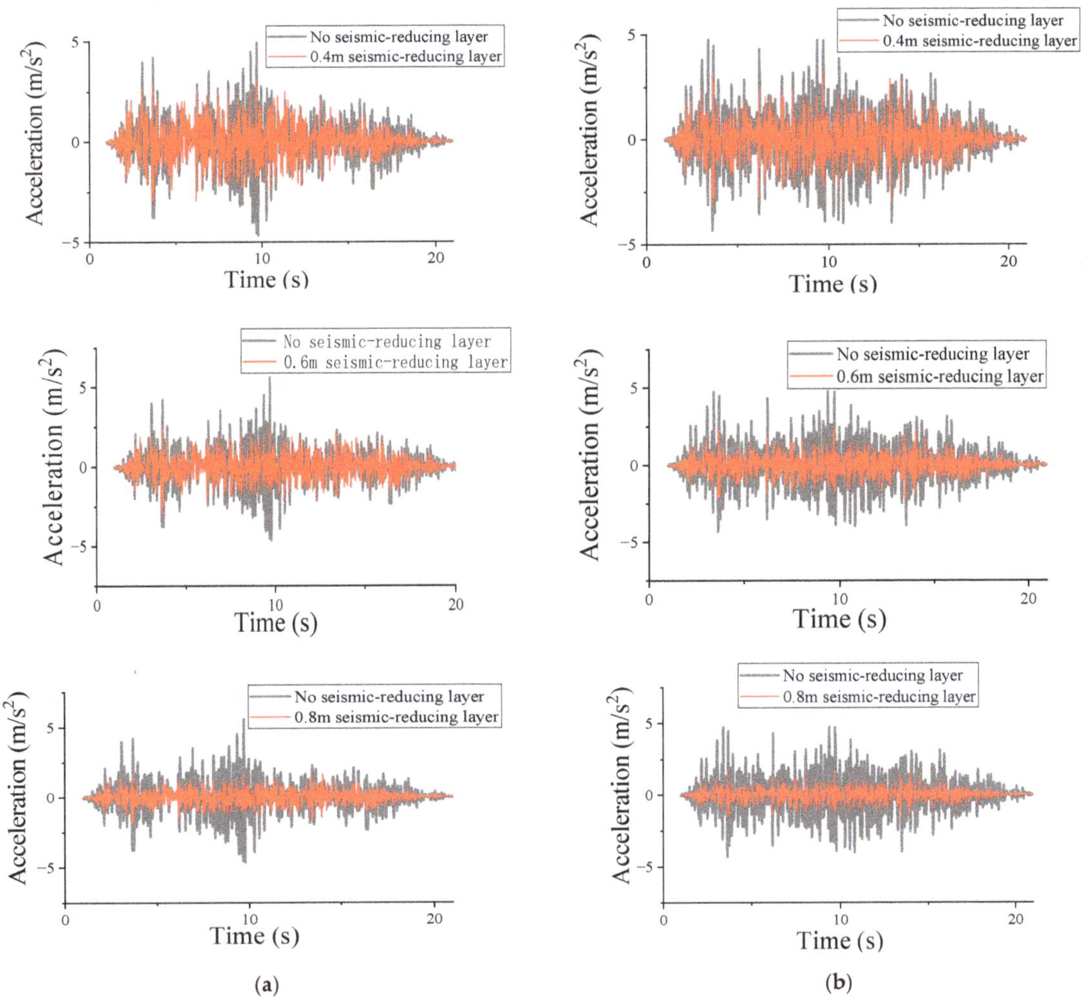

Figure 6. Peak acceleration–time history curve of an artificial ground motion record frame structure: (**a**) raft foundation and (**b**) strip foundation.

The following two points can be obtained from the qualitative analysis in Figure 6:

(1) Common points: under both foundation forms (raft foundation and strip foundation), the shape, duration, and peak acceleration–time history curve of frame structures with varying thicknesses of the rubber-particle-modified soil isolation layer are similar. The peak acceleration–time history curve of frame structures with a rubber-particle-modified soil isolation layer is smaller than that of frame structures without a rubber-particle-modified soil isolation layer.

(2) Different points: the peak value of the peak acceleration–time history curve of the frame structure gradually decreases with an increase in the thickness of the rubber-particle-modified soil isolation layer. This indicates that increasing the thickness of the rubber-particle-improved soil isolation layer effectively reduces the peak acceleration of the frame structure, enhancing its isolation effect. It is evident from the figure that the peak acceleration–time history curve of the frame structure with a 0.6 m and 0.8 m thick rubber-particle-modified soil isolation layer is significantly lower than that of the frame structure with a 0.4 m rubber-particle-modified soil isolation layer. However, the distinction between the peak acceleration–time history curves of the frame structures with a 0.6 m and 0.8 m thick rubber-particle-modified soil isolation layer is not pronounced.

Further quantitative analysis from Table 5 shows that as the thickness of the rubber-particle-modified soil isolation layer increases, the damping effect of a frame structure using the raft foundation improves, but the rate of improvement in the damping effect gradually decreases. The peak acceleration of the frame structures with rubber-particle-modified soil isolation layers of 0.4 m, 0.6 m, and 0.8 m thickness decreased by 46.5%, 51.65%, and 53.78%, respectively, compared to frame structures without a rubber-particle-modified soil isolation layer. Among them, the acceleration peak drop of the modified soil layer with a 0.6 m thickness increased by 5.15% compared to that with a 0.4 m thickness. Similarly, the peak acceleration reduction of the modified soil layer with a 0.8 m rubber particle layer increased by 2.13% compared to that of 0.6 m, and the reduction was not as significant as the former. In the case of the strip foundation, the damping effect of frame structure increased with an increase in the thickness of the layer. The amplitude of increase follows a similar pattern as that observed for the raft foundation. However, it is important to note that the damping effect was consistently smaller compared to the same thickness for the raft foundation.

Table 5. Comparison of peak acceleration of the structure vertex.

Type of Foundation	Peak Acceleration (m/s^2)		Reduction (%)	Peak Acceleration (m/s^2)		Reduction (%)	Peak Acceleration (m/s^2)		Reduction (%)
	No Seismic Isolation Layer Thickness	0.4 m Seismic Isolation Layer Thickness		No Seismic Isolation Layer Thickness	0.6 m Seismic Isolation Layer Thickness		No Seismic Isolation Layer Thickness	0.8 m Seismic Isolation Layer Thickness	
Raft foundation	4.8772	2.6089	46.5	4.8772	2.3577	51.65	4.8772	2.254	53.78
Strip foundation	4.9233	2.707	45.01	4.9233	2.4712	49.8	4.9233	2.367	51.92

4.2.2. Comparison of Column Height–Peak Absolute Displacement

The observated column illustrated in Figure 1 is also employed to investigate the influence of the rubber-modified soil isolation layer's thickness on the displacement of the frame structure. An observation point is designated every meter along column A, and the column height–absolute displacement peak curves of raft foundation and strip foundation frame structures with various isolation layer thicknesses are extracted under the condition that the input peak value is a 0.2 g artificial ground motion record, as shown in Figure 7.

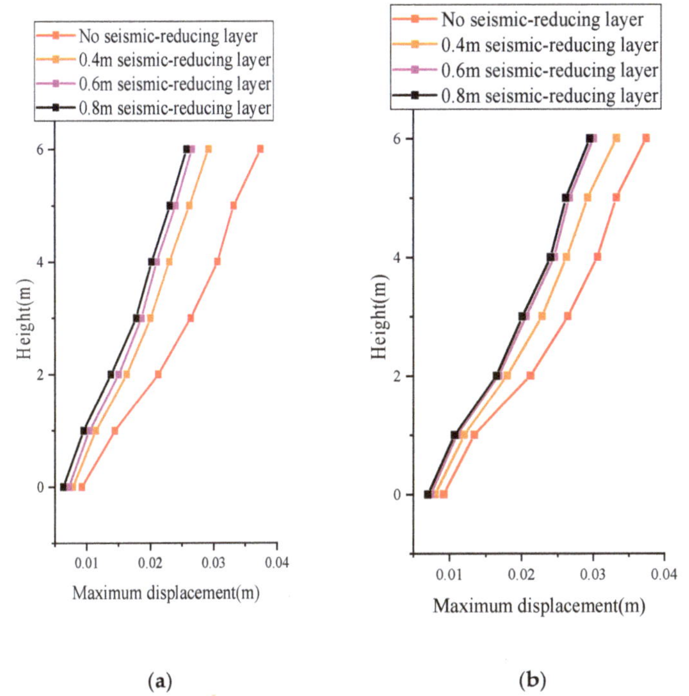

Figure 7. Column height–peak curve of absolute displacement: (**a**) raft foundation and (**b**) strip foundation.

The following two points can be obtained from the qualitative analysis in Figure 7:

(1) Common points: the trend in the column height–maximum displacement curve for each thickness of the rubber-particle-modified soil isolation layer is consistent across both foundation types, with displacement increasing with height.

(2) Different points: with the increase in the thickness of the rubber-particle-modified soil isolation layer, the lateral horizontal displacement at the same height gradually decreases. When the thickness of the rubber-particle-modified soil isolation layer is less than 0.6 m, the lateral horizontal displacement of the frame structure decreases significantly. However, when the thickness of the rubber-particle-modified soil isolation layer is greater than or equal to 0.6 m, the lateral horizontal displacement of the frame structure decreases slightly. The results demonstrate that increasing the thickness of the rubber-particle-modified soil isolation layer effectively reduces the lateral horizontal displacement of the frame structure. When the thickness of rubber-particle-modified soil isolation layer reaches 0.6 m, the damping effect approaches a limit, and further increasing the thickness of the rubber-particle-modified soil isolation layer only slightly reduces the lateral horizontal displacement of the frame structure.

To provide an intuitive expression of the structural displacement limit under earthquake conditions, this paper proposes the calculation of the structural displacement angle. The ultimate displacement of the structure is described by calculating the displacement angle of the columns within the structure. The calculation of the displacement angle of the column is shown in Equation (1). The maximum displacement angle of each column is presented in Table 6. It is evident from the table that the maximum displacement angle of the structure is significantly reduced under the working condition of a rubberized soil foundation.

$$\text{Displacement angle of the structure} = \frac{\text{Horizontal displacement at the top of the structure}}{\text{Height of structure}} \times 100\% \quad (1)$$

Table 6. Maximum displacement angle of the observed column.

Type of Foundation	Displacement Angle of the Structure (%)			
	No Seismic Isolation Layer Thickness	**0.4 m Seismic Isolation Layer Thickness**	**0.6 m Seismic Isolation Layer Thickness**	**0.8 m Seismic Isolation Layer Thickness**
Raft foundation	0.61	0.468	0.423	0.415
Strip foundation	0.62	0.486	0.434	0.428

By comparing the maximum displacement angle of columns under different working conditions, it can be concluded that the displacement angle of a frame structure without an isolation layer and with a 0.4 m isolation layer decreases significantly from 0.61% and 0.62% to 0.468% and 0.486%, respectively. This indicates that a rubber-modified soil isolation layer can effectively reduce the displacement angle, thus enhancing the isolation effect of the frame structure. The rubber-modified soil layer proves to be effective in mitigating the displacement response of the frame structure during earthquake events, resulting in a reduction of the frame structure's displacement under seismic conditions. The 0.6 m and 0.8 m rubber-modified soil isolation layers also exhibit a significant effect in reducing the displacement angle. However, the difference in the reduction of the displacement angle between these two isolation layer thicknesses is relatively small. In conclusion, a 0.6 m thick rubber-particle-modified soil isolation layer appears to be the optimal choice.

4.3. *Influence of Ground Motion Intensity on the Isolation Effect*

4.3.1. Acceleration Response Analysis of the Top Layer of the Structure

In this section, particle-modified soil isolation layers with thicknesses of 0 m and 0.4 m and raft foundations are selected to explore the influence of different ground motion intensities on the isolation effect of a rubber-modified soil isolation layer under the same ground motion record. The time history curves of frame structure vertex acceleration corresponding to each ground motion intensity are illustrated in Figure 8.

The following two points can be obtained from the qualitative analysis in Figure 8:

(1) Common points: when applying ground motion records of varying intensities to the frame structure, both with and without a 0.4 m thick rubber-particle-modified soil isolation layer, the shape, duration, and time of the peak acceleration–time history curve are similar. However, the peak acceleration of the frame structure with a rubber-particle-modified soil isolation layer is consistently lower than that of the frame structure without a rubber-particle-modified soil isolation layer.

(2) Different points: as the intensity of the input ground motion record increases, the peak acceleration of both the frame structure without the rubber-particle-modified soil isolation layer and the frame structure with the 0.4 m thick rubber-particle-modified soil isolation layer also increases. With an increase in the intensity of the input ground motion record, the frame structure with a rubber-particle-modified soil isolation layer experiences a smaller rise in peak acceleration compared to the structure without a rubber-particle-modified soil isolation layer. The disparity in vertex acceleration is most pronounced at the peak of peak acceleration. This suggests that a heightened ground motion intensity amplifies the peak acceleration of the frame structure. However, the frame structure with a rubber-particle-modified soil isolation layer exhibits lower peak acceleration and a superior damping effect.

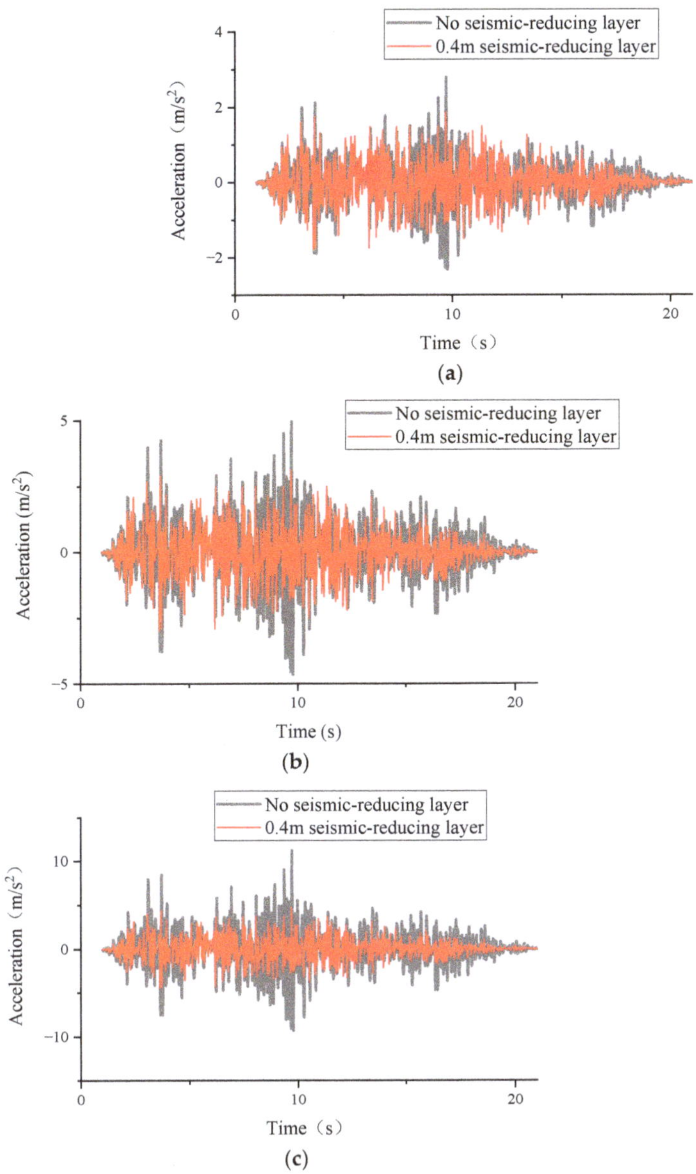

Figure 8. Comparison of the time history curve of frame structure vertex acceleration using an artificial ground motion record with different acceleration peaks: (**a**) 0.1 g; (**b**) 0.2 g; (**c**) 0.4 g.

Further quantitative analysis in Table 7 reveals that the peak acceleration of the frame structure, both without a rubber-particle-modified soil isolation layer and with a 0.4 m thick rubber-particle-modified soil isolation layer, increases with the rise in input seismic intensity. At input ground motion record intensities of 0.1 g, 0.2 g, and 0.4 g, the peak vertex acceleration of the frame structure with a 0.4 m thick rubber-particle-modified soil isolation layer decreases by 16.97%, 46.5%, and 51.03%, respectively, compared to the structure without a rubber-particle-modified soil isolation layer. The reduction in peak acceleration increases with an increase in the ground motion record intensity. The results indicate that as the ground motion record intensity increases, the frame structure with a 0.4 m thick rubber-particle-modified soil isolation layer exhibits a greater damping effect.

Table 7. Comparison of peak acceleration of frame structures under different ground motion intensities.

Ground Motion Record Intensity	Peak Acceleration of No Seismic Isolation Layer Thickness (m/s^2)	Peak Acceleration of 0.4 m Seismic Isolation Layer Thickness (m/s^2)	Reduction (%)
0.1 g	3.24	2.69	16.97
0.2 g	4.8772	2.6089	46.5
0.4 g	10.13	4.96	51.03

4.3.2. Column Height—Maximum Displacement and Maximum Shear Stress

The frame structure with a 0.4 m thick rubber-particle-modified soil isolation layer under a raft foundation is subjected to artificial ground motion records with acceleration peaks of 0.1 g, 0.2 g, and 0.4 g. The study focuses on the influence of ground motion records with varying peak acceleration on the isolation effect of the frame structure. The maximum displacement and maximum shear stress of the observed column are presented in Tables 8 and 9.

Table 8. Maximum displacement of observation column.

Peak Acceleration of Ground Motion Record	Maximum Displacement of Observation Columns in Structures without Isolation Layers (m)	Maximum Displacement of Observation Column for 0.4 m Isolation Layer Structure (m)	Reduction (%)
0.1 g	0.0364	0.0312	14.28
0.2 g	0.0423	0.0356	15.83
0.4 g	0.0485	0.0403	16.9

Table 9. Maximum shear stress of the observed column.

Peak Acceleration of Ground Motion Record	Maximum Shear Stress of Observed Column in Structures without Isolation Layers (Pa)	Maximum Shear Stress of Observed Column with 0.4 m Isolation Layer Structure (Pa)	Reduction (%)
0.1 g	22765	21674	4.79
0.2 g	26694	22348	16.28
0.4 g	32267	26549	17.72

It can be observed from Tables 8 and 9 that, as the peak acceleration of the input ground motion record acceleration increases to 0.1 g, 0.2 g and 0.4 g, the displacement of the frame structure column with a 0.4 m thick rubber-particle-modified soil isolation layer gradually increases by 14.28%, 15.83%, and 16.9%, respectively, compared to that without the rubber-particle-modified soil isolation layer. The reduction range of the maximum shear stress of the observed column gradually increased to 4.79%, 16.28%, and 17.72%, respectively. This suggests that with an increase in the ground motion record acceleration peak, the damping effect of the frame structure with a rubber-particle-modified soil isolation layer was enhanced.

4.4. Influence of Ground Motion Spectrum Characteristics on the Isolation Effect

Comparison of Fourier Spectrum Curves

A frame structure with a raft foundation is employed as a model to investigate the influence of seismic spectrum characteristics on seismic isolation. Taking the three ground motion records listed above (the Friuli (Italy) earthquake of 6 May 1976, the San Fernando, 1971 earthquake at Santa Anita Dam, and the artificial ground motion record), a peak acceleration of 0.2 g is applied to the ground motion record and fed into the finite element model. The optimal configuration of a 0.6 m isolation layer thickness and a raft foundation form is chosen to investigate the influence of ground motion spectrum characteristics on the isolation effect. The structure's peak acceleration–time history curve is extracted for each working condition and compared with the Fourier spectrum. The acceleration –time history curve and Fourier spectrum of each working condition are presented in Figure 9a–c.

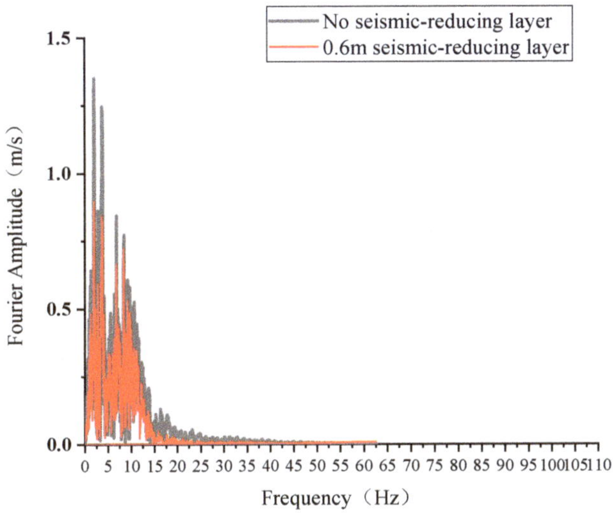

(a)

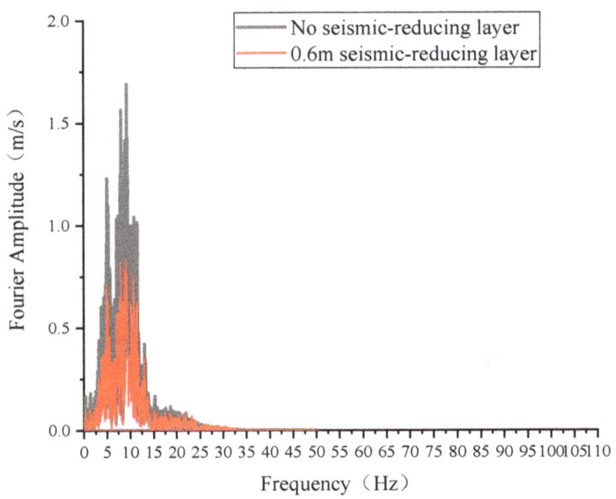

(b)

Figure 9. *Cont.*

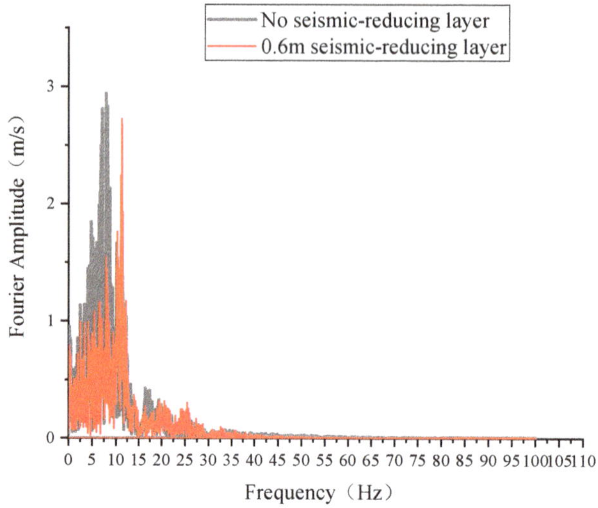

(**c**)

Figure 9. Fourier spectrum comparison of different types of ground motion records: (**a**) Friuli (Italy); (**b**) San Fernando; (**c**) artificial ground motion record.

The following two points can be drawn from Figure 9:

(1) Image comparison results: the Fourier spectrum of the structural response closely aligns with that of the input ground motion. However, there is a disparity between the Fourier spectrum of the input artificial ground motion record and the structural response. The fundamental frequency in the Fourier spectrum related to the vertex acceleration–time history of the frame structure is similar for ground motion records with comparable spectral characteristics. However, there is a discernible difference in the Fourier amplitude corresponding to the fundamental frequency. Specifically, the Fourier amplitude when subjected to artificial ground motion records is higher than the corresponding amplitude under the influence of the Friuli ground motion records and San Fernando ground motion records.

(2) The observed discrepancy may be attributed to the distinctive Fourier curve shape of the input artificial ground motion record, which diverges from the Fourier spectrum of the structural response. Due to absence of a rubber-particle-modified soil isolation layer, the frame structure experiences a heightened seismic response, accumulating internal damage and generating cracks. This, in turn, leads to a reduction in the natural frequency. Additionally, under the influence of the three ground motion records, both the Fourier spectrum of the structural response and the Fourier spectrum of the input ground motion decrease. This occurs because the presence of the rubber-particle-modified soil isolation layer creates a milder soil–structure interaction, resulting in reduced nonlinear residual deformation of the soil and a corresponding decrease in the amplitude of the fundamental frequency.

5. Conclusions

In this paper, a two-layer frame structure was utilized as an example to investigate the seismic isolation effects associated with frame structure with various foundation forms. Through dynamic time history analysis, the influences of base contact areas, the thickness of the rubber-modified soil layer, the intensity of ground motion records and the spectral characteristics of ground motion on the seismic isolation effect were compared. The primary research conclusions are as follows:

(1) Through a comparison of the peak acceleration–time history curve and the column height–maximum displacement curve of a frame structure with a raft foundation and a strip foundation, it is concluded that a larger contact area between the foundation and the rubber-modified soil layer, especially with thicknesses of 0.4 m, 0.6 m, and 0.8 m, leads to an enhanced isolation effect of the rubber-modified soil.
(2) In the frame structure with artificial ground motion records as input, by comparing the peak acceleration–time history curve and maximum column–displacement angle for four finite element models with rubber-particle-modified soil layers of 0 m, 0.4 m, 0.6 m, and 0.8 m thickness, it is concluded that as the thickness of the rubber-modified soil layer increases, the isolation effect of the structure improves. The increase in the isolation effect is no longer significant when the thickness exceeds 0.6 m.
(3) Different ground motion records with varying peak accelerations were applied to the finite element model. Once comparisons were made between the peak acceleration–time history curve of the structure, the column height-maximum displacement curve, and the column height–maximum shear stress curve, it was concluded that the peak acceleration of the frame structure increased with an increased peak acceleration of the ground motion records. Laying a rubber-particle-modified soil isolation layer can reduce the amplitude of the peak acceleration of the frame structure with an increase in the intensity of the input ground motion record.
(4) Three different types of ground motion records were utilized as inputs for the finite element models, each with the same peak acceleration. The Fourier spectrum of the frame structure's response to these three ground motion records was then compared with the Fourier spectrum curve of the input seismic motion. It was concluded that ground motion records with abundant low frequencies resulted in an intense structural dynamic response and higher Fourier amplitude. The soil–structure interaction was smoother when rubber particles were installed, and the Fourier spectrum amplitude of the structure was lower than that of the input seismic motion.

Author Contributions: S.C.: methodology, conceptualization, data curation, validation; Y.C.: reviewing, funding acquisition; D.C.: supervision, writing; Q.C.: reviewing; W.W.: methodology; J.L.: writing, data analysis. All authors have read and agreed to the published version of the manuscript.

Funding: This research was funded by the Natural Science Basic Research Plan in Shaanxi Province, (No. 2021-JQ-243), research on the dynamic characteristics of rubber-particle-improved ground bases and the key technology of structural shock absorption (grant number KJYF-2023-7GS-CG08), and Fundamental Research Funds for the Central Universities, CHD (No. 300102282201).

Data Availability Statement: The data used to support the findings of this study are included in the article.

Conflicts of Interest: The authors declare no conflict of interest.

References

1. Soltani, A.; Deng, A.; Taheri, A.; Sridharan, A. Swell-Shrink-Consolidation Behavior of Rubber-Reinforced Expansive Soils. *Geotech. Test. J.* **2019**, *42*, 761–788. [CrossRef]
2. Raeesi, R.; Soltani, A.; King, R.; Disfani, M.M. Field performance monitoring of waste tire-based permeable pavements. *Transp. Geotech.* **2020**, *24*, 100384. [CrossRef]
3. Akbarimehr, D.; Eslami, A.; Esmail, A. Geotechnical behaviour of clay soil mixed with rubber waste. *J. Clean. Prod.* **2020**, *271*, 122632. [CrossRef]
4. Akbarimehr, D.; Rahai, A.; Eslami, A.; Karakouzian, M. Deformation Characteristics of Rubber Waste Powder-Clay Mixtures. *Sustainability* **2023**, *15*, 12384. [CrossRef]
5. Saparudin, N.A.; Kasim, N.; Abu Taib, K.; Azahar, W.N.I.W.; Kasim, N.A.; Ali, M. Improvement of Problematic Soil Using Crumb Rubber Tyre. *IIUM Eng. J.* **2022**, *23*, 72–84. [CrossRef]
6. Sadek, D.M.; El-Attar, M.M. Structural behavior of rubberized masonry walls. *J. Clean. Prod.* **2015**, *89*, 174–186. [CrossRef]
7. Chen, Y.; Chai, S.Q.; Cai, D.B.; Wang, W.; Li, X.P.; Liu, J.H. Experimental study on shear mechanical properties of improved loess based on rubber particle incorporation and EICP technology. *Front. Earth Sci.* **2023**, *11*, 127012. [CrossRef]
8. Zhou, T.L.; Dong, C.S.; Fu, Z.P.; Li, S. Study on Seismic Response and Damping Performance of Tunnels with Double Shock Absorption Layer. *KSCE J. Civ. Eng.* **2022**, *26*, 2490–2508. [CrossRef]

9. Srijit, B.; Aniruddha, S.; Reddy, G.R. Performance of sand and shredded rubber tire mixture as a natural base isolator for earthquake protection. *Earthq. Eng. Eng. Vib.* **2015**, *14*, 683–693.
10. Hazarika, H.; Kohama, E.; Sugano, T. Underwater Shake Table Tests on Waterfront Structures Protected with Tire Chips Cushion. *J. Geotech. Geoenviron. Eng.* **2008**, *134*, 1706–1719. [CrossRef]
11. Alhan, C.; Gavin, H. A parametric study of linear and non-linear passively damped seismic isolation systems for buildings. *Eng. Struct.* **2004**, *26*, 485–497. [CrossRef]
12. D'Amato, M.; Gigliotti, R.; Laguardia, R. Seismic isolation for protecting historical buildings: A case study. *Front. Built Environ.* **2019**, *5*, 87. [CrossRef]
13. Amorosi, A.; Boldini, R.; Elia, R. Parametric study on seismic ground response by finite element modelling. *Comput. Geotech.* **2010**, *37*, 515–528. [CrossRef]
14. Schanze, E.; Schanze, E.; Gomez, M.; Lopez, A. Numerical study of the seismic response of an instrumented building with underground stories. *Appl. Sci.* **2021**, *11*, 3190. [CrossRef]
15. Adampira, M.; Derakhshandi, M.; Ghalandarzadeh, A. Experimental study on seismic response characteristics of liquefiable soil layers. *J. Earthq. Eng.* **2021**, *25*, 1287–1315. [CrossRef]
16. Cui, G.Y.; Wei, H.H.; Ning, M.Q.; Zhang, J.J. Analysis on influence of thickness of shock absorption layer on shock absorption effect of urban shallow-buried rectangular tunnel. *J. Saf. Sci. Technol.* **2020**, *16*, 137–142.
17. Li, S.T.; Bao, X.; Liu, J.B.; Wang, F.; Wang, D. Improvement research on the stability of explicit integration algorithms with 3D viscoelastic artificial boundary elements. *Eng. Comput.* **2023**, *40*, 494–513. [CrossRef]
18. Brunet, S.; Llera, D.L.C.J.; Kausel, E. Non-linear modeling of seismic isolation systems made of recycled tire-rubber. *Soil. Dyn. Earthq. Eng.* **2016**, *85*, 134–145. [CrossRef]
19. Quan, D.Z.; Chai, S.B.; Wang, Y.L.; Fan, Z.S.; Bu, Y.H. 3-D Numerical Simulation of Seismic Response of the Induced Joint of a Subway Station. *Buildings* **2023**, *13*, 1244. [CrossRef]
20. Zheng, Y.; Shi, X.L.; Wang, Y.H.; Zhang, Y.C.; Jin, N. Experiment Scheme Design for Rural Isolation Masonry Structure with Shaking Table. *Technol. Earthq. Disaster Prev.* **2016**, *11*, 297–305.
21. Cao, W.L.; Zhou, Z.Y.; Wang, Q.; Dong, H.Y.; Zhang, J.W. Experimental study on base vibration isolation and anti-seismic masonry structure in rural areas by shaking table test. *J. Vib. Shock* **2011**, *30*, 209–213.
22. *GB 55008-2021*; General Code for Concrete Structures. China Planning Publishing House: Beijing, China, 2021.
23. Du, X.L.; Zhao, M.; Wang, J.T. Artificial stress boundary conditions for near field wave simulation. *J. Mech.* **2006**, *1*, 49–56.
24. Du, X.L.; Zhao, M. Seismic response analysis method of arch dam based on viscoelastic boundary. *J. Water Conserv.* **2006**, *9*, 1063–1069.
25. Zhao, M.; Gao, Z.D.; Du, X.L.; Wang, J.J.; Zhong, Z.L. Response spectrum method for seismic soil-structure interaction analysis of underground structure. *Bull. Earthq. Eng.* **2019**, *17*, 5339–5363. [CrossRef]

Disclaimer/Publisher's Note: The statements, opinions and data contained in all publications are solely those of the individual author(s) and contributor(s) and not of MDPI and/or the editor(s). MDPI and/or the editor(s) disclaim responsibility for any injury to people or property resulting from any ideas, methods, instructions or products referred to in the content.

buildings MDPI

Article

Estimation Method for an In Situ Stress Field along a Super-Long and Deep-Buried Tunnel and Its Application

Qitao Pei [1], Xiaonan Wang [1,2], Lihong He [1], Lu Liu [1], Yong Tian [1] and Cai Wu [3,4,*]

[1] Wuhan Municipal Engineering Design & Research Institute Co., Ltd., Wuhan 430023, China; pqt011@163.com (Q.P.)
[2] Wuhan Urban Construction Investment Development Group Co., Ltd., Wuhan 430023, China
[3] School of Civil Engineering, Hubei Engineering University, Xiaogan 432000, China
[4] Changjiang River Scientific Research Institute of Changjiang Water Resources Commission, Wuhan 430010, China
* Correspondence: wucai@whut.edu.cn

Abstract: Aiming at some stress-induced failure phenomena in surrounding rock that occur during the construction of super-long and deep-buried tunnels, a method for estimating the in situ stress in the tunnels based on multivariate information integration is proposed, which uses a small amount of in situ stress measurement, stereographic projection technology, and a numerical simulation method. Firstly, by conducting a macroscopic analysis of the regional geological structure, topography, and pre-excavated small tunnels (such as exploration of adits and pilot tunnels), the strength of the tectonic stress field and the orientation of the principal stresses in the tunnel sections are preliminarily determined. Secondly, the reliability of the in situ stress measurement data were analyzed using full-space stereographic projection and the plane stress projection method. Then, some representative measurement points that reflected the distribution characteristics of in situ stress in the project area, on the whole, were determined. Thirdly, the finite difference (FDM) and multiple regression analysis (MRA) methods were used to inverse the in situ stress field in the project area. The proposed method was applied to a super-long and deep-buried tunnel project in Qinling, and the in situ stress distribution characteristics of the tunnel sections at different mileages were obtained. The results show that both the calculated principal stress values and the azimuth angle of the maximum horizontal principal stress are in good agreement with the measured ones, indicating that the method used in this study is reasonable. Finally, the typical surrounding rock failure phenomena encountered during the excavation of the project were investigated, and targeted treatment measures were proposed. The research results can provide references for support design and disaster management of surrounding rock in deep-buried long tunnels.

Keywords: super-long and deep-buried tunnels; in situ stress; back analysis; finite element method; numerical simulation

1. Introduction

With the rapid development of China's social economy and the implementation of the Western Development Strategy, significant progress has been made in tunnels and underground engineering, showing a trend in deep-buried, super-long, and extra-large development. The increasing number of tunnels in complex environments poses enormous challenges to the safe construction and healthy operation of engineering projects [1–3]. Engineering practice has shown that in situ stress is one of the important basic data requirements for geological environment and crustal stability evaluation and engineering design and construction, and it is also the fundamental force that causes deformation and failure when the environment in which the rock mass is located changes [4,5]. With the increasing attention paid to engineering safety, it has become an essential task to study the distribution characteristics of in situ stress in the project area in advance during the line selection, layout, design, and construction stages of super-long and deep-buried tunnels.

In situ stress testing is the most direct way to understand the in situ stress field in engineering areas. At present, the commonly used in situ stress testing methods include hydraulic fracturing (HF) [6], over-coring [7–9], hydraulic testing of pre-existing fractures (HTPF) [10], surface stress relief for rock mass, and so on [11]. However, during the implementation process, it was found that there are still the following shortcomings when using in situ stress testing. On the one hand, most super-long and deep-buried tunnels are subjected to the occurrence environment of long tunnels, large burial depths, complex geological structures, and stratum lithologies. Owing to high cost, long-term testing, and unfavorable on-site measurement conditions, it is difficult to carry out in situ stress testing for the entire line, and the results only reflect sporadic stress fields located near the limited measurement points. On the other hand, the application conditions of in situ stress testing methods are relatively strict, and it is difficult to fully meet this strict requirement in the actual testing process. For example, the over-coring method requires the tested rock mass to meet uniform, continuous, isotropic, and linear elastic conditions, while the HF method assumes principal stress along the vertical direction [8–10]. However, under extremely high-stress environments, core disking often occurs in boreholes, which greatly reduces the success rate of in situ stress testing [12,13]. Additionally, in high-stress environments, the testing equipment pressure is limited. Insufficient equipment pressure leads to the formation of wellbore cracks, making it difficult to implement in situ stress testing methods. The HTPF method assumes that the tested rock mass is uniform, continuous, and linearly elastic. When the distance between the tested fractures is greater than 50 m, a stress gradient value needs to be provided [10]. At present, due to limited measured data and geological information, studies typically use numerical simulation technology to invert the distribution characteristic of 3-D in situ stress fields for the entire research area [14–16]. However, the accuracy of the in situ stress field obtained using this method depends not only on the reliability and representativeness of the in situ stress testing data but also on numerous factors such as numerical models, boundary conditions, and geological parameters [17].

Numerous engineering practices have shown that a substantial amount of rock mass deformation and failure information is often revealed during the excavation process, although this phenomenon is harmful to engineering construction safety. However, from another perspective, failure phenomena provide an indirect basis for estimating the magnitude and direction of in situ stress. In recent decades, many scholars have conducted in-depth research on the relationship between testing methods, such as the borehole fracturing method [18–21], the borehole stress relief method [22–26], and the HF method [6,10], and in situ stress. Then, the magnitude and orientation of in situ stress based on failure phenomena and limited testing data can be estimated. With the improvement in computer technology and the continuous development of numerical simulation technology, the comprehensive use of in situ stress testing technology, rock excavation failure phenomenon, back analysis, and numerical simulation methods to obtain the distribution characteristics of the in situ stress field of deep underground projects, has gradually become a commonly used analysis method [27–30]. This method compensates for the problems of limited or unreliable in situ stress measurement data and provides a feasible approach for determining the 3-D in situ stress field of super-long and deep-buried tunnels. However, in the case of soils and soft rocks, the numerical models used to evaluate stresses and displacements must be based on unconfined conditions, not limited to oedometric parameters [31]. In addition, the back analysis method has been widely applied in computational simulations of tests (namely, direct analyses), sensitivity analyses for the optimal design of experiments, model reduction procedures, and other aspects [32,33]. At the same time, it plays an important role in the mechanical characterization of materials and the non-destructive assessment of possible damages to industrial plant components and civil engineering structures and infrastructures [34].

It can be seen that the current research on initial stress field inversion mostly focuses on how to improve the fitting degree between the calculated stress values at the measurement points and the measured ones, while the research on the effectiveness and representativeness of the measured data for subsequent back analysis is not in-depth enough. In addition, existing inversion methods lack recognition of failure phenomena in the surrounding rock during tunnel excavation. Thus, the obtained in situ stress field cannot match well with the failure characteristics of the surrounding rock. To address the above issues, a comprehensive analysis method that can estimate the 3-D in situ stress field of tunnels passing through multiple geological tectonic units based on incomplete measured stress data and geological information is proposed. The method is designed for use prior to the excavation of deep-buried long tunnels. Firstly, by conducting a macroscopic analysis of the regional geological structure, topography, and failure phenomena of pre-excavated small tunnels, the strength of the tectonic stress field and the orientation of the principal stresses in the tunnel sections are preliminarily determined. Secondly, by use of stereographic projection technology, the reliability of the in situ stress measured results is analyzed, and representative measurement points are selected that can reflect the distribution characteristics of in situ stress in the project area on the whole. On this basis, the FDM and MRA methods are used to complete a back analysis of the in situ stress field in the project area. The method above was directly applied to a super-long and deep-buried tunnel project in Qinling, and the in situ stress distribution characteristics of the tunnel sections at different mileages were obtained. Finally, the typical failure phenomena in the surrounding rock encountered during the excavation of the project were investigated, and targeted treatment measures were also proposed.

2. In Situ Stress Estimation Method for Super-Long and Deep-Buried Tunnels

During the construction of deep-buried tunnels, it is necessary to excavate some small tunnels (such as exploration, auxiliary, and pilot tunnels) in advance to expose stress-induced failure phenomena in the surrounding rock (such as peeling and sudden rock burst phenomena). Although stress-induced failure phenomena are harmful to construction safety, from another perspective, they also provide a new method to explore certain characteristics of in situ stress.

This study proposes a multivariate information integration method for estimating the in situ stress of deep-buried tunnels based on a small amount of stress measurement data and the pre-exposed stress-induced rock failure characteristics of small tunnels in advance. A flowchart for this method is shown in Figure 1. The basic principle and implementation process for the method are described in detail below:

2.1. Macroscopic Characteristic Analysis

Using an analysis of the tectonic history and current tectonic activity within the project region, the orientation of the tectonic stress was inferred as follows:

(1) Using a study on the geological age of the regional geological structure in the project area, the types of main control faults (such as normal, thrust, and strike-slip fault), and the relationship between faults in the near-field zone and tectonic stress, the orientation of the in situ stress field in the project area was preliminarily revealed.

(2) The relationship between the topography and tectonic stress in the project area was studied, and the influence characteristics of the topography on the distribution of in situ stress was preliminarily determined.

(3) In combination with the possible occurrence of failure phenomenon in the surrounding rock of the exploration adits, the strength of the structural stress field and the orientation of the principal stresses in the tunnel sections were preliminarily determined.

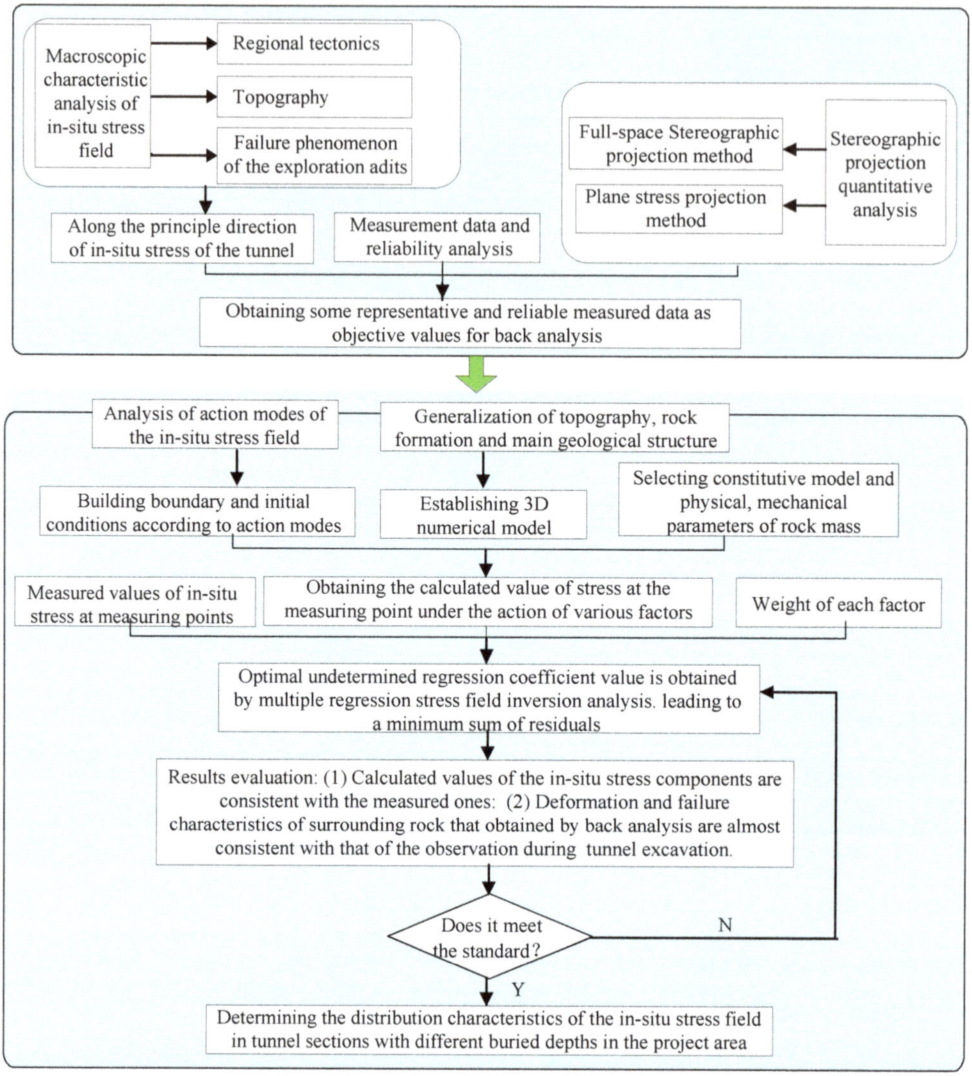

Figure 1. Flowchart showing the in situ stress estimation method for super-long and deep-buried tunnels.

2.2. Reliability Analysis and Selection of In Situ Stress Testing Results

Using the in situ stress measurement data, the distribution characteristics of the in situ stress field in the project area were preliminarily analyzed, and the azimuth and inclination angles of the principal stresses were described with stereographic projection technology. Subsequently, the potential deformation and failure characteristics of the surrounding rock of the tunnel were obtained according to the spatial relationship between the measured principal stress direction and tunnel axis. Within the same project area, the measured results should be consistent with the stress state in the macroscopic area. The reliability of the in situ stress measurement should be tested from the following aspects:

(1) When the measured results consider errors, such as rounding, they sometimes do not constitute a stress tensor. Therefore, it is necessary to analyze the complete stress tensor

(six components) from the measured values using stereographic projection technology [30]. The orthogonality of three principal stress directions was examined.

(2) Verification of consistency between the measured principal stress direction and macroscopic regional tectonic stress direction.

(3) Verification of the consistency between the measured stress-induced deformation and failure parts of surrounding rock and the inferred position using the spatial relationship between the principal stress and the tunnel axis.

(4) Combined with Section 2.1, the representative measurement data that can reflect the distribution characteristics of the in situ stress in the project area are selected for subsequent back analysis of the in situ stress field.

2.3. Multiple Regression Analysis Method (MRA) for an In Situ Stress Field

MRA of an in situ stress field is an important method for estimating the stress state in a project area. This method has been widely used in many large-scale underground projects [17,26,35]. Based on the theory of geomechanics, this method establishes a mathematical calculation model by dividing the main components of the in situ stress fields, such as the self-weight and tectonic stress field, and obtains the corresponding weights of each factor using multiple regression analysis so that the fitting error between the calculated and measured stress values at the test location is minimized. Based on current research results [35,36], there are four main types of action modes of the in situ stress field in this study. The impact of tectonic action on the in situ stress is achieved by applying unit stress or displacement at the boundary. The load and boundary constraints are plotted in Figure 2.

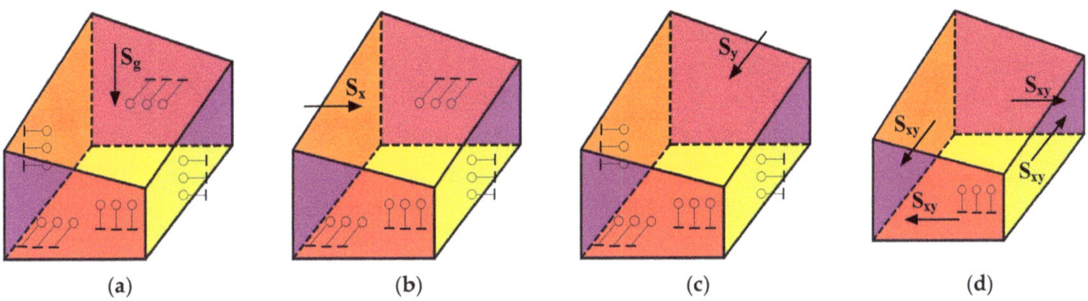

Figure 2. Schematic diagram showing applied loads and boundary constraints, (**a**) Self-weight action, (**b**) Compressive tectonic stress along the X-direction. (**c**) Compressive tectonic stress along the Y-direction. (**d**) Shear tectonic stress in the horizontal plane.

The calculation process for the MRA method is described as follows [37,38]:

Using the in situ stress field regression value $\hat{\sigma}_k$ as the dependent variable and the stress value $\sigma_k{}^i$ for each influencing factor as the independent variable, the form of the regression equation is as follows:

$$\hat{\sigma}_k = \sum_{i=1}^{n} L_i \sigma_k^i \tag{1}$$

where k is the sequence number of the observation points, L_i is the multiple regression coefficient for the corresponding independent variable, $\hat{\sigma}_k$ is the regression-calculated value for the kth observation point, $\sigma_k{}^i$ is the single-column matrix for the calculated value of the corresponding stress component, and n is the number of working conditions, $i = 1, 2, \ldots, n$.

Supposing that there are m observation points, each with six stress components, the residual sum of squares S_c can be obtained as follows:

$$S_c = \sum_{k=1}^{m} \sum_{j=1}^{6} \left(\sigma_{jk}^{*} - \hat{\sigma}_{jk}\right)^2 = \sum_{k=1}^{m} \sum_{j=1}^{6} \left(\sigma_{jk}^{*} - \sum_{i=1}^{n} L_i \sigma_{jk}^{i}\right)^2 \tag{2}$$

where $\sigma_{jk}{}^{*}$ and $\hat{\sigma}_{jk}$ are the measured data and calculated values of the stress component j at the observation point k, respectively, $\sigma_{jk}{}^{i}$ is the numerically calculated value of the stress component j at the observation point k under the i condition, and $l = 1, 2, 3, \ldots, 6$ refers to σ_{xx}, σ_{yy}, σ_{zz}, σ_{xy}, σ_{yz}, and σ_{zx}, respectively.

Based on the principle of the least square method, a set of L_i is obtained to minimize the residual sum of squares S_c, and the solution is unique. The least square method (also known as the least squares method) is a mathematical optimization technique. By minimizing the sum of squares of errors to find the best function that matches the data, it can easily estimate unknown data and minimize the sum of squares of errors between the calculated and measured results. The equation that makes S_c the minimum is:

$$\begin{bmatrix} \sum\limits_{k=1}^{m}\sum\limits_{l=1}^{6}(\sigma_{lk}^{1})^2 & \sum\limits_{k=1}^{m}\sum\limits_{l=1}^{6}\sigma_{lk}^{1}\sigma_{lk}^{2} & \cdots & \sum\limits_{k=1}^{m}\sum\limits_{l=1}^{6}\sigma_{lk}^{1}\sigma_{lk}^{n} \\ symmetry & \sum\limits_{k=1}^{m}\sum\limits_{l=1}^{6}(\sigma_{lk}^{2})^2 & \cdots & \sum\limits_{k=1}^{m}\sum\limits_{l=1}^{6}\sigma_{lk}^{2}\sigma_{lk}^{n} \\ & \vdots & & \vdots \\ & symmetry & & \sum\limits_{k=1}^{m}\sum\limits_{l=1}^{6}(\sigma_{lk}^{n})^2 \end{bmatrix} \begin{Bmatrix} L_1 \\ L_2 \\ \vdots \\ L_n \end{Bmatrix} = \begin{Bmatrix} \sum\limits_{k=1}^{m}\sum\limits_{l=1}^{6}\sigma_{lk}^{*}\sigma_{lk}^{1} \\ \sum\limits_{k=1}^{m}\sum\limits_{l=1}^{6}\sigma_{lk}^{*}\sigma_{lk}^{2} \\ \vdots \\ \sum\limits_{k=1}^{m}\sum\limits_{l=1}^{6}\sigma_{lk}^{*}\sigma_{lk}^{n} \end{Bmatrix} \quad (3)$$

Solving the system in Equation (3) and obtaining n undetermined regression coefficients $L = (L_1, L_2, \ldots, L_n)^{\mathrm{T}}$, the initial in situ stress at any point P in the calculation region can be obtained with superimposing the calculated values of the finite element in each working condition. That is:

$$\sigma_{jP} = \sum_{i=1}^{n} L_i \sigma_{jP}^{i} \quad (4)$$

where σ_{jP} is the regression-calculated value of the stress component j at any observation point P in the calculation domain and σ_{jP}^{i} is the calculated value of the stress component j at any observation point P in the calculation domain under the i condition.

According to Equation (4), the initial in situ stress field in the calculation region is treated as a linear superposition of the self-weight stress field and the tectonic stress field. To ensure that the obtained initial in situ stress field conforms to the actual distribution characteristic of the stress field, the reliability of the regressed in situ stress field was evaluated using two aspects: (1) the calculated values of the in situ stress component were verified to be basically consistent with the measured ones to ensure that the spatial points are consistent and (2) the consistency of the predicted deformation and failure parts in the surrounding rock of the tunnel under the calculated stress field, as well as the observed deformation and failure phenomena during the tunnel excavation, were verified to ensure that the in situ stress characteristics of the local area were consistent.

3. Project Overview: A Case Study

3.1. Project Introduction

The Hanjiang-to-Weihe River Water Diversion Project, which is called the South-to-North Water Diversion Project in the Shaanxi Province of China, aims to solve the water shortage problem in cities along the Weihe River coast in the central Shaanxi area. The Qinling water conveyance tunnel is a key channel for diverting water from the Hanjiang River to the Weihe River, with a maximum burial depth of about 2002 m, a length of 81.779 km, and an average annual water transmission of 1.505 billion cubic meters. The inlet of the tunnel has a bottom elevation of 537.17 m, and the exit has a bottom elevation of 510.0 m. This study focuses on the north section of the Qinling water conveyance tunnel, which is 41.779 km long (K40 + 000~K81 + 779). The location of the tunnel is shown in Figure 3.

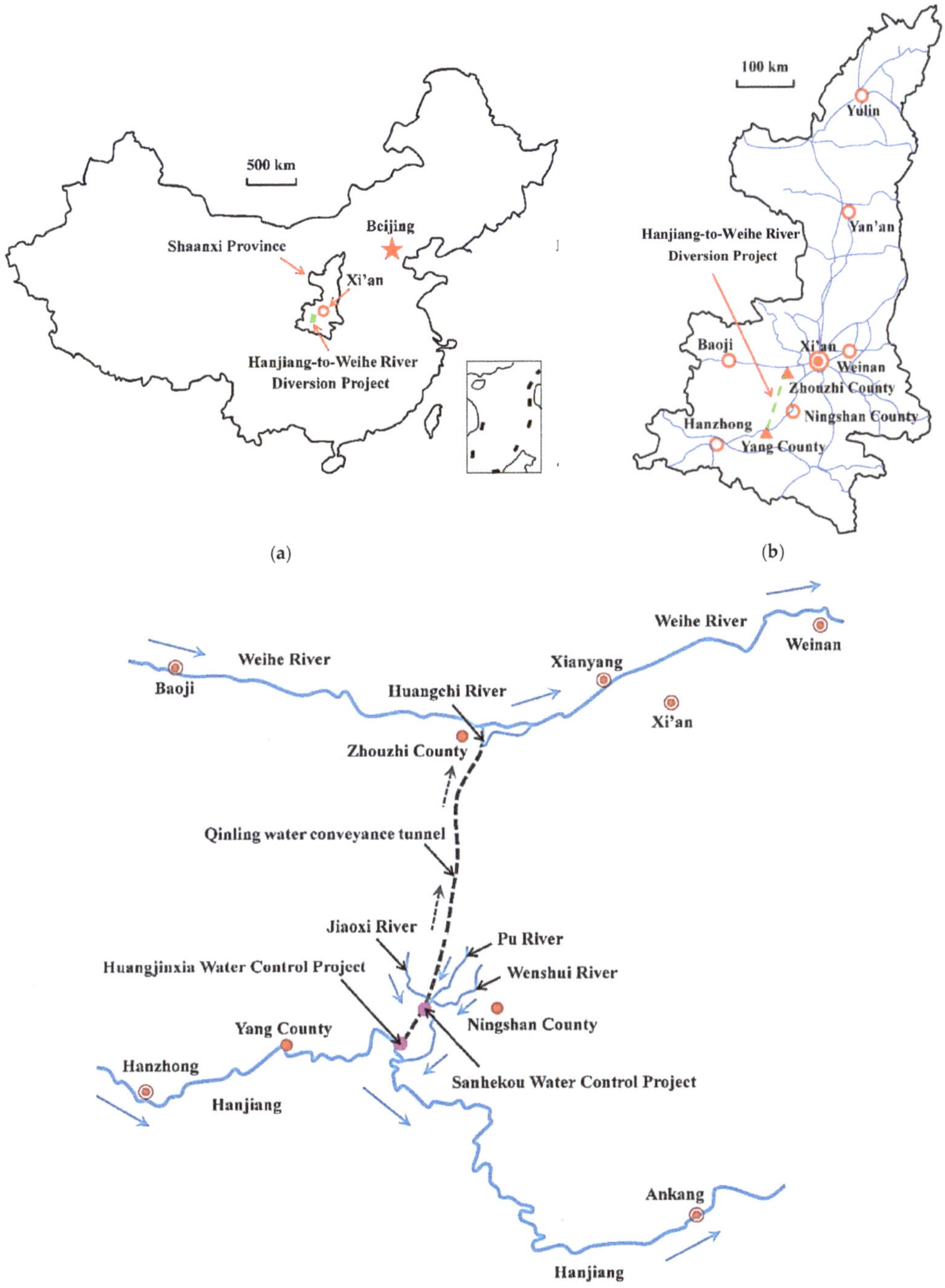

Figure 3. Location and layout of the Hanjiang-to-Weihe River Diversion Project. (**a**) Regional map of the project in China. (**b**) Geographic location of the project in Shanxi Province, China. (**c**) Diagram showing the project layout. (Modified from the Ref. [39]).

3.2. Geological Conditions

The project area is located in the northern section of the Qinling tunnel, which mainly includes the geomorphic units of the high and middle mountain areas of the Qinling Ridge, as well as the middle and low mountain areas of the northern Qinling Ridge. Under the influence of neotectonics, a relatively complex geomorphic unit was formed by long-term water erosion and cutting. The construction methods in the project area primarily include a TBM section and a drilling and blasting section. The geological conditions in the different sections are as follows:

(1) The TBM construction section is 25,164 m long (K40 + 000~K65 + 164), and its main lithology is granite, quartz diorite, metasandstone, phyllite, and hornblende schist. The rock mass is weakly weathered to slightly weathered, which is severely affected by geological structures. The joint fractures are relatively developed. The main faults that pass through are QF_4, f_8, f_9, f_{10}, f_{11}, f_{12}, f_{13}, f_{14}, f_{15}, f_{16}, f_{17}, QF_3, QF_{3-1}, QF_{3-2}, QF_{3-3}, QF_{3-4}, f_{18}, and f_{19}. The rock mass is relatively integrated, but the local stability of the surrounding rock is poor.

(2) The drilling and blasting construction section is 16,615 m long (K65 + 164~K81 + 779), and the main lithology is granite, granodiorite, granite porphyry, gneiss, hornblende schist, phyllite with sand, mica schist with chlorite schist, and marble with mica schist. The rock mass is weakly weathered to slightly weathered, which is severely affected by geological structures. The joint fractures are well-developed. The main faults that pass through are f_{21}, f_{22}, f_{23}, QF_2, f_{25}, and f_{26}. The integrity of the rock mass is poor, but the local stability of the surrounding rock is good.

The materials in the fault fracture zone are primarily composed of fault mud, breccia, Mylonite, and fragmented rock, which are loose and broken and contain water. The surrounding rock of the tunnel has poor self-stability, and its instability is prone to occur during construction.

4. In Situ Stress Analysis Using Multivariate Information Integration

4.1. Macro Characteristic Analysis

4.1.1. Analysis of the Regional Geological Structures

The Qinling water conveyance tunnel belongs to the Qinling Fold System in terms of tectonic units, and its geological structure mainly develops fold and fault structures. It passes through two regional major faults, namely, the Shangnan–Tangzhi fault, which strikes in the EW direction, shows compressive stress, and inclines in the N direction. The Fengzhen–Shanyang fault strikes in the NWW direction, shows tensile stress, and inclines in the N direction. Regional major faults have the characteristics of deep cutting, long extension, and large scale. The strike is mostly in the near EW direction, with a small number in the NE and NW directions, most of which are compressive. The distribution characteristics of major geological structures near the tunnel are shown in Figure 4. Some domestic scholars have deeply discussed the distribution characteristics of the tectonic stress field in this region using focal mechanisms, numerical analysis methods for microstructure, etc. It is believed that the tectonic stress field in this area is mainly extruded in the NW–SE direction [40]. This is basically consistent with the characteristics of the regional tectonic stress field in the North China Plate.

In summary, the main compressive stress direction of the modern tectonic stress field in the project area is generally in the NW or near EW direction, and there may be significant deviations in local areas due to the influence of topography, geological structure, and the ancient tectonic stress field.

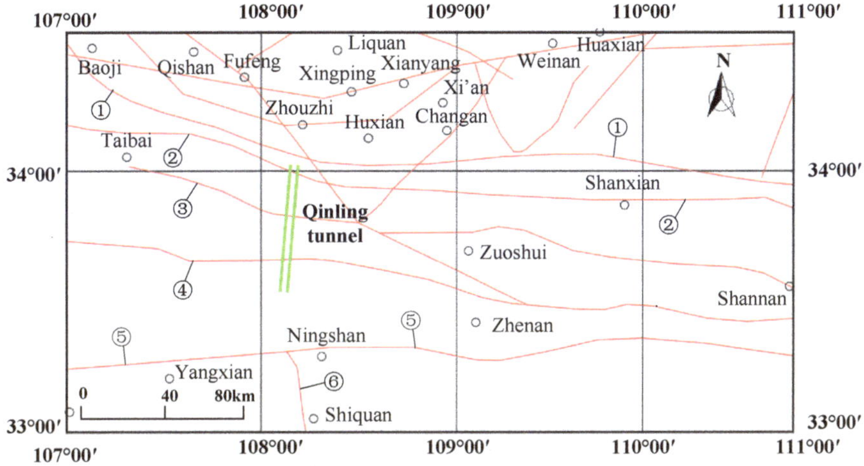

Figure 4. Distribution of major fault structures near the Qinling Tunnel project area. ① Baoji–Xieyuguan–Taiyao fault zone; ② Youfang–Huangtai fault zone; ③ Tangzhi–Shangnan fault zone; ④ Shanyang–Fengzhen fault zone; ⑤ Yangpingguan–Yangxian fault zone; and ⑥ Raofeng–Shiquan fault zone.

4.1.2. Analysis of the Topography in the In Situ Stress Field

Existing studies have shown that the distribution of in situ stress in river valleys generally includes a stress relief zone, stress concentration zone, and stress stability zone [41,42]. The elevation and maximum burial depth of the Qinling water conveyance tunnel project area are approximately 500 m–2500 m and 2002 m, respectively. The valley is relatively narrow, with a general width of 30–50 m and a natural slope of 40°–60°. The topographic and geomorphological features of the river valley along the tunnel are shown in Figure 5. The in situ stress testing indicates that the maximum horizontal principal stress direction is mainly in the NW direction outside the tunnel, but it is close to the EW direction inside the tunnel. The overall direction is between NW and EW, which is consistent with the regional structure and topography [43]. In addition, the lateral pressure coefficient of the maximum horizontal principal stress direction near the river valley outside the tunnel is generally greater than 2.0, indicating that the stress concentration phenomenon in the valley area is obvious. The tectonic stress field is strong, and the testing site is significantly affected by the terrain. The field test results are basically consistent with current research results.

Figure 5. Landforms in river valleys along the tunnel.

4.1.3. Relationship between Failure Characteristics of the Surrounding Rock and In Situ Stress

Recently, during the construction of the main and branch tunnels, a rock burst was observed. Also, instability in the surrounding rock is relatively common, mainly occurring in the fault fracture zones, weak structural planes, and joint dense zones. In some areas severely affected by geological structures, such as fault zones and lithological contact zones, it is easy to encounter large-scale collapses.

It can be seen that the distribution of in situ stress in the tunnel area is relatively complex. For the surrounding rock of the tunnel with large burial depths of hard and brittle rocks, the tectonic stress field has a strong effect. The stress concentration phenomenon is evident, and rock bursts are prone to occur. For soft rock areas with large burial depths, particularly in areas with fault structures, stress release is the main feature. Large deformation of the surrounding rock and rock mass sliding and collapse are more significant. Therefore, the characteristics of the in situ stress field in the project area are relatively complex, mainly formed by the superposition of self-weight stress and tectonic stress.

4.2. Analysis of In Situ Stress Testing Results

The in situ stress field test of the Qinling water conveyance tunnel mainly uses the HF method and 3-D hydraulic fracturing method. The HF method uses the method recommended by the International Society of Rock Mechanics for in situ stress testing. The principle underlying the 3-D hydraulic fracturing method and its implementation process are detailed in Ref. [43]. The layout of boreholes for in situ stress testing is shown in Figure 6. Among them, three deep boreholes, CZK-2, CZK-3, and SZK-2, were created using the HF method. The stress measurement results are shown in Table 1 and Figure 7, where σ_v is the vertical principal stress, σ_H is the maximum horizontal principal stress, and σ_h is the minimum horizontal principal stress. The 6# testing section used the 3-D hydraulic fracturing method, and the stress measurement results are shown in Table 2. The stress components and maximum horizontal principal stress results after coordinate system transformation are listed in Table 3.

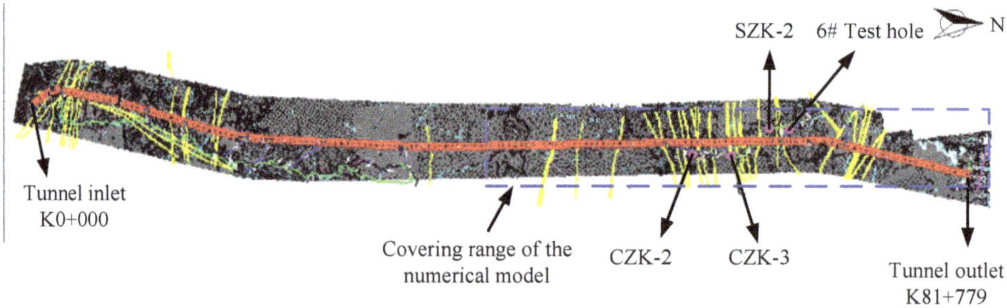

Figure 6. Layout of boreholes.

Table 1. The measured results for in situ stress in the Qinlin water conveyance tunnel.

Borehole	Measurement Point No.	Position	Depth (m)	σ_H (MPa)	σ_h (MPa)	σ_V (MPa)	σ_H/σ_V	σ_H Direction
CZK-2	1#	K50 + 390	353.0	18.11	11.26	9.34	1.94	N 52° W
	2#		507.80	18.53	11.68	13.44	1.38	N 47° W
	3#		548.75	18.53	12.68	14.52	1.28	/
	4#		597.10	19.51	13.35	15.80	1.23	/
	5#		631.15	20.04	13.89	16.70	1.20	N 37° W
	6#		646.85	20.49	14.14	17.12	1.20	/

Table 1. *Cont.*

Borehole	Measurement Point No.	Position	Depth (m)	σ_H (MPa)	σ_h (MPa)	σ_V (MPa)	σ_H/σ_V	σ_H Direction
CZK-3	7#	K53 + 940	265.50	9.12	6.10	7.03	1.30	/
	8#		275.40	10.21	6.70	7.29	1.40	N 40° W
	9#		285.40	7.31	5.30	7.55	0.97	/
	10#		310.40	9.06	6.54	8.21	1.10	N 43° W
	11#		334.40	7.79	5.78	8.85	0.88	/
	12#		341.40	8.36	5.85	9.03	0.93	/
	13#		354.40	12.49	8.47	9.38	1.33	N 46° W
	14#		363.40	13.08	8.56	9.62	1.36	/
SZK-2	15#	K64 + 315	126.40	17.77	10.27	3.54	5.02	N 16° E
	16#		195.40	18.96	10.96	5.47	3.47	N 9° E
	17#		225.15	19.26	11.26	6.30	3.06	/
	18#		238.20	19.89	11.39	6.67	2.98	N 16° W
	19#		251.20	22.02	12.52	7.03	3.13	N 33° W
	20#		270.90	27.71	16.71	7.59	3.65	N 35° W
	21#		284.25	29.85	16.85	7.96	3.75	/

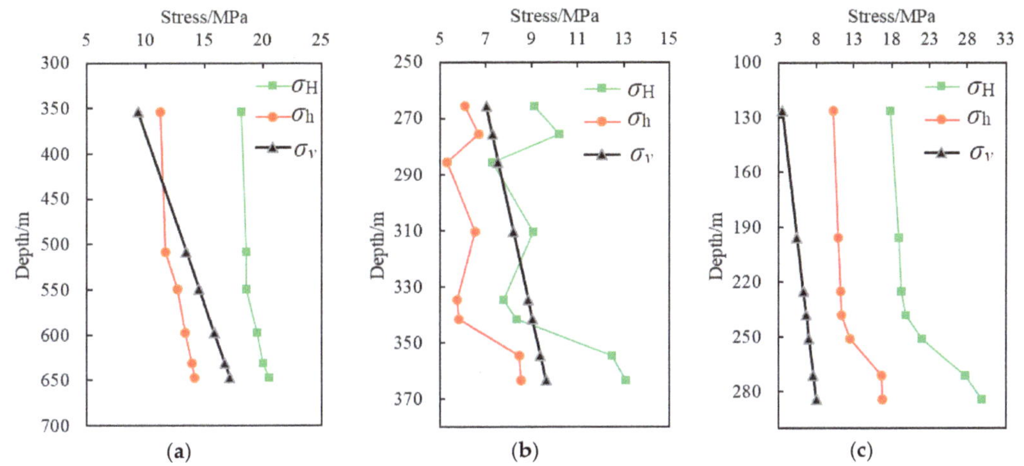

Figure 7. In situ stress test data using the HF method in the tunnel: (**a**) CZK-2; (**b**) CZK-3; and (**c**) SZK-2.

Table 2. In situ stress test data using the 3-D HF method at the 6# testing section.

Stress Order	First Principal Stress σ_1	Second Principal Stress σ_2	Third Principal Stress σ_3
Magnitude/MPa	25.4	15.5	14.1
Inclination/(°)	38	52	3
Azimuth/(°)	286	112	18

Note: position, K67 + 350; borehole burial depth, 1000 m.

Table 3. Stress component conversion results at the 6# test hole.

Type	Stress Components/MPa						Horizontal Principal Stress and Lateral Pressure Coefficient		
	σ_{xx}	σ_{yy}	σ_{zz}	σ_{xy}	σ_{yz}	σ_{zx}	σ_H/MPa	σ_h/MPa	$\lambda = \sigma_H/\sigma_Z$
Values	14.7	21.1	19.2	2.1	4.6	1.2	21.7	14.1	1.1

Note: position, K67 + 350; borehole burial depth, 1000 m.

As shown in Table 2 and Figure 7, in situ stress in the project area is dominated by horizontal tectonic stress, but there are certain differences in the distribution characteristics of stress among different boreholes. For boreholes CZK-2 and CZK-3, the relationship between the three principal stresses is basically $\sigma_H > \sigma_v > \sigma_h$, and the maximum principal stress is mainly horizontal. The predominant direction of the maximum horizontal principal stress is N 37° W to N 52° W, and the lateral pressure coefficient of the maximum horizontal principal stress (σ_H/σ_v) ranges from 0.88 to 1.94. However, for borehole SZK-2, the relationship between the three principal stresses is basically $\sigma_H > \sigma_h > \sigma_v$. The predominant direction of the maximum horizontal principal stress is N 9° E to N 35° W, and the lateral pressure coefficient of the maximum horizontal principal stress (σ_H/σ_v) ranges from 2.88 to 5.02.

In addition, to meet the requirements for quantitative analysis of in situ measured stresses, the so-called software 'GeoStress3D V1.0' was developed using OOP programming technology, the C++ language, and the open graphics library OpenGL. This method not only verifies the orthogonality of the three principal stresses direction but also determines the in situ stress distribution characteristics at measurement points on each projection plane using plane projection technology. Using the measurement point at the 6# testing section as an example, the stereographic projection of a unit vector of typical principal stress orientations and their projection stress ellipses on the XY, YZ, and YZ planes are shown in Figure 8a. Also, Figure 8b shows a plot of the deformation and failure characteristics of the surrounding rock of the tunnel. In the figure, the red circle ●, green square ■, and blue diamond ◆ stand for the unit vector of the orientation of the maximum, intermediate, and minimum principal stresses, respectively (same as below). The three numbers behind Sigma (i.e., Sigma1, Sigma2, and Sigma3) stand for the principal stress values, azimuth, and dip angles, respectively. It can be concluded that the long axial direction of the stress ellipse is inclined to the upper (or lower) reach of the river valley on the XY (or YZ) plane. On the XZ plane, the long axial direction is inclined to the inside mountain. Similarly, the projection features of the stresses at other testing points can also be interpreted.

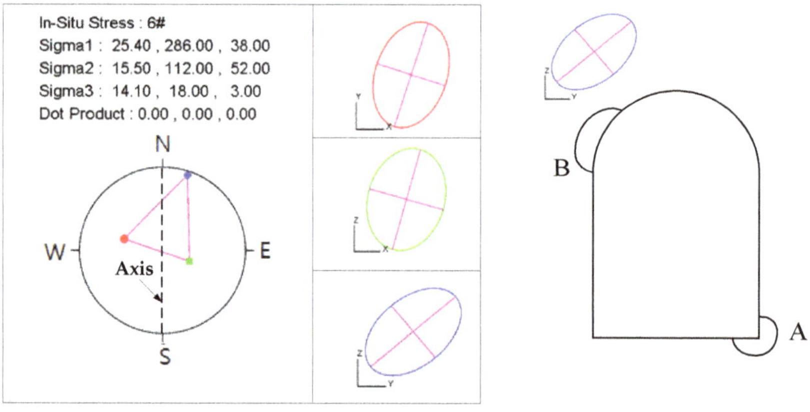

Figure 8. Sketch of stereographic projection at the measurement point in the 6# testing section of the Qinling water conveyance tunnel: (**a**) Sketch showing the stereographic projection of unit vectors of the principal stress orientations; (**b**) Deformation and failure position of surrounding rock.

Thus, the orientation of the maximum horizontal principal stress is orthogonal to the axis of the Qinling water conveyance tunnel (near the NS direction), which is relatively unfavorable for the stability of the cavern. During the actual construction process, some effective measures should be taken to reduce the impact of the secondary stress field on the deformation and failure of surrounding rock after tunnel excavation.

4.3. Selection of Representative Measured Data

Based on the above, it can be concluded that the in situ stress field in the project area has the following characteristics: (1) The in situ stress field is significantly affected by tectonic stress in the shallow part, and as the burial depth increases, the self-weight gradually dominates. (2) The orientation of the maximum horizontal principal stress is between NW and NWW. The lateral pressure coefficient of the maximum horizontal principal stress decreases with increasing burial depth, most of which are less than 1.5. (3) The projection of the measured stress tensor on each plane should conform to the characteristics shown in Figure 8.

It can be seen that the burial depth of borehole SZK-2 is relatively shallow, and the measured results are greatly affected by the topography and geomorphology. Moreover, the measurement points 9#–12# in borehole CZK-3 are significantly affected by local geological defects. Namely, the above measured results are not representative. Thus, ten measurement points, 1–8#, 13#, and 14#, can represent the overall characteristics of the in situ stress field in the project area, which are selected as the basic information for a subsequent back analysis of the in situ stress field.

4.4. MRA of the In Situ Stress Field

4.4.1. Computational Model and Mechanical Parameters for the Rock Masses

The numerical analysis method is widely recognized as an important approach for estimating the stress state in deep-buried tunnels. Here, the commercial software FLAC3D V4.0 was used for numerical calculation. The computational model corresponded to a region of the rock mass measuring 5000 m $\times$ 45,000 m $\times$ 3600 m (X $\times$ Y $\times$ Z, where the positive direction of the X-axis and Y-axis point to the east and north, respectively, and the Z-axis extends upwards from an elevation of -1000 m to the mountain top). The computational model and grid division based on regional geology, topography, and geomorphology are plotted in Figure 9a,b. In this model, seventeen typical strata and thirteen major faults were included; the main faults are shown in Figure 9c. In addition, a 3-D solid element (Solid45) was used to establish the finite element calculation model. Different types of strata and faults in the model were simulated using solid units, and the constitutive model was linear elastic. The bottom of the model was constrained, and the tectonic effects were simulated by applying horizontal compressive and shear loads around the model (see Section 2.3). The number of grid elements and nodes were 185,832 and 1,056,155, respectively. The physical and mechanical parameters of the rock masses are listed in Table 4.

Table 4. Physical and mechanical parameters of the rock mass.

Type of Rock Mass	Deformation Modulus/(GPa)	Poisson Ratio	Bulk Density/(kg/m^3)
Granite (γ_5)	37.05	0.20	26.6
Quartz diorite (δ_4)	24.30	0.22	27.0
Variable sandstone	12.67	0.28	26.0
Sericite thousand rocks, quartz metasandstone	7.66	0.31	26.5
Hornblende schist, sericite mudstone	15.48	0.27	26.5
Granite (γ_3)	24.40	0.21	26.5
Granodiorite, granite porphyry	23.33	0.22	26.9
Black cloud plagioclase gneiss	15.08	0.26	26.1
Mica schist, green clay, quartz schist	6.81	0.32	26.1
Dali rock sandwiched by mica schist	17.77	0.26	27.0
Mica quartz schist, chlorite schist	10.62	0.30	26.2
Faults	1.00	0.33	23.5

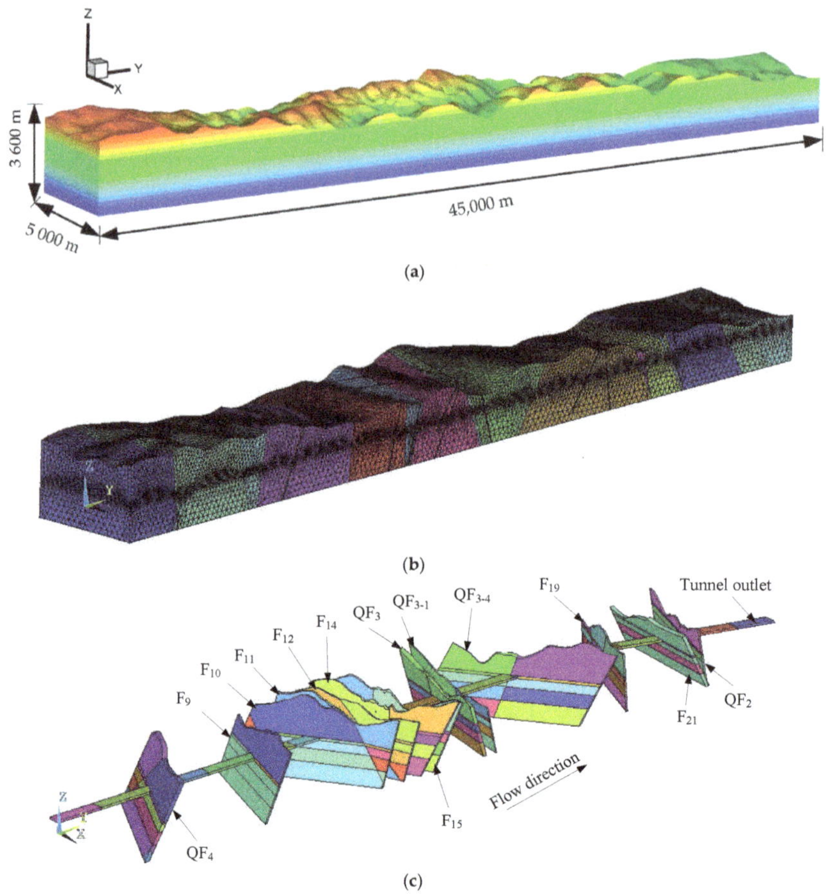

Figure 9. Computational model: (**a**) entity model; (**b**) mesh grid; and (**c**) major fractures.

4.4.2. Back Analysis of the In Situ Stress Field

Based on Section 2.3, the finite element calculation of regional ground stress field distribution under self-weight stress and three kinds of tectonic action was carried out, and the stress components at the corresponding depths of each borehole were extracted. According to the MRA, the regression coefficients for the four influencing factors affecting the in situ stress field of the Qinling water conveyance tunnel were obtained, and the initial in situ stress field regression equation was preliminarily established as:

$$\sigma_{cal} = 0.93S_g + 2.416S_x + 5.589S_y - 16.236S_{xy} \quad (5)$$

where σ_{cal} is the initial regression stress value; S_g is the self-weight stress value; S_x and S_y are the unit extrusion stresses applied along the horizontal boundary in the x and y directions, respectively; and S_{xy} is the unit-tangential stress applied along the x and y directions in the horizontal plane.

By comparing the calculation values with those measured (Table 5), it is found that the multiple correlation coefficient is 0.956, and the F-test observed value is 50.51. When the significance level is 0.05, the critical value $F_{0.05}$ (4, 24 − 4 − 1) is 2.93, namely, $F > F_{0.05}$. Thus, the precision of the back-analysis is satisfactory.

Table 5. Comparison between the measured data obtained using a field test and the calculated results using numerical simulation.

Boreholes	Measurement Point No.	Data Origin	σ_x /MPa	σ_y /MPa	σ_z /MPa	τ_{xy} /MPa	σ_H /MPa	σ_h /MPa	Azimuth /(°)
CZK-2	1#	Measured	15.51	13.86	9.34	−3.32	18.11	11.26	N 52° W
		Calculated	14.11	14.62	12.73	−1.63	16.01	12.72	N 40.52° W
	2#	Measured	14.75	15.46	13.44	−3.41	18.53	11.68	N 42° W
		Calculated	15.60	15.64	15.82	−1.50	17.11	14.12	N 44.59° W
	3#	Measured	15.30	15.91	14.52	−2.91	18.53	12.68	N 42° W
		Calculated	16.03	15.93	16.72	−1.45	17.43	14.53	N 43.99° W
	4#	Measured	16.11	16.75	15.80	−3.06	19.51	13.35	N 42° W
		Calculated	16.47	16.27	17.64	−1.42	17.80	14.95	N 43.03° W
	5#	Measured	16.12	17.81	16.70	−2.96	20.04	13.89	N 37° W
		Calculated	16.83	16.58	18.43	−1.39	18.10	15.31	N 42.38° W
	6#	Measured	16.44	18.19	17.12	−3.05	20.49	14.14	N 37° W
		Calculated	16.92	16.65	18.63	−1.39	18.18	15.40	N 42.23° W
CZK-3	7#	Measured	7.35	7.87	7.03	−1.49	9.12	6.10	N 40° W
		Calculated	9.44	11.22	6.57	−3.57	14.00	6.65	N 38.01° W
	8#	Measured	8.15	8.76	7.29	−1.73	10.21	6.70	N 40° W
		Calculated	10.19	12.70	7.40	−3.78	15.43	7.46	N 35.81° W
	13#	Measured	10.55	10.41	9.38	−2.01	12.49	8.47	N 44° W
		Calculated	10.31	13.26	9.50	−4.02	16.06	7.51	N 34.94° W
	14#	Measured	10.90	10.74	9.62	−2.26	13.08	8.56	N 46° W
		Calculated	10.22	13.32	10.21	−3.77	15.84	7.69	N 33.82° W

Figure 10 plots the relationship between the measured value of the principal stress and the calculated values at the measurement points. It can be seen that the measured values at most measurement points are numerically close to the calculated ones, and the absolute error of the stress component is mostly not more than 2 Mpa. The overall error is relatively small. In addition, the maximum horizontal principal stress azimuth at each measurement point is in good agreement with the measured data. The maximum absolute error of the principal stress azimuth is approximately 12°, generally not exceeding 6°. It can be concluded that the optimized back analysis method used in this paper is credible and behaves well when simulating the initial in situ stress field in the project area under complex conditions.

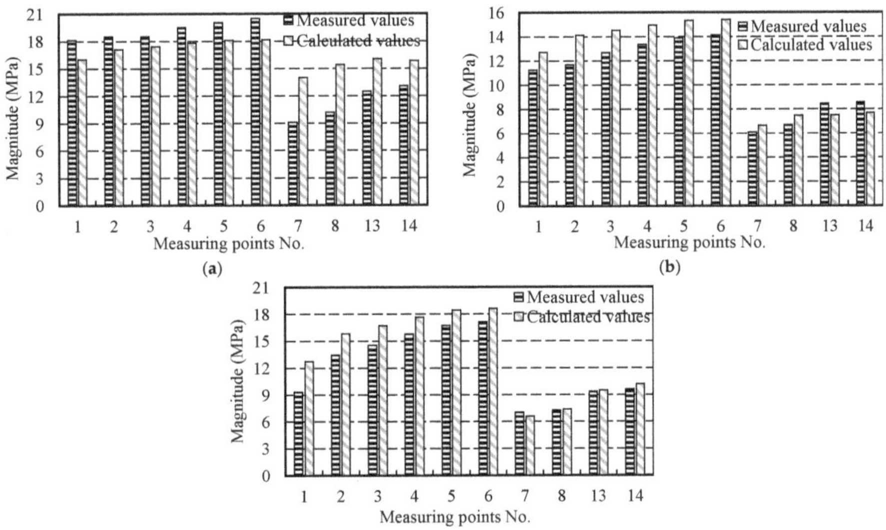

Figure 10. Comparative histogram showing the principal stresses between the measured and calculated values at the measurement points: (**a**) maximum horizontal principal stress; (**b**) minimum horizontal principal stress; and (**c**) vertical principal stress.

4.5. Distribution Characteristics of the In Situ Stress Field

Figure 11 shows the distribution of the three principal stress contours along the longitudinal section of the axis of the Qinling water conveyance tunnel. The distribution characteristics of the in situ stress field of the tunnel can be described as follows:

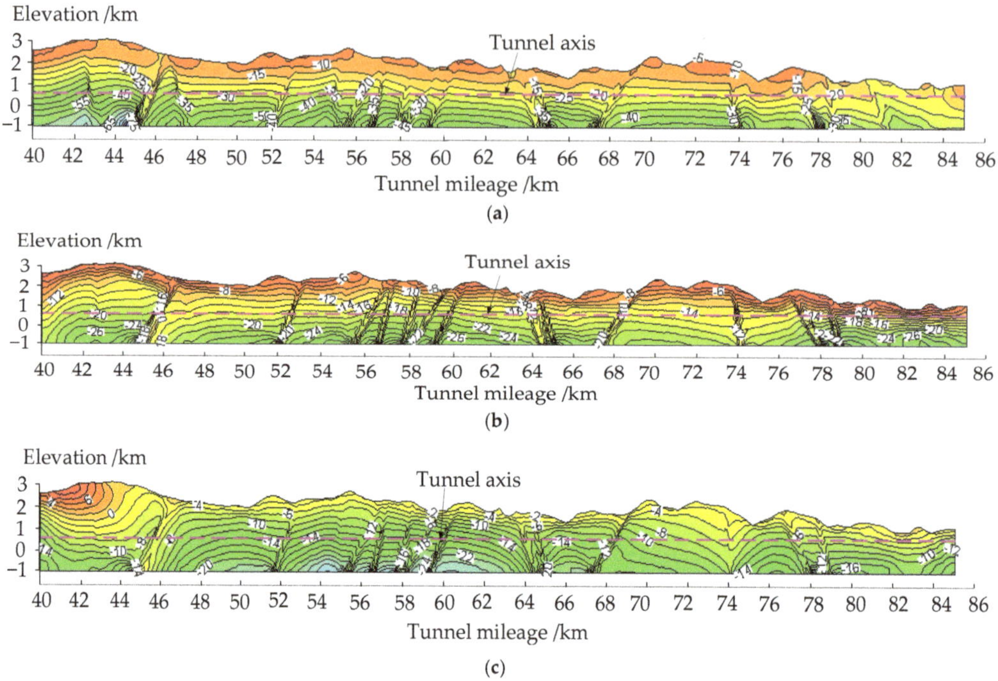

Figure 11. Distribution of the principal stress field at the axis of the Qinling water conveyance tunnel (unit: MPa): (**a**) the first principal stress; (**b**) the second principal stress; and (**c**) the third principal stress.

(1) The magnitude of in situ stress near the axis of the tunnel increases with the burial depth. The maximum burial depth is approximately 2000 m. The magnitude of the maximum principal stress is as high as 52.65 MPa.

(2) The tunnel has a long route that passes through different geological structural units, resulting in a very complex distribution of the in situ stress field in the surrounding rock. Specifically, when the burial depth of the tunnel is less than 1000 m, the in situ stress field at the tunnel elevation is mainly horizontal, that is, $\sigma_H > \sigma_Z > \sigma_h$. the lateral pressure coefficient of the maximum horizontal principal stress (σ_H/σ_v) ranges from 1.02 to 1.88. When the burial depth of the tunnel is greater than 1000 m, the maximum principal stress at the tunnel elevation is primarily owing to its own weight, that is, $\sigma_Z > \sigma_H > \sigma_h$. The lateral pressure coefficient of the maximum horizontal principal stress is between 0.7 and 1.1. In general, the lateral pressure coefficient of the principal stress gradually decreases with an increase in burial depth.

(3) Due to the influence of topography and faults, a certain stress concentration phenomenon occurs at the foot of slope. However, there are obvious stress release phenomenon in the fault zone and adjacent areas, and the maximum horizontal principal stress direction also undergoes a certain deviation. Furthermore, the fault and its surroundings show that the vertical principal stress in the rock mass is greater than that of the maximum horizontal principal stress, that is, $\sigma_Z > \sigma_H > \sigma_h$.

(4) The in situ stress in the fault zone changes sharply, and the stress contour appears to be a phenomenon of 'hump and mutation'. After reaching a certain distance, it tends to stabilize. Further analysis showed that at the end of the fault, each stress component and principal stress have evident stress concentrations. However, the position of the stress concentration zone is asymmetric at both ends of the fault, which is primarily caused by the difference in the angle between the boundary stress ratio of the fault and the maximum principal stress.

5. Failure Mechanism of the Surrounding Rock and Treatment Method

5.1. Description of Failure Characteristics during Tunnel Excavation

Figure 12 shows the geological profile along the axis of the Qinling water conveyance tunnel from K79 + 900 to K78 + 779. The burial depth of the tunnel in this section ranges from 183 m to 282 m, and more than ten large collapses occurred during its construction. The collapses are mainly located on the left arch and side wall. The height of the collapsed cavity is about 8 m, with a longitudinal extension of 4 m and a circumference of 6 m. After the collapse, the initial convergence deformation of the surrounding rock was large, reaching 10 mm/day. A typical failure image obtained in the surrounding rock is shown in Figure 13.

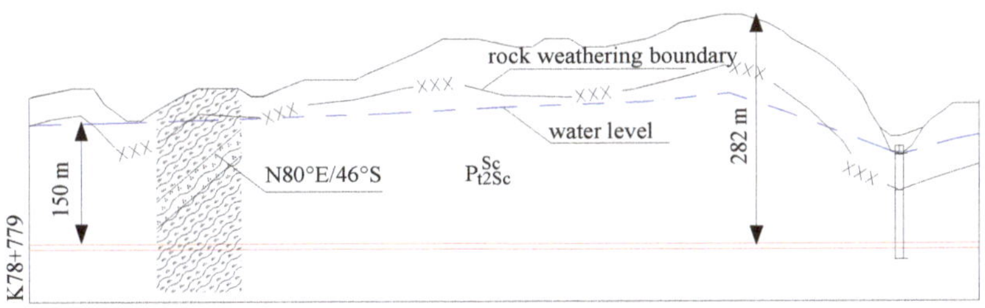

Figure 12. Geological profile along the axis of the Qinling tunnel from K79 + 900 to K78 + 779.

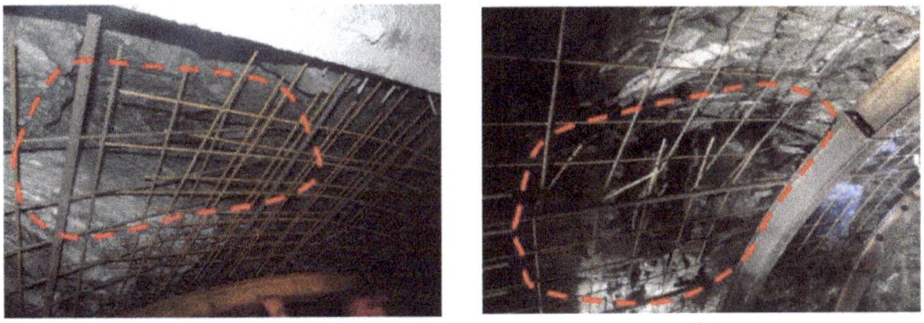

(**a**) (**b**)

Figure 13. Image showing a failure in the surrounding rock of the Qinling tunnel: (**a**) left arch and (**b**) left side wall.

5.2. Failure Mechanism Analysis

According to the geological data obtained during the excavation, the lithology of the locally damaged area is primarily mica quartz schist with local carbonaceous schist, high mica content, soft lithology, and joint development. Three groups of joints, i.e., straight, smooth, and no filling, are well developed, which can be described as follows:

(a) J1: N 20° W/60° S, spacing from 0.1 m to 0.2 m, extending approximately 1 m.

(b) J2: N 15° W/60° N, spacing from 0.3 m to 0.5 m, extending about 2 m.

(c) J3: N 30° E/55° N, spacing from 0.05 m to 0.1 m, extending about 1 m.

Due to the extremely smooth surface of mica schist, low internal friction angle, and poor shear resistance, unloading deformation occurred during the tunnel excavation. Plastic deformation subsequently occurred when exposed to water, which made it easy for the blocks to cause slip damage to the side wall. It can be concluded that the accident in this project was mainly owing to the collapse caused by the fragmentation of rock mass, soft lithology, and the formation of sliding blocks due to the cutting of the structural plane.

5.3. Targeted Treatment Method and Rationality Verification

5.3.1. Support Measures

Based on geological data and engineering experience, the parameters of Phase I support and lining in the surrounding rock are listed in Table 6. The corresponding cross-section of support is plotted in Figure 14.

Table 6. Support parameters of the surrounding rock.

Location	Spray C20 Concrete Thickness (cm)	Steel Arches		Secondary Lining	
		Specification	Spacing (m)	Grade	Thickness (cm)
Arches	21	I16	1.2	C30	40
Side walls	21			C30	40
Motherboard	/	/	/	C30	45

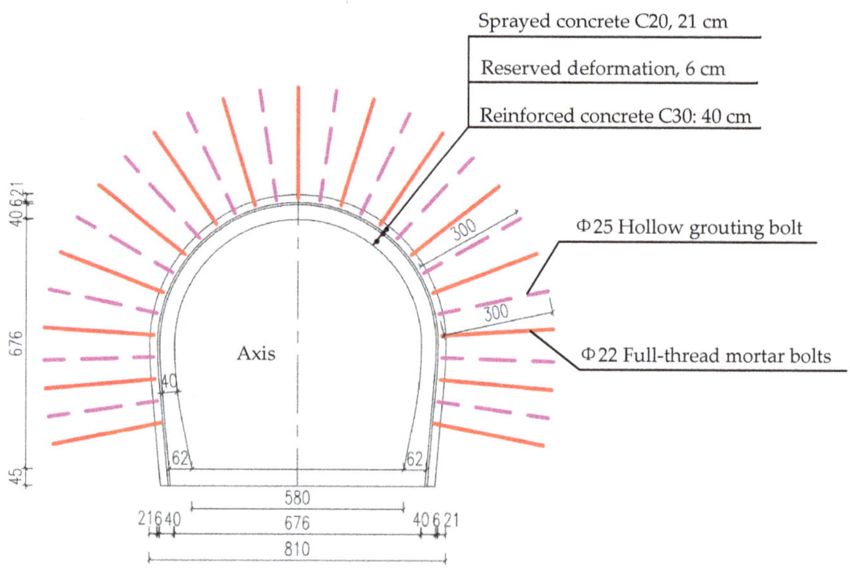

Figure 14. Cross-section of the Phase I support in the surrounding rock (unit: cm).

Owing to the fragmentation of surrounding rock, the development of fissure water, and poor stability, a collapse is prone to occur after excavation. Therefore, advanced small conduits with a diameter of 42 mm were added to this section of the tunnel. The small conduits are arranged within 120° of the arch, with a longitudinal spacing of 2.0 m/ring, a circumferential spacing of 30 cm, and an external insertion angle ranging from 5° to 10°. The collapsed area was backfilled with C20 spray concrete.

5.3.2. Numerical Simulation Analysis and Verification

To evaluate the feasibility and rationality of the collapse treatment method for surrounding rock, a three-dimensional finite element model (see Figure 15) was established based on the excavation and support conditions of the tunnel at the mileage from K79 + 900 to K78 + 779. In the model, surrounding rock, shotcrete, and secondary lining structures were simulated using 3-D solid element elements. Advanced small conduits and anchors were simulated using rod elements. Steel arches were simulated using beam elements. The mesh model contained 265,508 grid nodes and 142,612 elements.

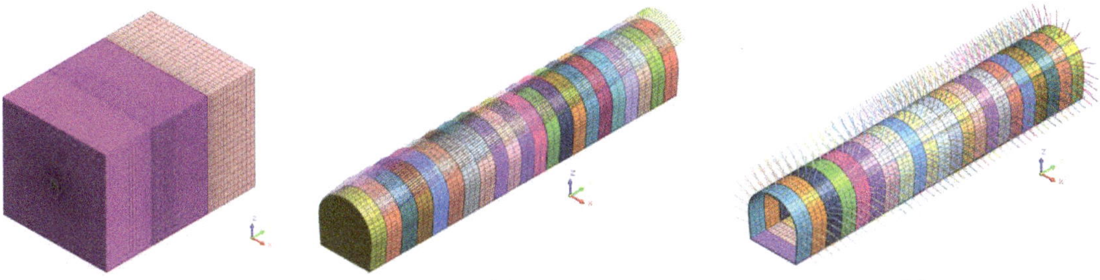

(a) (b) (c)

Figure 15. Computational model: (**a**) mesh model; (**b**) excavated tunnel and advanced small conduit model; and (**c**) tunnel system anchors with the shotcrete model.

Using the construction conditions and instability of the surrounding rock, as well as the stress estimation for the tunnel described in Section 4, a numerical simulation was carried out. The calculated results are as follows:

(1) Displacement distribution. Figures 16 and 17 show the distribution characteristics of the displacement field after tunnel excavation with untimely initial support and with timely initial support, respectively. In the case of untimely support, the maximum horizontal displacement at the tunnel sidewall can reach 96.7 mm, and the maximum settlement at the top arch is 44.0 mm. At this time, the soft rock is extremely vulnerable to damage. However, with timely support, the maximum horizontal displacement at the tunnel sidewall can reach 49.3 mm, and the maximum settlement at the top arch is 27.3 mm.

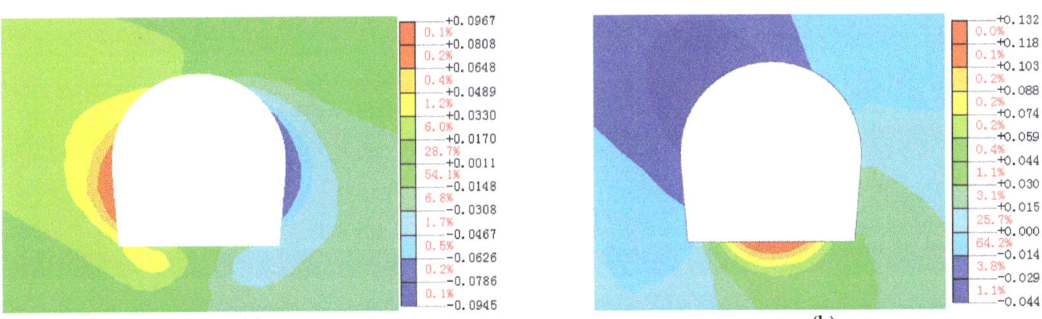

(a) (b)

Figure 16. Untimely initial support (unit: m): (**a**) horizontal deformation and (**b**) vertical deformation.

(2) In-stress stress distribution. Figure 18 shows the distribution characteristics of the first and third principal stresses after tunnel excavation. It can be seen that there is a phenomenon of stress concentration in the local area around the tunnel (the intersection of the arch foot and side wall with the bottom plate), and the stress concentration is mainly located in the left arch shoulder and side wall. The above results are basically consistent with the on-site failure phenomenon (see Figure 13), which are in good agreement with the inversion results in Section 4. Furthermore, the values of the major principal compressive stress and third principal stress lie in the ranges −0.23 to −13.2 and 0.26 to −5.74 MPa, respectively.

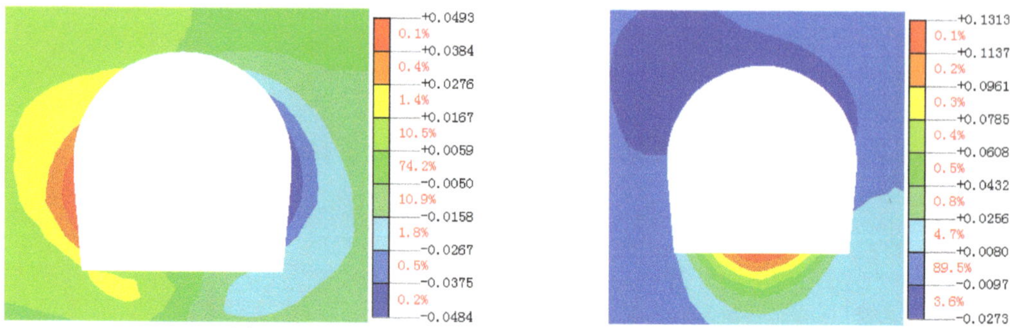

Figure 17. Timely initial support (unit: m): (**a**) horizontal deformation and (**b**) vertical deformation.

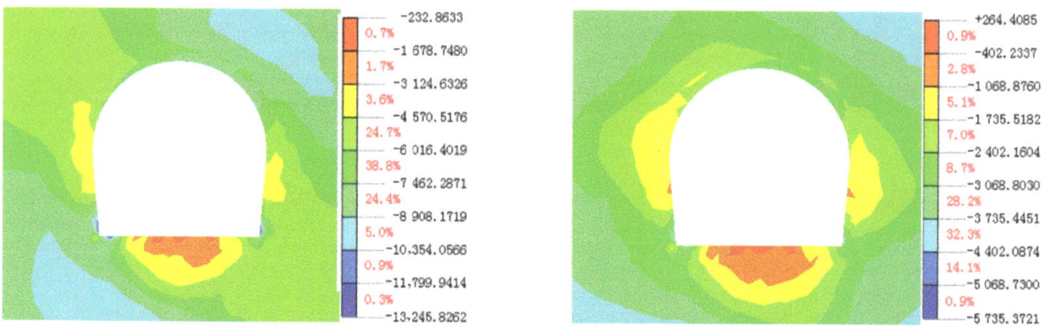

Figure 18. Schematic diagram showing the in situ stress field in the surrounding rock (unit: kPa): (**a**) the first principal stress and (**b**) the third principal stress.

5.4. Discussion

An in-depth analysis of the failure mechanism of surrounding rock during tunnel excavation at the mileage from K79 + 900 to K78 + 779 was carried out using a combination of the monitoring data, interpretation of rock failure characteristics, treatment measures for rock failure, and numerical simulations of the construction process and characteristics of the rock mass. We believe that the internal cause for the failure of the surrounding rock after the tunnel excavation was the fragmentation of rock mass, soft lithology, and the formation of sliding blocks due to the cutting of the structural plane. The external cause was improper construction and untimely initial support. After the tunnel excavation, it is not timely to carry out pre-grouting with advanced small conduit, bolting, and shotcreting support on the face of the tunnel, which results in unloading deformation of the surrounding rock. Then, plastic deformation occurs when encountering water, causing sliding damage to the side walls and arches. Further analysis showed that the sliding of the rock mass and the weakening of the mechanical properties of the structural plane caused a further adjustment of the stress field in the surrounding rock. In turn, the adjusted stress field led to further deformation of the structural plane. The complex interaction between the evolution of the rock mass structure and the adjustment of the stress field made the deformation of the surrounding rock exhibit certain time-effect characteristics, which was consistent with the maximum deformation of 10 mm/day observed during the tunnel excavation.

It can be concluded that for Class IV rock masses with fractured rock and relatively soft lithology, the failure of surrounding rock is more sensitive to the intensity of rock excavation disturbance. Therefore, in the subsequent construction process, the initial support measures such as adding an advanced small conduit with a diameter of 42 mm should be conducted at the face of the surrounding rock soon after the excavation. Meanwhile, the timely use of C20 sprayed concrete backfilling for collapsed areas should also be carried out. In that case, the impact of tunnel excavation on the stability of surrounding rock can be significantly reduced. Both the monitoring and numerical simulation indicated that during subsequent tunnel excavation, the stability of the surrounding rock was good, and the effectiveness of support measures was reasonably verified.

6. Conclusions

The contribution of this paper is the proposal of a method for in situ stress estimation of super-long and deep-buried tunnels based on multivariate information integration. The prominent features of this method include selecting the effectiveness and representativeness of the stress measurement data, identifying the failure phenomenon in surrounding rock during tunnel excavation, and verifying the results of the in situ stress inversion using multiple information types such as incomplete measurement data and on-site failure characteristics of the surrounding rock. This method was used to estimate the in situ stress in the sections of the super-long and deep-buried tunnel project in Qinling. Also, the failure mechanism and treatment measures of the surrounding rock are discussed. The main findings are as follows:

(1) The magnitude of the in situ stress near the axis of the Qinling water conveyance tunnel increased with the burial depth. The magnitude of the maximum principal stress was as high as 52.65 MPa, and the orientation of the maximum horizontal principal stress was mainly NW–EW.

(2) There were significant differences in the stress state along the deep-buried tunnel. When the burial depth of the tunnel was less than 1000 m, the in situ stress field at the tunnel elevation was mainly horizontal, that is, $\sigma_H > \sigma_Z > \sigma_h$. When the burial depth of the tunnel was greater than 1000 m, the maximum principal stress at the tunnel elevation was primarily owing to its own weight, that is, $\sigma_Z > \sigma_H > \sigma_h$.

(3) Compared with other areas, the fault zone and adjacent areas showed evident stress release characteristics, and the maximum horizontal principal stress direction also underwent a certain deviation. Furthermore, the fault and its surroundings showed that the vertical principal stress of rock mass was greater than that of the maximum horizontal principal stress, that is, $\sigma_Z > \sigma_H > \sigma_h$.

(4) The tunnel section from K79 + 900 to K78 + 779 of tunnel mileage collapsed in numerous locations during the construction process. The research showed that the initial support measures such as adding an advanced small conduit with a diameter of 42 mm should be conducted at the face of the fourth grade of surrounding rock soon after the excavation. The deformation of the surrounding rock after support will be significantly reduced, which will satisfy the construction safety requirements.

However, current in situ stress studies in underground engineering are mainly focused on the in situ stress field around deep hard rocks rather than soft rock tunnels. The in situ stress test method and theoretical verification near the large-deformation tunnel of soft rock are not yet mature and require further exploration.

Author Contributions: Q.P.: methodology, data curation, validation, and writing; X.W.: methodology and supervision; L.H.: investigation and supervision; L.L.: data curation and writing; Y.T.: data curation; C.W.: supervision and funding acquisition. All authors have read and agreed to the published version of the manuscript.

Funding: This research was funded by the National Science Foundation of China (No. 51609018), the Central Non-profit Scientific Research Fund for Institutes of China (No. CKSF2021431/CL), and the Key Project of the National Natural Science Foundation of China (No. 52239009).

Data Availability Statement: The data used to support the findings of this study are included in the article.

Conflicts of Interest: The authors declare no conflict of interest.

References

1. Lyu, H.M.; Sun, W.J.; Shen, S.L.; Zhou, A.N. Risk assessment using a new consulting process in fuzzy AHP. *J. Constr. Eng. Manag.* **2020**, *146*, 04019112. [CrossRef]
2. Jin, Y.F.; Yin, Z.Y.; Zhou, W.H.; Shao, J.F. Bayesian model selection for sand with generalization ability evaluation. *Int. J. Numer. Anal. Methods Geomech.* **2019**, *43*, 2305–2327. [CrossRef]
3. Yin, Z.Y.; Jin, Y.F.; Shen, J.S.; Hicher, P.Y. Optimization techniques for identifying soil parameters in geotechnical engineering: Comparative study and enhancement. *Int. J. Numer. Anal. Methods Geomech.* **2018**, *42*, 70–94. [CrossRef]
4. Cai, M.F. *Principle and Techniques of In-Situ Stress Measurement*; Science Press: Beijing, China, 2000.
5. Fu, C.; Wang, W.; Chen, S. Back analysis study on initial geostress field of dam site for Xiluodu hydropower project. *Chin. J. Rock Mech. Eng.* **2006**, *25*, 2305–2312.
6. Bjarnason, B. Hydro Fracturing Rock Stress Measurements in the Baltic Shield. Licentiate Thesis, Lulea University, Luleå, Sweden, 1986.
7. Kim, K.; Franklin, J.A. Suggested methods for rock stress determination. *Int. J. Rock Mech. Min. Sci. Geomech. Abstr.* **1987**, *24*, 53–73.
8. Sjöberg, J.; Christiansson, R.; Hudson, J.A. ISRM suggested methods for rock stress estimation d Part 2: Overcoring methods. *Int. J. Rock Mech. Min. Sci.* **2003**, *40*, 999–1010. [CrossRef]
9. Sjöberg, J.; Klasson, H. Stress measurements in deep boreholes using the Borre (SSPB) probe. *Int. J. Rock Mech. Min. Sci.* **2003**, *40*, 1205–1223. [CrossRef]
10. Haimson, B.C.; Cornet, F.H. ISRM suggested methods for rock stress estimation-Part 3: Hydraulic fracturing (HF) and/or hydraulic testing of pre-existing fractures (HTPF). *Int. J. Rock Mech. Min. Sci.* **2003**, *40*, 1011–1020. [CrossRef]
11. Liu, Y.F.; Yin, J.M.; Liu, Y.K. *Geostress Measurement Methods and Engineering Application*; Hubei Science and Technology Press: Wuhan, China, 2014.
12. Ljunggren, C.; Chang, Y.; Janson, T.; Christiansson, R. An overview of rock stress measurement methods. *Int. J. Rock Mech. Min. Sci.* **2003**, *40*, 975–989. [CrossRef]
13. Zheng, M.Z.; Li, S.J.; Yao, Z.; Zhang, A.D.; Xu, D.P.; Zhou, J.F. Core discing characteristics and mitigation approach by a novel developed drill bit in deep rocks. *J. Cent. S. Univ.* **2020**, *27*, 2822–2833. [CrossRef]
14. Shukla, P.K.; Singha, D.K.; Sain, K. Modeling of in-situ horizontal stresses and orientation of maximum horizhontal stress in the gas hydrate-bearing sediments of the Mahanadi offshore basin. *Geomech. Geophys. Geo-Energ. Geo-Resour.* **2022**, *8*, 90. [CrossRef]
15. Liu, J.; Ding, W.; Yang, H.; Wang, R.; Yin, S.; Li, A.; Fu, F. 3D geomechanical modeling and numerical simulation of in-situ stress fields in shale reservoirs: A case study of the lower Cambrian Niutitang formation in the Cen'gong block, South China. *Tectonophysics* **2017**, *712*, 663–683. [CrossRef]
16. Krietsch, H.; Gischig, V.; Evans, K.; Doetsch, J.; Dutler, N.O.; Valley, B.; Amann, F. Stress measurements for an in-situ stimulation experiment in crystalline rock: Integration of induced seismicity, stress relief and hydraulic methods. *Rock Mech. Rock Eng.* **2019**, *52*, 517–542. [CrossRef]
17. Xu, D.P.; Huang, X.; Jiang, Q.; Li, S.J.; Zheng, H.; Qiu, S.L.; Xu, H.S.; Li, Y.H.; Li, Z.G.; Ma, X.D. Estimation of the three-dimensional in situ stress field around a large deep underground cavern group near a valley. *J. Rock Mech. Geotechn. Eng.* **2021**, *13*, 529–544. [CrossRef]
18. Haimson, B.C.; Lee, H. Borehole breakouts and compaction bands in two high-porosity sandstones. *Int. J. Rock Mech. Min. Sci.* **2004**, *41*, 287–301. [CrossRef]
19. Leeman, E.R. The measurement of stress in rock: Part I: The principles of rock stress measurements. *J. S. Afr. Inst. Min. Metall.* **1964**, *65*, 45–81.
20. Zoback, M.D.; Barton, C.A.; Brudy, M.; Castillo, D.A.; Finkbeiner, T.; Grollimund, B.R.; Moos, D.B.; Peska, P.; Ward, C.D.; Wiprut, D.J. Determination of stress orientation and magnitude in deep wells. *Int. J. Rock Mech. Min. Sci.* **2003**, *40*, 1049–1076. [CrossRef]
21. Lund, B.; Zoback, M.D. Orientation and magnitude of geostress to 6.5 km depth in the Baltic Shield. *Int. J. Rock Mech. Min. Sci.* **1999**, *36*, 169–190. [CrossRef]
22. Stacey, T.R. Contribution to the mechanism of core discing. *J. S. Afr. Inst. Min. Metall.* **1982**, *82*, 269–274.
23. Kaga, N.; Matsuki, K.; Sakaguchi, K. The geostress states associated with core disking estimated by analysis of principal tensile stress. *Int. J. Rock Mech. Min. Sci.* **2003**, *40*, 653–665. [CrossRef]
24. Jiang, Q.; Feng, X.T.; Xiang, T.B.; Su, G.S. Rockburst characteristics and numerical simulation based on a new energy index: A case study of a tunnel at 2500 m depth. *Bull. Eng. Geol. Environ.* **2010**, *69*, 381–388. [CrossRef]
25. Funato, A.; Ito, T. A new method of diametrical core deformation analysis for in-situ stress measurements. *Int. J. Rock Mech. Min. Sci.* **2017**, *91*, 112–118. [CrossRef]
26. Zhang, C.Q.; Feng, X.T.; Zhou, H. Estimation of in-situ stress along deep tunnels buried in complex geological conditions. *Int. J. Rock Mech. Min. Sci.* **2012**, *52*, 139–162. [CrossRef]
27. Matsuki, K.; Kaga, N.; Yokoyama, T.; Tsuda, N. Determination of three dimensional in-situ stress from core disking based on analysis of principal tensile stress. *Int. J. Rock Mech. Min. Sci.* **2004**, *41*, 1167–1190. [CrossRef]
28. Bell, J.S. Practical methods for estimating geostresses for borehole stability applications in sedimentary basins. *J. Pet. Sci. Eng.* **2003**, *38*, 111–119. [CrossRef]

29. Kim, H.; Xie, L.; Min, K.B.; Bae, S.; Stephansson, O. Integrated in situ stress estimation by hydraulic fracturing, borehole observations and numerical analysis at the EXP-1 borehole in Pohang, Korea. *Rock Mech. Rock Eng.* **2017**, *50*, 3141–3155. [CrossRef]
30. Feng, X.T.; Hudson, J.A. *Rock Engineering Design*; CRC Press/Balkema: Leiden, The Netherlands, 2011.
31. Rodríguez, C.A.; Rodríguez-Pérez, Á.M.; López, R.; Hernández-Torres, J.A.; Caparrós-Mancera, J.J. A Finite Element Method Integrated with Terzaghi's Principle to Estimate Settlement of a Building Due to Tunnel Construction. *Buildings* **2023**, *13*, 1343. [CrossRef]
32. Buljak, V.; Cocchetti, G.; Cornaggia, A.; Garbowski, T.; Maier, G.; Novati, G. Materials mechanical characterizations and structural diagnoses by inverse analyses. In *Handbook of Damage Mechanics*; Springer: New York, NY, USA, 2015; pp. 619–642.
33. Chai, S.B.; Liu, H.; Song, L.; Li, X.P.; Fu, X.D.; Zhou, Y.Q. Static pressure and dynamic impact characteristics of filled jointed rock after frozen-thaw cycle damage. *Front. Ecol. Evol.* **2023**, *11*, 1222676. [CrossRef]
34. Maier, G.; Bolzon, G.; Buljak, V.; Garbowski, T.; Miller, B. Synergic Combinations of Computational Methods and Experiments for Structural Diagnoses. In *Computer Methods in Mechanics. Advanced Structured Materials*; Kuczma, M., Wilmanski, K., Eds.; Springer: Berlin/Heidelberg, Germany, 2010; Volume 1. [CrossRef]
35. Zhang, Z.; Gong, R.; Zhang, H.; Lan, Q.; Tang, X. Initial ground stress field regression analysis and application in an extra-long tunnel in the western mountainous area of China. *Bull. Eng. Geol. Environ.* **2021**, *80*, 4603–4619. [CrossRef]
36. Zhou, C.; Yin, J.M.; Dong, Z.H.; Liu, Y.K.; Han, X.Y.; Zhang, X.H. The partitioned inversion method of initial stress field of extra-long tunnel considering the direction of boundary load. *Chin. J. Rock Mech. Eng.* **2022**, *41*, 2725–2734.
37. Zhang, Q.; Xiang, W.; Yu, X. Back analysis of initial geo-stress field in underground powerhouse of Shuangjiangkou Hydropower Station. *China Civil Eng. J.* **2015**, *48*, 86–96.
38. Zheng, C.M.; Zhu, W.S.; Sun, G.F. The application of the least square method in back analysis of 3D initial geostress. In Proceedings of the 2009 International Conference on Artificial Intelligence and Computational Intelligence, Shanghai, China, 7–8 November 2009; Volume 11, pp. 603–607.
39. Liu, F.; Tang, C.A.; Ma, T.; Tang, L. Characterizing rockbursts along a structural plane in a tunnel of the Hanjiang-to-Weihe river diversion project by microseismic monitoring. *Rock Mech. Rock Eng.* **2019**, *52*, 1835–1856. [CrossRef]
40. Xu, W.L. A study on tectonic stress field on the middle section of northern Qinling mountain in Shanxi province. *Seismol. Geol.* **1991**, *13*, 161–172.
41. Tan, C.X.; Wang, R.J.; Sun, Y. Numerical modelling estimation of the 'tectonic stress plane' (TSP) beneath topography with quasiU-shaped valleys. *Int. J. Rock Mech. Min. Sci.* **2004**, *41*, 303–310. [CrossRef]
42. Liu, Y.Q.; Li, H.B.; Pei, Q.T.; Yu, C.; Luo, C.W.; Yang, F.W. Study of distribution regularities of in situ stress field in steep and narrow river valleys. *Chin. J. Rock Mech. Eng.* **2011**, *30*, 2435–2443.
43. Zhou, C.H.; Yin, J.M.; Ding, X.L.; Ai, K. Complex measurements of geo-stress in deep diversion tunnel and research on distribution characteristic of regional stress field in Qinling mountain. *Chin. J. Rock Mech. Eng.* **2012**, *31*, 2956–2964.

Disclaimer/Publisher's Note: The statements, opinions and data contained in all publications are solely those of the individual author(s) and contributor(s) and not of MDPI and/or the editor(s). MDPI and/or the editor(s) disclaim responsibility for any injury to people or property resulting from any ideas, methods, instructions or products referred to in the content.

buildings

MDPI

Article

Study on the Seismic Response Characteristics of Shield Tunnels with Different General Segment Assembly Methods and Widths

Shuaifa Zhang [1,2], Zhihua Gao [1], Sui Wang [2,*], Bin Chen [2] and Chaozeng Mao [2,3]

1 School of Civil Engineering, Chang'an University, Xi'an 710064, China; 19836108089@163.com (S.Z.); zhgao@chd.edu.cn (Z.G.)
2 School of Civil and Transportation Engineering, Ningbo University of Technology, Ningbo 315211, China; chenbin.nb@163.com (B.C.); chaozengmao@gmail.com (C.M.)
3 School of Faculty of Architectural, Civil Engineering and Environment, Ningbo University, Ningbo 315211, China
* Correspondence: wangsui10610@163.com

Abstract: Shield tunnels assembled with general ring segments are widely used in urban areas. Segment assembly methods and widths cause changes in the mechanical properties of the structure and influence the seismic response of shield tunnels. To investigate the influence of the assembly method and width of the general ring segment on the seismic performance of a shield tunnel, a three-dimensional refined soil–structure dynamic interaction finite element model of the shield tunnel was established based on ABAQUS, and the mechanical response and joint deformation of the general ring lining under seismic loads were studied. The simulation results show the following: (i) The overall deformation of the tunnel lining is not significantly affected by the assembly method, and the difference is only 5.24% under a 0.4 g earthquake. (ii) The seismic responses of general ring tunnels with different assembly methods are quite different, and the mechanical properties of the shield tunnel assembled with the straight assembly method are better than those of the shield tunnel assembled with staggered joints, but the deformation of the structure is larger. Under the action of a 0.1 g earthquake, the radial force, circumferential force, and bending moment of the staggered 90° assembly tunnel are respectively reduced by 13.6%, 11.1%, and 17.8% compared with the staggered 45° assembly structure, but the maximum intra-opening deformation increases by 0.19, 0.58, and 2.4 mm, respectively. (iii) The internal force distribution of the bolts is controlled by the deformation of the joint; compared with the CF90 and TF assembled tunnels, the mechanical properties and deformation characteristics of the CF45 and CF90 assembled tunnels are more reasonable. (iv) The extrados and intrados joint opening deformation and shear dislocation of the 1.2 m wide general ring segment under the staggered assembly increase by 1.2 mm and 1.03 mm, respectively, compared with the 1.5 m wide segment, while the radial force, circumferential force, and bending moment are reduced by 24.4%, 36.5%, and 41.7%, respectively, indicating that the seismic performance of the shield tunnel with a segment width of 1.5 m is better than that of the shield tunnel with a width of 1.2 m.

Keywords: assembly type; general segment; segment width; seismic response; bolted joint; segment joint

1. Introduction

With the rapid development of urban scale and economic construction and the rapid growth in population in metropolitan areas, the rapid transit system, built underground, has become an important tool of transportation, and subways have become the main basic transportation facilities in many cities to alleviate surface traffic saturation [1]. The area involved in the construction of rail transit in mainland China exceeded 9000 km at the end of 2021 [2]. The tunnel, as a kind of underground structure, is mainly affected by interaction with the surrounding soil during the action of an earthquake, while the load on the aboveground structures mainly comes from the inertial force generated by acceleration. There are differences in the mechanisms of earthquake damage between tunnel and aboveground structures. In general, metro shield tunnels have better seismic performance as a result of the interaction of the surrounding soil [3]. Thus, researchers have long thought that subways are less threatened by earthquakes. With the development of underground structures, more and more cases of underground structures being damaged by earthquakes have been paid attention to [4,5]. The seismic performance of underground structures has attracted the attention of more and more scholars as the views of people on the safety of underground structures have gradually changed. Situated at the intersection of the world's two major seismic belts, the Pacific and the Eurasian, China is a country with frequent earthquakes, and most of the cities are located in or close to earthquake zones. At the same time, most of the extensive and complex subway systems that China has built are in large cities. In previous seismic disasters, including the 2008 Wenchuan Earthquake [6], the 1995 Kobe Earthquake in Japan [7], and the 1999 Chi-chi earthquake [8], it was observed that the tunnel lining was damaged to varying degrees.

Most of the coastal cities in southeast China have soft-soil tunnels, which have a low modulus of elasticity and poor soil quality compared with mountain tunnels. This is of central importance for seismic design in the case of metro systems, considering their significance for life safety and economy and that they are generally more vulnerable in soft-soil ground conditions [9]. With regard to the seismic resistance of shield tunnels in areas with soft soil, researchers have conducted a range of investigations, mainly focusing on the dynamic response of tunnel structures under complex soft-soil ground conditions with different earthquake intensities and site conditions and the corresponding seismic design measures [10–12]. However, the tunnel lining structure is simplified to a homogeneous circumferential ring in the above study, only the overall performance of the segment ring is taken into account, and the weak link of the shield tunnel connection is not taken into account, which is somewhat different from the actual structure. In fact, prefabricated segments were used in the construction of the tunnel and were bolted together on site, and the forces of the segment blocks are transmitted through the joint bolts under seismic loading, leading to changes in the mechanical failure mechanism of the shield tunnel [13], which have a significant impact on the overall seismic performance and dynamic characteristics of the tunnel [14]. Segment joints are also a vulnerable part of the structure when subjected to seismic loads, which will cause major deterioration of the structure in forms such as lining cracks, misaligned segment joints, and water leakage under the action of an earthquake [15]. Some scholars are now studying the effect of the discontinuity of shield tunnel tube joints on the seismic performance of shield tunnels [16], showing that the presence of joints makes a large difference in the internal force distribution and deformation at the joints of shield tunnels with the homogeneous circular ring model, leading to the redistribution of internal forces and deformation. The general ring segment is increasingly used in the construction of urban subways and large sections of underwater shield tunnels due to its advantages of great versatility, high assembly efficiency, accurate line control, and easy deviation correction [17], and the existing shield tunnel segment ring assembly method includes straight assembly and staggered assembly. Except for some soft ground represented by Shanghai, China, where the straight joint assembly method is used, most cities in China use the staggered assembly method [18]. The overall stiffness of a staggered assembly segmental structure is higher than that of a straight-assembly structure, and the deformation is also relatively small. Studies, including the model test and numerical simulation, have shown that the assembly method of general ring segments has a significant influence on the development of the internal force and deformation of shield tunnels [19–22]. Wang [23] studied the influence of a change in tunnel diameter on the performance of the joints and the stress state of the concrete, and the change in tunnel diameter was proposed as an evaluation parameter to estimate the performance of soft-soil tunnel structures. Liu [6,24] studied through a full-scale experiment that the failure mode of straight assembly depends on the bolt structure performance of the longitudinal joint, while staggered assembly is influenced by interaction between the rings. However, most of the above studies have focused on the influence of different segment assembly modes on the deformation development and failure process of general ring segments under static loading. The failure mechanism of the underground structure due to soil interaction is more complicated, but few studies have focused on the dynamic response of segment ring assembly methods to tunnel structures under dynamic loads such as earthquake loads, train vibrations, etc. Zhao [25] carried out seismic analysis on a shield tunnel in a soft-soil area and simulated an earthquake disaster affecting the shield tunnel using the ABAQUS three-dimensional discontinuous model. The simulation results accurately reproduced the mechanical characteristics of the segments and joints. Wang [26] established a staggered assembly general ring tunnel model to study the effects of the staggered assembly method and the straight assembly method on the amount of longitudinal joint tensioning and the damage law of the internal support of a shield tunnel. Gou [27] investigated the adaptability of the staggered assembly method and the direct assembly method for shield tunneling in ground fissures in Xi'an, China. Chen [28] used a beam spring model to investigate the seismic performance of a tunnel by using the reactive displacement method, and the results showed that the assembly angle has a significant effect on the internal forces of the tunnel structure. Zhang [29] mainly investigated the effect of the position of the capping block on the internal force of a segmental tunnel under a seismic wave.

In summary, many results have been obtained regarding the seismic response of shielded tunnels using the staggered assembly method, while the research on the seismic performance of shield tunnels constructed with different assembly methods mainly focuses on the structural mechanical properties. The joint deformation laws and mechanical behavior at the joints caused by different assembly methods of general ring segments under different seismic intensities are less involved, and the analysis of the effect of the width of the general ring segment is also less involved. According to this, a method based on dynamic time history analysis is adopted in the present study, and a discontinuous three-dimensional surrounding rock–structure interaction model of shield tube sheet joints was established. Ningbo Railway Transit Line 6 was selected as the engineering background, and the EI-Centro seismic wave was selected for the time history response analysis of the shield tunnel. The changes in the lining internal force, joint deformation, and bolt internal force of segments caused by different widths of general ring segments and assembling methods are studied. This paper provides a reference for the seismic design of shield tunnels in soft-soil areas.

2. Engineering Research Background

Ningbo Railway Transit Line 6 is an east–west urban line with a total length of about 56 km, in which the overburden depth of the shield tunnel interval is 10~26 m, and this paper adopts the middle burial situation of the tunnel (burial depth 16.4 m). According to the geological situation traversed by Railway Transit Line 6, the shield tunnel was selected to pass through three complex soil layers of muddy silty clay, silty fine sand, and silty clay. The soil layers in which the tunnel is located are, from top to bottom, muddy silty clay, silty fine sand, muddy silty clay, and silty clay. The shield tunnel structure is located in an area with a wide range of muddy silty clay and sandy soil, and the top and bottom of the arch are partially in soft soils, while near the waist of the arch is sandy soil, and the soft layers are in close contact with the middle hard layer. Figure 1 shows the soil profile and location of the tunnel in this example.

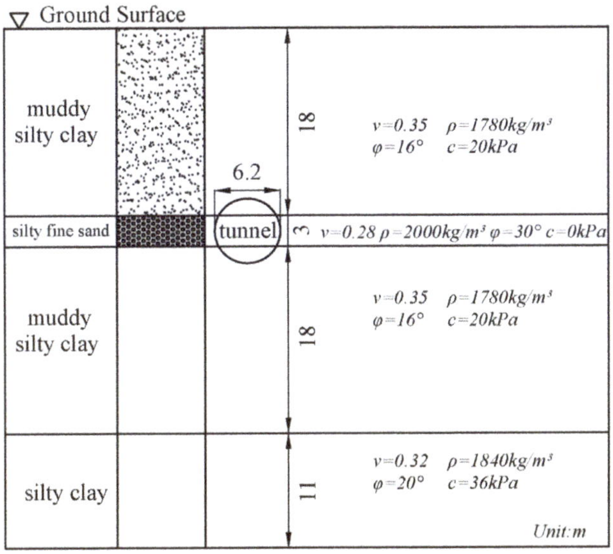

Figure 1. Description of soil profile.

3. Numerical Model of Soil–Shield Tunnel Interaction

Based on the ABAQUS FEA simulation platform, a shield-segment three-dimensional discontinuous-contact finite–infinite element coupling model to analyze the seismic response of shield tunnels was constructed. The size of the model was selected according to the properties of the layered soil and the influence of the number of model elements on the calculation efficiency. Therefore, the size of the model is 80 m × 50 m × 14.4 m, and the buried depth of the center of the tunnel is 19.5 m. Considering the properties of some soft-soil layers are similar in the area where this project is located and the changes in parameters are minor, the soil layers are appropriately simplified based on the properties of the soil layers. The dynamic elastic modulus and dynamic shear modulus of the soil are converted from the weighted average shear wave velocity of each soil layer measured on site in the survey report. To ensure the accuracy and efficiency of dynamic calculation, the grid size of the soil layer model should be smaller than 1/10 to 1/8 of the minimum seismic wavelength [30]. The central soil grid size is 0.8 m, and the maximum size is 1.4 m to meet these conditions. There were 62,344 elements and 67,963 nodes in the finite element model developed; the dynamic calculation model is shown in Figure 2.

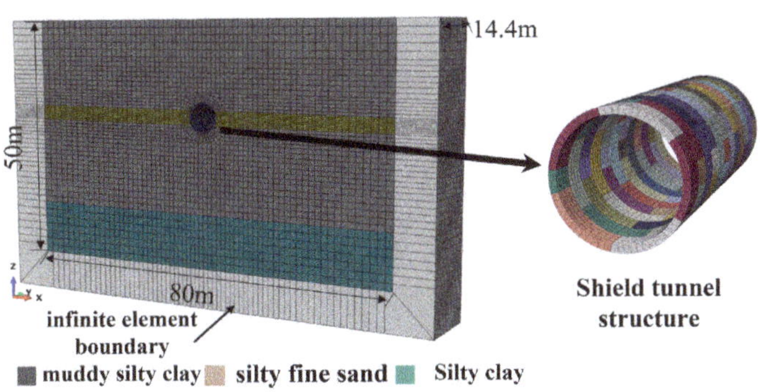

Figure 2. Dynamic calculation model of soil layers and tunnel structure.

The outer radius of the tunnel segment ring is 6.2 m, the inner radius is 5.5 m, and the segment thickness is 0.35 m. It is assumed that the tunnel lining consists of six precast concrete segments made of C50 concrete, one capping block (F, 20° of the center of the circle), two adjacent blocks (L1, L2, circular angle 67.5°), and three standard blocks (B1, B2, circular angle 68.75°), as shown in Figure 3. The maximum width is 1218.6 mm, the minimum width is 1181.4 mm, and the average width is 1200 mm. There are certain bolted connections between different segment blocks and segment rings in the shield tunnel structure that reduce the overall structural stiffness of the shield tunnel and affect the internal force distribution and deformation of the shield tunnel. The circumferential joints of the tunnel ring were connected by two segment blocks with a pair of Grade M30 steel bolts. In addition, the longitudinal tunnel joint was connected by two rings through a number of sixteen Grade M30 steel bolts evenly distributed, as shown in Figure 4. The linear brick solid elements are used to model soils, using reduced integration (C3D8R) to reasonably facilitate the computational efficiency of soil behavior [31].

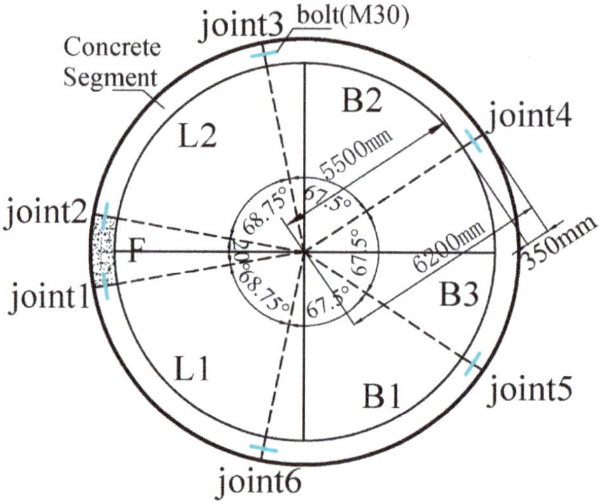

Figure 3. Schematic diagram of the division of the general ring segments.

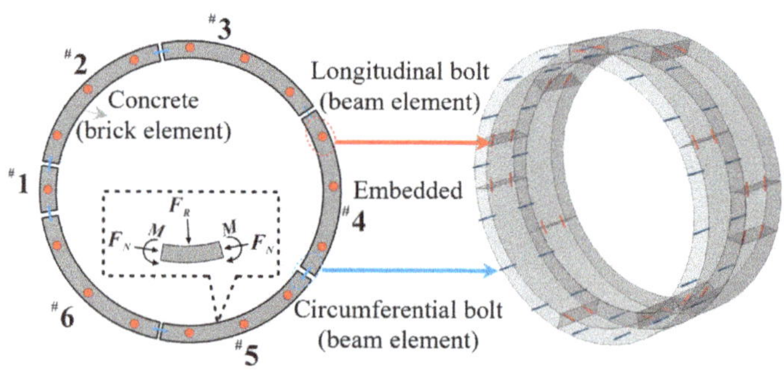

Figure 4. Schematic diagram of shield tunnel unit form and connection mode and positive direction of tunnel lining internal force.

The T3D2 truss element was used to simulate the internal reinforcement of the tunnel segment and embedded in the concrete segment to realize the mechanical properties of reinforced concrete, and the shield tunnel was modeled using the C3D8R linear brick element with an embedded linear beam element B31. The concrete segment structure and tunnel internal reinforcement adopt the elastic constitutive model; the beam element corresponding to the steel bolt location adopts the bifold reinforcement model; and the Mohr–Coulomb constitutive model was used to simulate the surrounding soil layer. The physical and mechanical performance of each soil layer and shield tunnel is shown in Table 1.

Table 1. Material parameters of numerical model.

Materials	Thickness (m)	Density (kg/m^3)	Poisson's Ratio	Elastic Modulus (MPa)	φ	Cohesion (kPa)
Muddy silty clay	0~18	1780	0.35	77	16°	20
Silty fine sand	18~21	2000	0.28	300	30°	0
Muddy silty clay	21~39	1780	0.35	84	16°	20
Silty clay	39~50	1840	0.32	300	20°	36
Concrete	-	2500	0.2	3.45×10^4	-	-
Bolt	-	7800	0.2	2.10×10^5	-	-

4. Surrounding Rock–Tunnel Interaction, Boundary Conditions, and Input Motions

To consider the soil–tunnel contact relationship at the interface, the shield tunnel segments and soil are usually set as interaction contact surfaces to account for contact discontinuity between the shield tunnel and the surrounding soil medium and any possible compression, separation, or sliding of these structural components from the surrounding soil under seismic excitation [32]. Contact surfaces consisting of master and slave surfaces were defined to implement the contact pair method. In this paper, the soil deposit surface around the tunnel was considered as the master surface, and the outer surface of the shield tunnel segment was regarded as the slave surface. Additionally, the contact between different tunnel segment blocks needs to be taken into consideration, and there is no requirement to purposefully distinguish the master and slave surfaces between tunnel segments because the material and mesh of the master and slave surfaces are identical. To simulate the interaction in the normal direction of the surfaces, the hard-contact algorithm was used so that pressure can only be transmitted when the master and slave surfaces are in contact with each other. The tangential behavior of the interfaces followed the classical frictional contact model, and according to previous studies, the coefficient of friction can be taken as $\mu = 0.5$ [33]. CIN3D8 (eight-node linear unidirectional infinite brick) elements were used to describe the infinite area, which allowed the energy reflected by the seismic wave on the truncation boundary to be absorbed by the model. The role of infinite element boundaries is different in static analysis and dynamic analysis [14]. Therefore, the following steps were used to perform static–dynamic boundary transformation and seismic response analysis: during the static analysis step, the gravity load was applied, the static boundary condition was that the fixed boundary was used in all directions at the bottom of the soil deposit, and the roller boundary, which allowed vertical displacement and prevented horizontal displacement, was used on the side boundary to balance the initial stress of the model to obtain the static stress field of the soil in the static analysis, and the method of model change was used to simulate the excavation process and to obtain the initial stress field after excavation. However, during the seismic analysis, when the seismic motion was input, the original static roller boundaries at the side boundaries were replaced by dynamic infinite element boundaries (one-way CIN3D8 element dynamic infinite element boundaries) [31]. In this study, the El Centro wave in the United States was selected as the input seismic wave, and the acceleration peaks were adjusted to 0.1 g, 0.2 g, and 0.4 g, respectively. The acceleration time history of the seismic wave after peak adjustment is shown in Figure 5. The effective duration takes the first 15 s before reaching 10% of the peak acceleration, and the processed seismic waves were imposed from the bottom boundary of the model.

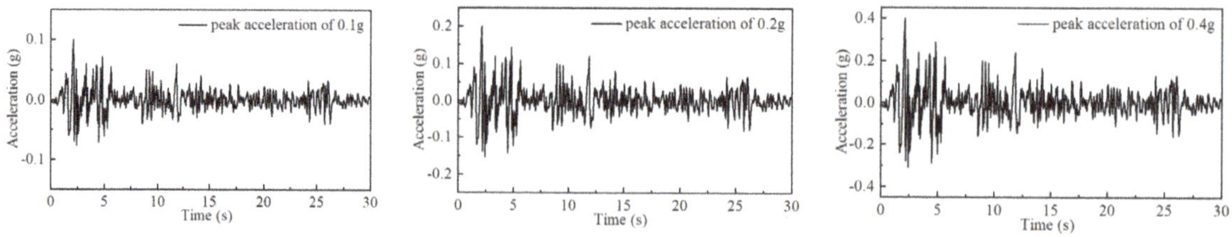

(a) (b) (c)

Figure 5. Time history of seismic wave after peak adjustment: (**a**) 0.1 g, (**b**) 0.2 g, (**c**) 0.4 g.

5. Dynamic Response Characteristics of Shield Tunnel under Different Assembly Methods

In this section, the seismic responses of shield tunnels with four assembly methods during different seismic intensities (the peak accelerations of the input seismic motions are 0.1 g, 0.2 g, and 0.4 g, respectively) are analyzed from the aspects of tunnel internal force and deformation response and the mechanical behaviors of segmental joints.

5.1. FEM of Shield Tunnel under Different Assembly Methods

In order to study the influence of the general ring assembly methods on the seismic response of the shield tunnel structure, four kinds of assembly methods were used: one straight assembly (TF) and three staggered joint assemblies (CF). The staggered joint assemblies were 22.5, 45, and 90 degrees along the central axis of the tunnel. The angles between two adjacent ring capping blocks were 45, 90, and 180 degrees, respectively, recorded as CF45, CF90, and CF180, respectively. The schematic diagram of the assembly method and the specific scheme are shown in Table 2.

Table 2. Assembly methods of segments.

Assembling Method	Schematic Diagram	Instructions
CF45	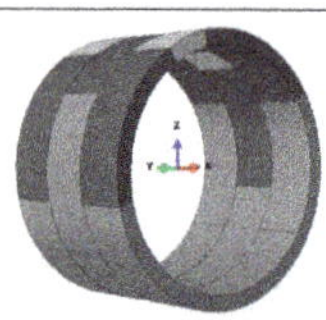	The front and rear proximity rings rotate 45 degrees relative to the middle ring
CF90	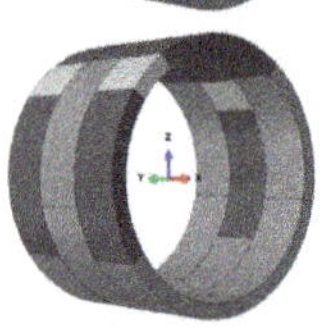	The front and rear proximity rings rotate 90 degrees relative to the middle ring
CF180	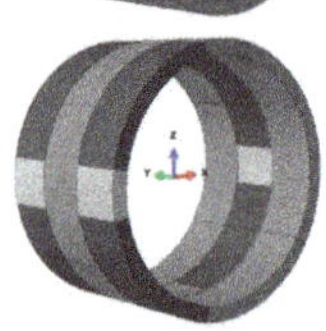	The front and rear proximity rings rotate 180 degrees relative to the middle ring
TF	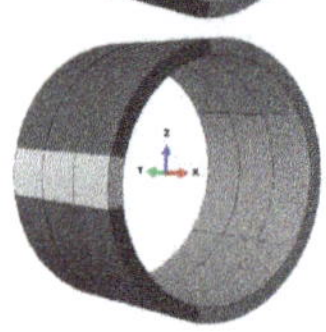	Straight assembly

5.2. Numerical Investigations and Discussions

5.2.1. Deformation of Shield Tunnel Structure with Different Assembly Methods due to Seismic Loads

Shear deformation of the shield tunnel section is one of the important indicators reflecting the seismic response characteristics of the structure and should be considered. The shear deformations defined here are the ratio of the relative horizontal displacement between the top and bottom of the arch to the tunnel diameter. Figure 6 shows the influence of the assembly method on the shear deformation under different seismic loads. It can be seen that the shear deformation of the shield tunnel does not vary significantly depending on the assembly method, and the assembly method has little effect on the overall deformation of the structure. Under the action of an earthquake with 0.1 g peak acceleration, there is only a 0.03% difference between the shear deformation of condition CF90, which is the maximum, and that of condition TF, which is the minimum. The difference between the maximum of shear deformation (CF90) and the minimum (TF) under the action of 0.4 g seismic loading is 5.24%, indicating that the tunnel structure and the surrounding soil and seismic medium are moving simultaneously and that the overall deformation of the shield tunnel is controlled by the relative displacement of the surrounding soil, not by the assembly method.

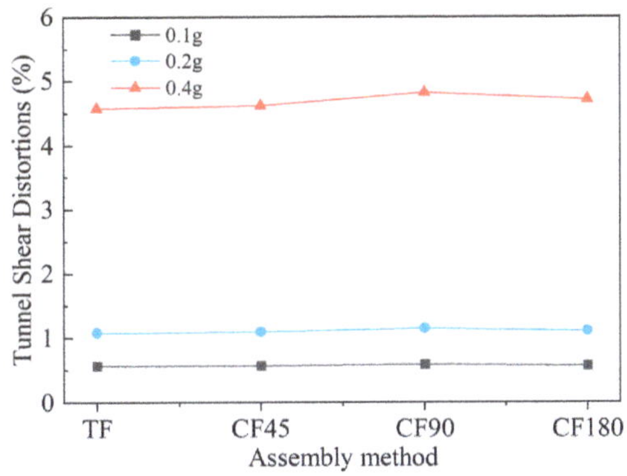

Figure 6. Maximum shear deformation of shield tunnel under different assembly methods.

5.2.2. Mechanical Behavior of Shield Tunnel Joints under Different Assembly Methods

The phenomenon of earthquake damage to the structure of shield tunnels in the past indicates that despite the high strength of the tunnel section, the damage is slight, but the mechanical properties of the joints of the shield tunnel are relatively weak [34]. Therefore, the deformation indicator of segment joints was used to evaluate the seismic performance of shield tunnels. Figure 7 shows the time history curves of the extrados opening, intrados opening, and shear dislocation at each joint of the general ring segment under different assembly methods at 0.1 g seismic loading. It can be seen that due to the shear deformation caused by the relative displacement between the soil layers during the transmission of the seismic wave, the joints are compressed, opened, and displaced due to the relative slip and separation between the shield segments. The deformation of the segment joints is unevenly distributed, with some joints deforming heavily under tensile forces and others deforming poorly under compression. The time histories at the joint alternate between positive and negative, and the location of the peak joint deformation varies as the assembly method changes. As an illustration of the extrados opening, the maximum extrados opening joint deformation occurs at segment Joint 2 for the assembly methods TF and CF180, while that of CF45 and CF90 occurs at segment Joint 4. The different positions of the capping block also change due to an alteration in the assembly method, resulting in different degrees of joint deformation. The extrados joint opening of method TF is the maximum, which is 0.27 mm, the intrados joint opening of method CF90 is the maximum, which is 0.52 mm, and the shear dislocation of method CF45 is the maximum, which is 0.16 mm. Figure 8 shows the maximum deformation of the extrados opening, the intrados opening, and the shear dislocation of the shield tunnel segment joints in the four assembly methods under different seismic loadings. It can be seen that the joint deformations of the assembly methods CF45 and CF180 are relatively similar under 0.1 g and 0.2 g earthquakes. Compared with the minimum value of CF180, the maximum deformations of the CF90 assembly increased by 0.19, 0.55, and 2.6 mm, respectively, under the three seismic intensities. This is even higher than the assembly method TF, with lower overall stiffness. Meanwhile, the maximum joint opening deformations under the action of 0.4 g earthquakes of the TF and CF90 assembly methods significantly exceed the maximum limit values recommended by the Chinese Code for design of shield tunnel engineering (3 mm), while the joint deformations of the CF45 and CF180 staggered assembly methods are relatively close and still within the limit, which can ensure their safety and watertightness. The deformation of the segment joint is greatly affected by the position of the capping block, and it should be avoided that the location of the capping block is at the spandrel of the tunnel, as the deformation of the joint is greatly improved.

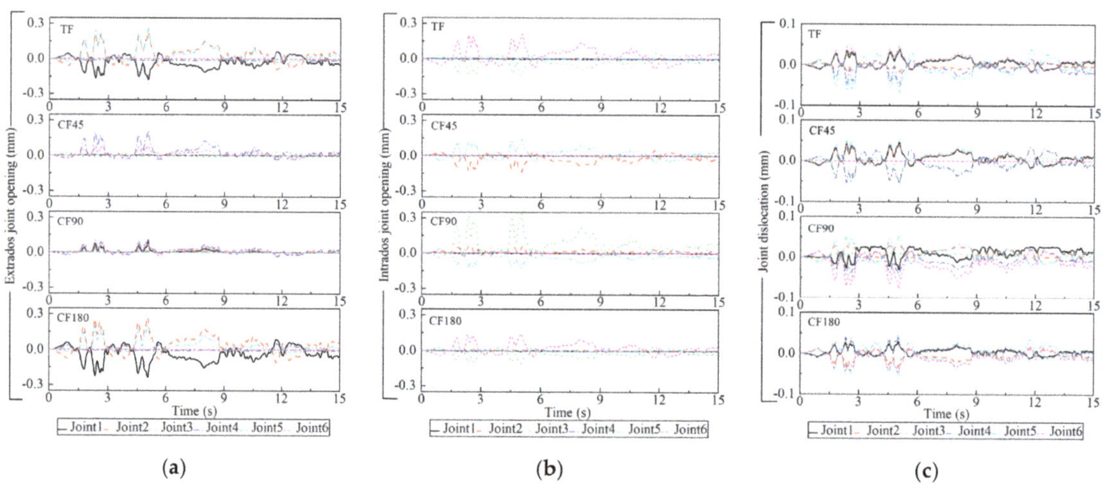

Figure 7. Time histories of segment joint deformation under different assembly methods with the input 0.1 g seismic loading: (**a**) extrados opening, (**b**) intrados opening, (**c**) shear dislocation.

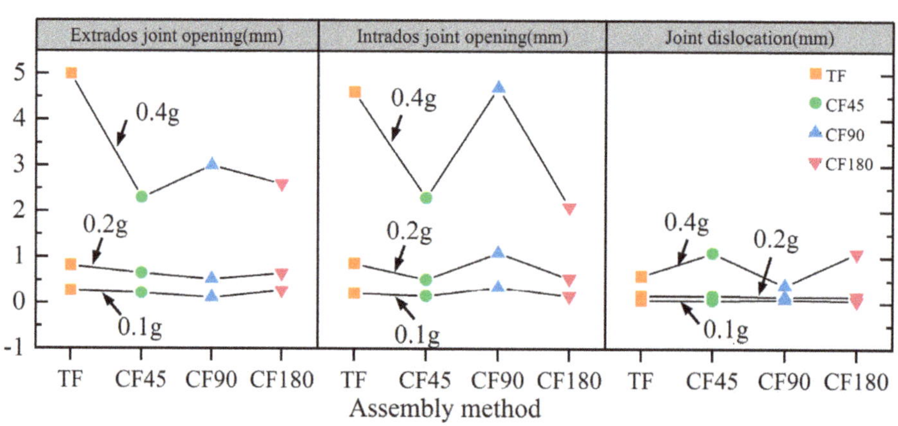

Figure 8. Maximum deformation of segment joints under different assembly methods.

Deformation of the segment leads to the opening, compression, and dislocation of the joint, which affect the internal force distribution of the joint bolts. Figure 9 shows the time histories of the axial and shear forces of each joint bolt of the four general ring segment assembly structures under the action of a seismic wave with a peak value of acceleration of PGA = 0.1 g. It can be seen that the time histories of the axial and shear forces are basically consistent with the change law of the corresponding joint deformation. The position of the maximum axial force of the bolt corresponds to the maximum opening deformation of the joint, and the position of the maximum shear force of the bolt corresponds to the maximum dislocation of the joint, which further indicates that the internal force distribution of the bolts is controlled by the deformation of the joint. Figure 10 shows the peak axial and shear forces of the joint bolts in the four assembly methods under different seismic loadings. Using 0.1 g seismic loading as an example, the maximum bolt axial force of CF90 is 55 kN, which is 13 kN higher than the 42 kN obtained by using the straight assembly method, and comparing the other two staggered assembly methods, CF45 and CF180 increased by 31 kN and 21 kN, respectively. The shear force difference in the four assembly methods is relatively close under 0.1 g and 0.2 g earthquake action, which indicates that under this earthquake intensity, the difference in the assembly method of the general ring segment has negligible influence on the bolt shear force. However, under the action of a 0.4 g earthquake, the bolt shear force of the CF45 and CF180 assembly methods is obviously more developed than that of TF and CF90.

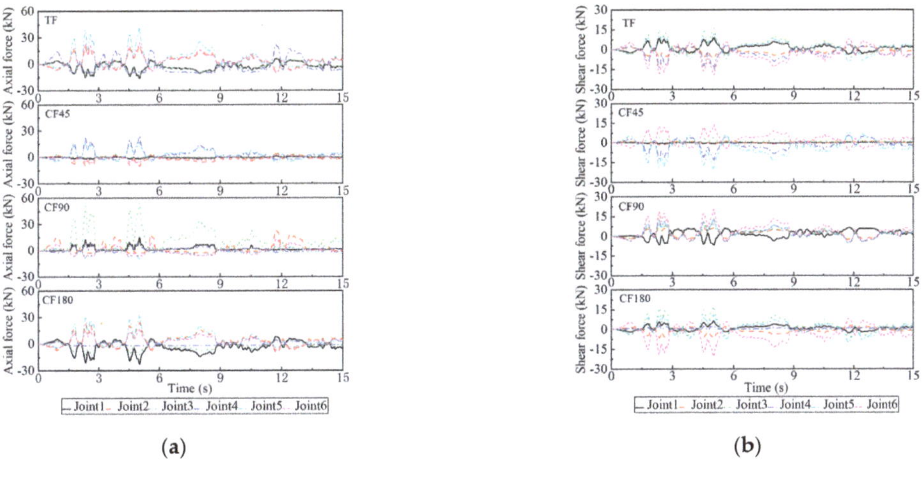

Figure 9. Internal force curves of joint bolts under different assembly methods with the input 0.1 g seismic loading: (**a**) axial force, (**b**) shear force.

Figure 11 shows the internal force envelopes of shield tunnels with four assembly methods under the action of a seismic wave with a peak value of acceleration of 0.1 g (the direction of the internal forces of the segment is indicated in Figure 4). It can be seen that the distribution of internal force is not greatly affected by the assembly method, and the maximum circumferential internal force and maximum circumferential bending moment occur within the range of ±45° from the tunnel axis. The main reason for this is that the different assembly methods of the general ring segments change the stiffness of the shield tunnel structure, but the distribution of additional stresses on the segment structure caused by the same seismic wave and site conditions is the same. However, due to the change in joint position caused by the position of the capping block under different assembly methods, the structural stiffness of the ring joint is weakened, and the deformation of the joint under the earthquake causes the redistribution of the internal force value. The difference is mainly due to the difference in structural stiffness among the joints and segments caused by the assembly method, and low stiffness has better adaptability to deformation. In addition, due to the different positions of the capping blocks, the deformation of the segment joints under the earthquake is significantly different, and the release of internal forces at the joints is also an important reason for the difference in internal forces. The circumferential axial force and bending moment of the CF90 method are smaller to varying degrees near the capping blocks, which indicates that the assembly method has changed the numerical law of the internal force of the lining structure. The maximum internal force of the shield tunnel under different assembly methods is shown in Figure 12, taking a 0.1 g earthquake as an example under different assembly methods. The maximum radial axial force of CF45 is 126.9 kN, and the minimum of CF90 is 112.6 kN, with a difference of 13.6%; the maximum circumferential axial force of CF45 is 98.3 kN, and the minimum of CF90 is 88.2 kN, with a difference of 11.1%; and the maximum circumferential bending moment of CF45 is 38.3 kN·m, and the minimum of CF90 is 32.5 kN·m, with a difference of 17.8%. Meanwhile, as the intensity of the earthquake loading rises, the internal force envelope law changes little, but the difference is more significant.

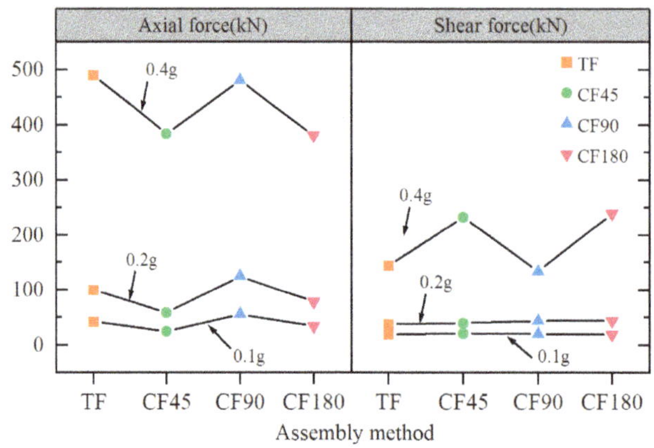

Figure 10. Maximum internal force of bolts with different assembly methods.

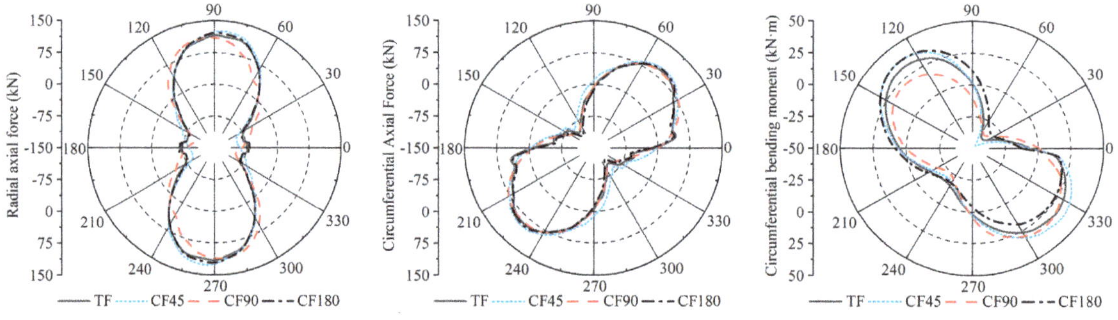

Figure 11. Envelope diagram of tunnel internal force under different assembly methods with the input 0.1 g seismic loading: (**a**) radial axial force: FR, (**b**) circumferential axial force: FN, (**c**) circumferential bending moment: M.

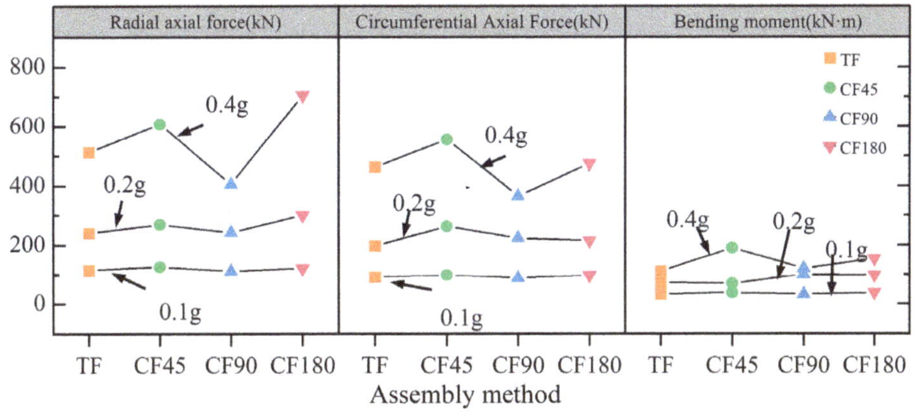

Figure 12. Maximum internal force of tunnel under different assembly methods.

Considering the structural stress characteristics and based on previous research experience [32], the maximum principal stress (MAPS) and the minimum principal stress (MIPS) were selected for stress characteristic analysis. MAPS is primarily used to evaluate the tensile failure location of the tunnel, while MIPS is primarily used to evaluate the compression failure location of the tunnel. Figure 13 shows nephograms of stress of each general segment assembly method under a small earthquake. In each assembly method, the MAPS of method CF90 is the maximum, which is 3.94 MPa; the MIPS of method CF180 is the maximum, which is 7.63 MPa. Under an earthquake intensity of 0.1 g, the segment structures for all assembly methods did not exceed the tensile and compressive damage limits of concrete, and the structure is in an elastic working state. Table 3 shows the MAPS and MIPS of four shield tunnel assembly methods under different earthquake action. With the increase in earthquake intensity, the MAPS and MIPS of the shield tunnel gradually increase, which is consistent with the seismic response law of general underground structures. In the same case, the MIPS of the tunnel structure changes in a small range, and the MAPS of the tunnel structure using the TF assembly method is smaller than those of the other three staggered assembly methods, which may be due to the lower stiffness.

Table 3. MAPS and MIPS of four shield tunnel assembly methods under earthquake action.

Assembly Method	Seismic Wave	MAPS (MPa)	MIPS (MPa)
TF	PGA = 0.1 g	2.54	7.25
CF45		3.82	6.89
CF90		3.98	7.02
CF180		3.94	7.63
TF	PGA = 0.2 g	3.98	8.70
CF45		6.17	8.11
CF90		5.98	8.25
CF180		5.89	8.70
TF	PGA = 0.4 g	8.85	20.06
CF45		19.1	19.64
CF90		19.6	19.64
CF180		19.2	20.01

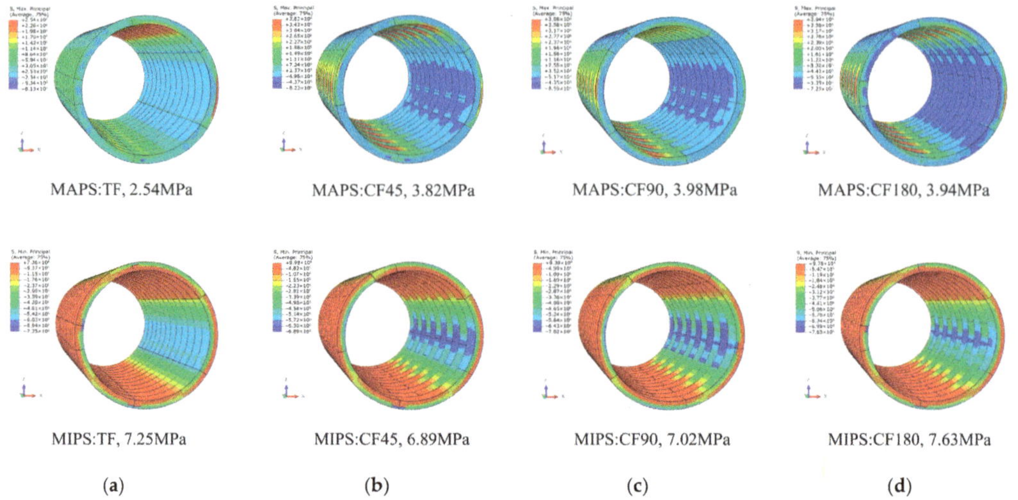

Figure 13. The maximum mechanical results of different assembly methods: (**a**) TF; (**b**) CF45; (**c**) CF90; (**d**) CF180.

5.2.3. Evaluation and Discussion of Seismic Performance

Under the conditions of 0.1 g and 0.2 g earthquake loadings, the safety and watertight performance of the tunnel lining can be guaranteed. Table 4 summarizes the seismic performance analysis of the four general ring assembly methods under the action of 0.4 g earthquake conditions; the maximum joint deformation of CF90 and TF far exceeds the code limit, the maximum bolt stress exceeds the yield point, and the maximum stress variation is in a wide range. The seismic performance of the shield tunnels assembled using CF90 and TF cannot meet the design requirements. However, the mechanical properties and deformation characteristics of the tunnels assembled using CF45 and CF180 are more reasonable, and the tunnel structure is in a safe working condition. In summary, the seismic performance of the CF45 and CF180 assembled tunnels can meet safety and water resistance requirements.

Table 4. Seismic performance evaluation of shield tunnel under different assembly methods.

Research Object	Mechanical Property	Joint Deformation	Bolt Internal Force	Stress Property	Seismic Performance Evaluation
Assembly method	Radial force/ Circumferential force/ Bending moment	Extrados opening/ Intrados opening/ Dislocation	Axial force/ Shear force	MAPS/ MIPS	The methods TF and CF90 do not meet the seismic design requirements
TF CF45 CF90 CF180	706 kN (CF90)/556 kN (CF45)/190 kN·m (CF45)	5 mm (TF)/ 4.7 mm (CF90)/ 1.1 mm (CF180)	490 kN (TF)/ 239 kN(CF180)	19.6 MPa (CF90)/ 20.6 MPa (TF)	

6. Dynamic Response Characteristics of Shield Tunnel under Different Segment Widths

The change in the segment width causes the overall stiffness of the shield tunnel to change, and the decrease in the width increases the discontinuity of the structure. In this section, general segments of different widths under two assembly methods are selected for analysis under the peak acceleration of 0.4 g earthquakes.

6.1. FEM of Shield Tunnel under Different Segment Widths

In order to study the influence of the width of the general ring segment on the seismic performance of the shield tunnel structure, two kinds of segment widths were used. General segments of 1.2 m and 1.5 m widths under the two assembly methods of straight and 180-degree assembly, recorded as TF1.2, CF1.2, and TF1.5, CF1.5, respectively, are shown in Figure 14.

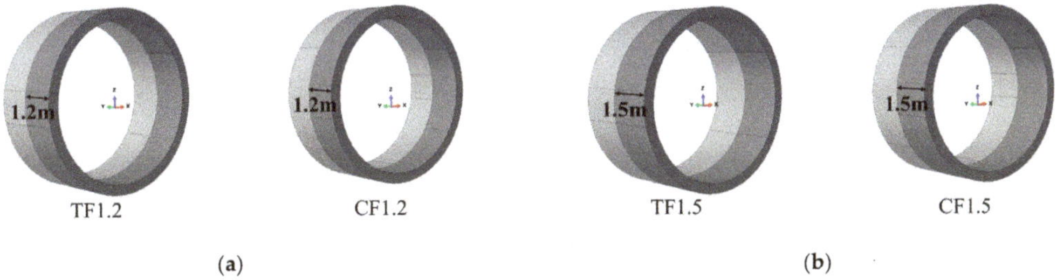

Figure 14. Schematic diagram of general ring segments with different widths: (**a**) segment width: 1.2 m; (**b**) segment width: 1.5 m.

6.2. Numerical Investigations and Discussions

6.2.1. Mechanical Behavior of the Shield Tunnel Joints under Different Segment Widths

Figure 15 shows the time histories of the maximum joint deformation of the general ring segments with widths of 1.5 m and 1.2 m. The residual deformation of the structure is relatively large after the 0.4 g earthquake excitation; the large irreversible plastic deformation of the soil layer is the main reason for this. Meanwhile, the deformation of joints with a width of 1.5 m is significantly smaller than that of joints with a width of 1.2 m, which is mainly due to the integrity of the tunnel, structural rigidity, and resistance to deformation increasing with increasing segment width.

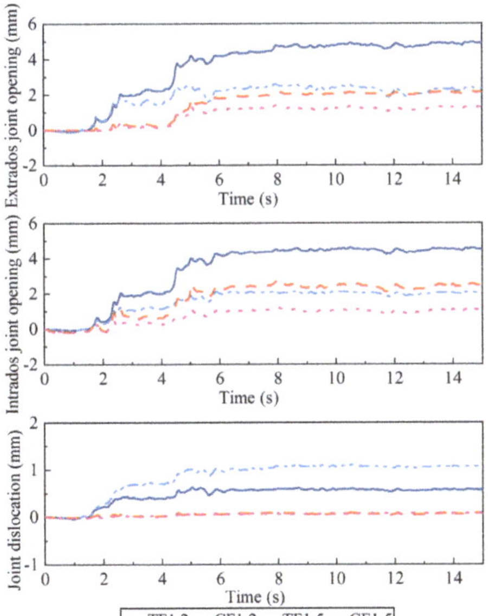

Figure 15. Maximum deformation time history curves of segment joints under different widths.

The maximum axial force and shear force of the shield tunnel joint bolts under different widths are shown in Figure 16. It can be seen that the internal force distribution of the joint bolts is still controlled by the deformation law of the joint. The bolt axial force of the condition TF1.2 is the maximum, which is 490 kN, and that of the condition CF1.5 is the minimum, which is 200 kN. The bolt shear force of the condition CF1.2 is the maximum, which is 239 kN, and that of the condition TF1.5 is the minimum, which is 73 kN. Meanwhile, the bolt axial and shear forces of the 1.5 m wide general ring tunnel are reduced compared with those of the 1.2 m wide ring whether the straight assembly method or staggered assembly method is used. The internal force of the bolt of TF1.2 exceeds the yield strength of 400 kN. According to the development law of the maximum axial force and shear force of bolts with different general ring segment widths, the strength of the 1.2 m wide segment connecting bolts should be increased accordingly to prevent possible damage.

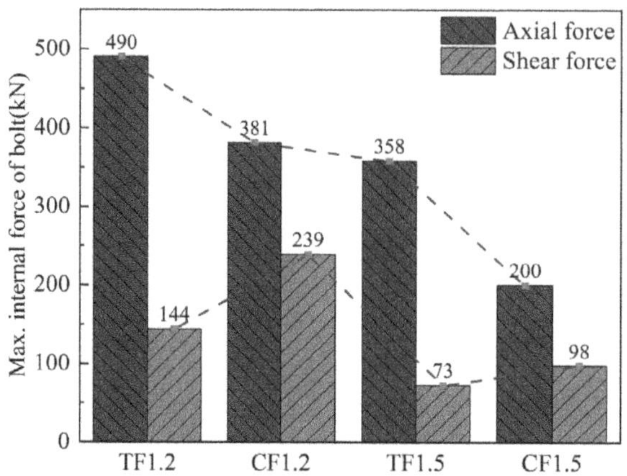

Figure 16. Maximum internal force of joint bolts under different widths.

6.2.2. Mechanical Behavior of Segments under Different Segment Widths

Figure 17 shows the internal force envelope diagrams of two kinds of general segment width shield tunnels under the earthquake intensity of 0.4 g. It can still be seen that the internal forces of the lining in a straight assembly method are smaller than those of a staggered assembly method. Considering the internal force law of tunnels using the staggered assembly method under two kinds of width conditions as an example, the maximum radial axial force with a width of 1.5 m is 878.5 kN, and that with a width of 1.2 m is 705.8 kN; the maximum circumferential axial force with a width of 1.5 m is 649.6 kN, and that with a width of 1.2 m is 476.1 kN; the circumferential bending moment with a width of 1.5 m is 218.1 kN·m, and that with a width of 1.2 m is 154 kN·m. In terms of specific values, the radial axial force, circumferential axial force, and circumferential bending moment differ by 24%, 36.5%, and 41.7%, respectively, which means that the change in segment width has a significant influence on the mechanical properties of the tunnel structure. The main reason for this is that increasing the segment width can enhance the stiffness of the shield tunnel and reduce the discontinuity between the segment ring and the ring, increasing the overall stiffness of the structure and reducing the deformation capacity. Notably, the deformation of the structure is opposite to the law of the internal force: the joint deformation and bolt internal force of the 1.5 m wide segment are both reduced, but the internal force of the structure is significantly increased. Therefore, the need to determine how to coordinate the structural stiffness of the tunnel makes its internal force and deformation particularly important.

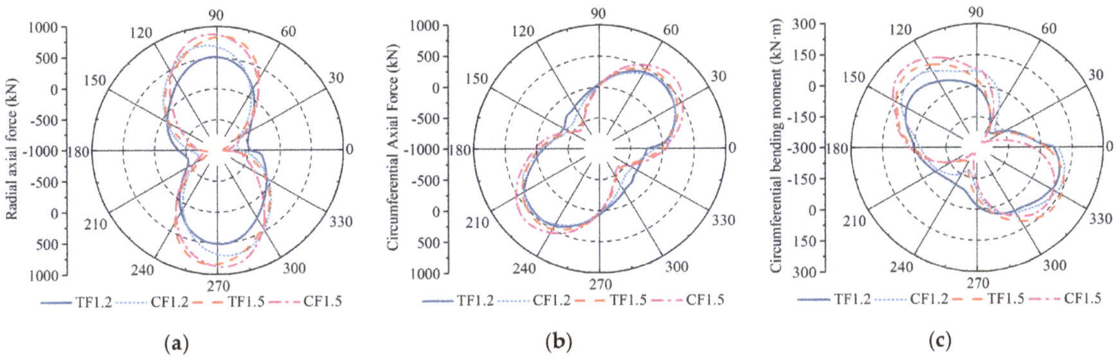

Figure 17. Envelope diagram of internal force of segments under different widths: (**a**) radial axial force: FR, (**b**) circumferential axial force: FN, (**c**) circumferential bending moment: M.

The stresses of each ring are approximately similar, so the cloud diagram of the MAPS and MIPS of the fifth ring of the shield tunnel change is given as shown in Figure 18. The maximum MAPS value for the CF1.2 operating condition is 16.7 MPa, and the minimum MAPS value for the TF1.2 operating condition is 8.63 MPa; the maximum MIPS value for the TF1.2 operating condition is 19.5 MPa, and the minimum MIPS value for the TF1.5 operating condition is 18.4 MPa. It can be seen that as the segment width increases, the MAPS of the segment ring under the staggered and straight assembly methods decreases with the increase in width, while the MIPS does not change significantly.

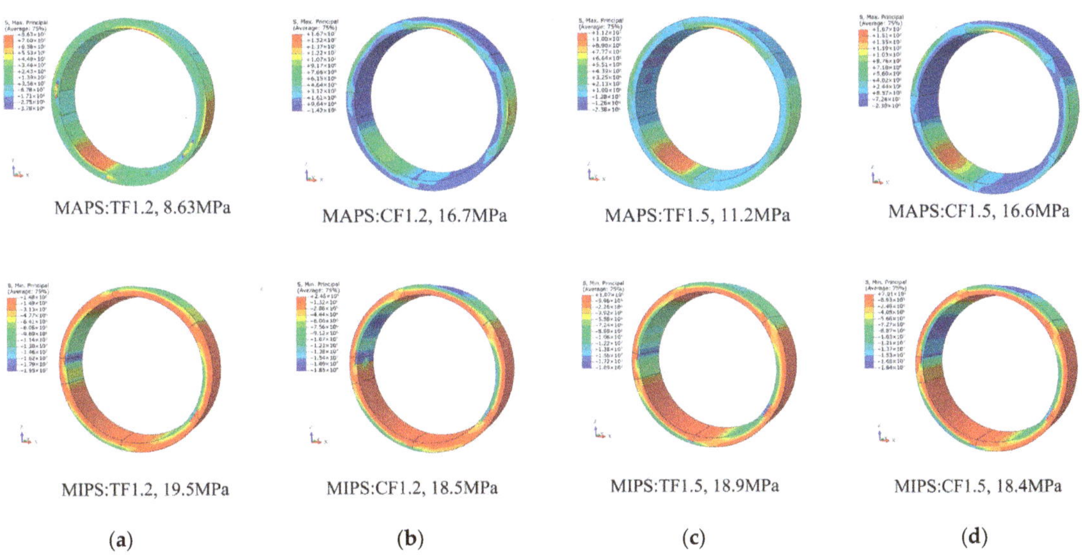

Figure 18. MAPS and MIPS of shield tunnel under different widths: (**a**) TF1.2; (**b**) CF1.2; (**c**) FT1.5; (**d**) CF1.5.

6.2.3. Analysis and Discussion of Seismic Performance

Table 5 summarizes the seismic performance evaluation of general ring tunnels of two widths under 0.4 g earthquake conditions. Except for TF1.2, the maximum joint deformations of TF1.2, TF1.5, and CF1.5 did not exceed the specification limit, and the maximum bolt stress was less than the yield point. The mechanical properties and overall stiffness of the tunnel are enhanced with the increase in the width of the general segment ring, and the deformation and internal force characteristics of the segment joints are significantly improved. To sum up, the segment width should be appropriately increased to reduce the discontinuity between the segment rings to increase the seismic resistance of the tunnel while ensuring the construction requirements are met.

Table 5. Seismic performance evaluation of shield tunnels under different segment widths.

Research Object	Mechanical Property	Joint Deformation	Bolt Internal Force	Stress Property	Seismic Performance Evaluation
Segment Width	Radial force/ Circumferential force/ Bending moment	Extrados opening/ Intrados opening/ Dislocation	Axial force/ Shear force	MAPS/ MIPS	
TF1.2 CF1.2 TF1.5 CF1.5	905 kN (TF1.5)/ 582 kN (CF1.5)/ 280 N·m (CF1.5)	5 mm (TF1.2)/ 4.6 mm (TF1.2)/ 1.1 mm (CF1.2)	490 kN (TF1.2)/ 239 kN (CF1.2)	16.7 MPa (CF1.2)/ 19.5 MPa (TF1.2)	Shielded tunnel with a 1.5 m width segment is better than the 1.2 m wide shielded tunnel in terms of seismic performance

7. Conclusions

In this paper, taking the sandy soil interlayer site in the coastal soft-soil area of Ningbo as the research background, a three-dimensional contact discontinuous finite element–infinite element coupling model of a soil–shield tunnel interaction was established by using ABAQUS, and the impact of the general ring segment assembly methods and the width on the shield tunnel was studied. The internal force response of the structure and the deformation and mechanical properties of the segment joints were also studied. The following conclusions can be reached:

(1) Under the action of 0.1 g, 0.2 g, and 0.4 g earthquakes, the shear deformation of the lining structure with different assembly methods is relatively close, and the difference is only 5.24% under the action of 0.4 g earthquake loading. The overall deformation of the lining structure is not greatly affected by the assembly method, but is mainly controlled by the forced displacement of the formation.

(2) The joint deformation of the segment joints and the internal force distribution of the connecting bolts are heterogeneous, and the location of the maximum joint deformation will change depending on the assembly method. The joint deformation and bolt internal force under the condition of staggered assembly at 90° are more obvious than those of the other three assembly methods, and the internal force distribution of the tunnel structure is less affected by the assembly method. The internal force of the lining is relatively large near the 45° angle with the vault, and the internal force decreases at the joint deformation, which changes the numerical law of the internal force of the tunnel structure.

(3) During the action of 0.4 g earthquake intensity, the deformation of the joints and the internal force of the bolts of shield tunnels assembled with straight assembly and staggered 90°assembly exceed the limit, while the other two staggered joint assembly methods can maintain safety and waterproof performance. One should avoid placing the capping block on the spandrel for the staggered assembly method from the perspective of earthquake resistance.

(4) Under the conditions of staggered assembly, the maximum difference in internal force between 1.5 m width and 1.2 m width exceeds 40%, indicating that the internal force characteristics of the structure are significantly affected by changes in the structural stiffness due to changes in segment width.

Author Contributions: S.Z.: methodology, conceptualization, data curation, software, writing; Z.G.: conceptualization, formal analysis, supervision; S.W.: conceptualization, writing, supervision, Funding acquisition; B.C.: conceptualization, investigation, supervision; C.M.: software, visualization. All authors have read and agreed to the published version of the manuscript.

Funding: The work presented in this paper was sponsored by the Supported by Systematic Project of Key Laboratory of New Technology for Construction of Cities in Mountain Area (No. LNTCCMA-20200104), the Systematic Project of Guangxi Key Laboratory of Disaster Prevention and Structural Safety (No. 2019ZDK005), the Ningbo Natural Science Foundation (2019A610394), the Ningbo Public Welfare Science and Technology Planning Project (No. 2019C50012). These financial supports are gratefully acknowledged.

Data Availability Statement: The data used to support the findings of this study are included in the article.

Conflicts of Interest: The authors declare no conflict of interest.

References

1. Shen, Y.; Zhang, D.; Wang, R.; Li, J.; Huang, Z. Sbd-k-medoids-based long-term settlement analysis of shield tunnel. *Transp. Geotech.* **2023**, *42*, 101053. [CrossRef]
2. Xu, L.; Guo, J.; Xu, C.; Chen, R.; Lin, J. Study on Dynamic Response of Soil Layer at the Bottom of Subway Shield Tunnel under Seismic Action. *Geotech. Geol. Eng.* **2022**, *41*, 1635–1646. [CrossRef]
3. Masanori, H. *Dynamic Behaviors of Underground Structures during Earthquakes and Earthquake-Resistant Design;* Springer: Berlin/Heidelberg, Germany, 2014; pp. 229–273.
4. Yang, L.; Xu, C.; Du, X. Causal Analyses of Different Degree of Earthquake Damage Occurred on Daikai Subway Station and Its Running Tunnels during Kobe Earthquake. *J. Disaster Prev. Mitig. Eng.* **2020**, *40*, 326–336.
5. Liu, R.; Zhu, Z. Review of Earthquake Damage Prediction for Underground Structures. *China Earthq. Eng. J.* **2020**, *42*, 1349–1360.
6. Liu, X.; Bai, Y.; Yuan, Y.; Mang, H.A. Experimental Investigation of the Ultimate Bearing Capacity of Continuously Jointed Segmental Tunnel Linings. *Struct. Infrastruct. Eng.* **2016**, *12*, 1364–1379. [CrossRef]
7. Du, X.-L.; Li, Y.; Xu, C.-S.; Lu, D.-C.; Xu, Z.-G.; Jin, L. Review on Damage Causes and Disaster Mechanism of Daikai Subway Station During 1995 Osaka-Kobe Earthquake. *Yantu Gongcheng Xuebao/Chin. J. Geotech. Eng.* **2018**, *40*, 223–236.
8. Weng, M.-C.; Wu, J.-H.; Hwang, J.-H.; Chigira, M.; Massey, C. Preface to the Special Issue of Geo-Hazards Induced by the 1999 Chi-Chi Earthquake, Taiwan: Lessons Learned and Progress in Two Decades. *Eng. Geol.* **2022**, *297*, 106505. [CrossRef]
9. Huang, Z.K.; Pitilakis, K.; Tsinidis, G.; Argyroudis, S.; Zhang, D.M. Seismic Vulnerability of Circular Tunnels in Soft Soil Deposits: The Case of Shanghai Metropolitan System. *Tunn. Undergr. Space Technol.* **2020**, *98*, 103341. [CrossRef]
10. Gong, G.; Liang, J.; Ba, Z.; Xu, A.; Yan, Q.; Wang, Z. Seismic Transverse Time-History Analysis of Shield Tunnel in Complex Soft Soil. *Tianjin Daxue Xuebao (Ziran Kexue Yu Gongcheng Jishu Ban)/J. Tianjin Univ. Sci. Technol.* **2019**, *52*, 106–112.
11. Dong, S.H.; Zhang, X.Y.; Jia, C.X.; Li, S.Q.; Wang, K. Study on Seismic Response and Vibration Reduction of Shield Tunnel Lining in Coastal Areas. *Sustainability* **2023**, *15*, 4185. [CrossRef]
12. Khan, M.A.; Sadique, M.R.; Harahap, I.H.; Zaid, M.; Alam, M.M. Static and Dynamic Analysis of the Shielded Tunnel in Alluvium Soil with 2d Fem Model. *Transp. Infrastruct. Geotechnol.* **2022**, *9*, 73–100. [CrossRef]
13. Arnau, O.; Molins, C. Theoretical and Numerical Analysis of the Three-Dimensional Response of Segmental Tunnel Linings Subjected to Localized Loads. *Tunn. Undergr. Space Technol.* **2015**, *49*, 384–399. [CrossRef]
14. Shen, Y.; Zhong, Z.; Li, L.; Du, X.; El Naggar, M.H. Seismic Response of Shield Tunnel Structure Embedded in Soil Deposit with Liquefiable Interlayer. *Comput. Geotech.* **2022**, *152*, 105015. [CrossRef]
15. Zhang, W.-J.; Cao, W.-Z. Mechanical and Waterproof Performances of Joints of Shield Tunnels with Large Cross-Section under Earthquakes. *Yantu Gongcheng Xuebao/Chin. J. Geotech. Eng.* **2021**, *43*, 653–660.
16. Zhu, T.; Wang, R.; Zhang, J.-M. Seismic Response Analysis of Shield Tunnels in Liquefiable Soils. *Yantu Gongcheng Xuebao/Chin. J. Geotech. Eng.* **2019**, *41*, 57–60.
17. Wu, H.; He, C.; Yan, Q.; Feng, K.; Cheng, T. A Study on Curve Fitting Algorithm of Curved Shield Tunnels Assembled by Universal Wedge Segments and Its Application. *Tiedao Xuebao/J. China Railw. Soc.* **2016**, *38*, 90–98.
18. Wang, S.; Shen, X.; He, X.; Yao, J. A Model Test for the Mechanical Property and Failure Mode of Lining Segments with Different Assembly Types of Shield Tunnel. *China Civ. Eng. J.* **2017**, *50*, 114–124.
19. Blom, C.B.M.; Van Der Horst, E.J.; Jovanovic, P.S. Three-Dimensional Structural Analyses of the Shield-Driven 'Green Heart' Tunnel of the High-Speed Line South. *Tunn. Undergr. Space Technol.* **1999**, *14*, 217–224. [CrossRef]
20. Li, W.; He, C. Study on Mechanical Behavior and Controlling Assembling Modes of Universal Segment Lining for Shield Tunnel. *Tiedao Xuebao/J. China Railw. Soc.* **2007**, *29*, 77–82.
21. Teachavorasinskun, S.; Chub-uppakarn, T. Influence of Segmental Joints on Tunnel Lining. *Tunn. Undergr. Space Technol.* **2010**, *25*, 490–494. [CrossRef]

22. Feng, K.; He, C.; Su, Z. Prototype Test on Failure Characteristics of Segmental Lining Structure for Nanjing Yangtze River Tunnel. *Xinan Jiaotong Daxue Xuebao/J. Southwest Jiaotong Univ.* **2011**, *46*, 564–571.
23. Wang, R.-L.; Zhang, D.-M. Mechanism of Transverse Deformation and Assessment Index for Shield Tunnels in Soft Clay under Surface Surcharge. *Yantu Gongcheng Xuebao/Chin. J. Geotech. Eng.* **2013**, *35*, 1092–1101.
24. Liu, X.; Zhang, Y.; Wang, R. Discussion on Deformation and Failure of Segmental Metro Tunnel Linings. *China Civ. Eng. J.* **2020**, *53*, 118–128.
25. Zhao, W.-S.; He, X.-Z.; Chen, W.-Z.; Yang, J.-P.; Wang, H.; Yuan, J.-Q. Method for Analyzing Seismic Response of Shield Tunnel and Its Application. *Yantu Lixue/Rock Soil Mech.* **2012**, *33*, 2415–2421.
26. Wang, C.; Tang, P.; Zhuang, H.; Yang, M. Seismic Performance of Large Shield Tunnel under the Yangzi River with Considering the Staggered Joints. *J. Nat. Disasters* **2021**, *30*, 116–123.
27. Gou, Y.; Huang, Q.; Wang, L.; Yan, Y.; Jia, S. Study on Structural Behaviors and Adaptability of Shield Tunnel in the Ground Fissure Environment. *Railw. Stand. Des.* **2020**, *64*, 117–125.
28. Chen, H.-W. Seismic Response Analysis of Shield Tunnel Considering Joint Effect. *Urban Rapid Rail Transit* **2018**, *31*, 78–85.
29. Zhang, W.; Lu, Q.; Zhang, G.; Li, H. Study on Dynamic Response of Shield Tunnel Ring Structure by Cap Location. *Chin. J. Undergr. Space Eng.* **2020**, *16*, 588–595+609.
30. Zhuang, H.; Hu, Z.; Wang, X.; Chen, G. Seismic Responses of a Large Underground Structure in Liquefied Soils by Fem Numerical Modelling. *Bull. Earthq. Eng.* **2015**, *13*, 3645–3668. [CrossRef]
31. Sharari, N.; Fatahi, B.; Hokmabadi, A.S.; Xu, R. Impacts of Pile Foundation Arrangement on Seismic Response of Lng Tanks Considering Soil–Foundation–Structure Interaction. *J. Perform. Constr. Facil.* **2022**, *36*, 04021110. [CrossRef]
32. Tao, L.; Ding, P.; Lin, H.; Wang, H.; Kou, W.; Shi, C.; Li, S.; Wu, S. Three-Dimensional Seismic Performance Analysis of Large and Complex Underground Pipe Trench Structure. *Soil Dyn. Earthq. Eng.* **2021**, *150*, 106904. [CrossRef]
33. Zhuang, H.-Y.; Ren, J.-W.; Wang, R.; Miao, Y.; Chen, G.-X. Elasto-Plastic Working States and Seismic Performance Levels of Frame-Type Subway Underground Station with Two Layers and Three Spans. *Yantu Gongcheng Xuebao/Chin. J. Geotech. Eng.* **2019**, *41*, 131–138.
34. Pakbaz, M.C.; Yareevand, A. 2-D Analysis of Circular Tunnel against Earthquake Loading. *Tunn. Undergr. Space Technol.* **2005**, *20*, 411–417. [CrossRef]

Disclaimer/Publisher's Note: The statements, opinions and data contained in all publications are solely those of the individual author(s) and contributor(s) and not of MDPI and/or the editor(s). MDPI and/or the editor(s) disclaim responsibility for any injury to people or property resulting from any ideas, methods, instructions or products referred to in the content.

buildings

MDPI

Article

Characteristics and Mechanism of Large Deformation of Tunnels in Tertiary Soft Rock: A Case Study

Dengxue Liu [1], Shuling Huang [1,*], Xiuli Ding [1], Jianjun Chi [2] and Yuting Zhang [1]

1 Key Laboratory of Geotechnical Mechanics and Engineering of Ministry of Water Resources, Changjiang River Scientific Research Institute, Wuhan 430010, China; liudx@mail.crsri.cn (D.L.)

2 China Water Resources Beifang Investigation, Design and Research Co., Ltd., Tianjin 300222, China

* Correspondence: huangsl@mail.crsri.cn

Abstract: During the excavation of a water-conveyance tunnel in Tertiary soft rocks in China, significant deformation of the surrounding rocks and damage to the support were observed. Substantial horizontal deformation, reaching magnitudes of meters, was observed in the right side wall after a certain period of tunnel excavation. Extensive investigations, including field surveys, monitoring data analysis, laboratory tests, and numerical simulations, were conducted to understand the underlying mechanisms of this large deformation. The section of the tunnel with large deformation consisted of Tertiary sandy mudstone, mudstone interbedded with marl, and glutenite. Laboratory tests and mineral composition analysis revealed that the sandy mudstone and mudstone interbedded with marl exhibited low strength, which was closely related to the water content of the rock specimens. The compressive strength gradually decreased with increasing water content, and when the water content of mudstone interbedded with marl reached 26.96%, the uniaxial compressive strength decreased to only 0.24 MPa. Additionally, sandy mudstone and mudstone interbedded with marl contained a significant amount of hydrophilic minerals, with montmorillonite constituting 30% and 34% of the two rock samples, respectively. The tunnel passed beneath a perennially flowing gully, and a highly permeable glutenite layer was present in the middle of the tunnel. This resulted in groundwater seepage from the inverted arch during excavation, leading to the softening effect on the mudstone interbedded with marl in the lower part of the tunnel. Through numerical simulation and back-analysis techniques, the varying degrees of softening induced by groundwater were quantitatively analyzed in the surrounding rocks on the left and right sides. The study revealed that the large deformation of the tunnel was triggered by two factors: the plastic flow caused by tunnel excavation under the low strength of the surrounding rocks and the softening effect of groundwater. The damage to the support system was primarily attributed to the squeezing and swelling deformation of the surrounding rocks and the non-uniform deformation between different rock layers.

Keywords: large deformation; tunnel; tertiary soft rock; squeezing and swelling deformation; incompatible deformation

1. Introduction

Large deformation is an engineering geological damage that is frequently encountered in tunnel construction. Ding et al. [1] defined the large deformation of surrounding rocks in tunnels as deformation that is significantly greater than normal and such that it may trigger adverse effects. The normal level of deformation in tunnels refers to the deformation that occurs in the surrounding rock of a tunnel section with non-adverse geological conditions after considering conventional support measures. Generally, when the deformation of the tunnel is at a normal level, the surrounding rock will not undergo damage caused by deformation. The first documented serious large deformation of surrounding rocks occurred in the Simplon tunnel in Switzerland that was completed in 1906. Significant deformation and damage occurred to surrounding rocks during excavation of the tunnel, attracting extensive attention among engineers [2]. Afterwards, large deformation of surrounding rocks to different extents also happened during the construction of the Arlberg expressway tunnel and Tauern tunnel in Austria [3,4], the Enasan tunnel in Japan [5], and the Jiazhuqing tunnel [6], Muzhailing tunnel [7], Wushaoling tunnel [8], and Hadapu tunnel [9] in China. The large deformation was also accompanied by damage to the supports and even engineering disasters such as failure of the surrounding rocks in the tunnels. These tunnels shared characteristics of a low strength of surrounding rocks and high ground stress, which resulted in large deformation of surrounding rocks and enduring deformation during construction.

The large deformation of surrounding rock after tunnel excavation may be attributed to either the tectonically high stress-dominated external environment or the low strength of rock masses. In the former case, the stress redistribution resulting from tunnel excavation under high ground stress conditions can induce significant secondary stress disturbances in the surrounding rock, leading to substantial deformations in soft rock formations as well as thin layered and anisotropic medium–hard rock formations [10,11]. In the latter case, it generally pertains to tunnel excavation taking place within soft rock formations [12–16]. Soft rocks are characterized by a low uniaxial compressive strength (UCS). According to the recommendation of the International Society for Rock Mechanics (ISRM), rocks with a UCS of less than 25 MPa are classified as soft rocks. Soft rocks are commonly related to specific lithology, including siltstone, mudstone, and diabase, which have UCSs far lower than intact hard rocks, such as granite, gneiss, limestone, and marble. Soft rocks have strong rheological properties and low strength [17], so stress redistribution in surrounding rocks lasts for a long time after excavating a tunnel in soft rocks: deformation of surrounding rocks generally lasts for several weeks to several months [18], and, on occasion, beyond one year [15]. Considering the high frequency and hazard of large deformation during the construction of tunnels in soft rocks, numerous scholars have investigated the mechanism underlying large deformations during construction [19–23]. Their research on the large deformation of surrounding rocks of tunnels excavated in soft rocks shows that squeezing and swelling are the main modes of deformation of soft rocks. The squeezing deformation of surrounding rocks is a result of yielding of rock caused by excavation-induced stress redistribution beyond the strength of the rock, and it is mainly dependent on the rock strength and overburden thickness, that is, the ground stress. Wood [19] evaluated the stability of tunnels by virtue of the ratio of the UCS of rocks to the overburden stress. After similarity investigation and analysis of more than 20 tunnels with large squeezing deformation, Aydan et al. [12,21] found that tunnels constructed in soft rocks are commonly subjected to squeezing failure when satisfying the following conditions: the ratio of the UCS of rocks to the overburden stress is less than two and the tangential strain in the side walls of tunnels exceeds 1%. In the Standard for Engineering Classification of Rock Mass (GB50218-2014) in China [24], the $R_c/\sigma_{\max}$ index (R_c represents the saturated uniaxial compressive strength of rock, $\sigma_{\max}$ represents the maximum initial stress in the direction vertical to the tunnel axis) is used to evaluate the behavior during tunnel excavation. When $R_c/\sigma_{\max}$ is less than 4, spalling or extremely significant deformation and even large deformation are likely to occur in the rock surrounding a tunnel constructed in soft rock, and the deformation may persist. The swelling deformation of tunnels built in soft rocks refers to deformation due to the swelling of some minerals in the surrounding rocks in reaction with water. Chen [20] found that the volumetric swelling of surrounding rocks after reaction with groundwater is the main cause for deformation in Jinchuan deposit in China. In fact, it is generally impossible to separate squeezing and swelling in the excavation of tunnels in soft rocks, and a large deformation of surrounding rocks caused by squeezing or swelling alone is very rare.

To reveal the mechanism behind such a large deformation of tunnels built in soft rocks, many scholars generally performed comprehensive analysis from both mechanical and geological perspectives [15,18,25,26]. Bian et al. [16] studied the occurrence mechanism of large deformation for the Huangjiazhai tunnel in China by combining engineering analysis, laboratory analysis, and micro-analysis. Li et al. [27] analyzed the large deformation mechanism of the Jiaoding tunnel in layered rock strata and proposed treatment measures based on field tests and theoretical analysis. However, the mechanism underlying the large deformation of a specific project generally calls for targeted research because of the complex geological conditions and ground stress at specific sites. The present research focuses on the significant deformation of surrounding rock and the failure of supporting structures observed during the excavation of a tunnel in Tertiary soft rock in China. Various methods, including field geological investigation, analysis of deformation monitoring data during tunnel excavation, laboratory experiments, and numerical simulations, were employed to analyze the characteristics and mechanisms of the large deformations in the soft rock tunnel.

2. Geology and Topography

2.1. Topography

The water-conveyance tunnel has a total length of 41.823 km, maximum burial depth of 2268 m, design flow 70 m^3/s, and diameter of 6.9 m [28]. It is designed as a free-flow tunnel and constructed using a drill and blast-heading method. The tunnel section studied here is in the chainage range of K3 + 700 to K5 + 500, beneath a hilly area where the ground elevation is 1420 to 1560 m. The tunnel section is buried at the depth of 140 to 300 m. The mountains in the area are relatively flat, with a natural slope angle of 25° to 35°. The tunnel section traverses beneath a gully with a width of 150 to 200 m. The topography in the area of the tunnel section is shown in Figure 1.

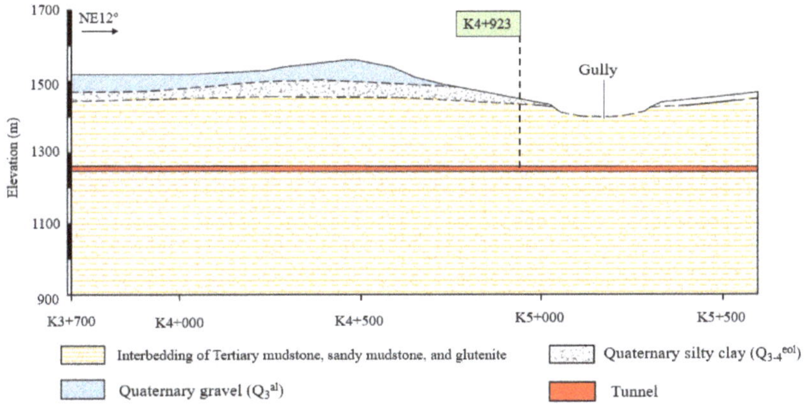

Figure 1. Geological profile of the water-conveyance tunnel (A part of the water conveyance tunnel section from K3 + 700 to K5 + 500).

2.2. Strata and Lithology

Quaternary silty clay ($Q_{3-4}{}^{eol}$) and gravel ($Q_3{}^{al}$) form the main overburden in the region. The geological prospecting by drilling boreholes on the tunnel line shows that interbedding of Tertiary mudstone, sandy mudstone, and glutenite mainly occurs below the overburden (Figure 1). Glutenite, also known as conglomerate, is characterized by its coarse-grained nature, consisting of rounded or angular fragments of various sizes and types of rock cemented together by a finer-grained matrix.

Figure 2 illustrates the lithological distribution revealed at the tunnel face (K4 + 923) after excavation of the tunnel. Three types of rocks are found on the cross-section: namely, mudstone interbedded with marl, glutenite, and sandy mudstone. The upper portion of the cross-section predominantly consists of a reddish–brown, thick to ultra-thick sandy mudstone. This particular rock type exhibits weak to medium argillaceous and calcareous cementation. It has a natural moisture content of 18.03%. Situated in the middle of the cross-section is a medium-thick bedded, gravelly glutenite. This glutenite also displays a reddish–brown hue, which is accompanied by weak cementation and inadequate sorting of gravels. It exhibits a relatively moist condition and contains approximately 15% to 20% gravel, with grain sizes ranging from 5 to 10 mm and a maximum grain size of 30 mm. The particle sizes within the glutenite are unevenly distributed. The lower portion of the cross-section is occupied by mudstone interbedded with marl. This rock type primarily exhibits a reddish–brown color interspersed with off-white marl layers. The mudstone interbedded with marl displays weak argillaceous and calcareous cementation. Its natural moisture content is 15.48%.

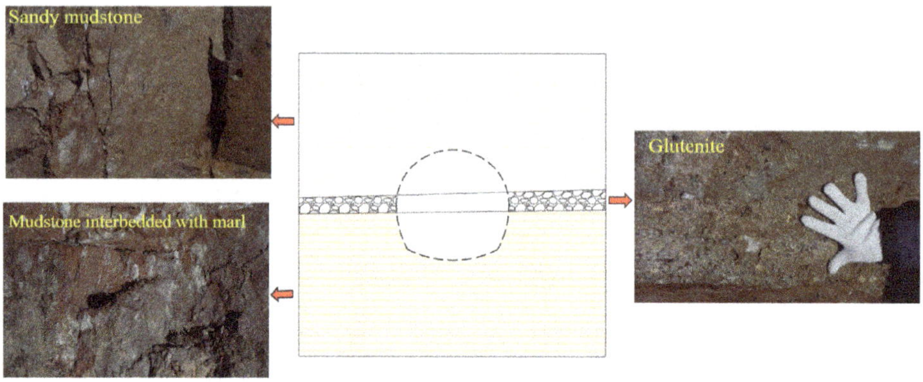

Figure 2. Lithology revealed by excavation at the K4 + 923 cross-section.

2.3. Geological Structures

The geological prospecting results prior to construction show that there is no large fault in the region. Three fractures are mainly developed. ① The first is a fracture with occurrence of NW320°~325°SW∠30°~45° with straight and smooth fracture planes that is opened by 1 to 2 mm. It has striated mirror faces and is not filled; these lie at a spacing of 0.3 to 0.8 m. ② The second is a fracture with occurrence of NE25°~30°NW∠45°~50° with straight and smooth fracture planes with scratches opened by 0.5 to 1.0 mm. The fracture is not filled and has a spacing of 0.5 to 1.0 mm. ③ The third is a fracture with occurrence of NE45°~50°NW∠25°~30° with straight and smooth fracture planes opened by 0.5 to 1 mm. It is not filled and has a spacing greater than 1 m.

2.4. Hydrogeology

The tunnel section studied in the research traverses beneath a gully that has a width of 150 to 200 m at a vertical distance of 138 m from the tunnel. There are ice and snow meltwater from nearby mountains and atmospheric precipitation flow throughout the year in the gully. The design flood flows for the twenty-year and fifty-year return periods are 142 and 164 m^3/s, respectively. The real scene of the gully is shown in Figure 3. The annual precipitation over the area is 350.2 mm, and the precipitation mainly occurs from April to September every year. During the tunnel excavation process, certain surrounding rocks within the tunnel section were observed to undergo water seepage, which was identified by the on-site geological engineer as being associated with the water source originating from the nearby gully.

Figure 3. Real picture of gully.

3. Large Deformation during Tunnel Excavation and Analysis of Corresponding Monitoring Data

3.1. Large Deformation during Tunnel Excavation

The tunnel was constructed using the new Austrian tunneling method [29] with the implementation of effective drainage measures throughout the tunnel construction process. It was excavated in two benches, and the footage of each cyclic excavation is no longer than 250 mm. Based on the empirical evidence obtained from field construction, the average daily advance rate was observed to range from 1 to 1.5 m, which corresponds to a total of four to six cycles of simultaneous excavation of the upper and lower benches. The primary support system employed for the tunnel encompassed a composite approach involving the implementation of hollow grouting anchor bolts, steel arches, mesh reinforcement, and shotcrete. Subsequently, a reinforced concrete structure was utilized for the secondary lining. The detailed support parameters of the tunnel section are summarized in Table 1, and the excavation section size is shown in Figure 4.

Table 1. Rock support measures.

Support Stage	Support Measures	Detailed Parameters
Advance support	Forepole: advanced steel pipe	Location: upper 160° range of tunnel roof; Diameter of pipes: 42 mm; Length of pipes: 5 m; Spacing of pipes: 0.35 m
Primary support	Steel arch	HW150-type; 0.5 m spaced steel bars with 22 mm diameter are used to connect the steel arch; feet-lock bolts are used to fix the steel arch
	Anchor bolt	Hollow grouting anchor bolt at 1.25 m × 1.25 m spacing and 3.5 m length
	Shotcrete	0.2 m thick with steel net
Secondary support	Reinforced concrete lining	0.6 m thick

In mid to late September 2019, large deformation occurred in the surrounding rocks during the excavation of the tunnel section K4 + 904 to K4 + 924. Monitoring data show that the maximum horizontal deformation at the deformation monitoring point on the right-hand wall on the cross-section at K4 + 923 was 990.6 mm, invading the construction space of the secondary lining. Then, the bottom of the tunnel section was also subject to sever heave. On 23 October 2019, longitudinal cracks were found on the lining in the tunnel section (K4 + 904.3 to K4 + 913.3) where a lining had been applied to the inverted arch; on 3 November 2019, the inverted arch was cracked and lifted upwards in the unlined tunnel section (K4 + 913.3 to K4 + 924). Meanwhile, support in the tunnel section with large deformation of surrounding rocks also suffered from a series of damaging events such as cracking of the shotcrete and yielding of its steel arch supports, posing a threat to the stability and long-term safety of the surrounding rock-support system in the construction and operating periods of the tunnel. Figure 5 shows the consequences of the large deformation and damage to the primary support in the tunnel section.

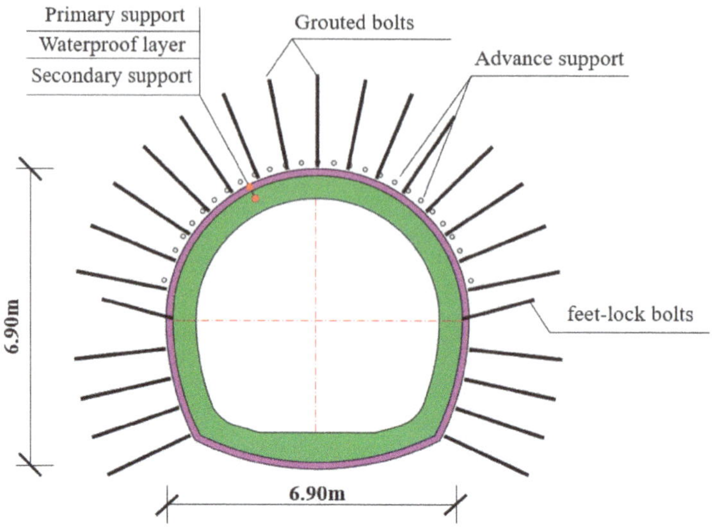

Figure 4. Tunnel cross-sectional dimensions.

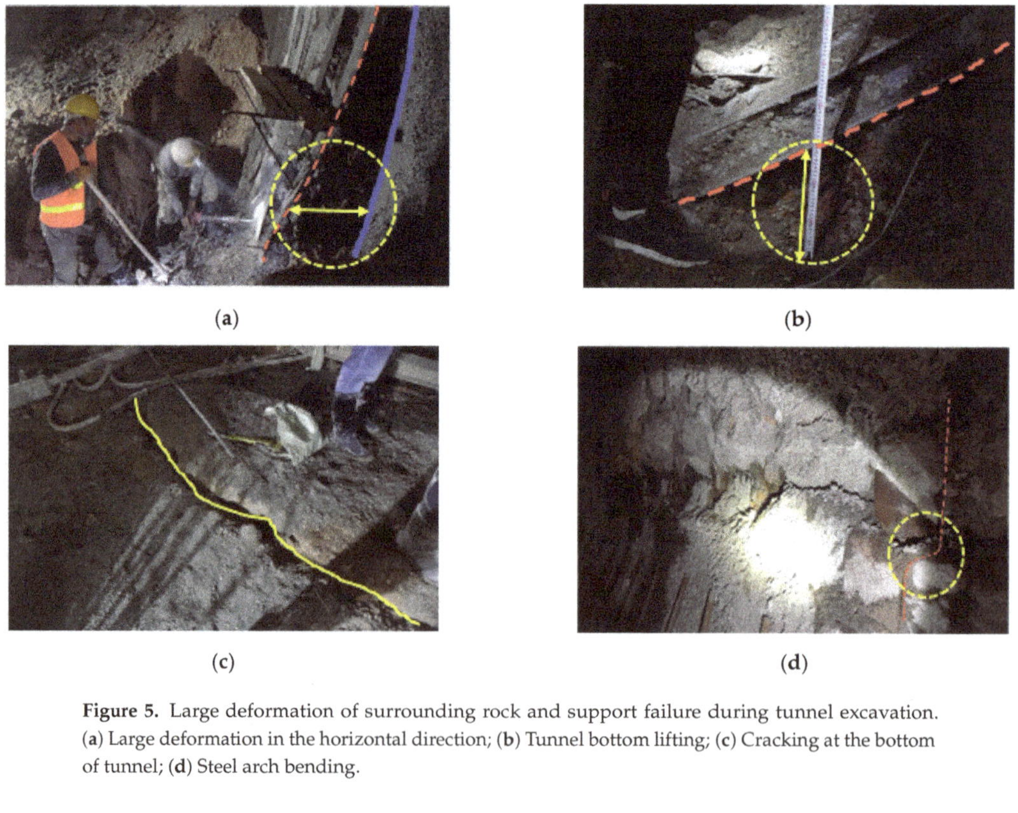

Figure 5. Large deformation of surrounding rock and support failure during tunnel excavation. (**a**) Large deformation in the horizontal direction; (**b**) Tunnel bottom lifting; (**c**) Cracking at the bottom of tunnel; (**d**) Steel arch bending.

3.2. Analysis of Monitoring Data

The cross section K4 + 923 under monitoring is within the tunnel section K4 + 904 to K4 + 924 where large deformation occurred. The monitoring results of the cross-section include those at points B and C in the upper bench and points D and E in the lower bench. The points in the upper bench of the cross-section were monitored from 12 September 2019. Figure 6 shows the horizontal deformation curves of surrounding rocks on the cross-section at K4 + 923. Figure 7 shows the horizontal deformation rate curves of surrounding rocks on the cross-section at K4 + 923.

The deformation curves at monitoring points in the upper bench of the cross-section basically change in a consistent trend. The deformation rate always peaked on the second day of monitoring, which was followed by a gradual decrease. The deformation of surrounding rocks tended to gradually converge by 18 September 2019 (the 16th day of monitoring). By 30 November 2019 (the last time the monitoring points were used to collect data in the upper bench), the cumulative horizontal deformation at point B on the left spandrel was 332.2 mm, and the horizontal deformation rate was 0.2 mm/d; the cumulative horizontal deformation at point C on the right spandrel was 368.4 mm, and the horizontal deformation rate was 0.1 mm/d.

Points in the lower bench on the cross-section at K4 + 923 were monitored from 17 September 2019. The deformation rate at point D on the left-hand wall reached the peak on the second day of monitoring, with a horizonal deformation rate of 38.1 mm/d, which then gradually declined. By 2 October 2019 (the 15th day of monitoring), the deformation of surrounding rocks tended to converge. After starting to monitor the deformation at point E on the right-hand wall, the horizontal deformation inward to the tunnel increased constantly and reached a peak of 65.3 mm/d on 27 September 2019 (the 10th day of monitoring). Then, the rate of deformation gradually declined, and the deformation of surrounding rocks tended to convergence by 4 October 2019 (the 17th day of monitoring). On 18 October 2019, the deformation rates at points D and E in the lower bench tended to increase. The horizontal deformation rate at point D on the left-hand wall was 10.1 mm/d while that at point E on the right-hand wall was 16.2 mm/d on 26 October 2019. Thereafter, the deformation rates at the two points gradually decreased. By 30 November 2019, the cumulative horizontal deformation at point D on the left-hand wall was 530.7 mm with the horizontal deformation rate 1.4 mm/d. The cumulative horizontal deformation at point E on the right-hand wall was 990.6 mm by 19 November 2019, and the horizontal deformation rate was 4.4 mm/d.

According to the above analysis, the deformation curves at different positions in the upper bench of the cross-section K4 + 923 follow similar trends: the deformation rates both reached the peak on the second day of monitoring and then gradually declined, tending to converge on the 16th day of monitoring. In addition, the deformations on the left and right-hand sides in the upper bench are similar. After the first convergence of deformations at two points on the left and right-hand walls in the lower bench of the cross-section, the deformation rates increased significantly on 18 October 2019. According to the distortion of steel arches observed in the field during this time period, it is speculated that because the steel arches were under too much load and therefore locally yielded, the primary support failed to bear the pressure of surrounding rocks. This finally led to the large increase in deformation rates after the first convergence of deformations at points on left and right-hand walls in the lower bench. In addition, the horizontal deformation of the right-hand wall in the lower bench was significantly larger than that of the left-hand wall.

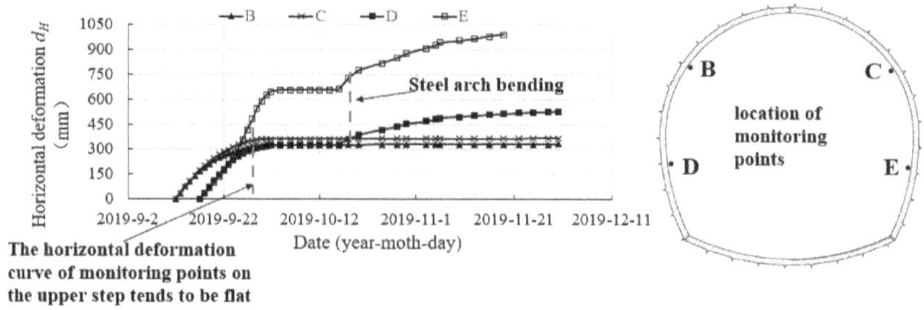

Figure 6. Horizontal deformation curves of surrounding rocks at monitoring points on the cross-section at K4 + 923.

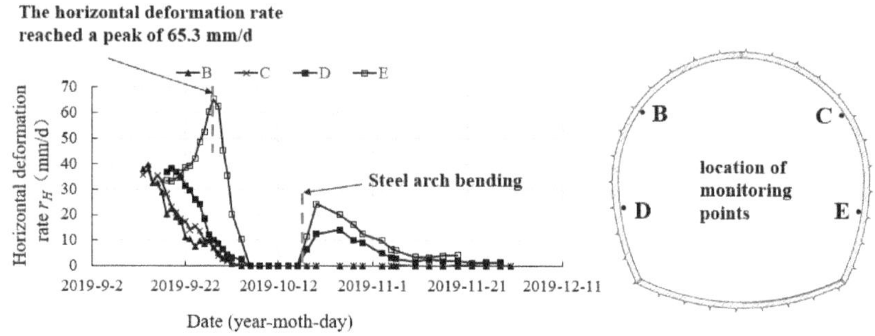

Figure 7. Horizontal deformation rate curves of surrounding rocks at monitoring points on the cross-section at K4 + 923.

4. Investigation of Causes of the Large Deformation

4.1. Rock Strength

The mudstone interbedded with marl and sandy mudstone exposed in the tunnel section with large deformation were sampled in the field, allowing measurement of the physico-mechanical properties of the rocks. The glutenite was not successfully sampled because it was too loose to allow intact specimen recovery. After being sampled in situ, the rock surfaces were sealed with paraffin immediately, and then the samples were wrapped using multiple layers of transparent tape and plastic film to ensure the intactness of samples and decrease water loss during transportation. The rock samples taken from the field were machined into cuboidal specimens measuring 50 mm × 50 mm × 100 mm (Figure 8a,b) and then immersed in water and air-dried to obtain specimens of different water contents. An RMT-301 (RMT-401) rigid servo-motor testing system for rocks was used to test the UCS of rock specimens, and the test results are listed in Table 2. According to the test results, except for some of the two types of rock specimens that have a UCS value between 5 and 15 MPa, the UCS of others is less than 5 MPa, so they are classified as soft to extremely soft rocks. The water content significantly affects the strength of the two types of rocks. To be specific, the UCSs of mudstone interbedded with marl and sandy mudstone decrease to different extents with the increasing water content; at a given water content, mudstone interbedded with marl has a lower UCS than sandy mudstone.

In the natural state, the morphology of the two samples subjected to failure under uniaxial compression is shown in Figure 9. From the observations of the failure states of the two rock samples under uniaxial loading, the two rock samples are shear failures under uniaxial stress. The fracture surface crosses the entire specimen diagonally, creating a complete fracture crack. Additionally, there are localized micro-cracks present on the end faces of the specimen.

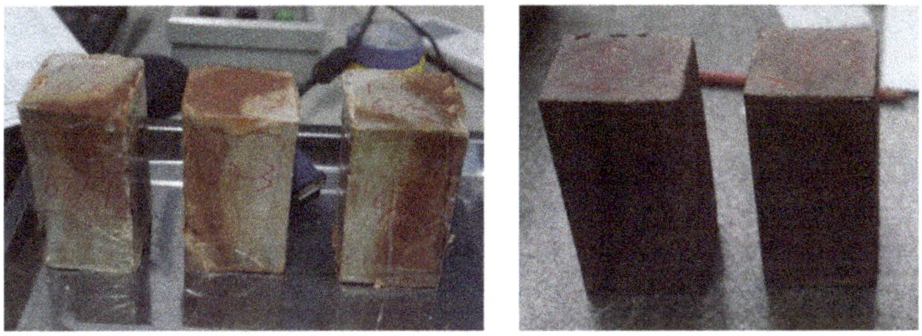

(a) (b)

Figure 8. Specimens of standard size: (**a**) Mudstone interbedded with marl and (**b**) Sandy mudstone.

Table 2. Results of the uniaxial compression tests on mudstone interbedded with marl and sandy mudstone.

Lithology	Treatment	Water Content C_W (%)	Elastic Modulus E (MPa)	UCS (MPa)
Mudstone interbedded with marl	Untreated	15.48	88	0.84
	Air-dried for 4 days	9.48	149	1.44
	Immersed in water for 1.5 h	26.96	12	0.24
Sandy mudstone	Untreated	18.03	654	3.72
	Air-dried for 4 days	6.99	835	8.54
	Immersed in water for 2 h	18.21	432	2.76
	Immersed in water for 4 h	26.10	219	1.40
	Immersed in water for 24 h	36.65	126	0.58

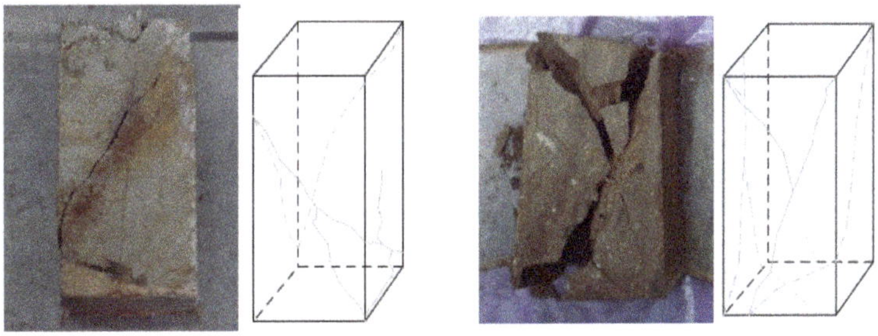

(a) (b)

Figure 9. Uniaxial compression failure form of rock samples in the natural state: (**a**) Mudstone interbedded with marl with UCS of 0.84 MPa and (**b**) Sandy mudstone with UCS of 3.72 MPa.

4.2. Swelling of Rocks

The swelling of rocks rich in clay minerals exposed to water is an important cause of the large deformation. We employed the X-ray diffraction (XRD) technique to analyze the composition of minerals in the rock samples. The testing and inspection were conducted in accordance with the Chinese Specification of testing quality management for geological laboratories (DZ/T0130-2018) [30] and the Joint Committee on Powder Diffraction Standards. The results are shown in Table 3. The minerals in the specimens of mudstone interbedded with marl are dominated by montmorillonite and quartz, which separately account for 34% and 30%, followed by illite, calcite, and feldspar. Calcite and montmorillonite are the dominant minerals in the specimens of sandy mudstone and they separately account for 45% and 30%; they are followed by illite, quartz, and feldspar. Both rock samples are enriched with clay minerals such as montmorillonite and illite, which confer upon them the characteristic of swelling.

Table 3. Mineral compositions of mudstone interbedded with marl and sandy mudstone.

Lithology	Mineral Content C_M (%)				
	Montmorillonite	Quartz	Illite	Calcite	Feldspar
Mudstone interbedded with marl	34	30	16	15	5
Sandy mudstone	30	10	10	45	5

The swelling tests (Table 4) indicate that the free swelling percentage and swelling pressure of mudstone interbedded with marl are 4.07% to 6.38% and 0.41 to 0.55 MPa, respectively; those of sandy mudstone are separately 1.31% to 1.51% and 0.22 to 0.34 MPa. It is evident that the mudstone interbedded with marl exhibits significantly higher swelling percentage and swelling pressure compared to sandy mudstone. The basis for the swelling tests is in accordance with the Chinese code for rock tests in water and hydropower projects [31].

Table 4. Results of the swelling test on mudstone interbedded with marl and sandy mudstone.

Lithology	Free Swelling Percentage PE_s (%)	Swelling Pressure P_s (MPa)
Mudstone interbedded with marl	4.07 to 6.38	0.41 to 0.55
Sandy mudstone	1.31 to 1.51	0.22 to 0.34

Meanwhile, a cubic specimen of mudstone interbedded with marl and that of sandy mudstone were selected and immersed in water to investigate their hydrophilicity and observe the changes in their appearance after immersion for 24 h. As shown in Figure 10, mudstone interbedded with marl absorbed water and rapidly swelled after being exposed to water, showing argillation and softening characteristics. When exposed to water, mudstone interbedded with marl underwent physical and chemical reactions, which resulted in the swelling, breakage, and decomposition of the rock block. This was also mainly because mudstone interbedded with marl contains some clay minerals that have high hydrophilicity. After being immersed in water for 24 h, the rock block changed obviously in the shape and was argillated, losing almost all load-bearing capacity. As for sandy mudstone, only a thin muddy layer was formed on the surface of the specimen after being exposed to water, while the specimen did not change greatly in the appearance and retained its original shape. Some fragments spalled upon pinching the specimen manually.

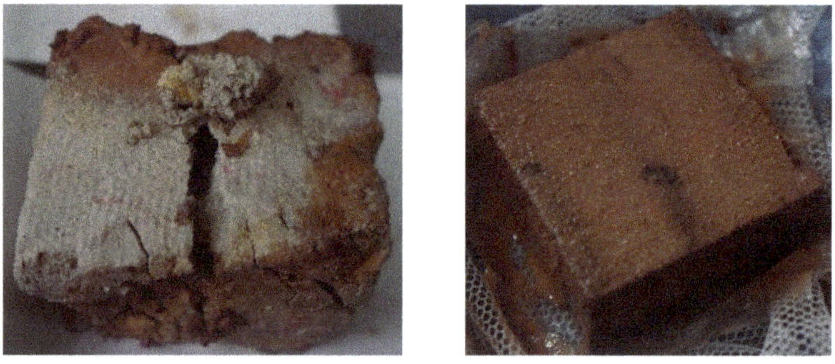

(a) (b)

Figure 10. Appearances of the two types of rocks after being immersed in water for 24 h: (**a**) Mudstone interbedded with marl and (**b**) Sandy mudstone.

4.3. Groundwater

The large deformation of tunnels is closely related to activities of groundwater. The tunnel face was basically dry in the initial stage of excavation, and local areas of the glutenite layer were moist. After 7 to 10 days, the glutenite layer had become saturated, which was accompanied by water seepage. After one month, groundwater began to seep from the tensile fractures occurring due to the uplift of the inverted arch at K4 + 914 (Figure 11). It was speculated that water in the tunnel was from the gully, although mudstone can be regarded as impermeable, whose permeability coefficient is generally 10^{-9}~10^{-7} cm/s [32]. The strata exposed in the middle of the tunnel face were a glutenite layer (NW275°SW∠15°~25°) of high water permeability at an angle of 83° with the tunnel axis (NE12°). According to the direction of the tunnel axis and the occurrence of the glutenite layer exposed during excavation, the end of the glutenite layer is located just beneath the gully, thus forming a channel allowing the migration of groundwater. The formation of plastic zones in the excavation process caused changes in the seepage field, which guided groundwater to flow to the tunnel [33]. Based on the strength tests of the two types of rock samples and the water immersion test, it can be observed that the presence of groundwater has a significant softening effect on the mudstone interbedded with marl beneath the glutenite layer. This softening effect results in a considerable reduction in the strength of the rock mass. Furthermore, the presence of groundwater induces seepage forces within the rock mass. As the water flows through the rock mass, it exerts pressure on the rock grains, causing them to displace and facilitating the deformation of the rock mass. This seepage force acts as an additional mechanism for weakening the rock mass and reducing its strength. Overall, the combined effects of groundwater softening and seepage forces contribute to the significant reduction in the strength of the mudstone interbedded with marl beneath the glutenite layer. These findings highlight the importance of considering groundwater conditions in the stability analysis and design of engineering projects involving such rock masses.

Figure 11. Seepage of groundwater from the inverted arch near K4 + 914.

4.4. Ground Stress

Ground stress also has important influences on the deformation of surrounding rocks. In the Chinese Code for Investigation of Geotechnical Engineering (GB50021-94) [34], the level of the initial stress field is evaluated using the $R_c/\sigma_{\max}$ ratio. $R_c/\sigma_{\max}$ is the ratio of the UCS of saturated rocks to the maximum principal stress. It is equivalent to using 0.25 as the critical value of σ_{cm}/P_0 to distinguish rocks under medium and severe squeezing [35]. In the above code, the critical value of $R_c/\sigma_{\max}$ is set to 4 to distinguish extremely high ground stress from high ground stress by summarizing geological data from multiple sources. When $R_c/\sigma_{\max}$ is less than 4, a region is regarded as bearing extremely high ground stress.

However, the measured ground stress data collected in tunnels show that the initial ground stress field of the strata where a tunnel is excavated always has a lateral pressure coefficient not equal to unity: that is, the vertical initial stress is not equal to the horizontal one. Therefore, the mechanical responses of surrounding rocks and distribution of the secondary stress field caused by excavation of the tunnel also differ. On this basis, our research group proposed the criterion for large squeezing deformation considering the tectonic stress and excavation disturbance, as given by Equation (1) [1].

$$\begin{cases} SSR = \frac{R}{(3-\lambda)\sigma_{0\max}} < 1 & \lambda \leq 1 \\ SSR = \frac{R}{(3-\frac{1}{\lambda})\sigma_{0\max}} < 1 & \lambda > 1 \end{cases} \quad (1)$$

where SSR represents the calculated strength–stress ratio of surrounding rocks; R is the UCS of rocks; λ denotes the lateral pressure coefficient of ground stress; and $\sigma_{0\max}$ represents the initial maximum principal stress.

During the project planning phase, geostress tests in the region were conducted using the hydraulic fracturing method and stress relief method. The horizontal maximum principal stress σ_H is dominant in the elevation range of the tunnel in the region. The vertical stress σ_z is similar to the horizontal minimum stress σ_h. The quantitative relationship of the three principal stresses can be expressed as $\sigma_H = 1.2\sigma_z$ and $\sigma_H > \sigma_h \approx \sigma_z$. The burial depth of the tunnel section under investigation is 190 m, so the vertical initial ground stress on the tunnel section is about 4.5 MPa by estimating that the overburden above the tunnel has an average bulk density of 2.38 $\mathrm{g/cm^3}$. Then, the horizontal maximum principal stress is about 5.4 MPa. According to the laboratory test results, the UCS of mudstone interbedded with marl is about 0.84 MPa. Then, the SSR of surrounding rocks is calculated to be 0.072, which is much less than unity, so it is judged that large squeezing deformation will occur after excavation of the tunnel.

5. Numerical Simulations

Numerical simulations can be an important means of analyzing problems pertaining to rock engineering. A corresponding numerical model was established according to geological conditions revealed after excavation of the tunnel to quantify the softening effect of groundwater on mudstone interbedded with marl below the glutenite layer.

5.1. Initial Simulation Conditions

5.1.1. Methods and Model

The commercial software FLAC^{3D} was used to conduct numerical simulation. Figure 12a shows the numerical model established based on the geological conditions revealed by excavation of the cross-section at K4 + 923, and Figure 12b displays the mesh generated around the tunnel. Hexahedral meshing was applied to the whole simulation model, forming 3548 elements and 7156 nodes. The model was in a range with the horizontal and vertical lengths both of 150 m. It is worth noting that the burial depth of the cross-section at K4 + 923 of the tunnel is 190 m, while the thickness of the overburden above the tunnel in the model is 75 m. Therefore, geostatic stress of overlying strata with a thickness of 115 m was also applied to the model in the simulation process. The boundary conditions for the numerical model are as follows: the left and right vertical sides are subject to horizontal restrictions only, while the bottom boundary is subject to both horizontal and vertical restrictions.

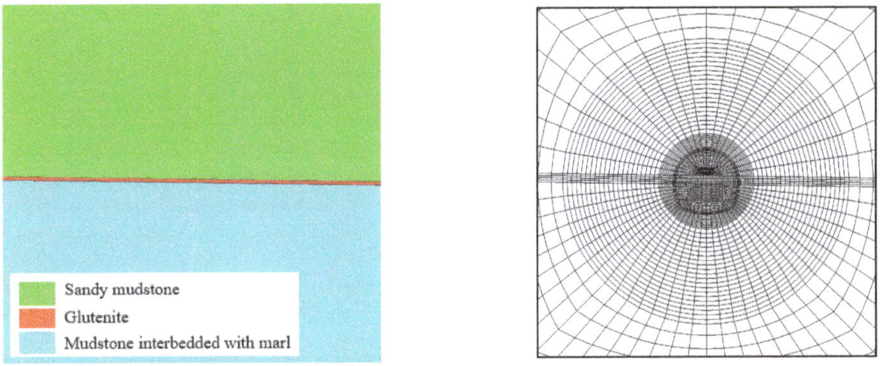

(**a**) (**b**)

Figure 12. Numerical simulation model and the mesh: (**a**) Numerical simulation model and (**b**) The mesh around the tunnel.

5.1.2. Mechanical Parameters of Rocks and Initial Ground Stress Field

The mechanical parameters of the three types of rock masses were provided based on results of laboratory tests and general geological judgment (Table 5). The ideal elasto-plastic constitutive model and Mohr–Coulomb yield criterion were used for simulations.

According to the measurements of in situ stress, the lateral pressure coefficients λ_H and λ_h of the horizontal maximum and horizontal minimum principal stresses were determined to be 1.2 and 1.0, respectively. Then, the initial in situ stress field of the simulation mesh could be obtained based on the two values.

Table 5. Suggested mechanical parameters of the rock masses.

Lithology	Unit Weight γ (kN/m^3)	Shear Strength		Tensile Strength t_s (MPa)	Elastic Modulus E (GPa)	Poisson's Ratio ν
		Cohesion c (MPa)	Friction Angle φ (°)			
Sandy mudstone	21.0	0.30	26.56	0.15	0.65	0.39
Glutenite	21.0	0.20	28.81	0.10	0.50	0.39
Mudstone interbedded with marl	21.0	0.28	24.23	0.14	0.09	0.39

5.1.3. Support Measures of the Tunnel Considered in the Simulation

The support measures implemented during the tunnel excavation are outlined in Table 1. For the numerical simulation, distinct elements, namely cable, beam, and solid elements, were employed to represent the anchor bolts, steel arch, and shotcrete, respectively. The cable and beam elements are structural components integrated within FLAC^{3D}. It is important to note that the simulation was conducted using a two-dimensional model; thus, the advance support measures specified in Table 1 were not considered in the simulation.

5.2. Analysis of Simulation Results

Figure 13 shows the horizontal deformation of rocks after excavation of the tunnel based on rock parameters in Table 1. The maximum horizontal deformation of the tunnel is found in the middle–upper part of the left and right-hand walls, and the horizontal deformation at this position in the right-hand wall is 366.5 mm. Figure 14 illustrates the distribution of plastic zones. The depths of plastic zones in the vault and side walls of the tunnel are 5.7 and 8.6 m, respectively. Plastic zones are dominated by shear plastic zones, with a small area of shear-tension plastic zones distributed in shallow surrounding rocks of the tunnel. The deformation of surrounding rocks obtained by numerical simulations was compared with that measured during excavation of the tunnel, which indicated that the two are consistent in the upper bench. However, the measured deformation of surrounding rocks in the lower bench is much greater than that attained by numerical simulation. This is because the softening effect of groundwater on the lower mudstone interbedded with marl was not considered in the numerical simulation. As mentioned in Section 4.3, the formation of plastic zones during the excavation induces changes in the seepage field, guiding groundwater to flow toward the tunnel, which thus significantly decreases the strength of mudstone interbedded with marl below the glutenite layer. This is also the cause for the significant increase in the deformation of surrounding rocks in the lower bench. Therefore, the rock mass parameters of mudstone interbedded with marl after encountering groundwater shall be obtained through back-analysis based on the monitoring data from the lower bench of the tunnel.

For simplicity, only the weakening effect of groundwater on the elastic modulus of rock mass is considered in the back-analysis process. The range of the surrounding rock involved in the back-analysis is the plastic zone of mudstone intercalated with marl, as shown by the black box line in Figure 14. According to results of back-analysis, the elastic modulus of softened mudstone interbedded with marl in the left-hand wall is 0.012 GPa, while that of the softened mudstone interbedded with marl in the right-hand wall is 0.007 GPa. Figure 15 shows the horizontal deformation of surrounding rocks after excavation of the tunnel when considering the softening effect of groundwater on mudstone interbedded with marl. The maximum horizontal deformation of the left-hand wall in the lower bench is 518.5 mm, while that of the right-hand wall is 992.2 mm, which is consistent with the measured deformation. Differential deformations are observed in the lower bench's side walls, which are attributed to varying degrees of groundwater-induced softening effects on the mudstone interbedded with marl. Field surveys further reveal a relatively higher moisture content in the right-hand wall compared to the left, providing corroborative evidence for the aforementioned explanation.

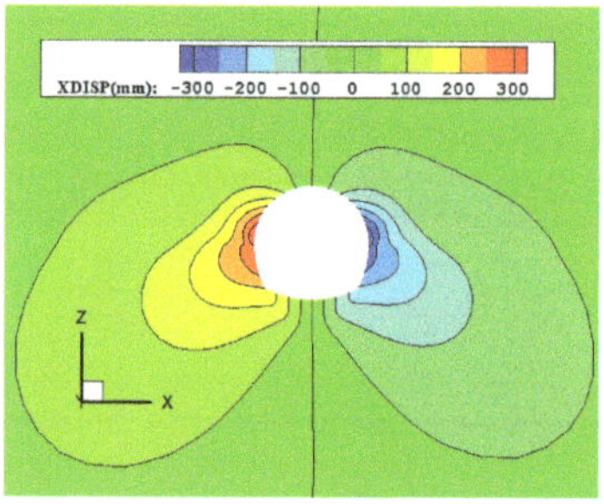

Figure 13. Horizontal deformation of surrounding rocks of the tunnel when not considering the softening effect of groundwater.

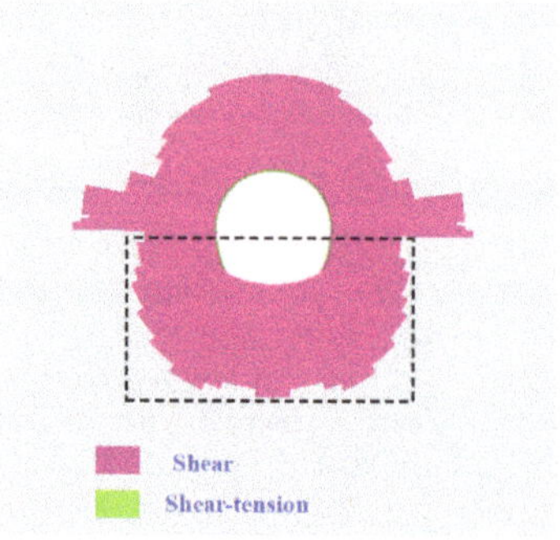

Figure 14. Plastic zones after excavation of the tunnel.

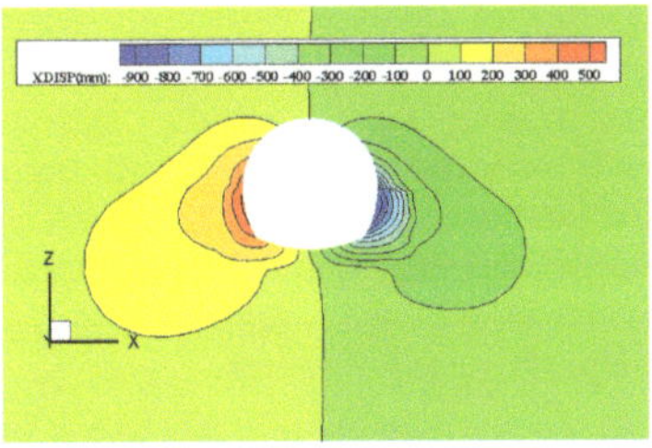

Figure 15. Horizontal deformation of surrounding rocks of the tunnel when considering the softening effect of groundwater.

According to the deformation value of surrounding rock obtained by numerical simulation, the tunnel outline after deformation is fitted, as shown in Figure 16. By comparing the tunnel outlines before and after the deformation of the surrounding rock, it can be seen that the results of the numerical simulation can well reproduce the large deformation and the damage of the supporting structure during the tunnel excavation.

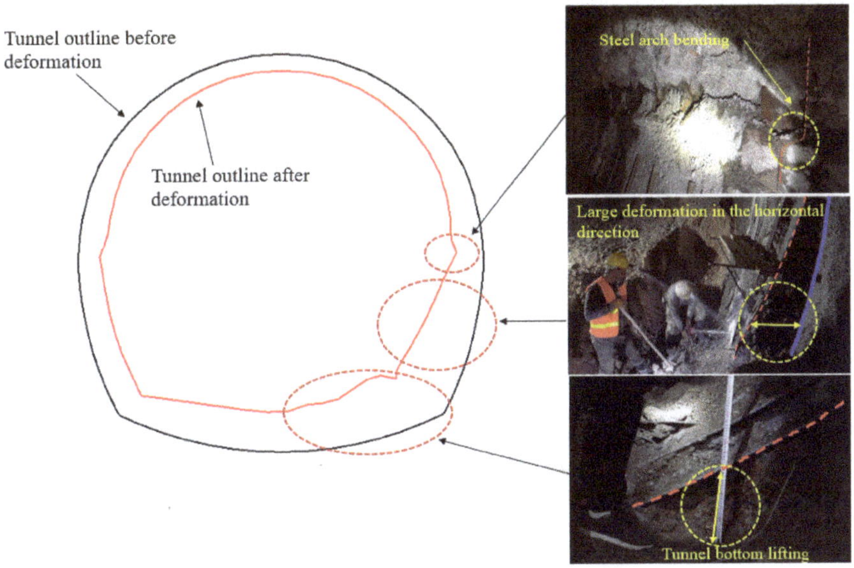

Figure 16. Tunnel outlines before and after the deformation of the surrounding rock.

6. Discussion on the Mechanism of Large Deformation

6.1. Plastic Flow of Soft Rocks

Plastic flow caused by excavation is regarded as one of the most important factors that induce a large deformation of surrounding rocks of tunnels constructed in soft rocks [17]. Excavation of a tunnel alters the stress state of surrounding rocks from typical three-dimensional compression to a two-dimensional state and even induces tensile stresses. The excavation also increases the tangential stress while reducing the radial stress and inducing large shear stresses within a certain depth range. Once the stress reaches the yield plane, the rocks enter the plastic state and undergo plastic deformation in the plastic zones [36].

Deformation in plastic flow can occur at various scales, namely minor, severe, and extreme, depending on factors such as the primary structure of surrounding rocks, ground stress, and rock strength. The tunnel section is primarily composed of mudstone interbedded with marl, sandy mudstone, and glutenite, which have UCSs lower than 25 MPa, as determined by test results. These rocks are classified as soft rocks. Furthermore, the ratio of the strength of the surrounding rocks to the ground stress in the tunnel section is relatively small, indicating a region with extremely high ground stress. The stress redistribution resulting from tunnel excavation significantly surpasses the strength of the surrounding rocks, leading to their plastic yielding and substantial plastic deformation. The numerical analysis results also provide strong evidence that when adopting the Mohr–Coulomb yield criterion, the depth of the plastic zone in the side wall reaches 8.6 m. Therefore, the entire cross-section at K4 + 923 exhibits an extremely severe squeezing deformation.

6.2. Softening Effect of Groundwater on Rocks

Results of laboratory tests indicate that the UCS values of mudstone interbedded with marl and sandy mudstone decrease to different extents with increasing water content, reflecting the softening effect of water on rocks. Softening is a phenomenon that changes over time and it reduces the shear strength of the rock. Reduction in the strength can be represented as descent of the rupture envelop with the decrease in non-linearity at low stress [37]. XRD results indicate that mudstone interbedded with marl and sandy mudstone exposed during excavation of the tunnel both contain clay minerals such as montmorillonite and illite, so these rocks are of high hydrophilicity. Although mudstone has a low coefficient of permeability, the glutenite layer with its high water permeability and fractures caused by excavation unloading provide conditions for the seepage of groundwater. Groundwater flows to mudstone along these fractures, thus softening the mudstone. Terzaghi [38] described the relationship between the internal softening and stability of a cutting slope constructed in a hard, cracked clay. It is worth noting that fractures and unloading both play an important role in the process.

Figure 17 depicts the mechanism through which the groundwater's softening effect facilitates significant deformation of the tunnel's surrounding rocks. The process is explained as follows: the tunnel section from K4 + 904 to K4 + 924 is close to the gully and the excavation of the tunnel changes the seepage path of the groundwater. The glutenite layer, with its high water permeability in the middle of the tunnel, becomes the preferential pathway for the migration of groundwater, so that the water content or permeability of surrounding rocks in the middle and lower parts of the tunnel increases significantly. Meanwhile, the surrounding rocks are unloaded and relaxed due to excavation of the tunnel, resulting in the initiation, propagation, and coalescence of fractures (Figure 9). This further increases the permeability of the rocks and significantly changes the moisture conditions therein. Surrounding rocks are argillated and softened when exposed to water, show significantly decreased strength, and swell to some extent, leading to large plastic deformation. In particular, the mudstone interbedded with marl in the middle and lower parts of the tunnel undergoes greater plastic deformation under the influences of the glutenite aquifer. Therefore, the mechanism of occurrence of the large deformation of the rock surrounding the tunnel section from K4 + 904 to K4 + 924 is closely related to the strata structure and lithological distribution of the section as well as water recharging from the gully.

In the tunnel section where the large deformation occurred, the mechanical performance of mudstone interbedded with marl below the glutenite layer is more significantly deteriorated compared with that of the sandy mudstone above the glutenite layer under the combined influences of the lithological difference and groundwater. The difference in the mechanical performance of surrounding rocks below and above the glutenite layer generates an uneven load on the primary support (as evinced by the larger deformation in the lower side walls than in the upper part). This also induces local buckling of the streel arch at the glutenite layer.

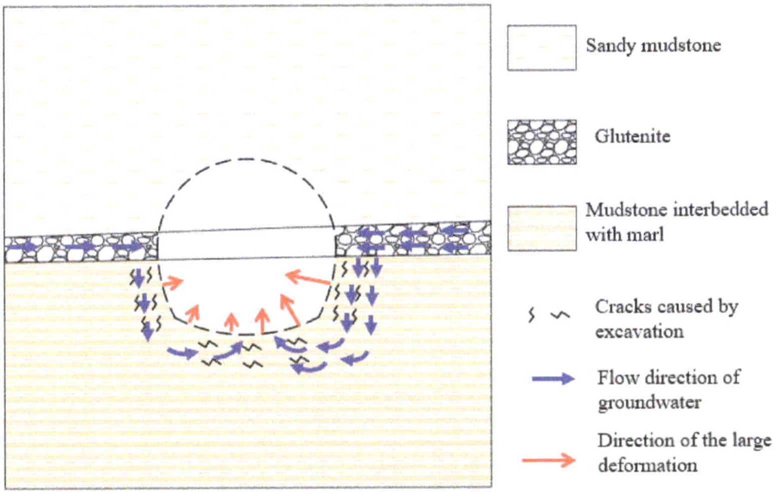

Figure 17. Diagrammatic sketch for the mechanism of large deformation in the tunnel.

7. Conclusions

Based on the comprehensive analysis of field surveys, deformation monitoring data, laboratory tests, and numerical simulations, the following conclusions can be drawn regarding the significant large deformation and damage to support observed during the excavation of a tunnel in Tertiary soft rocks in China.

(1) The large deformation of the surrounding rocks in the tunnel is primarily caused by two factors: plastic flow induced by tunnel excavation under low rock strength and the softening effect of groundwater on the surrounding rocks.
(2) The occurrence mechanism of the large deformation in the rocks surrounding the tunnel is closely related to the strata structure, lithological distribution of the tunnel section, and water recharge from the gully.
(3) The damage to the support structure in the tunnel is primarily driven by the combined effects of squeezing and swelling deformation of the surrounding rocks and non-uniform deformation between different rock layers.

In conclusion, the findings from this study highlight the importance of considering the strata structure, lithological distribution, and groundwater influence when assessing the potential for large deformation and damage to support during tunnel excavation in Tertiary soft rocks. These insights can inform future tunnel construction projects and help mitigate the risks associated with such geological conditions.

Author Contributions: D.L.: methodology, conceptualization, data curation, validation, writing. S.H.: conceptualization, reviewing, funding acquisition, supervision. X.D.: conceptualization, supervision. J.C.: data curation, resources. Y.Z.: data curation, reviewing. All authors have read and agreed to the published version of the manuscript.

Funding: This research was funded by the Fundamental Research Funds for Central Public Welfare Research Institutes (Grant No. CKSF2021457/YT and CKSF2021715/YT) and the National Natural Science Foundation of China (Grant No. 51979008).

Data Availability Statement: The data used to support the findings of this study are included in the article.

Conflicts of Interest: The authors declare no conflict of interest.

References

1. Ding, X.L.; Zhang, Y.T.; Huang, S.L.; Chi, J.J.; Zhang, C.J.; Liu, D.X. Large deformation mechanism of surrounding rock masses of tunnels, prediction method of squeezing large deformation and its application. *Chin. J. Rock Mech. Eng.* **2023**, *42*, 521–544.
2. Fox, F. The boring of the simplon tunnel, and the distribution of temperature that was encountered. *Proc. R. Soc. Lond. Ser. A* **1905**, *76*, 29–33.
3. Enrich, H.A.C.K.L. Experiences from the construction of the Tauern Tunnel. *Rock Mech. Felsmech Mec. Roches* **1974**, *6*, 90.
4. John, M. Construction of the Arlberg expressway tunnel tube. *Tunn. Tunn. Int.* **1980**, *12*, 45–50.
5. Ito, Y. Design and construction by NATM through Chogiezawa fault zone for Enasan Tunnel on central motorway. *Tunn. Undergr.* **1983**, *14*, 7–14.
6. Gao, S.J. Main countermeasures against the serious deformation of surrounding rock in Jiazhuqing Tunnel. *Mod. Tunn. Technol.* **1998**, *1*, 52–56.
7. Liu, G.; Zhang, F.Y.; Li, X.Z.; Yang, Z.C. Research on large deformation and its mechanism of Muzhailing tunnel. *Chin. J. Rock Mech. Eng.* **2005**, *24* (Suppl. S2), 5521–5526.
8. Qing, S.H.; Huang, R.Q. Study on the large deformation characteristics of the soft rocks in Wushaoling tunnel. *Mod. Tunn. Technol.* **2005**, *42*, 7–14.
9. Zhao, S.Q.; Lin, A.N.; Yan, X.J. Large deformation analysis and control for vertically-oriented slate in Hadapu Tunnel. *Mod. Tunn. Technol.* **2011**, *48*, 5.
10. Song, S.W.; Feng, X.M.; Liao, C.G.; Cai, D.W.; Liu, Z.X.; Yang, Y.H. Measures for controlling large deformations of underground caverns under high in-situ stress condition—A case study of Jinping I hydropower station. *J. Rock Mech. Geotech. Eng.* **2019**, *8*, 605–618.
11. Tian, Y.; Shu, X.Y.; Tian, H.M.; He, L.K.; Jin, Y.; Huang, M. Effect of horizontal stress on the mesoscopic deformation and failure mechanism of layered surrounding rock masses in tunnels. *Eng. Fail. Anal.* **2023**, *148*, 107226.
12. Aydan, Ö.; Akagi, T.; Kawamoto, T. The squeezing potential of rock around tunnels: Theory and prediction with examples taken from Japan. *Rock Mech. Rock Eng.* **1996**, *29*, 125–143. [CrossRef]
13. Hoek, E. Big Tunnels in Bad Rock. *J. Geotech. Geoenviron. Eng.* **2001**, *127*, 726–740.
14. Yassaghi, A.; Salari-Rad, H. Squeezing rock conditions at an igneous contact zone in the Taloun tunnels, Tehran-Shomal freeway, Iran: A case study. *Int. J. Rock Mech. Min. Sci.* **2004**, *42*, 95–108. [CrossRef]
15. Wang, M.Y.; Zhang, N.; Li, J.; Ma, L.J.; Fan, P.X. Computational method of large deformation and its application in deep mining tunnel. *Tunn. Undergr. Space Technol.* **2015**, *50*, 47–53.
16. Bian, K.; Liu, J.; Liu, Z.P.; Liu, S.G.; Ai, F.; Zheng, X.Q.; Ni, S.H.; Zhang, W. Mechanisms of large deformation in soft rock tunnels: A case study of Huangjiazhai Tunnel. *Bull. Eng. Geol. Envinron.* **2019**, *78*, 431–444.
17. Yoshida, N.; Nishi, M.; Kitamura, M.; Adachi, T. Analysis of mudstone deterioration and its effect on tunnel performance. *Int. J. Rock Mech. Min. Sci.* **1997**, *34*, 353.e1–353.e19. [CrossRef]
18. Brox, D.; Hagedorn, H. Extreme deformation and damage during the construction of large tunnels. *Tunn. Undergr. Space Technol.* **1999**, *14*, 23–28. [CrossRef]
19. Wood, A.M.M. Tunnels for roads and motorways. *Q. J. Eng. Geol. Hydrogeol.* **1972**, *5*, 111–126. [CrossRef]
20. Chen, Z.J. The mechanical problems for the long-term stability of underground galleries. *Chin. J. Rock Mech. Eng.* **1982**, *1*, 1–20.
21. Aydan, O.; Akagi, T.; Kawamoto, T. The squeezing potential of rocks around tunnels: Theory and prediction. *Rock Mech. Rock Eng.* **1993**, *26*, 137–163.
22. He, M.C.; Jing, H.H.; Sun, X.M. *Engineering Mechanics of Soft Rock*; Beijing Science Press: Beijing, China, 2002.
23. Li, G.F.; He, M.C.; Zhang, G.F.; Tao, Z.G. Deformation mechanism and excavation process of large span intersection within deep soft rock roadway. *Min. Sci. Technol.* **2010**, *20*, 28–34. [CrossRef]
24. *GB50218-2014*; Standard for Engineering Classification of Rock Mass. China Planning Press: Beijing, China, 2014.
25. Dalgic, S. Tunneling in squeezing rock, the Bolu tunnel, Anatolian motorway, Turkey. *Eng. Geol.* **2002**, *67*, 73–96. [CrossRef]
26. Khanlari, G.; Meybodi, R.G.; Mokhtari, E. Engineering geological study of the second part of water supply Karaj to Tehran tunnel with emphasis on squeezing problems. *Eng. Geol.* **2012**, *145*, 9–17. [CrossRef]
27. Li, J.Q.; Wang, Z.F.; Wang, Y.Q.; Chang, H.T. Analysis and countermeasures of large deformation of deep-buried tunnel excavated in layered rock strata: A case study. *Eng. Fail. Anal.* **2023**, *146*, 107057.
28. Liu, D.X.; Chi, J.J.; Ding, X.L.; Huang, S.L.; Zhang, Y.T. Discussion on Lining Cracking Mechanism and Prevention Measures of A Water Conveyance Tunnel. *Mod. Tunn. Technol.* **2020**, *57*, 852–858.
29. Anagnostou, G.; Kovári, K. The New Austrian Tunnelling Method: A Review of Recent Advances and Applications. *Tunn. Undergr. Space Technol.* **2017**, *69*, 139–153.
30. *DZ/T0130-2018*; The Specification of Testing Quality Management for Geological Laboratories. China Planning Press: Beijing, China, 2018.
31. *SL/T264-2020*; Code for Rock Tests in Water and Hydropower Projects. China Water Power Press: Beijing, China, 2020.
32. Wang, T.T.; Huang, T.H. An experience of tunnelling in mudstone area in southwestern Taiwan. *Tunn. Undergr. Space Technol.* **2002**, *17*, 425–436. [CrossRef]
33. Fattah, M.Y.; Shlash, K.T.; Salim, N.M. Effect of Reduced Ko Zone on Time Dependent Analysis of Tunnels. *Adv. Civ. Eng.* **2011**, *2011*, 963502.
34. *GB50021-94*; Standard for Investigation of Geotechnical Engineering. China Planning Press: Beijing, China, 1994.

35. Hsiao, F.Y.; Wang, C.L.; Chern, J.C. Numerical simulation of rock deformation for support design in tunnel intersection area. *Tunn. Undergr. Space Technol.* **2009**, *24*, 14–21.
36. Chai, S.B.; Liu, H.; Song, L.; Li, X.P.; Fu, X.D.; Zhou, Y.Q. Static pressure and dynamic impact characteristics of filled jointed rock after frozen-thaw cycle damage. *Front. Ecol. Evol* **2023**, *11*, 1222676.
37. Yoshida, N.; Adachi, T. FE analysis of time-dependent instability of cut slopes in clay shale. *Int. J. Rock Mech. Min. Sci. Geomech. Abstr.* **1993**, *30*, A255.
38. Terzaghi, K. Stability of slopes in natural clay. In Proceedings of the 1st International Conference of Soil Mechanics and Foundations, Cambridge, MA, USA, 22–26 June 1936; pp. 161–165.

Disclaimer/Publisher's Note: The statements, opinions and data contained in all publications are solely those of the individual author(s) and contributor(s) and not of MDPI and/or the editor(s). MDPI and/or the editor(s) disclaim responsibility for any injury to people or property resulting from any ideas, methods, instructions or products referred to in the content.

buildings

MDPI

Article

Mechanical Behavior of Fully Grouted Rock Bolts in Hydraulic Tunnels Subjected to Elevated Ground Temperatures

Haibo Jiang [1], Shuangxi Li [1], Qinglin Li [2] and Juncai Xu [3,*]

[1] College of Hydraulic and Civil Engineering, Xinjiang Agriculture University, Urumqi 830052, China
[2] College of Water Conservancy & Architectural Engineering, Shihezi University, Shihezi 832003, China
[3] Department of Civil Engineering, Case Western Reserve University, Cleveland, OH 44106, USA
* Correspondence: juncai.xu@gmail.com

Abstract: In this study, the mechanical behavior of fully grouted rock bolts in hydraulic tunnels subjected to elevated ground temperatures was investigated. A differential equation for axial displacement of the rock bolt was formulated, which considers the force equilibrium of infinitesimal bolt segments and the stress transfer mechanism at the anchor–rock interface. The distribution functions for axial stress within the bolt and the interfacial shear stress were obtained by solving the differential equation, which incorporated the displacement of the surrounding rock mass as a parameter. This study showed that the effectiveness of the bolt–shotcrete support system decreases over time, considering the displacement relaxation rate of the surrounding rock mass. The mechanical model's variation laws at 20 °C, 50 °C, and 80 °C were summarized by integrating the thermal deformation equation for material parameters, and the numerical simulation results were compared and analyzed. The findings revealed that the bond strength between the rock bolt and the rock mass diminishes as the temperature of the surrounding rock increases, leading to a reduction of interfacial shear stress at both extremities of the bolt. Moreover, the maximum axial force within the bolt escalates as the neutral point migrates farther from the tunnel wall.

Keywords: rock bolts; hydraulic tunnels; elevated temperatures; interfacial shear stress; bond strength

1. Introduction

With the rapid advancement of large-scale construction projects, numerous deeply buried underground projects, such as mountain tunnels, hydropower projects, and diversion tunnels, have been executed. Nevertheless, in recent years, concerns surrounding rock mass stability and associated hazards have become increasingly significant [1–5]. The complex geological environment and elevated ground temperatures present substantial technical challenges for engineering construction [6,7]. High ground temperatures can induce thermal stress in tunnels, potentially damaging the lining and causing incomplete hydration of concrete, thereby reducing its strength [8,9]. Furthermore, elevated ground temperatures can alter the rock's microstructure and increase its porosity, leading to changes in its physical and mechanical properties [10–13]. As tunnel construction continues to expand globally, advanced deep tunnel support technologies have rapidly evolved as a new category of support technology that combines resistance with yield [14–18]. These technologies encompass advanced stress release techniques, such as yield stress control, graded support, and reinforcing support. It has been developed to address the challenges posed by complex geological environments and high ground temperatures in the construction of deeply buried underground projects. Deep tunnel support technology aims to enhance the stability of rock masses and prevent disasters by controlling the deformation of the surrounding rock and minimizing damage to the tunnel lining, thus ensuring the safety and reliability of underground engineering construction.

In order to meet the demands of deep tunnel support technology, numerous studies have focused on understanding and optimizing the mechanical behavior of rock bolts in various geological conditions. These efforts have led to the development and refinement of theoretical models, such as the neutral point theory, as well as innovative simulation methods and experimental approaches, all aimed at improving tunnel stability and safety in challenging environments. The neutral point theory for bolts, introduced by Freeman and Wang et al. [19,20], has gained considerable recognition and has progressively evolved into a comprehensive theoretical system [21–26]. Wen et al. utilized theoretical analysis to determine the distribution function for axial force and shear stress along the bolt's length, demonstrating that the selection of initial support time significantly influences the stability of the surrounding rock and the anchoring efficacy of the bolt [27]. Zhou et al. simulated fully grouted bolts using an implicit anchor column unit and derived the continuous displacement of the surrounding rock through a shear slip model for the anchorage interface. They employed interpolation fitting to obtain distribution functions for shear stress and axial force at the anchorage body interface [28]. Liu et al. applied the finite difference method to discern the distribution laws for axial force within the anchorage body and interfacial shear stress along the bolt's length [29]. Additionally, Liu revealed the anchoring mechanism under the interaction between jointed rock masses and bolts, suggesting an optimization method for anchoring jointed rock masses [30]. Yuan et al. developed a novel simulation method for bolts within bolt-surrounding rock assemblies, reflecting the stress distribution characteristics of bolts along their axial length [31]. Zhao et al. formulated a hyperbolic tangent function model for interfacial shear stress and shear displacement in anchorage bodies, subsequently analyzing the distribution characteristics of axial force and interfacial shear stress in bolts [32]. Zhang Yiming et al. proposed various techniques for crack evolution and modeling in brittle materials, encompassing the Global Cracking Elements Method (GCEM) and the cracking elements method for dynamic brittle fracture [33–35]. Bolts effectively enhance rock–soil body strength and improve stability to control tunnel deformation. However, due to the complex forces in rock–soil media and diverse geological environments, force analysis and calculation design for bolts in specialized environments, such as high ground temperatures, lack corresponding theoretical support. Semi-theoretical and semi-empirical design methods often result in significant discrepancies between computational analysis models and actual engineering implementations [36–38].

There exists an urgent need to devise robust theoretical frameworks that are capable of supporting the analysis and design of rock bolts in specialized environments, such as elevated ground temperatures. The development of precise and dependable theoretical models is essential for bridging the gap between computational analysis and real-world engineering applications. The present study investigates the mechanical behavior of fully grouted rock bolts in hydraulic tunnels subjected to high ground temperature conditions. The main contributions of this work are as follows:

- The influence of bolt–shotcrete support timing on the anchorage effect was investigated.
- The impact of temperature change on the anchoring effect was studied.
- The axial force distribution and shear stress of the bolt along the bolt length under high ground temperatures were investigated.

The paper is structured as follows: Initially, drawing from existing research on fully grouted bolt theoretical models, the authors derive distribution functions for shear stress and axial force at the anchorage interface, incorporating the displacement of tunnel surrounding rock as a parameter. Subsequently, the impact of bolt–shotcrete support timing on the anchoring effect is thoroughly examined through the introduction of a displacement release rate. Next, the distribution of axial force and shear stress in bolts along their length under high ground temperature conditions is analyzed, accounting for the effects of temperature changes on the physical properties of the surrounding rock and anchorage system. Ultimately, the theoretical validity of the proposed model is corroborated through rigorous numerical simulation.

2. Analysis of Stress Distribution in Fully Grouted Bolts

This investigation focuses on a 4.1 km-long tunnel section with an elevated ground temperature within a hydraulic tunnel in Xinjiang. Constructed using drilling and blasting methods, the tunnel's rock mass consists primarily of mica quartz schist interbedded with graphite schist. The tunnel remains dry, exhibiting no groundwater leakage. The overlying strata thickness is approximately 560 m, and longitudinal ventilation is employed to reduce the temperature within the tunnel.

As tunnels undergo excavation, stress redistribution prompts the surrounding rock to displace toward the tunnel. Following the installation of bolts within the surrounding rock, deformation is constrained by the bolts. Deformation is significant near the cavity wall and gradually decreases with increasing distance until reaching a point of no deformation. Consequently, after anchoring, shear stress pointing toward the cavity wall is transmitted to the bolt through the bonding material between the surrounding rock and the bolt at one end near the cavity wall. In contrast, at the other end far from the cavity wall, the bolt experiences shear stress directed away from the cavity wall. The bolt body is subject to opposing shear stresses, and a neutral point with zero shear stress emerges at a specific location along the bolt. The bolt segment extending from the cavity wall to the neutral point is referred to as the drawing segment, whereas the portion from the neutral point to the far end of the cavity wall is known as the anchorage segment, as depicted in Figure 1.

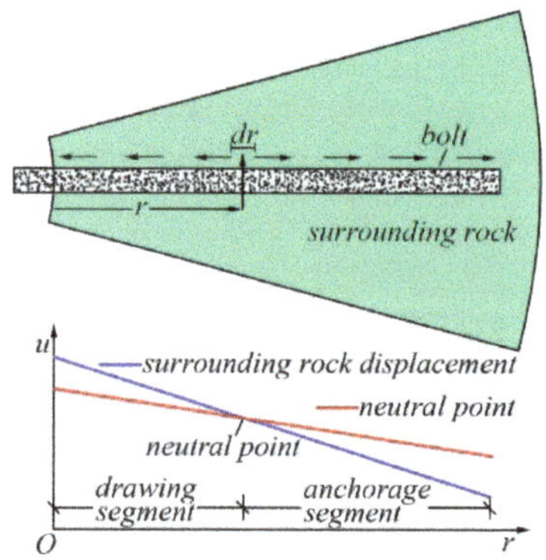

Figure 1. The force and displacement of the bolt in the rock mass.

2.1. Basic Assumptions

In the construction sequence and process of fully grouted rock bolts, holes are drilled into the cavity wall following tunnel excavation. Bonding materials, such as mortar or resin anchoring agents, are then injected before steel bars are implanted. Upon solidification of the bonding material, a connection is formed between the bolt and the surrounding rock, resulting in an integrated anchorage-bearing system.

Theoretical analysis of this anchorage process and the force transmission characteristics of the system necessitates several assumptions:

(1) The anchorage body, comprising the bolt and bonding material, remains elastic under external loads, and only experiences axial deformation in the bolt.

(2) The bolt and bonding material maintain a complete bond without damage, such as slip failure or cracking. Moreover, the thinness of the bonding material justifies disregarding changes in shear stress on its inner and outer sides.

2.2. Stress Distribution of the Fully Grouted Bolt

Based on the above assumptions, an anchoring micro-element (dr), was obtained in order to study the interaction between the bolt and the surrounding rock, as shown in Figure 2.

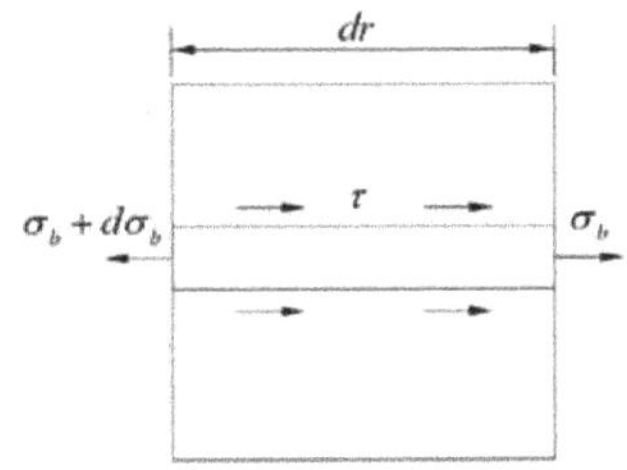

Figure 2. Anchoring micro-element.

The deformation caused by tunnel excavation generates deformation shear stress (τ) on the anchorage interface. By maintaining force balance, the following equation was derived:

$$dN(r) = Ad\sigma_b = -\pi D\tau dr \tag{1}$$

where *N(r)* is the axial force at the corresponding micro-element of the bolt, *N*; *A* is the cross-sectional area of the bolt, m^2; *D* is the diameter of the bolt, *m*; and σ_b is the axial stress of the bolt, MPa.

Assuming that the axial displacement of the bolt is u_b, directed towards the cavern, the axial stress distribution of the micro-element of the bolt can be expressed as:

$$\sigma_b = -E_b \cdot \varepsilon_b = -E_b \frac{du_b}{dr} \tag{2}$$

where E_b is the elastic modulus of the bolt, and ε_b is the axial strain of the micro-element of the bolt.

Substituting Equation (1) into Equation (2) yields the following expression of displacement and interfacial shear stress of the bolt:

$$\frac{d^2u_b}{dr^2} = -\frac{4}{E_bD}\tau \tag{3}$$

Considering the assumption that the change in shear stress between the inner and outer sides of the bonding material is negligible, the shear stress in the anchorage layer of the bolt can be obtained as:

$$\tau = \frac{G_n}{h_n}(u_s - u_b) \tag{4}$$

where G_n is the shear modulus of the bonding material, GPa; h_n is the thickness of the bonding material, *m*; and u_s is the radial displacement of the rock and soil surrounding the bolt after anchoring support.

Combining Equations (3) and (4), the differential equation of the axial displacement of the fully grouted bolt can be obtained as:

$$u_b'' - a^2u_b + a^2u_s = 0,\ a^2 = \frac{4G_n}{E_bDh_n} \tag{5}$$

Upon solving the aforementioned differential equation, the following equation is obtained:

$$u_b = C_1e^{ar} + C_2e^{-ar} - \frac{ae^{ar}}{2}\int e^{-ar}u_sdr + \frac{ae^{-ar}}{2}\int e^{ar}u_sdr \tag{6}$$

In order to derive the expression for the axial displacement of the bolt, it is essential to determine the radial displacement of the surrounding rock in the anchoring area. By substituting the expressions for the radial displacement of the surrounding rock in different zones, the axial displacement of the bolt can be calculated. Following tunnel excavation, the surrounding rock typically comprises plastic, elastic, and undisturbed zones, sequentially from the cavity wall. According to the Drucker–Prager criterion, the radius of the plastic zone post-tunnel excavation is given by [39]:

$$R_p = r_0 [\frac{(P + k/3\alpha)(1 - 3\alpha)}{P_1 + k/3\alpha}]^{\frac{1-3\alpha}{6\alpha}} \tag{7}$$

where r_0 is the tunnel excavation radius, m; P is the original rock stress, MPa; P_1 is the support resistance, MPa; and k and α are the material constants related to the cohesion (c) of surrounding rock and the internal friction angle (φ), respectively [40].

$$k = \frac{6c \cdot \cos\varphi}{\sqrt{3}(3 + \sin\varphi)}, \ \alpha = \frac{2\sin\varphi}{\sqrt{3}(3 + \sin\varphi)} \tag{8}$$

The boundary displacement of the elastic and plastic zones of the tunnel under anchoring support is [41]:

$$u_p = \frac{3\alpha(P + \mathrm{k}/3\alpha)R'_p}{2G} \tag{9}$$

Assuming that the volume of the surrounding rock in the plastic zone remains constant, the total displacement of the tunnel's surrounding rock influenced by anchoring support is:

$$u_0 = \frac{R'_p u_p}{r_0} = \frac{3\alpha(P + \mathrm{k}/3\alpha)R'^2_p}{2Gr_0} \tag{10}$$

Taking into account the space effect and construction sequence of the tunnel, the displacement of the tunnel wall after anchoring and support is:

$$u_2 = u_0 - u_1 \tag{11}$$

where u_1 is the released displacement of the surrounding rock before anchoring support, expressed by the displacement release rate λ, $\lambda \in [0, 1]$. Consequently, Equation (11) can be rewritten as:

$$u_2 = u_0 - \lambda u_0 = (1 - \lambda)u_0 \tag{12}$$

The support resistance of the shotcrete is [42]:

$$P_1 = k_c u_2, \ k_c = \frac{2G_c(2r_0h_c - h_c^2)}{r_0[(2 - 2v_c)r_0^2 - 2r_0h_c + h_c^2]} \tag{13}$$

where v_c and G_c are the Poisson's ratio and shear modulus of the shotcrete layer, respectively, and h_c is the thickness of the shotcrete. After anchoring support, the radial displacement (u_s) of the surrounding rock in the anchoring area is:

$$u_s = \frac{r_0 u_2}{r} \tag{14}$$

Substituting Equation (14) into Equation (6) yields the following equations:

$$u_b = C_1e^{ar} + C_2e^{-ar} - \frac{ar_0u_2e^{ar}}{2}\int \frac{e^{-ar}}{r}dr + \frac{ar_0u_2e^{-ar}}{2}\int \frac{e^{ar}}{r}dr \tag{15}$$

$$\frac{du_b}{dr} = C_1ae^{ar} - C_2ae^{-ar} - \frac{a^2r_0u_2e^{ar}}{2}\int \frac{e^{-ar}}{r}dr - \frac{a^2r_0u_2e^{-ar}}{2}\int \frac{e^{ar}}{r}dr \tag{16}$$

$$\int \frac{e^{-ar}}{r}dr = \ln|r| + \sum_{i=1}^{\infty} \frac{(-ar)^i}{i \cdot i!} \tag{17}$$

$$\int \frac{e^{ar}}{r}dr = \ln|r| + \sum_{i=1}^{\infty} \frac{(ar)^i}{i \cdot i!} \tag{18}$$

When the independent variable (r) is large, the solutions provided by Equations (17) and (18) produce large errors. In this paper, through continuous integration by parts and subsequent series expansion, more accurate results can be obtained by considering the first finite term in the calculation.

$$\int \frac{e^{-ar}}{r}dr = -\frac{e^{-ar}}{ar} - \int \frac{e^{-ar}}{ar^2}dr \approx -\frac{e^{-ar}}{ar}\sum_{i=0}^{\infty}\frac{(-1)^i i!}{(ar)^i} \tag{19}$$

$$\int \frac{e^{ar}}{r}dr = \frac{e^{ar}}{ar} - \int \frac{e^{ar}}{ar^2}dr \approx \frac{e^{ar}}{ar}\sum_{i=0}^{\infty}\frac{i!}{(ar)^i} \tag{20}$$

According to engineering practice, the axial stress at both ends of the bolt is zero.

$$\left.\frac{du_b}{dr}\right|_{r=r_0} = 0,\ \left.\frac{du_b}{dr}\right|_{r=r_0+l} = 0 \tag{21}$$

By substituting Equation (21) into Equation (16), C_1 and C_2 can be obtained as follows:

$$C_1 = \frac{e^{-ar_0}}{2(e^{2al}-1)}\left[\sum_{i=0}^{\infty}\frac{((-1)^i-1)i!}{a^i r_0^{i+1}} - e^{ar_0}\sum_{i=0}^{\infty}\frac{((-1)^i-1)i!}{a^i(r_0+l)^{i+1}}\right] \tag{22}$$

$$C_2 = \frac{e^{a(r_0+l)}}{2(e^{2al}-1)}\left[e^{ar_0}\sum_{i=0}^{\infty}\frac{((-1)^i-1)i!}{a^i r_0^{i+1}} - \sum_{i=0}^{\infty}\frac{((-1)^i-1)i!}{a^i(r_0+l)^{i+1}}\right] \tag{23}$$

By substituting the constants C_1 and C_2 into Equations (1) and (4), the distribution functions of the axial force and shear stress of the fully grouted bolt at room temperature can be obtained as:

$$N(r) = aAE_b\left[-C_1e^{ar} + C_2e^{-ar} - \frac{1}{2}\sum_{i=0}^{\infty}\frac{((-1)^i-1)i!}{a^i(r)^{i+1}}\right] \tag{24}$$

$$\tau(r) = -\frac{G_n}{h_n}\left[C_1e^{ar} + C_2e^{-ar} - \frac{r_0u_2}{r} + \frac{r_0u_2}{2}\sum_{i=0}^{\infty}\frac{((-1)^i+1)i!}{a^i(r)^{i+1}}\right] \tag{25}$$

2.3. Support Timing on Anchoring Effect and Bolt Stress

Determining the displacement release rate of the surrounding rock during different excavation steps in the numerical analysis of engineering excavation processes is a complex problem. It depends on factors such as the geological conditions, engineering excavation depth, supporting structure design, excavation method, supporting construction time, and auxiliary construction measures. A displacement release coefficient can be introduced to prevent unbalanced internal forces caused by excavation from being loaded onto an excavation step all at once. It is a simplified numerical analysis method that simulates unit changes sequentially across multiple construction phases. After tunnel excavation, the displacement generated when reaching secondary equilibrium without anchorage is called the total displacement (u) of the surrounding rock. Anchorage is applied before the surrounding rock reaches total displacement. To facilitate the analysis, the displacement release rate (λ) is introduced, and it represents the rate at which stored energy is released through displacement of materials.

Case 1: The tunnel had a radius (r_0) of 1.5 m and an initial geostress (P_0) of 6.49 MPa. The rock–soil body had an elastic modulus (E_r) of 7.1 GPa, a Poisson's ratio (ν_r) of 0.28, an internal friction angle (φ) of 37°, and cohesion (c) of 1.1 MPa. The shotcrete thickness (h_c) was 0.15 m, and C20 concrete was used with an elastic modulus (E_c) of 12 GPa and a Poisson's ratio (ν_c) of 0.167. The bolt had an elastic modulus (E_b) of 200 GPa, a diameter (D) of 250 mm, and a length (l) of 2 m, whereas the mortar anchoring layer had a thickness (h_n) of 10 mm. When the displacement release rate was at values such as 0.3, 0.5, or 0.7, theoretical calculations for the plastic zone radius and shotcrete resistance are shown in Table 1, whereas the shear stress and axial force on bolts are shown in Figures 3 and 4, respectively.

Table 1. Plastic zone radius and shotcrete resistance at displacement release rates of 0.3, 0.5, and 0.7.

Displacement Release Rate λ	Plastic Zone Radius/m	Shotcrete Resistance/MPa
0.3	3.09	3.24
0.5	3.26	2.60
0.7	3.53	1.81

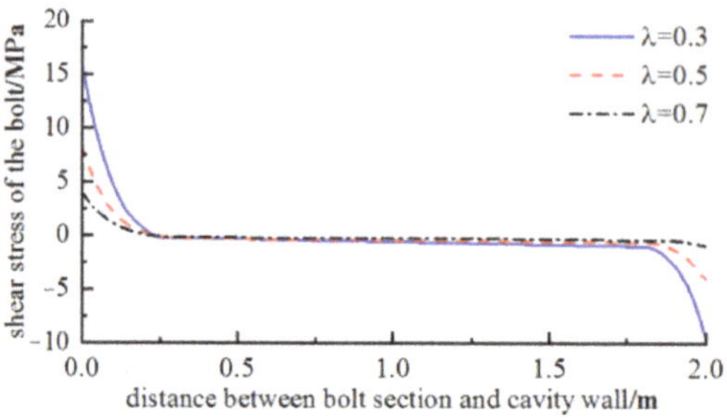

Figure 3. Curves of shear stress distribution at different displacement release rates (λ).

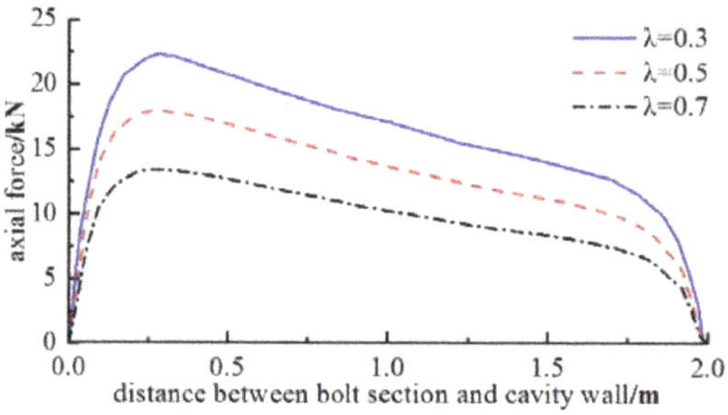

Figure 4. Axial force distribution at different release rates (λ).

Under different displacement release rates, both positive and negative shear stress values occur at the anchorage body interface. The segment with positive values is known as the drawing segment, with shear stress pointing towards the cavity wall. In contrast, the segment with negative values is the anchoring segment, with shear stress pointing towards the deep part of the surrounding rock. At the neutral point, where shear stress is zero, the axial force of the bolt is at its maximum.

According to Table 1, Figures 3 and 4, the plastic zone radius of the surrounding rock and the force borne by the anchoring structure vary with the timing of the bolt–shotcrete support. The study analyzed three different release rates, showing that when λ was 0.3, the axial force on the bolt and the shear stress on the interface between the bolt and surrounding rock were at their highest. At a distance of 0.28 m from the cavity wall, the maximum axial force on the bolt was 22.3 kN. The maximum shear stress at the cavity wall was 16.2 MPa, whereas the maximum shotcrete resistance was 3.24 MPa. The plastic zone had a radius of 3.09 m, the smallest observed in the study.

Conversely, when λ was 0.7, the axial force on the bolt and the shear stress on the interface between the bolt and surrounding rock were at their lowest. The maximum axial force was 13.1 kN, occurring at 0.25 m from the cavity wall, which was only 59% of that at $\lambda = 0.3$. The maximum shear stress was 4.5 MPa at the cavity wall. The shotcrete resistance was the lowest at 1.81 MPa, 44.1% less than that at $\lambda = 0.3$. The radius of the plastic zone at $\lambda = 0.7$ was the largest, measuring 3.53 m, 0.44 m more than that of $\lambda = 0.3$. These results suggest that a large displacement release rate diminishes the interaction between the bolt and the surrounding rock mass, leading to almost complete deformation of the surrounding rock. At this point, the bolt–shotcrete support becomes insignificant, and the surrounding rock bears greater deformation and load, making it more prone to loosening and damage. Choosing the appropriate time for bolt–shotcrete support after tunnel excavation ensures that the anchoring structure and surrounding rock form a common loading body. This can fully adjust and improve the surrounding rock's stress state, limit the plastic zone's expansion, and enhance the overall safety of the structure.

3. Effects of High Ground Temperature on the Mechanical Properties of Anchorage Systems

3.1. Introduction of the Thermal Deformation Equation of Material Parameters

The primary impact of high temperature on the anchoring system is the alteration of material mechanical properties in each component. Consequently, temperature changes generate thermal stress on both the surrounding rock and the bolt. The theoretical calculation primarily considers the temperature's effect on the materials' mechanical parameters. The current research summarizes the changing laws of mechanical properties of various materials with temperature through model tests and field monitoring. One reference fitted the empirical formula of the elastic modulus and Poisson's ratio of the surrounding rock up to 200 °C using experimental data [43]:

$$E_r{}^T = (1 - 1.33 \times 10^{-5}T - 4.22 \times 10^{-6}T^2 - 9.64 \times 10^{-9}T^3)E_r \quad (26)$$

$$\nu_r{}^T = (0.0023T + 0.94)\nu_r \quad (27)$$

where E_r and ν_r denote the elastic modulus and Poisson's ratio of surrounding rock at room temperature, and T is the temperature difference. As the temperature increases, the elastic modulus decreases gradually, whereas Poisson's ratio decreases first and then increases.

Reference measures of the value of the elastic modulus of concrete with temperature changes were obtained, and the fitted relationship between 20 °C and 600 °C is expressed as [44]:

$$E_c^T = (0.982 - 0.0014T)E_c \quad (28)$$

Reference studies of the mechanical properties of steel bars at high temperatures were utilized and gave the piecewise function of the elastic modulus of steel bars with changing temperature. The relationship within 370 °C can be expressed as [45]:

$$E_b^T = (1 - 4.86 \times 10^{-4}T)E_b \quad (29)$$

By substituting the above relations into Equations (24) and (25), the distribution equations of axial force and shear stress of the bolt along the bolt length at high temperatures can be obtained as follows:

$$N^T(r) = aAE_b^T\left[-C_1e^{ar} + C_2e^{-ar} - \frac{1}{2}\sum_{i=0}^{\infty}\frac{((-1)^i - 1)i!}{a^i(r)^{i+1}}\right] \quad (30)$$

$$\tau^T(r) = -\frac{G_n^T}{h_n}\left[C_1e^{ar} + C_2e^{-ar} - \frac{r_0u_2}{r} + \frac{r_0u_2}{2}\sum_{i=0}^{\infty}\frac{((-1)^i + 1)i!}{a^i(r)^{i+1}}\right] \quad (31)$$

In Case 1, the displacement release rate was set to $\lambda = 0.5$. The mechanical parameters at various temperatures were utilized for the theoretical calculations. Table 2 displays the theoretical calculation results of the plastic zone radius and the shotcrete's support counterforce. Figures 5 and 6 illustrate the theoretical calculation of the bolt's shear stress and axial force, respectively.

Table 2. Theoretical values of plastic zone radius and shotcrete resistance at 20 °C, 50 °C, and 80 °C.

Temperature/°C	Plastic Zone Radius/m	Shotcrete Resistance/MPa
20	3.26	2.60
50	3.29	2.54
80	3.30	2.51

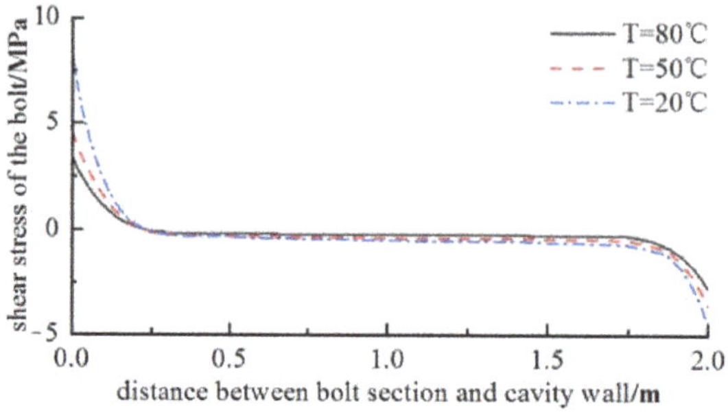

Figure 5. Curves of Shear Stress Distribution at Different Temperatures.

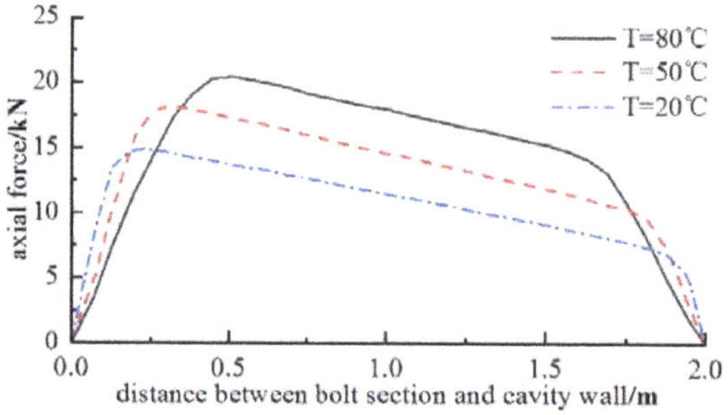

Figure 6. Curves of axial force distribution at different temperatures.

According to Table 2, Figures 5 and 6, the plastic zone radius of the surrounding rock enlarged, whereas the shotcrete resistance decreased with increasing temperature. This result indicates a decrease in the elastic modulus of the surrounding rock with increasing temperature. Specifically, the lower the elastic modulus, the weaker the rock, and the greater the deformation of the surrounding rock. Under such conditions, the surrounding rock bears more load, making it more susceptible to loosening and damage, resulting in a weaker anchoring effect of the shotcrete.

At a temperature of 20 °C, the shear stress of the anchorage body interface reached its highest value, peaking at 8.1 MPa, whereas the maximum axial force of the bolt was 17.9 kN, which occurred 0.28 m from the cavity wall. In contrast, at 80 °C, the shear stress reached its minimum, measuring only 3.4 MPa, which is only 42% of that at 20 °C. The bolt experienced a maximum axial force of 22.3 kN, 25% greater than the value at 20 °C, and the force occurred at a distance of 0.34 m from the cavity wall. As the temperature increases, the bonding ability between the bolt and the surrounding rock weakens. Consequently, the maximum shear stress value at the cavity wall decreases, the position of the neutral point moves away from the cavity wall, the length of the drawing segment increases, and the axial force of the bolt increases. Therefore, high-temperature environments require bolts with greater strength.

3.2. Model Validation

In order to validate the theoretical analysis, a comparative analysis was conducted using ANSYS numerical software. After tunnel excavation, the surrounding rock's displacement and stress increased as it approached the cavern, and the influence range was three to five times the cavern diameter. Considering that the bolt will also disturb the surrounding rock during support, the calculation range was set to three times the length of the bolt plus the cavern diameter (35 m × 35 m). Displacement in the X direction constrained the model's left and right boundaries, whereas displacement in the Y direction constrained the lower boundary. The surrounding rock adopted the elastic–plastic constitutive model and the D–P yield criterion, and the other material parameters remained the same as in Case 1.

Figure 7 illustrates the model and the finite element model's geometrical dimensions. During meshing, the thermal simulation unit of the surrounding rock used plane55 and plane42 as structural calculation units, beam3 element for the shotcrete layer, and link1 element for the bolt. The bolt and surrounding rock had good bonding, and a common node model connected them. The simulation model divided a total of 3550 units. The indirect coupling method and the "element birth and death" technology simulated the excavation process during the calculation. The axial force of the bolt located at the top of the arch was analyzed, and its values were extracted at intervals of 0.25 m. Figure 8 displays the simulated bolt axial force value and the difference between the theory and simulation.

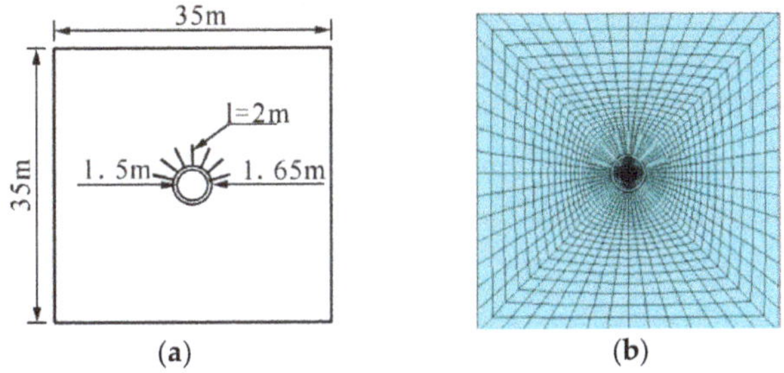

Figure 7. Simulation model: (**a**) geometric model; (**b**) finite element model.

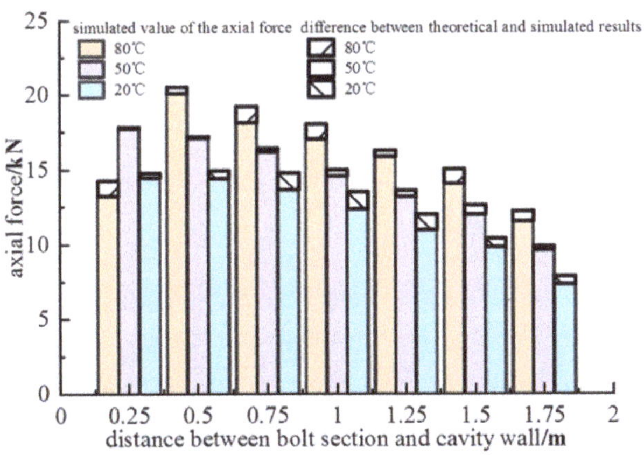

Figure 8. The simulated value of the axial force of the bolt.

At simulated temperatures of 20 °C and 50 °C, the maximum axial force of the bolt occurred at 0.25 m, with values of 17.74 kN and 14.47 kN, respectively. When the temperature reached 80 °C, the maximum axial force was 20.09 kN, occurring at 0.5 m. As the simulated temperature increased, the drawing segment of the bolt lengthened, and the neutral point moved away from the cavity wall, which is consistent with the results obtained from the theoretical analysis. These findings validate the computational model's accuracy in this paper. Furthermore, the simulated axial force values of the bolt align with the theoretical values, with a maximum difference of 7.7%, which is within an acceptable range. The difference between the two values may be due to the fact that the theoretical analysis only considered the effect of temperature change on the physical properties of the surrounding rock and anchoring structure. Conversely, the numerical simulation accounts for the impact of temperature stress caused by temperature change. However, temperature stress minimized the bolt's stress at normal and low temperatures, but significantly impacted the stress at high temperatures. Additionally, temperature stress is a secondary influencing factor under different temperature conditions. Figure 9 displays the plastic strain paths, whereas Figure 10 illustrates the plastic strain variation in different paths.

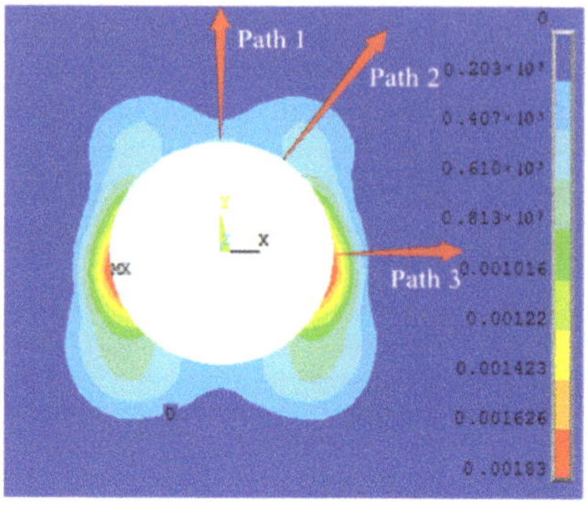

Figure 9. Paths of plastic strain in the plotting contour.

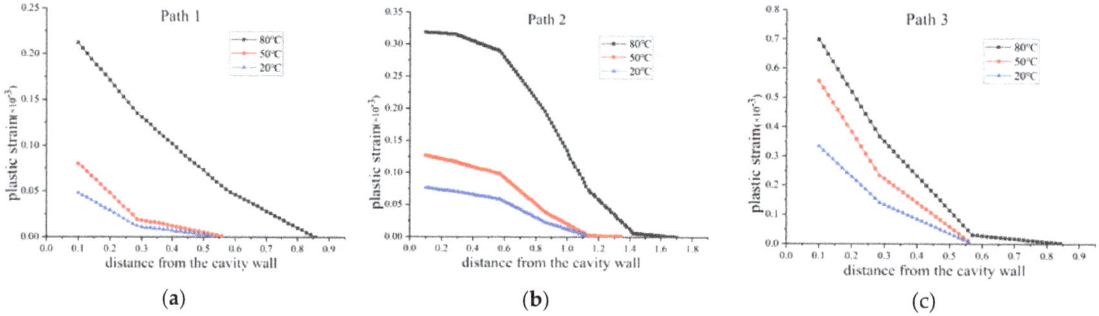

Figure 10. Paths of plastic strain in the plotting contour.

Based on Figure 10, the plastic strain at a particular location increases as the temperature of the surrounding rock increases, leading to a larger plastic zone. Additionally, the plastic strain is inversely proportional to the distance from the cavity wall. The plastic strain growth and changes in the plastic zone on Path 1 and Path 2 were significantly higher within a temperature range of 50 °C to 80 °C than within a temperature range of 20 °C to 50 °C. This suggests that the plastic zone at the vault and spinner is more sensitive to temperature changes at medium to high temperatures than at normal and low temperatures. The plastic strain variation with temperature on Path 3 indicates that the plastic strain at the arch waist increases proportionally with an increase in temperature, implying a certain regularity in the growth of the plastic strain at the arch waist. Further information can be found in Figure 9. The simulation results are consistent with the theoretical analysis, demonstrating the computational model's accuracy. However, the theoretical and simulated values differ due to the theoretical analysis only considering the effect of temperature change on the physical properties of the surrounding rock and anchoring structure. In contrast, the numerical simulation also considers the effect of temperature stress caused by temperature change.

4. Conclusions

In this study, the investigation focused on the distribution of shear stress and axial force on fully grouted rock bolts at the anchorage interface, which was attributed to deformation of the surrounding rock. The primary parameter considered in this research was the displacement of the tunnel's surrounding rock. Furthermore, the study explored the effects of the displacement release rate on the anchorage efficacy of bolt–shotcrete support timing, along with the impact of temperature variations on the physical properties of the surrounding rock and anchorage system. This was achieved by analyzing the axial force and shear stress distribution along the bolt length under elevated ground temperatures. The theoretical findings were substantiated by numerical simulations. The key conclusions drawn from this study include:

1. Lower displacement release rates result in more effective bolt–shotcrete support. With reduced displacement release rates, bolt–shotcrete support can fully adapt and improve the stress state of the surrounding rock, restricting plastic zone expansion in the surrounding rock and ensuring comprehensive structural stability.
2. In high-temperature environments, the bond strength between bolts and rock mass weakens as the temperature of the surrounding rock increases. This results in small interfacial shear stress and greater maximum axial force on bolts. At 80 °C, the maximum axial force increased by 25% compared to normal temperatures, and the neutral point moved away from the cavity wall.
3. The proposed theoretical model was validated through numerical simulations, which demonstrated that temperature fluctuations primarily affect the distribution of bolt axial force by impacting the anchoring system's pertinent material parameters. Additionally, temperature stress exerts a more significant influence on bolt stress at elevated temperatures than at normal temperatures.
4. This study establishes a mechanical model of fully grouted bolts in tunnels under high geothermal conditions, investigates the mechanical properties of fully grouted bolts in hydraulic tunnels under such environments, derives the distribution function of stress and interfacial shear stress of bolts, and verifies theoretical soundness using numerical calculations. However, a notable gap remains between computational analysis models and real-world engineering scenarios. To bridge this gap, the development of a more precise and dependable theoretical model is essential.

Although this paper primarily addresses deep, high-geothermal hydraulic tunnel engineering, the findings can also be applied to other underground projects facing high-geothermal challenges, such as high-temperature tunnels and deep geothermal mining. The research outcomes hold significance for analogous engineering endeavors. Future research should focus on refining the theoretical model presented in this study in order to better capture the intricate relationships between temperature, displacement, and the material properties in anchorage systems. This can be achieved through the development of advanced numerical simulations, model validation via experimental investigations, and the exploration of novel methodologies for optimizing the design and functionality of anchoring systems in geotechnical engineering. Expanding upon the conclusions of this study, additional research directions can be identified to further advance understandings of the mechanical behavior of fully grouted rock bolts under diverse conditions. To complement the theoretical and numerical analyses, it would be beneficial to conduct experimental studies on fully grouted rock bolts within controlled environments. This would aid in verifying the accuracy and reliability of the established models. Such investigations should concentrate on recreating various ground temperature and displacement scenarios and evaluating the performance of different rock bolt materials and designs.

Author Contributions: H.J. and S.L.: conceptualization, methodology, software, and validation; S.L., Q.L. and J.X.: formal analysis, investigation, data curation, and writing—original draft preparation; J.X. and H.J.: conceptualization, writing—review and editing, visualization, and funding acquisition. All authors have read and agreed to the published version of the manuscript.

Funding: This work was financially supported by the National Natural Science Foundation of China (Grant No. 51769031, 52179130) and the Regional Innovation Guidance Plan project of the XPCC (Grant No. 2021BB004).

Institutional Review Board Statement: Not applicable.

Informed Consent Statement: Not applicable.

Data Availability Statement: The data presented in this study are available on request from the corresponding author.

Acknowledgments: We are grateful to two anonymous reviewers for their constructive comments.

Conflicts of Interest: The authors declare no conflict of interest. The funders had no role in the design of the study; in the collection, analyses, or interpretation of data; in the writing of the manuscript; or in the decision to publish the results.

References

1. Yu, Y.; Chen, B.; Wang, F.; Wang, J.; Ke, D. Zonal disintegration of surrounding rock in deep underground cave based on force analysis of rock bolts. *Chin. J. Rock Mech. Eng.* **2017**, *37*, 1629–1640. [CrossRef]
2. Wang, H.; Li, Y.; Ni, Q.; Utili, S.; Jiang, M.; Liu, F. Analytical solutions for the construction of deeply buried circular tunnels with two liners in rheological rock. *Rock Mech. Rock Eng.* **2013**, *46*, 1481–1498. [CrossRef]
3. Xu, J.; Ren, Q.; Shen, Z. Sensitivity analysis of the influencing factors of slope stability based on LS-SVM. *Geomech. Eng.* **2017**, *13*, 447–458. [CrossRef]
4. Sakhno, I.; Liashok, I.; Svitlana, S.; Oleksandr, I. Method for controlling the floor heave in mine roadways of underground coal mines. *Min. Miner. Depos.* **2022**, *16*, 1–10. [CrossRef]
5. Sakhno, I.; Isayenkov, O.; Rodzin, S. Local reinforcing of footing supported in the destroyed rock massif. *Min. Miner. Depos.* **2017**, *11*, 9–16. [CrossRef]
6. Cao, G.; Liu, H. Complex Geological Conditions and Detection Technology of Tunnel. In *Three-Dimensional Exploration Technology of Tunnel Geology*; Springer: Berlin/Heidelberg, Germany, 2022; pp. 11–28.
7. Jiang, H.; Sun, H.; Shi, K.; Xu, J. Stability Analysis of the Surrounding Rock-Lining Structure in Deep-Buried Hydraulic Tunnels Having Seepage Effect. *Sustainability* **2022**, *14*, 16586. [CrossRef]
8. Xu, D.; Zhang, B.; Zubin, A.; Bu, X.; Pan, H.; Chen, S. Spatial-temporal evolution principle of temperature field in a high-temperature geothermal highway tunnel. *Ain Shams Eng. J.* **2023**, *14*, 101965. [CrossRef]
9. Kang, F.; Li, Y.; Tang, C.; Li, T.; Wang, K. Numerical Study on Thermal Damage Behavior and Heat Insulation Protection in a High-Temperature Tunnel. *Appl. Sci.* **2021**, *11*, 7010. [CrossRef]
10. Lee, Y.-K.; Pietruszczak, S. A new numerical procedure for elasto-plastic analysis of a circular opening excavated in a strain-softening rock mass. *Tunn. Undergr. Space Technol.* **2007**, *23*, 588–599. [CrossRef]
11. Borgia, A.; Oldenburg, C.M.; Zhang, R.; Pan, L.; Daley, T.M.; Finsterle, S.; Ramakrishnan, T. Simulations of CO_2 injection into fractures and faults for improving their geophysical characterization at EGS sites. *Geothermics* **2017**, *69*, 189–201. [CrossRef]
12. Yan, J.; He, C.; Wang, B.; Meng, W. Influence of high geotemperature on rockburst occurrence in tunnel. *Rock Soil Mech.* **2017**, *40*, 1543–1550. [CrossRef]
13. Gudala, M.; Govindarajan, S.K.; Yan, B.; Sun, S. Numerical investigations of the PUGA geothermal reservoir with multistage hydraulic fractures and well patterns using fully coupled thermo-hydro-geomechanical modeling. *Energy* **2022**, *253*, 124173. [CrossRef]
14. Barla, G.; Bonini, M.; Semeraro, M. Performance monitoring and analysis of a yield-control support system in squeezing rock. In *Rock Mechanics in Civil and Environmental Engineering*; CRC Press: Boca Raton, FL, USA, 2010. [CrossRef]
15. Moritz, B. Yielding elements—Requirements, overview and comparison/Stauchelemente—Anforderungen, berblick und Vergleich. *Geomech. Tunn.* **2011**, *4*, 221–236. [CrossRef]
16. Li, C.C.; Doucet, C. Performance of D-Bolts Under Dynamic Loading. *Rock Mech. Rock Eng.* **2012**, *45*, 193–204. [CrossRef]
17. Krykovskyi, O.; Krykovska, V.; Skipochka, S. Interaction of rock-bolt supports while weak rock reinforcing by means of injection rock bolts. *Min. Miner. Depos.* **2021**, *15*, 8–14. [CrossRef]
18. Liu, Y.; Xia, C.; Wu, F.; Xu, C.; Deng, Y. A combined support technology of long and short bolts of soft rock tunnels under high ground stresses. *Chin. J. Rock Mech. Eng.* **2019**, *39*, 105–114. [CrossRef]
19. Freeman, T.J. Behavior of Fully Bonded Rock Bolts in the Kielder Experimental Tunnel. *Tunn. Tunn.* **1978**, *10*, 37–40. [CrossRef]
20. Wang, M.; He, X.; Zhang, Y. Mechanical model of full-length anchored bolt and its application. *Met. Mine* **1983**, *4*, 24–29.

21. Li, C. Study on the loading and deformation of tunnel segments in soft clay with consideration for the soil mass rheological characteristics. *Geotech. Geol. Eng.* **2018**, *37*, 673–682. [CrossRef]
22. Kargar, A.R. An analytical solution for circular tunnels excavated in rock masses exhibiting viscous elastic-plastic behavior. *Int. J. Rock Mech. Min. Sci.* **2019**, *124*, 104128. [CrossRef]
23. Wu, F.; Wang, Z.; Xia, C.; Deng, Y.; Liu, Y. Analytical design method for yielding pipe length of yielding bolt. *Tunn. Constr.* **2019**, *39*, 119–124. [CrossRef]
24. Tao, L.; Hongbin, C.; Yonggang, W.; Jian, Z.; Minnian, Z. Analysis on the influence of the lateral pressure coefficient on the bolt in the tunnel system with high ground stress. *Chin. J. Undergr. Space Eng.* **2020**, *16*, 437–441.
25. Zhao, L. Study on stress distribution and influencing factors of grouting anchor in tunnel surrounding rock. *J. Hebei Univ. Eng.* **2021**, *38*, 63–68. [CrossRef]
26. Li, P.; Huang, J.; Chen, K.; Tong, L. Statistical analysis on temporal and spatial characteristics of the axial force of anchor bolts in tunneling. *Mod. Tunn. Technol.* **2021**, *58*, 227–236. [CrossRef]
27. Wen, J.; Zhang, Y.; Wang, C. Study of mechanical model of fully grouted rock bolt's anchorage interface in tunnel surrounding rock. *Rock Soil Mech.* **2022**, *34*, 1645–1651. [CrossRef]
28. Hao, Z.; Ming, X.; Chen, J. Study of anchoring mechanism and analysis of anchoring effect of fully grouted rock anchor in large-scale underground caverns. *Rock Soil Mech.* **2016**, *37*, 1503–1511.
29. Liu, G.; Xiao, M.; Chen, J.; Zhou, H. Stress analysis method of fully grouted rock bolt in underground caverns. *J. Huazhong Univ. Sci. Technol.* **2017**, *45*, 113–116. [CrossRef]
30. Liu, C.; Li, Y. Research progress in bolting mechanism and theories of fully grouted bolts in jointed rock masses. *Chin. J. Rock Mech. Eng.* **2018**, *37*, 1856–1872. [CrossRef]
31. Yuan, Y.-H.; Xiao, M.; Chen, J.-T. A method for simulating stress distribution along fully grouted anchor. *Rock Soil Mech.* **2018**, *39*, 1908. [CrossRef]
32. Zhao, B.; Jin, J.; Wang, G.; Minghua, H.; Xin, T. Nonlinear Analysis on Mechanical Behavior of Fully Grouted Bolt in Tunnels. *Met. Mine* **2022**, *51*, 85. [CrossRef]
33. Zhang, Y.; Zhuang, X. Cracking elements method for dynamic brittle fracture. *Theor. Appl. Fract. Mech.* **2019**, *102*, 1–9. [CrossRef]
34. Zhang, Y.; Mang, H.A. Global cracking elements: A novel tool for Galerkin-based approaches simulating quasi-brittle fracture. *Int. J. Numer. Methods Eng.* **2020**, *121*, 2462–2480. [CrossRef]
35. Zhang, Y.; Huang, J.; Yuan, Y.; Mang, H.A. Cracking elements method with a dissipation-based arc-length approach. *Finite Elem. Anal. Des.* **2021**, *195*, 103573. [CrossRef]
36. Owaid, K.; Hamdoon, A.; Matti, R.; Saleh, M.; Abdelzaher, M. Waste Polymer and Lubricating Oil Used as Asphalt Rheological Modifiers. *Materials* **2022**, *15*, 3744. [CrossRef]
37. Abdelzaher, M.A.; Farahat, E.M.; Abdel-Ghafar, H.M.; Balboul, B.A.A.; Awad, M.M. Environmental Policy to Develop a Conceptual Design for the Water–Energy–Food Nexus: A Case Study in Wadi-Dara on the Red Sea Coast, Egypt. *Water* **2023**, *15*, 780. [CrossRef]
38. Benjeddou, O.; Ravindran, G.; Abdelzaher, M.A. Thermal and Acoustic Features of Lightweight Concrete Based on Marble Wastes and Expanded Perlite Aggregate. *Buildings* **2023**, *13*, 992. [CrossRef]
39. Sun, Z.-P.; Gao, Z.-N.; Meng, X.-R. Permeability Coefficient of Damage Zone of Surrounding Rock Based on Drucker-Prager Criterion. *J. Yangtze River Sci. Res. Inst.* **2013**, *30*, 26. [CrossRef]
40. Peng, C.; Bohou, X. Matching method of DP yield criterions to MC based on value of internal frictional angle. *Chin. Q. Mech.* **2012**, *33*, 269–274. [CrossRef]
41. Kun, W.; Guangmin, Z.; Xiangrui, M. The plastic analaysis of surrounding rock of roadway based on Drucker-Prager yield criterions of MC yield criterion. *Safe Coal Mines* **2013**, *44*, 67–70. [CrossRef]
42. Fu, C.; Zhou, H.; Chen, S. Equivalent mechanical model of joined rockmass reinforced by shotcrete lining and its application. *Rock Soil Mech.* **2009**, *30*, 1967–1973. [CrossRef]
43. Zhang, G. Analysis of bonding mechanism of wholly grouted anchor under high ground temperature environment. Master's Thesis, Southwest Jiaotong University, Chengdu, China, 2018.
44. Qin, L.; Song, Y.; Wang, Y.; Zhang, Z.; Yu, C. Testing research of mechanics characteristics of concrete affected by high temperature. *Concrete* **2004**, *5*, 9–11.
45. Ma, Y. Thermal and Mechanical Properties of Steel and Concrete under High Temperature. *Sci. Technol. West China* **2011**, *10*, 37–39. [CrossRef]

Disclaimer/Publisher's Note: The statements, opinions and data contained in all publications are solely those of the individual author(s) and contributor(s) and not of MDPI and/or the editor(s). MDPI and/or the editor(s) disclaim responsibility for any injury to people or property resulting from any ideas, methods, instructions or products referred to in the content.

buildings

MDPI

Article

Model Test and Numerical Simulation of Two Typical Close-Fitting Pile–Wall Integrated Structures in Deep Excavation

Changjiang Wu [1,2], Wuqin Shen [2,*], Ying Xu [3,*] and Guowei Wei [4]

1 School of Mechanics and Civil Engineering, China University of Mining & Technology, Xuzhou 221116, China; wucj2008@ntu.edu.cn
2 School of Transportation and Civil Engineering, Nantong University, Nantong 226019, China
3 School of Mines, China University of Mining & Technology, Xuzhou 221116, China
4 Jiangsu Zhongnan Construction Industry Group Co., Ltd., Nantong 226100, China
* Correspondence: 2118310011@stmail.ntu.edu.cn (W.S.); 15921360161@163.com (Y.X.)

Abstract: Compared to conventional support methods, the close-fitting pile–wall integration technique features a minimized construction spacing between the retaining pile and the basement retaining wall. This approach leverages the pile stiffness to minimize the wall thickness and enhance underground space utilization. However, it currently lacks significant discussions and measured data about the interaction laws between the pile and the wall. The model test and finite element method (FEM) are employed to study the deformation and internal force interaction laws of two typical close-fitting pile–wall integrated structures, and a comparison with conventional design is conducted. Furthermore, this study separately investigates the impact of sensitivity factors, specifically the pile–wall stiffness ratio and floor plate stiffness, on both structures during the basement construction and serviceability stages. The test results can closely match the numerical simulation. The study results reveal that the wall impacts the bending moment of the pile to some extent. The internal force in the wall is significantly influenced by the lateral deformation of the pile and the floor plate. Compared to conventional designs, this structure significantly reduces the bending moment of the wall, particularly in the composite structure. Additionally, the analysis of sensitivity factors reveals their considerable influence on the pile–wall interaction.

Keywords: pile–wall integrated structure; model test; numerical simulation; deep excavation; retaining structure

1. Introduction

Shanghai, China's largest international economic center, is situated in a typical soft soil region [1]. With the swift development of the urban economy, large-scale development of underground spaces has been brought to the forefront, leading to numerous deep excavation projects [2,3]. In soft soil areas with high underground water levels, excavation support projects typically employ a large number of bored piles as temporary retaining structures. In conventional support methods, the retaining piles are regarded merely as temporary support during the excavation stage. The external unit pressure is considered to directly act on the wall without considering the supportive role of the retaining piles during the serviceability stage, resulting in considerable waste [4].

Pile–wall integration technology combines the basement wall and the retaining pile. By fully utilizing the bending resistance of the pile and reducing the space between the pile–wall, this approach significantly reduces the basement wall thickness [5], expands the basement area, and offers economic and energy-saving benefits [6]. However, the current pile–wall integration technical standards are relatively crude, distributing the load borne by the pile and wall solely based on their relative bending stiffness without considering their combined interaction. This technique requires high verticality and quality of the retaining piles and involves more complex construction processes. Moreover, existing monitoring data for excavations primarily focus on the excavation stage, lacking sufficient monitoring data and research on the pile's durability [7] during the basement construction and serviceability stage, as well as adequate practical case support. Consequently, this technology is generally adopted passively only when space within the retaining structure is insufficient. Therefore, "pile-wall integration" remains a technology that is not widely implemented in deep excavation practice, with a scarcity of experimental data and related research support. Understanding the characteristics and interaction mechanisms of pile–wall integration is of significant importance.

Many scholars have investigated the behavior and response of support structures in excavation projects through model tests. Their efforts to enhance the stability and safety of deep excavations primarily focus on designing retaining structure parameters, optimizing support methods, analyzing deformation development patterns, and analyzing the stress distribution [8–13]. However, in complex model tests, current analytical research mainly focuses on the excavation stage, with little attention given to the serviceability stage. Model tests can be time-consuming and less repetitive, and numerical simulation offers a quick and convenient method. Many scholars have conducted comprehensive analyses using this method on the internal forces and deformation patterns, three-dimensional spatial effects, and the impact of external soil parameters and excavation depth on retaining structures such as secant piles, steel sheet piles, pile groups, and retaining walls. They are dedicated to optimizing the design and construction methods of retaining structures through simulation, thereby providing a thorough analysis of the geotechnical engineering construction process [14–22]. Furthermore, the combination of numerical simulation and model tests can complement each other. Model experiments are better suited to reflect engineering characteristics, while numerical simulation can analyze influencing parameters that are difficult to achieve in model experiments [23,24]. Combining both methods allows for more in-depth research on engineering characteristics [11,25–30]. Unfortunately, in current research, there is almost no discussion on the interaction mechanism between retaining piles and the retaining walls of the underground main structure.

The updated Shanghai "Technical Code for Excavation Engineering" (DG_TJ 08-61-2018) [31] recognizes the new design concept of "pile-wall integrated structure", which is classified as the close-fitting type and the force-transferring plate type (the separated type). The close-fitting type is usually divided into superimposed and composite structures. The close-fitting type, with a small gap between the piles and walls, achieves the stress transmission and interaction between pile and wall by employing concrete grouting in the gap after cutting hair on the retaining structure or after pre-burying connections in the piles. On the other hand, the separated type, with a bigger gap, achieves the interaction by setting several force-transferring plates between the pile wall, resulting in a lower stress transfer efficiency than the former. Up to now, there has been little research on the concept of pile–wall integration. Additionally, the interaction mechanism between the pile and wall is complex, so applying the previous excavation research achievements directly to the pile–wall integration engineering practice is hard.

In this paper, a study is conducted on the two typical close-fitting piled-wall integrated structures using FEM and model tests. Within the study, model tests specially designed for the pile–wall integrated structure are conducted to be benchmark data for the FEM simulation, and then models of superimposed and composite structures are established to analyze the deformation and the development law of internal forces of them. Moreover, the impact of sensitivity factors, including the pile–wall stiffness ratio and horizontal plate stiffness, is investigated during basement construction and serviceability stages. The research results are of significant reference value for the design and construction of close-fitting pile–wall integrated structures, especially in formulating excavation plans and optimizing structural schemes of this structure in the deep excavation.

2. Physical Model Test Results

2.1. Model Test Design

This study focuses on the two types of close-fitting pile–wall integrated structures. The profile diagram of both structures is depicted in Figure 1. Figure 1a depicts the superimposed structure (SS), which typically employs concrete grouting between the pile and wall after cutting hair on the excavation side of the retaining structure or after pre-burying connection reinforcement in the piles. It enables horizontal force, shear force, and bending moment transfer between pile and wall. Figure 1b depicts the composite structure (CS), where the gap between the piles and walls is filled with lining materials, enabling only horizontal force transfer. In cases where the lining material possesses a certain thickness or a soft interlayer, the force-transferring plate must be added to the position of the floor plate to ensure proper load transfer.

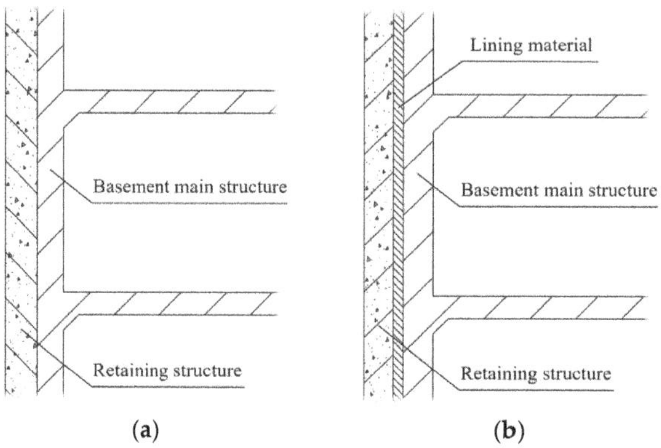

Figure 1. Two types of close-fitting pile–wall integrated structures (SS). (**a**) Superimposed structure, and (**b**) composite structure (CS).

The model test simulated the basic cross-section of the prototype, and it will be the reference for the FEM numeric simulation. The schedule of grouting between pile and wall was adopted, and it makes the model pile and wall bind together, inevitably allowing bending moments and shear forces to be transmitted between them. This aligns with the force-transmission characteristics of the superimposed structure (SS) making it suitable for the schedule of the model test. The study of the composite structure (CS) is carried out after the finite element simulation result is compared with the test to verify the reasonableness.

The excavation depth is about 10 m. During the excavation stage, two inner supports are installed at elevations of −0.5 m and −6 m. The structural profile of the model test is depicted in Figure 2. The tests were designed to comprehensively investigate the entire process of excavation, including the deformation and development of internal forces in the pile and the wall of the basement construction and serviceability stage.

The test stages and corresponding practical engineering references are depicted in Figure 3. The design of Test Stage 8 (serviceability stage) is somehow different from the actual construction. During the serviceability stage, it is difficult to simulate the transition from active soil pressure during the excavation stage to static soil pressure. We used an extra load to the outer ground surface to simulate the variation of the earth's pressure.

The experiments consist of the following test stages: Test Stage 1: Initial excavation to the first support bottom. Test Stage 2: Installation of the first support and the second excavation to the second support bottom. Test Stage 3: Installation of the second support and the third excavation to the foundation plate bottom. Test Stage 4: Installation of the foundation plate and removal of the second support. Test Stage 5: Installation of the basement retaining wall, grouting the gap between the pile wall to the elevation of the B_1 plate, and subsequent installation of the B_1 plate after the grout solidifies. Test Stage

6: Removal of the first support. Test Stage 7: Grouting the gap between the pile wall to the elevation of the B_0 floor plate and subsequent installation of the B_0 floor plate after the grout solidifies. Test Stage 8: Gradual loading of the ground outside the excavation with intensity from 0.2 to 0.6 kPa (equivalent to 10~30 kPa in the prototype).

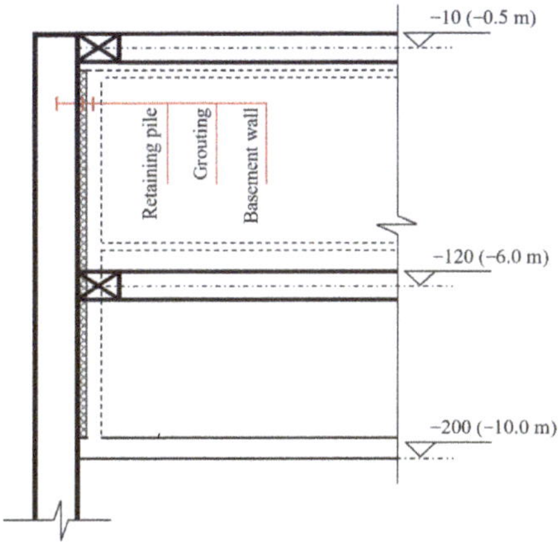

Figure 2. Section profile of underground structure for model test.

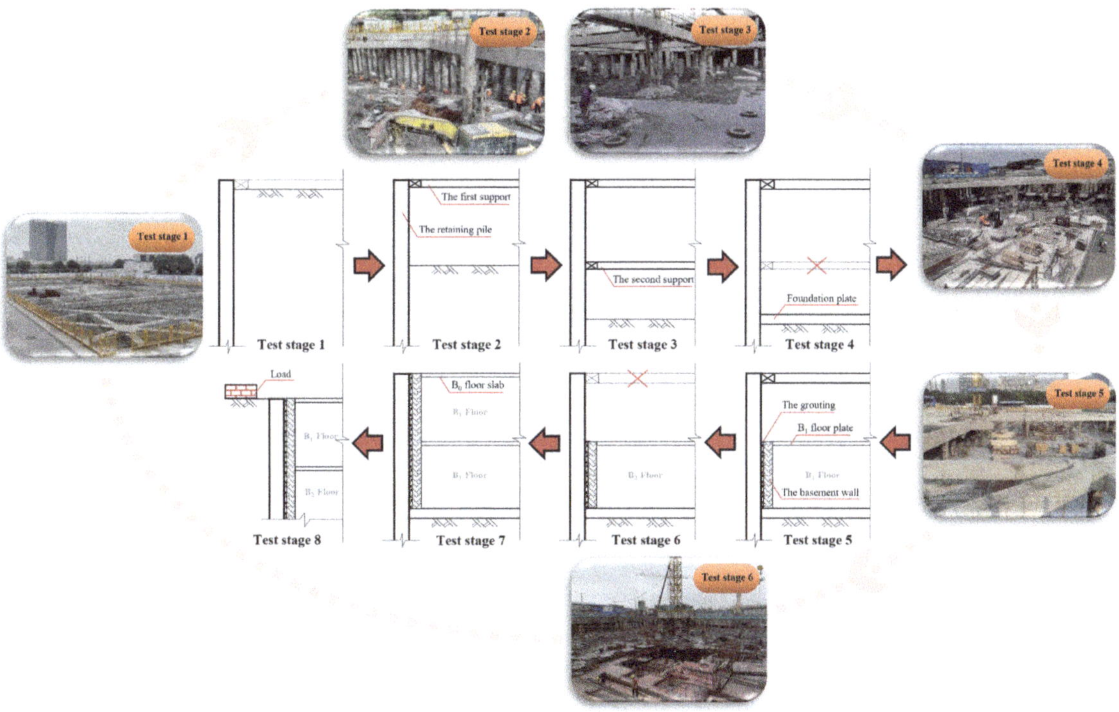

Figure 3. Test stages.

2.2. Design of Test Chamber and Selection of Similitude Ratio

The model test chamber is 1.0 m in length, 1.0 m in height, and 0.5 m in width. The test chamber is encased in 15 mm thick tempered glass panels with no top cover. The bottom plate of the model frame is a steel plate of 10 mm in thickness. On the upper section of a short side wall, six detachable stainless steel plates, measuring 500 × 50 × 10 mm, are used as a substitute for tempered glass. Furthermore, the exterior of the tempered glass panels is fortified with 5 mm thick angle steel to provide the model box with adequate rigidity, controlling chamber deformation and maintaining a planar strain state during testing. The schematic diagram of the model box can be seen in Figure 4. The similar model tests in this study were conducted under the gravity acceleration of 1 g. The geometric similitude ratio (n) of 50 is determined based on the modulus similarity relationship and the clearance size of the test chamber.

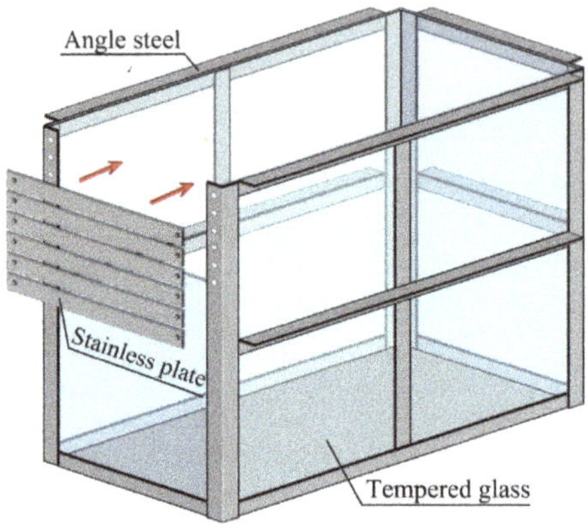

Figure 4. Test chamber.

2.3. Preparation of Similar Model Soil Materials

Orthogonal experiments are a common and effective method for configuring proper geotechnical materials [32–34]. In the model test, to embody the characteristics of the water-rich sand soil layer, four materials are selected as components. The fine sand and barite powder (1250 mesh) were employed as aggregates, while bentonite served as the binder, and laundry detergent (composed mainly of non-ionic surfactants) acted as an auxiliary material within the mixture. Based on these materials, a four-factor and four-level orthogonal table L16 (4^5) was designed to perform 16 sets of experiments. Subsequently, following the mixing and test soil sample preparation, direct shear and consolidation tests were conducted to measure the parameters of the model soil under each mixture ratio. Ultimately, the mass ratio of fine sand to barite powder to bentonite to water to laundry detergent was determined as 1:0.11:0.24:0.27:0.002. With this ratio, the physical and mechanical parameters of the similar model soil were $\gamma = 18.2$ kN·m^{-3}, $Es = 0.33$ MPa, c = 0.1 kPa, and $\varphi = 26.2°$, which matches the target parameters range presented in Table 1. The prototype parameters are based on the soil layers from an excavation project in Shanghai.

Table 1. Physical and mechanical parameters of prototype and model soil.

	Unit Weight γ (kN·m^{-3})	Water Content ω (%)	Compressive Modulus Es (MPa)	c (kPa)	φ (°)
Prototype	17.4~19.6	20~30	4~15	0~15	15~33
Expected	17.4~19.6	20~30	0.08~0.3	0~0.3	15~33
Model soil	18.2	20	0.33	0.1	26.2

2.4. Selection of Materials for Underground Structures

As the deep excavation retaining structure, the retaining pile mainly bears horizontal load and flexural deformation, so its bending deformation should be the main design objective in the 1 g model test. According to the similitude principle, the geometric similitude ratio n, and the stress similitude ratio N [8], the 1 g model test similitude relationship was obtained, as shown in Table 2.

Table 2. The 1 g model test similitude relationship.

Physical Quantity	Similitude Relationship	Prototype to the Model Ratio (for $N = n = 50$)
Acceleration	1	1
Linear dimension	n	50
Stress	$N = n$	50
Deformation	n	50
Bending stiffness	Nn^4	50^5
Moment	Nn^3	50^4

To choose appropriate materials, the specimen was placed on the multifunctional mechanical testing machine and a three-point bending test was conducted, as presented in Figure 5a. The load–displacement curve was obtained from the electronic test data generated during the three-point bending test, as depicted in Figure 5b.

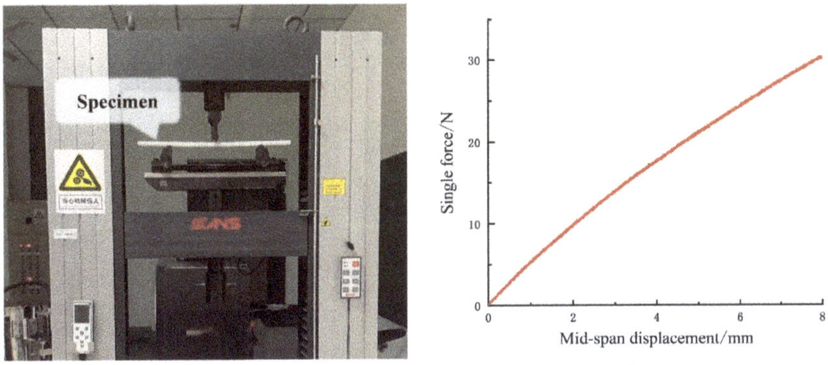

(a) **(b)**

Figure 5. Modulus of elasticity test. (**a**) Three-point flexural test, and (**b**) displacement versus load curve.

The elastic modulus E (MPa) can be determined by utilizing Formula (1), a material mechanics equation, which relates the concentrated load p (N) and the deflection f (mm) of a simply supported beam at the mid-span. This calculation requires values of the support distance l on the test bench, measured in millimeters (mm), and the sectional moment of inertia I of the tested material specimen, measured in square millimeters (mm^4).

$$f = \frac{pl^3}{48EI} \tag{1}$$

Table 3 presents the measured physical properties of three test materials, acrylic, PP, and PE. The elastic modulus of the PE (polyethylene) material closely matches the target value, and it is designated as a similar material in this experiment. However, PE, being a white, opaque material, exhibits inadequate incision smoothness during thin sheet processing. To ensure the accuracy of the grouting height and the better conjunction between the floor plate and the wall model, clear acrylic material is employed for the basement retaining wall. Furthermore, translucent PP (polypropylene) material is used for the floor plate. Tables 4 and 5 list the dimensions and mechanical properties of the various structural model, as well as their similitude ratios and relevant materials.

Table 3. Material parameters of the structural similarity model.

Material	Acrylic	PP	PE
Diameter (mm)	16	16	16
Elastic modulus E (GPa)	3.02	1.39	0.84

Table 4. Similitude ratio of bending structural model.

	Size (mm)	Length (mm)	Bending Stiffness E_sI (MPa·m^4)
Prototypical retaining pile	Φ 800	30,000	603.18
Model retaining pile	Φ 16	600	2.70×10^{-6} (PE)
Target similitude ratio	50	50	50^5
Practical similitude ratio	50	50	46.8^5
Prototypical basement wall	δ 500	10,000 (height)	312.5
Model basement wall	δ 6	160	1.09×10^{-6} (Acrylic)
Target similitude ratio	50	50	50^5
Practical similitude ratio	83	62.5	49.2^5

Φ means diameter; δ means thickness.

Table 5. Similitude ratio of the compressive structural model.

	Size (mm)	Length (mm)	Compressive Stiffness $EsAs{\cdot}L^{-1}$ (MPa·m^2·m^{-1})
Prototypical inner support	a 800	20,000 (section)	960
Model inner support	Φ 16	400	0.42 (PE)
Target similitude ratio	50^2	50	50^2
Practical similitude ratio	50^2	50	47.8^2
Prototypical foundation plate	δ 800	20,000 (section)	1200
Model foundation plate	δ 16	400	0.67 (PE)
Target similitude ratio	50	50	50^2
Practical similitude ratio	50	50	42.3^2
Prototypical floor plate	δ 200	20,000 (section)	300
Model floor plate	δ 2	400	0.17 (PP)
Target similitude ratio	50	50	50^2
Practical similitude ratio	100	50	46.5^2

Φ means diameter; δ means thickness; a means side length. The length of the prototype member refers to the basic cross-section of the prototype.

2.5. Instrumentation and Arrangement

The deformation of the retaining pile top is monitored utilizing a dial gauge displacement meter. Strain gauges are affixed to the surface of the pile body for monitoring purposes. Before installation, epoxy resin is applied to safeguard the surface of the strain gauges. The arrangement of strain monitoring points on the pile body is illustrated in Figure 6a, with two monitoring cross-sections. Similarly, the monitoring of internal forces on the external wall is also conducted by attaching strain gauges. Two monitoring points are designated, as presented in Figure 6b, with two monitoring cross-sections. The locations of the strain monitoring cross-section, along with an overview of the model box, are shown in Figure 6c.

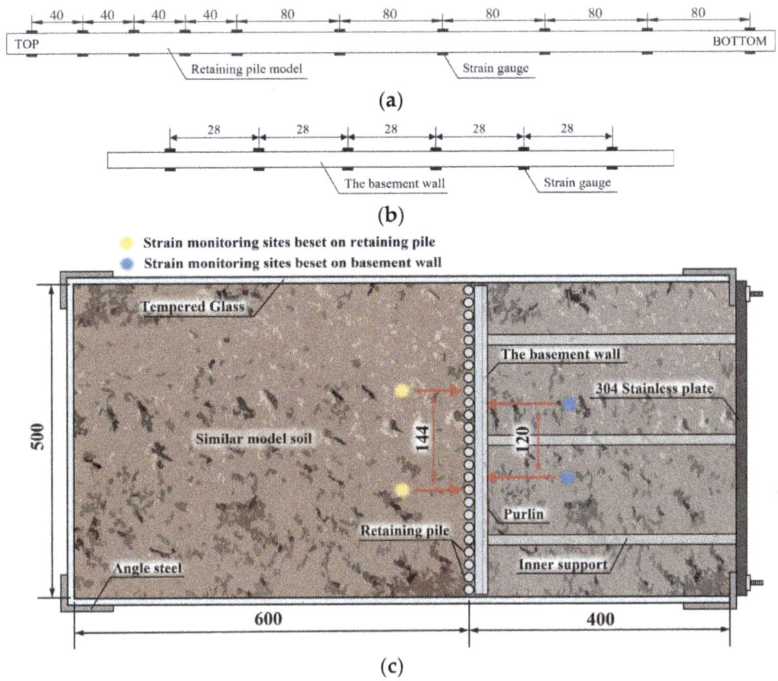

Figure 6. Scheme of strain monitoring points (unit: mm). (**a**) Arrangement of pile strain monitoring points, (**b**) arrangement of wall strain monitoring points, and (**c**) overview of strain monitoring sites.

2.6. Analysis of Strain Data

2.6.1. Deformation Data Inverse Analysis

In model tests, it is challenging to directly measure the lateral deformation of the pile body. The indirect derivation is possible by considering the relationship between the strain and deflection. The changes in deformation at different strain gauge point locations on the pile body are considered continuous variables. Alternatively, discrete variables can be employed to approximate these changes. The strain-tilt monitoring method suggested by Pei [35] can serve as a reference. The pile is considered a continuous beam with zero deformations and strain as boundary conditions at the bottom. The finite difference method is employed to calculate the lateral deformation of the retaining pile. Based on the strain gauge point locations, this method discretizes the retaining pile into multiple nodal elements. Based on the approximate differential equations for the beam deflection curves satisfied by the deflection of the retaining wall piles, the difference equations were developed in the following matrix form:

$$\frac{D}{d^2}\begin{bmatrix} 1 & -2 & 1 & 0 & \cdots & \cdots & 0 \\ 0 & \ddots & \ddots & \ddots & \ddots & \ddots & \vdots \\ \vdots & \ddots & \ddots & \ddots & \ddots & \ddots & \vdots \\ \vdots & \ddots & \ddots & \ddots & \ddots & \ddots & \vdots \\ \vdots & \ddots & \ddots & \ddots & \ddots & \ddots & 0 \\ 0 & \cdots & \cdots & 0 & 1 & -2 & 1 \end{bmatrix}\begin{bmatrix} \omega_0 \\ \omega_1 \\ \vdots \\ \vdots \\ \omega_n \\ \omega_{n+1} \end{bmatrix} = \begin{bmatrix} \varepsilon_1 \\ \varepsilon_2 \\ \vdots \\ \vdots \\ \vdots \\ \varepsilon_n \end{bmatrix} \tag{2}$$

where D is the diameter of the model's retaining pile and d is the distance between two adjacent strain measurement points (real or virtual). The parameter n is the number of strain measurement points, while the parameters ω and ε are the deflection and strain value of the strain measurement point, respectively. The virtual strain measurement points are introduced at the midpoint of sparsely distributed real points, which the average value of adjacent real points is taken as.

Based on the conditions of $\omega_0 = \omega_t$, $\omega_{n+1} = 0$, and the invertibility of the square matrix, the matrix can be rewritten as follows:

$$\begin{bmatrix} \omega_1 \\ \omega_2 \\ \vdots \\ \vdots \\ \vdots \\ \omega_n \end{bmatrix} = \frac{d^2}{D} \begin{bmatrix} -2 & 1 & 0 & \cdots & \cdots & 0 \\ 1 & -2 & 1 & 0 & \ddots & \vdots \\ \vdots & \ddots & \ddots & \ddots & \ddots & \vdots \\ \vdots & \ddots & \ddots & \ddots & \ddots & \vdots \\ \vdots & \ddots & \ddots & \ddots & \ddots & 0 \\ 0 & \cdots & \cdots & 0 & 1 & -2 \end{bmatrix}^{-1} \begin{bmatrix} \varepsilon_1 - \frac{D}{d^2}\omega_t \\ \varepsilon_2 \\ \vdots \\ \vdots \\ \varepsilon_n \end{bmatrix} \quad (3)$$

where ω_t is the pile-top deformation, which is measured. The equation above provides a direct calculation method for the lateral deformation of the strain measurement points on the pile model.

2.6.2. Bending Moment Data Inverse Analysis

The expression (Formula (4)) for the relationship between the bending moment (M) and the neutral section strain difference of the pile can be derived using the relationship between the curvature (κ) and the curvature radius (ρ). The same approach can also be applied to calculate the internal forces of exterior walls in underground structures.

$$M = EI \cdot \kappa = \frac{EI}{\rho} = \frac{(\varepsilon_1 - \varepsilon_2)}{D} EI \quad (4)$$

2.7. Main Steps of the Model Test

(1) The fine-grained sand, barite powder, and bentonite are weighed according to the proportions determined by the orthogonal experiment (Section 2.3). The materials are mixed evenly, and then the required amount of water and detergent is gradually added and stirred uniformly. The previously prepared model clay is placed in a bucket and sealed with a cover to prevent water evaporation, as depicted in Figure 7.

Figure 7. Preparation of model soil. (**a**) Production of model soil, and (**b**) preservation of model soil.

(2) As it is shown in Figure 8. After preparing the model soil, the sidewalls of the test chamber are lubricated with white Vaseline, and the model soil is filled to the elevation slightly above the pile bottom. The model of the retaining pile fitted with strain gauges is positioned and inserted. Then, the remaining model soil is layered and compacted up to the pile top. PE film is attached to the surface of the model soil to prevent water

evaporation. The model soil is allowed to settle for 1 to 2 days until it stabilizes. Then, the dial gauge deformation meter and connection of the strain gauges to the data-acquisition system (DH3818Y static stress–strain testing and analysis system) will be installed.

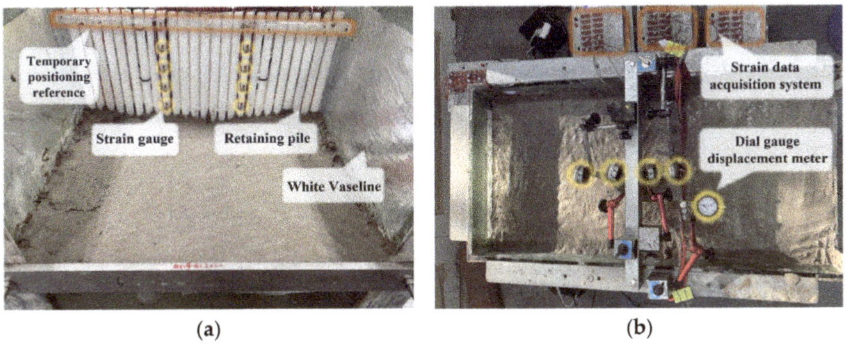

Figure 8. Model structure burial and monitor system installation. (**a**) Retaining pile model burial, and (**b**) monitoring system installation and commissioning.

(3) The data acquisition system is activated and the test process is initiated. We allow a period of 0.5 to 1.5 h before proceeding to the next test condition to ensure system stabilization. An overview of the test stages is shown in Figure 9.

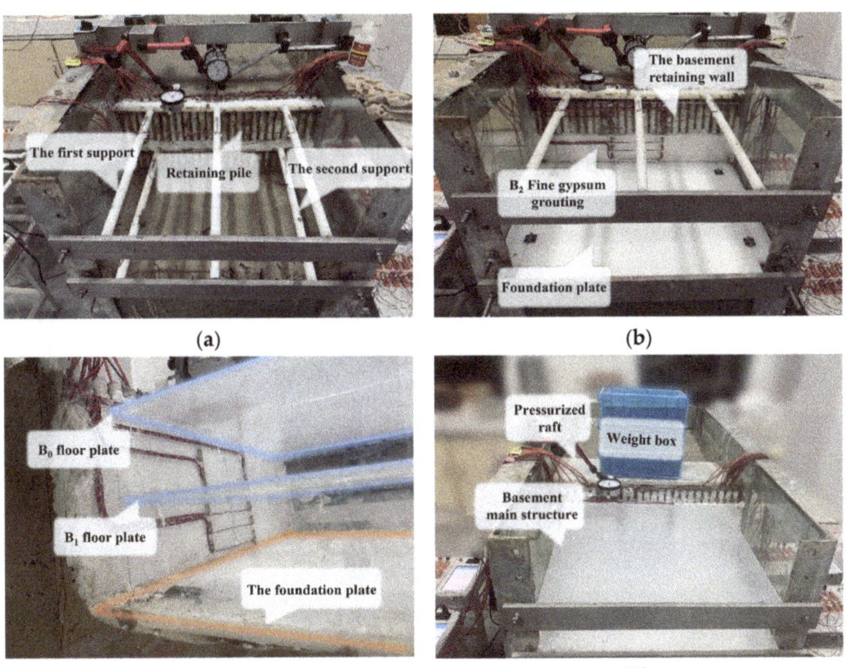

Figure 9. Overview of test stages. (**a**) Test Stage 3, (**b**) Test Stage 4, (**c**) Test Stage 7, and (**d**) Test Stage 8.

2.8. Model Test Results and Analysis

2.8.1. Analysis of Lateral Deformation in Retaining Piles

Figure 10 displays the lateral deformations, which are converted to the prototype according to the similitude relationship presented in Table 2. The following is evident in the diagram: (1) the pattern of lateral deformations of the retaining piles exhibits a characteristic "bulging" shape under various working conditions [4,36–38]; (2) the maximum pile lateral deformation is approximately 60 mm, which corresponds to 0.6% of the maximum excavation depth; (3) in Test Stage 1, the excavation deformation changes are minimal due to the shallow initial excavation depth; (4) in Test Stage 2, the pile's deformation significantly increases with the excavation depth, and the maximum lateral deformation occurs 3 m below the excavation face (−9 m elevation); (5) in Test Stage 3, as the excavation reaches the bottom, the maximum deformation section continues to move downward, approximately 3 m below the excavation face (−13 m elevation); (6) in Test Stage 4, after removing the second support, the lateral deformation of the pile body above the excavation face increases to some extent, while it decreases below the excavation face because of the support from the foundation plate; (7) in Test Stage 6, the removal of the first support causes a relatively bigger increase in the pile lateral deformation within the B_1 floor depth range, and a smaller increase within the B_2 floor depth range because of the formation of the B_2 floor main structure; (8) in Test Stage 8, the influence of the pile–wall integrated structure results in the pile lateral deformation remaining essentially unchanged, with a slight decrease within the B_2 floor depth range. Overall, it appears that the deformation of the pile develops very little because of the influence of the integrated structure during Stages 6 and 8.

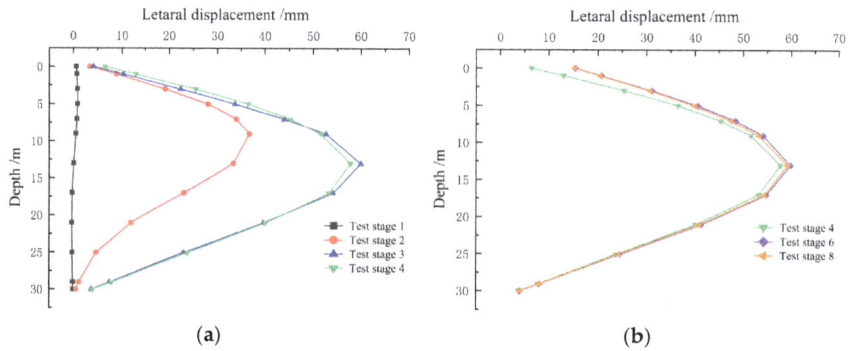

Figure 10. Lateral deformation of the pile (SS). (**a**) Lateral deformation of the pile in Test Stages 1 to 4, and (**b**) lateral deformation of the pile in Test Stages 4 to 8.

2.8.2. Analysis of Bending Moment in Retaining Piles

Figure 11 displays the pile's bending moments, which are converted to the prototype. The diagram illustrates the following: (1) The maximum pile's moment is about 770 kN·m, which occurs in Test Stage 3. (2) In Test Stage 1, the initial excavation reaches a shallow depth, and the pile body's internal force experiences minimal variation. (3) In Test Stage 2, with an increasing excavation depth, the pile's moment significantly increases. This reveals the law with an initial increase and subsequent decrease followed by an increase again, along with the depth. The contra-flexure point occurs at about 12 m under the excavation face (−18 elevation). (4) In Test Stage 3, as excavation to the excavation bottom, the position of the cross-section with the maximum moment continues to move downward, and the maximum value continues to increase. Compared with Test Stage 2, the pile body's moment above the depth of 9 m decreases. However, the pile's moment increases consistently below the depth of 9 m. (5) In Test Stage 4, the pile's moment above the depth of 9 m increases after the second support's removal. However, it decreases to some degree within the depth of 9 to 17 m because of the support of the foundation plate. (6) In Test Stage 6, within the floor B_2 depth range, because of the formation of the pile–wall integrated structure, the pile's moment above the excavation bottom decreases when the first support is removed. (7) In Test Stage 8, the formation of the pile–wall integrated structure is completed. With the load exacted outside the excavation, the pile body's moment slightly increases within the depth range of the B_1 floor. It slightly decreases adjacent to the foundation plate within the B_2 floor depth range due to the support provided by the foundation plate. Overall, as depicted in Figure 11, the section with the maximum moment depth is consistently located below the excavation face. Upon completion of the pile–wall integrated structure, the pile body's moment demonstrates a clear decreasing trend.

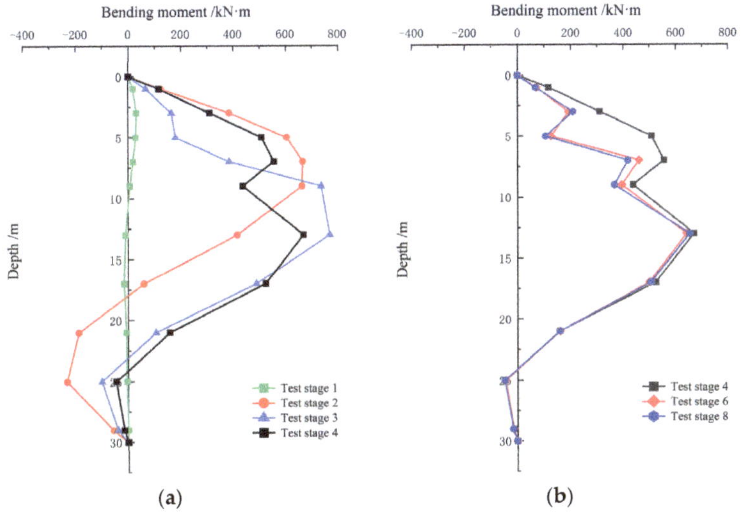

Figure 11. Bending moment of the pile (SS). (**a**) Excavation stage with Stage 4 and (**b**) basement construction stage.

2.8.3. Analysis of Internal Forces in Basement Retaining Wall

Figure 12 displays the basement retaining wall's bending moments converted to the prototype. The figure illustrates the following: (1) In Test Stage 6, the formation of the pile–wall integrated structure is only achieved within the B_2 floor depth range. Upon the removal of the first support, the retaining piles undergo rotation toward the excavation side around this location, prompted by the elastic support offered by the B_1 plate, resulting in the deformation of the basement retaining wall. At this stage, within the floor B_2 depth range, the exterior wall endures compression on the side of the excavation and tension on the outer side, leading to a significant negative moment. (2) In Test Stage 8, the pile–wall integrated structure is formed. When the load outside the excavation is conducted, the consequent active earth pressure rises. This leads to the gradual decrease in the B_2 floor wall negative moment and the increase in the B_1 floor wall positive moment.

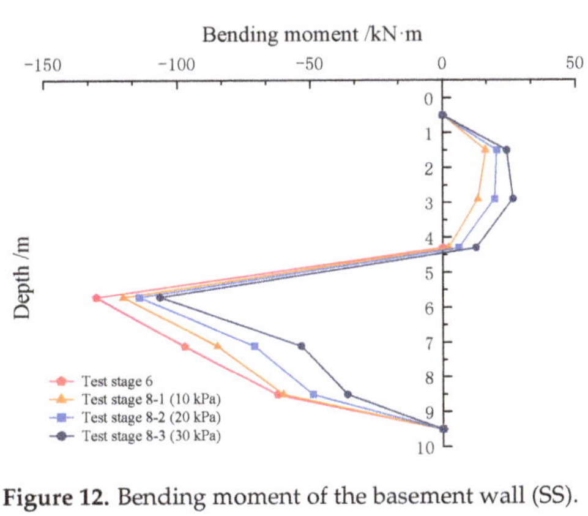

Figure 12. Bending moment of the basement wall (SS).

3. Numerical Simulation Results

3.1. Model Description and Details

Figure 13 presents the overview and calculation profile diagrams of the finite element model, respectively. The geotechnical finite element software Midas GTS NX (2022fea) was utilized for conducting numerical simulations in this study. The model has dimensions of 130 m in length, 130 m in width, and a height of 50 m. The excavation has an excavation area of 70 × 70 m and a depth of 10 m. The model is subjected to the following boundary conditions: the top surface is free, the four side surfaces have normal constraints, and the bottom surface is constrained in all directions. In total, the model is composed of 68,617 elements and 49,255 nodes.

The soil layer parameters refer to the same excavation project in Shanghai as the model test. The hardening soil model (HSS model) is taken as the soil constitutive model and is divided using a hexahedral mixed-mesh method. Tables 6 and 7 provide detailed soil layer information and calculation parameters.

Table 6. Soil layer parameters.

Soil Layer	Unit Weight γ (kN·m^{-3})	Compressive Modulus *Es* (MPa)	*c* (kPa)	φ (°)	μ
① Mucky silty clay	17.4	3.15	10	16.5	0.32
② Mucky clay	16.8	2.30	12	8.5	0.36
③$_1$ Silty clay	17.8	4.10	15	18.5	0.33
③$_2$ Sandy silt	18.2	11.82	4	33.5	0.32

Table 7. Soil layer parameters of the HSS constitutive model.

Soil Layer	E_{oed}^{ref}(MPa)	E_{50}^{ref} (MPa)	E_{ur}^{ref} (MPa)	*m*	G_0^{ref} (MPa)	$\gamma_{0.7}$
① Mucky silty clay	2.84	3.40	22.72	0.8	72.70	2.7×10^{-4}
② Mucky clay	2.41	2.89	19.30	0.8	54.03	2.7×10^{-4}
③$_1$ Silty clay	5.74	6.89	34.45	0.8	59.58	2.7×10^{-4}
③$_2$ Sandy silt	5.63	6.75	33.75	0.8	118.13	2.7×10^{-4}

The FEM model of the underground main structure is established as follows: the superimposed structure (abbreviated as SS) typically employs the pre-embedded reinforcement in the piles or the concrete grouting between the pile–wall after chiseling the excavation side of the retaining structure to make the pile–wall integrated structure. In this configuration, the piles and walls can transfer the horizontal force, shear force, and bending moment between them. Considering the load-transfer characteristic, the concrete grouting is modeled as a 3D elastic unit (with a thickness of 50 mm), as depicted in Figure 14a. The composite structure (CS) is filled with liner material between the pile and wall. Due to the liner material's inability to withstand tensile stress, a compression-only elastic link unit is used instead of a concrete grouting 3D unit, as depicted in Figure 14b.

The underground structures are constructed using C30 concrete material, with a unit weight of 25 kN·m^{-3}, an elastic modulus of 30 GPa, and a Poisson ratio of 0.2. The crown beams and internal supports employ 1D beam elements, whereas the basement retaining wall employs 2D elastic plate elements. Based on the principle of equivalent stiffness of the cross-section, the retaining piles (Φ800@1050) are converted into 2D elastic plate elements with a calculated equivalent thickness of 613 mm. The equation representing the equivalent transformation is as follows:

$$\frac{1}{12}(D+t)h^3 = \frac{1}{64}\pi D^4 \tag{5}$$

where *D* is the diameter of the pile (mm), *t* is the pile clearance (mm), and *h* is the equivalent plate thickness.

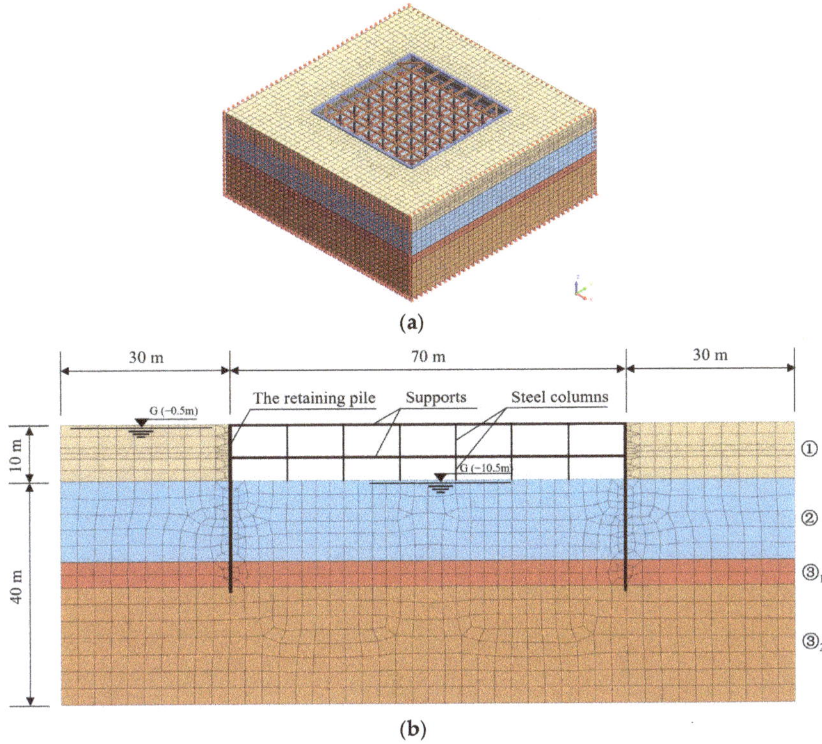

Figure 13. Finite element model. (**a**) Overview of meshing and (**b**) calculation profile diagram.

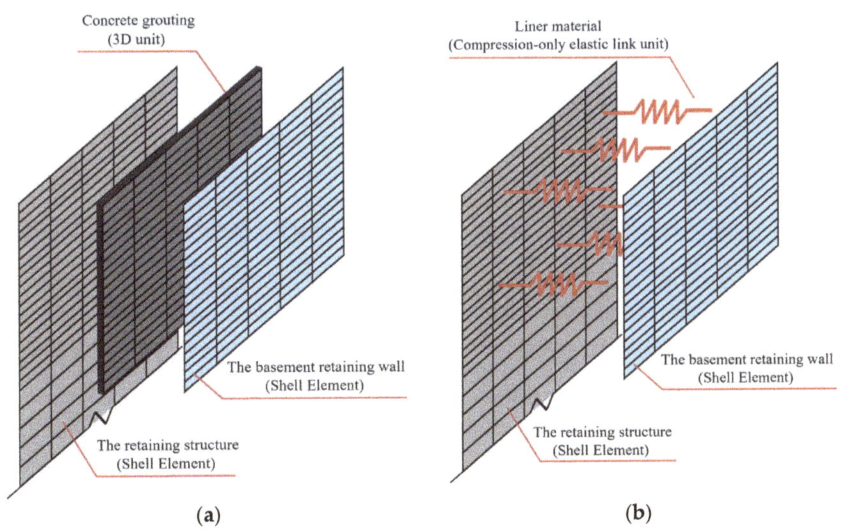

Figure 14. FEM modeling of the main underground structure. (**a**) Superimposed structure (SS) and (**b**) composite structure (CS).

Figure 15 presents the excavation process based on the model test in the FEM analysis. The analysis stages are as follows: Stage 0 involves blunting the retaining structures and balancing the earth's ground stresses, setting the groundwater level below the ground surface by −0.5 m. Stage 1 involves installing the first support, followed by excavation to the position of the second support. Stage 2 involves installing the second support and continuing the excavation to the bottom of the excavation. Stage 3 involves casting the foundation plate and removing the second support. Stage 4 involves installing the B_2 floor basement retaining walls and the floor plate of the B_1 floor. Stage 5 involves removing the first support. Stage 6 involves constructing the B_1 floor basement retaining walls and the floor plate of the B_0 floor. Stage 7 involves progressively overloading the excavation exterior from 10 to 30 kPa.

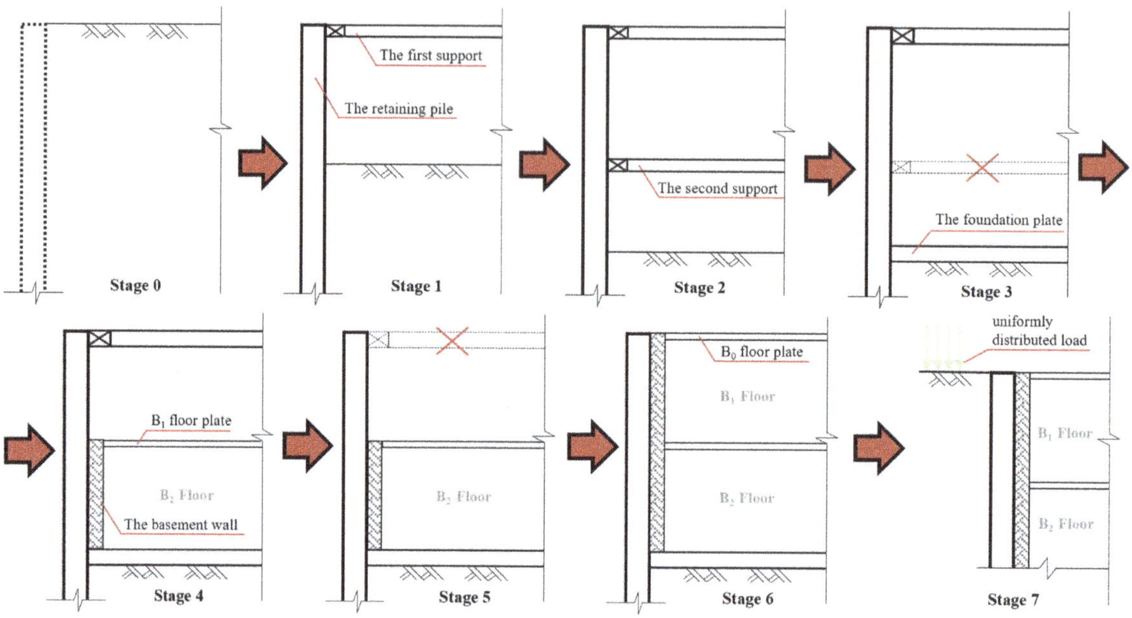

Figure 15. Scheme of FE simulation.

3.2. *Analysis of the Lateral Deformation of the Pile*

3.2.1. Excavation Stage

Figure 16a displays the lateral deformation of the piles at various excavation stages in the finite element simulation. Each incremental excavation depth is 2 m. It is discernible in the diagram that the presence of the first support restricts the pile top displacement during the excavation stages. The pile top displacement's development essentially halts at a depth of −6 m (at which point the second support is completed), followed by a rebound phenomenon occurring at an excavation depth of −10 m. This rebound exhibits a magnitude of approximately 30% relative to the maximum pile top displacement during the excavation stage. The pile's deformation forms a characteristic "bulging" shape, with the magnitude of deformation increasing as the excavation depth increases. The depth of the maximum deformation section descends in tandem with the excavation progress. It reaches approximately 3 m below the excavation face, matching the engineering laws of the soft soil region. In Stage 1, with the first excavation reaching down to −6 m, the maximum deformation is approximately 25 mm. In Stage 2, which includes further excavation to the bottom at the elevation of −10 m, the maximum deformation is around 62 mm, equivalent to 0.62% of the total excavation depth. The pile bottom displacement escalates in tandem with an increase in the excavation depth, while the development rate demonstrates relative constancy governed by the properties of the bearing stratum soil.

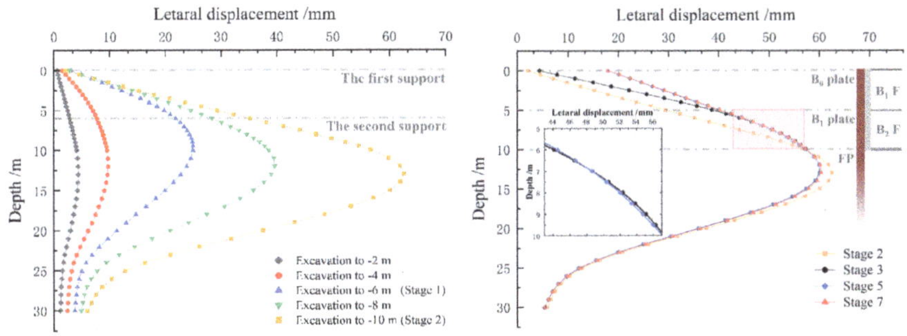

(a) (b)

Figure 16. Lateral deformation of the pile (SS). (**a**) Excavation stage and (**b**) basement construction stage.

3.2.2. Basement Construction Stage

Figure 16b displays the pile's lateral deformation during the basement construction in the finite element simulation. In Stage 3, removing the second support has a minimal impact on the pile's deformation. Due to the support from the foundation plate, the deformation above the excavation face increases, while it decreases beneath the excavation face. The resulting secondary deformation is approximately 15% of the maximum deformation throughout the excavation stage.

In Stage 5, removing the first support results in a significant top displacement of the pile. It is worth noting that, in conventional excavation support structures, the removal of the first support will lead to large deformation of the retaining structure above the excavation face [1,37–39]. However, within the pile–wall integrated structure, the maximum pile lateral deformation within the B_2 floor depth range changes very little. Because the B_2 floor main structure is completed at this time, additionally, the deformation of the pile body even exhibits a certain amount of rebound within this depth range. This is because of the elastic support imparted by the B_1 floor plate; the pile body (within the basement depth range) rotates toward the excavation side. However, since the B_2 floor has formed a pile–wall integrated structure, the amplitude of this rebound is relatively small.

In Stage 7, an external overload of 30 kPa is applied. By this stage, the basement construction has been completed, forming a fully integrated structure with the pile and wall. Therefore, the external overload has a negligible effect on the pile's lateral deformation. Due to the support provided by the floor plate and the foundation plate, the pile body deformation remains roughly unchanged. In general, there is a very slight overall deformation of the pile body towards the excavated area, accounting for approximately 0.5% of the previous stage deformation, which is attributed to the influence of the pile–wall integrated structure.

3.3. Comparative Analysis of Model Test and FEM Simulation Results

The preparation of the model test and the configuration of soil layers for the FEM simulation was conducted in adherence to an excavation project with distinctive soft soil conditions prevalent in Shanghai. Meanwhile, due to the relatively high water content of the model soil, observations and pore pressure measurements during the model tests indicate that the groundwater level consistently remains high. This aligns well with engineering conditions in high-water-level soft soil areas. The FEM simulation incorporates a higher groundwater level to match these conditions. This approach ensured a close resemblance between the model and the prototype's soil and underground water conditions, thereby enhancing the relevance and applicability of the ensuing analyses. Based on the similitude principle (refer to Table 2), measured results were converted to the prototype to verify the computed results as follows.

3.3.1. Comparison of Lateral Deformations of Retaining Pile

Figure 17a,b show the comparison of the pile lateral deformations in the excavation stage to the bottom (−10 m of elevation) and the basement serviceability stage, respectively. The diagram demonstrates that the pile lateral deformation distribution patterns from the model test (measured results) align closely with the FEM (computed results). In both cases, the maximum lateral deformation occurs around 3 m below the excavation face (−13 m of elevation). Moreover, the lateral deformation of the lower pile body retracts slower in the test results compared to the FEM results. This is because of the better properties of the bearing stratum from actual projects, while the sole stratum is taken in the test.

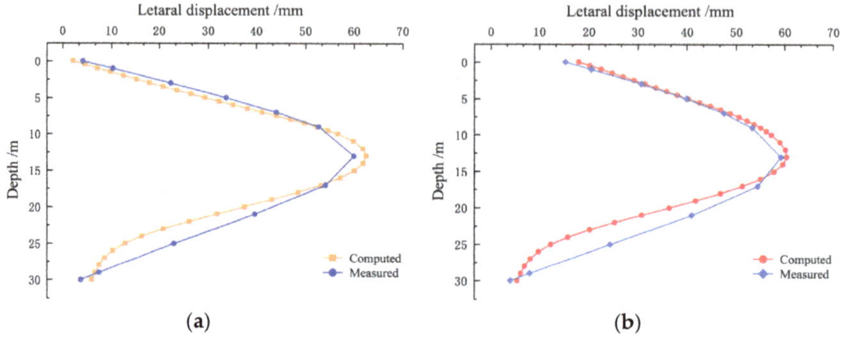

Figure 17. Lateral deformation of the pile comparison of the model test and FEM. (**a**) Excavation to −10 m and (**b**) basement serving stage (SS).

3.3.2. Comparison of Internal Forces in the Retaining Piles

Figure 18a,b show the comparison of the pile's bending moments in the excavation stage to the bottom and the basement serviceability stage, respectively. The diagram demonstrates that the pile's moment in the measured results exhibits a similar distribution pattern to the computed results. However, a notable difference exists in the moment distribution pattern of the bottom part of the pile. This can be attributed to the disparity in properties of the bearing stratum between the FEM and the model test.

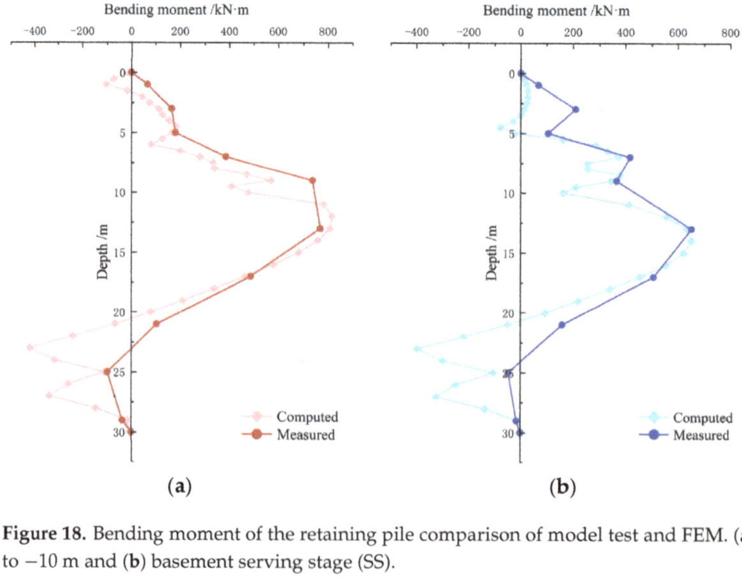

Figure 18. Bending moment of the retaining pile comparison of model test and FEM. (**a**) Excavation to −10 m and (**b**) basement serving stage (SS).

3.3.3. Comparison of Internal Forces in the Exterior Walls of the Basement

Figure 19 presents the comparison of the wall's bending moments in the basement serviceability stage. The diagram reveals a similar wall moment distribution pattern between the model test and FEM results. The load outside the excavation could lead to active earth pressure. So, in the upper part, there is a positive moment. The lower part experiences a significant negative moment due to the effect of the rotation and rebound tendency of piles. In both cases, the maximum moments occur around −6.5 m of elevation. However, there is still a slight discrepancy in the wall's moment upper and lower part distribution ratio. This discrepancy can be attributed to the size effects and errors in the test. Additionally, assembly errors prevent the grouting thickness between the pile and the wall from reaching the ideal thickness of around 1 mm. Despite controlling the gypsum grouting to a smaller thickness as much as possible, it still has an impact.

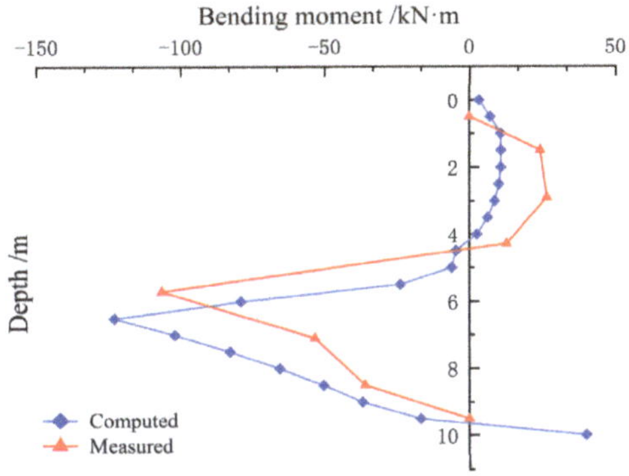

Figure 19. Bending moment of the wall comparison of model test and FEM (SS).

The deformation and internal force development of the close-fitting pile–wall integrated structure (SS) simulated by the model test exhibits strong consistency with the results of finite element simulation. Therefore, the model test and FEM numerical simulation conducted in this study are considered scientifically effective and reliable. The subsequent FEM simulation of the composite structure, which is based on the superimposed structure with slight modifications, introducing compressive-only elastic links between the pile and wall, should obtain the same evaluation.

3.4. Internal Force Analysis of Pile–Wall Integrated Structures

Figure 20 displays the bending moment in Stages 5 and 7 of the superimposed and composite structures. During basement construction and serviceability stages, the two structures' internal force is analyzed.

3.4.1. Commonality Analysis of the Two Structures

In engineering practice, the load pressure within the range of 2~12 m outside the excavation should not exceed 20 kPa. However, in this FEM simulation, a loading pressure of 30 kPa is selected to observe clearer patterns.

Regarding the effects of the overload on the pile–wall structure, there is general consistency between the two structures. The effect on the pile body below the excavation face is minimal, while that above the excavation face increases. The increase in the pile body's bending moment within the B_1 floor depth range is significantly greater than that within the B_2 floor. Because of the integrated structure, the wall and pile work together. The B_1 wall bears part of the active soil pressure generated by the overload, causing a tensile moment on the excavation side of the wall. On the other hand, the moment of the B_2 wall generally develops in the positive bending moment direction. However, the foundation plate's support causes an increase in the negative moment near its position.

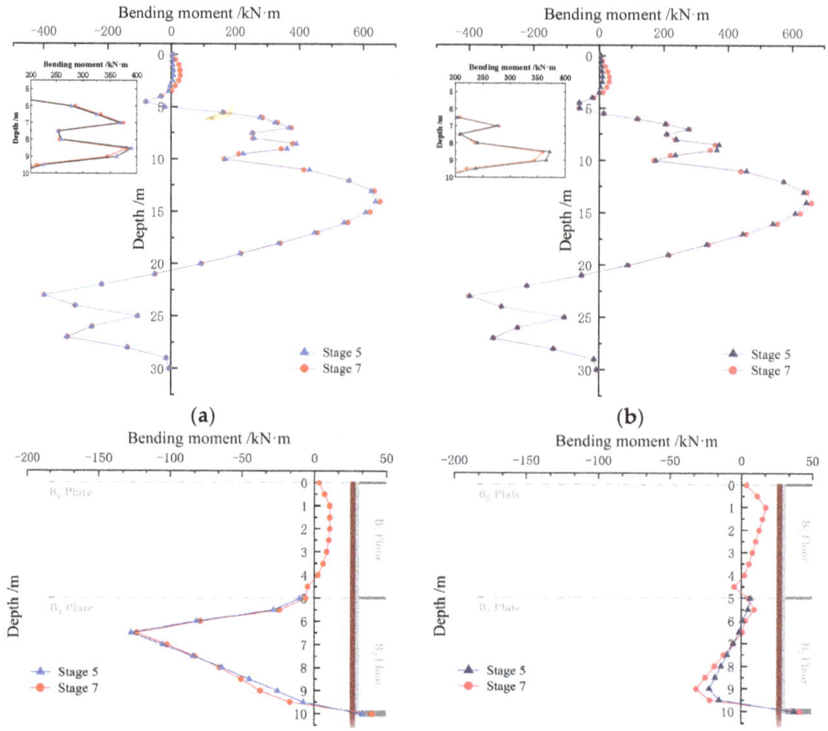

Figure 20. Bending moment of integration structures. (**a**) Bending moment of the pile (SS), (**b**) bending moment of the pile (CS), (**c**) bending moment of the basement wall (SS), and (**d**) bending moment of the basement wall (CS).

3.4.2. Main Difference Comparison of the Two Structures

For the retaining piles, the development of the pile body's bending moment is generally consistent within the B_1 floor depth range for both structures. Within the B_2 floor depth range, the pile body's moment is significantly smaller under the composite structure because the pile rotating tendency is unresisted by the wall due to the tensile stress inactivation between the pile, lining, and wall.

Regarding the basement retaining wall, the wall's maximum bending moment is slightly larger at the upper position within the floor B_1 depth range under the composite structure. This is because the superimposed structure has a more direct and complete connection with the pile wall, resulting in a more uniform transmission and distribution of the active earth pressure outside the excavation. The composite structure, within the B_1 floor depth range, has a weaker redistribution effect of the active earth pressure transmitted by the pile body due to the presence of the lining. And there is slightly more pressure for the upper part of the wall to bear, which illustrates a more concentrated moment distribution. Within the B_2 floor depth range, the appearance of the negative moment in the wall is mainly due to the restraining effect caused by the rotation and rebound tendency of the retaining piles toward the excavation site and the deformation of the wall itself. In the case of composite structures, the tensile stress between the pile and wall is inactive. The negative moment exists by self-deformation within the B_2 floor depth range, and it is approximately 1/4 of that of the superimposed structure.

3.5. Comparison with the Conventional Design Method

In conventional design theory, during the basement-serviceability stage, the design of the walls is calculated using the continuous-beam method, typically without considering the presence of retaining piles. The basement walls bear the full earth pressure. The excavated area's soil layer is mucky silty clay, with a unit weight γ = 17.4 kN/m^3 and a lateral earth pressure coefficient k_0 = 0.72. The thickness of the basement wall is 500 mm.

Figure 21 presents the wall's calculated values using the continuous-beam method in the serviceability stage. Due to the irregular load transfer from the external load to the wall, conventional design usually adopts a conservative approach by assuming a rectangular, uniformly distributed load. Based on the magnitude of the bending moment at the most unfavorable section, the superimposed structure is approximately 50% of the conventional method, and the composite structure is about 20%. Therefore, it is evident that adopting a close-fitting pile–wall integrated structure, especially a composite structure, can significantly minimize the design values of basement walls, thereby minimizing the wall thickness.

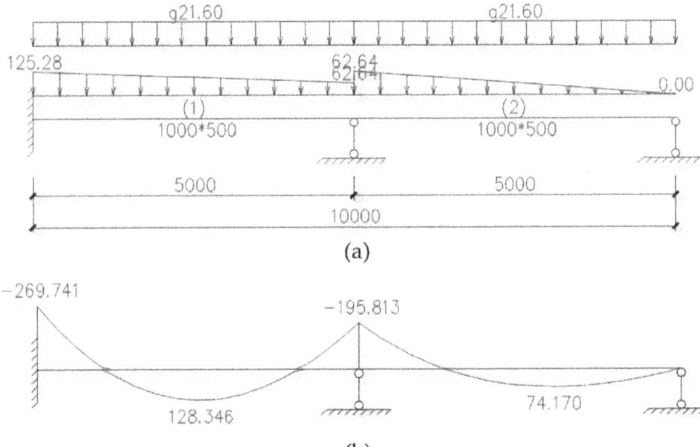

Figure 21. The conventional design methods of the basement retaining wall. (**a**) Load diagram of the continuous-beam method (unit: kN), (**b**) bending moment envelope diagram (unit: kN·m).

3.6. Analysis of Sensitivity Factors in the Pile–Wall Interaction Structures

3.6.1. Influence Analysis of the Pile–Wall Stiffness Ratio

Figure 22 displays the bending moment distribution of the pile and wall under different stiffness ratios. Based on conventional design principles, the wall thickness of double-floor basements in Shanghai typically ranks between 400 and 600 mm. Consequently, the wall thicknesses of 400, 500, and 600 mm are, respectively, used for different stiffness ratios.

Figure 22a,b reveal a similar pattern in the moment variations of the exterior walls under different pile–wall stiffness ratios for both structures. Increasing the wall thickness significantly impacts both the positive and negative moments. Despite the increase in earth pressure outside the excavation from the basement-construction stage to the serviceability stage, the integrated structure lateral deformation remains minimal. This results in only a slight impact on the pile's moment. However, the thickness and stiffness increase in the wall leads to a significant increase in their moment.

Figure 22c,d reveal that the variation in wall thickness has minimal influence on the pile's moments beneath the excavation face. And its effect within the basement depth range is negligible. Within the depth range of the B_1 floor, the pile's moment decreases with the increased wall thickness because the pile–wall structure in this region collectively withstands the active earth pressure. Within the depth range of the B_2 floor, the moment of the pile increases with the wall thickness due to the resistance to the pile rotation and rebound tendency under the superimposed structure. However, the tension stress between the pile and the wall is inactive under the composite structure, resulting in the pile's moment remaining relatively constant within the B_2 floor depth range.

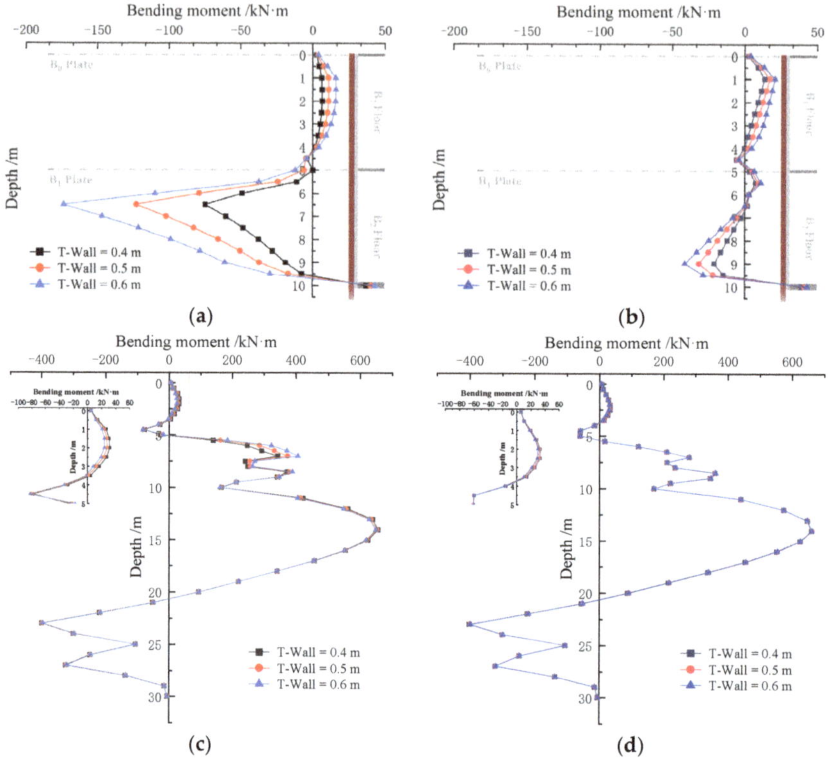

Figure 22. Bending moment of the integrated structure with different pile–wall stiffness ratios. (**a**) Bending moment of the basement wall (SS), (**b**) bending moment of the basement wall (CS), (**c**) bending moment of the pile (SS), and (**d**) bending moment of the pile (CS).

3.6.2. Analysis of the Impact of Horizontal Floor Plate Stiffness

Figure 23 presents different stiffness ratios (0.25, 0.33, and 0.42) chosen to examine the bending moment distribution of the pile and wall under varying floor plate stiffness conditions. The usual foundation plate thickness of double-floor basements in Shanghai ranks between 600 and 800 mm, while the intermediate floor plate thickness ranks between 120 and 250 mm. Consequently, the stiffness ratio ranks between 0.15 and 0.42. To emphasize the impact of floor plate stiffness variation on the integrated structure, a fixed thickness of 600 mm is selected for the foundation plate.

Figure 23a,b reveal similar patterns in the wall's bending moment variations with different horizontal floor plate stiffnesses for both structures. The impact of floor plate stiffness on the moment is significant. As the horizontal floor plate stiffness increases, the moment adjacent to the B_0 plate gradually increases within the B_1 floor depth range. However, the moment near the B_1 plate transitions to a negative moment. This occurs because the gradual increase in horizontal floor plate stiffness reduces the displacement at the plate position. It increases the moment borne by the wall near the plate, resulting in an upward shift of the positive moment distribution. In the B_2 floor depth range, the moment exhibits a decreasing trend. This effect occurs because the increase in floor plate stiffness reduces the pile lateral deformation, leading to a decrease in interaction with the basement retaining wall and subsequently reducing the moment.

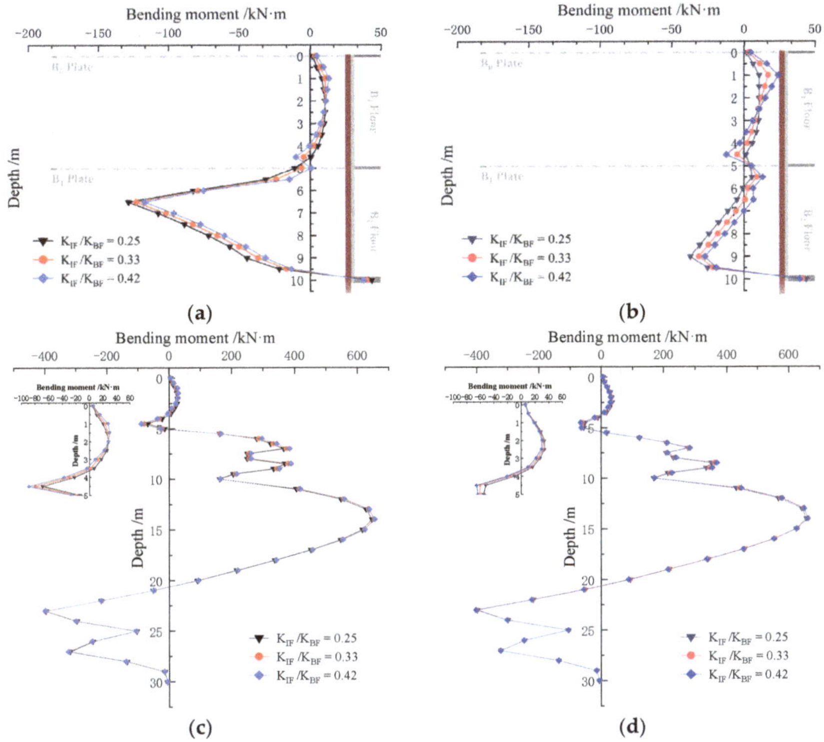

Figure 23. Bending moment of the integrated structure with different horizontal plate stiffnesses. (**a**) Bending moment of the basement wall (SS), (**b**) bending moment of the basement wall (CS), (**c**) bending moment of the retaining pile (SS), and (**d**) bending moment of the retaining pile (CS).

Figure 23c,d reveal similar patterns in the pile's moment variations with different horizontal floor plate stiffness for both structures. The effect of varying horizontal floor plate stiffnesses on the moment beneath the excavation face is negligible, as is its impact within the basement depth range. For the superimposed structure, as the horizontal floor plate stiffness increases, the moment adjacent to the B_0 plate gradually increases within the B_1 floor depth range. However, it exhibits a negative moment trend near the B_1 plate, presenting a similar moment variation of the basement retaining wall because of the high pile–wall integration under the superimposed structure. For the composite structure, the moment of this range solely decreases with the increase in the plate stiffness. This is because the interaction between the pile and wall is only influenced by horizontal compressive stress in this structure. As the horizontal floor plate stiffness increases, the pile lateral deformation within the B_1 floor depth range decreases, resulting in a reduction in the moment. Within the B_2 floor depth range, the moment slightly rises with the increase in the plate stiffness under the superimposed structure. This is due to the restriction of the B_1 plate on the deformation rebound caused by the pile rotation tendency. The presence of tensile stress is inactive between the pile and wall under the composite structure. Hence, the moment within the B_2 floor depth range remains essentially unchanged.

4. Conclusions

This paper presents a systematic study of the force and deformation mechanisms, along with the development of internal force laws, for two typical types of close-fitting pile–wall integrated structures. This study employs a combination of model tests and finite element numerical simulations. The following observations and conclusions are presented herein:

(1) The computed results align well with the measured results, validating the reliability of FEM and mode test employed in this study, which means a certain significant reference value for simulating and analyzing close-fitting pile–wall integrated structures.
(2) A comparative analysis reveals the following: in the superimposed structure, the direct connection between the pile and wall (through rebar and cement) causes the wall to be significantly influenced by the pile's deformation, experiencing varying degrees of tension and compression at different stages. This leads to a better stress distribution and structural integrity, enhancing overall engineering stability. In the composite structure, the developed wall internal forces are relatively small, allowing for a reduced wall thickness in design. Additionally, the presence of the lining layer facilitates basement waterproofing construction.
(3) A sensitivity factor analysis reveals the following: different pile–wall stiffness ratios have a slight effect on the pile body's bending moment above the excavation face but a significant impact on the wall's bending moment. Different horizontal floor plate stiffness has a slight impact on the pile body's bending moment above the excavation face but causes notable changes in the wall's internal forces adjacent to the plate location. Both factors should be carefully considered to achieve a better stress distribution and structural performance.
(4) The analysis of the integrated structure indicates the following: for the pile, its bending moment is significantly reduced after the construction of the basement retaining wall compared to the excavation stage, but it still bears substantial forces during the serviceability stage. For the wall, the bending moment distribution is significantly influenced by the floor plate stiffness in both structures. However, due to the utilization of the pile's stiffness, the computed and measured bending moments are much lower than those in conventional design methods under the same conditions, especially for the composite structure.
(5) In a close-fitting pile–wall integrated structure, the pile's maximum bending moment occurs during the excavation stage, with its section consistently located below the excavation face, similar to conventional structures. However, the moment distribution on the wall differs significantly from conventional design methods, especially under a superimposed structure (a significant negative bending moment occurs on the B_2 wall). Therefore, the design of the retaining structure should primarily focus on the internal forces developed beneath the excavation face during this stage. Additionally, the design of the basement retaining wall should account for the combined action of the horizontal floor plate support and the retaining pile.
(6) For deep excavation, applying a close-fitting integrated structure can significantly save space between the pile and wall and reduce the cost of the basement retaining wall by fully utilizing retaining piles that are otherwise wasted. Additionally, this approach improves structural performance with minimal additional cost.

Author Contributions: Conceptualization, C.W.; methodology, C.W. and W.S.; software, W.S.; validation, Y.X., C.W. and W.S.; formal analysis, C.W. and W.S.; investigation, C.W., W.S. and G.W.; writing—original draft preparation, W.S. and C.W.; visualization, W.S.; writing—review and editing, C.W., Y.X. and W.S.; supervision C.W. and Y.X.; resources, G.W.; project administration, G.W. and C.W.; funding acquisition C.W. All authors have read and agreed to the published version of the manuscript.

Funding: The work described in this paper was jointly supported by the Jiangsu Natural Science Foundation Youth Project (Grant No. BK20180954) and the Postgraduate Research & Practice Innovation Program of the School of Transportation and Civil Engineering, Nantong University (Grant No. NTUJTXYGI2205).

Data Availability Statement: The original contributions presented in the study are included in the article, further inquiries can be directed to the corresponding authors.

Conflicts of Interest: Author Guowei Wei is employed by the Jiangsu Zhongnan Construction Industry Group Co., Ltd. The remaining authors declare that the research was conducted in the absence of any commercial or financial relationships that could be construed as a potential conflict of interest.

References

1. Shi, J.; Liu, G.; Huang, P.; Ng, C.W.W. Interaction between a large-scale triangular excavation and adjacent structures in Shanghai soft clay. *Tunn. Undergr. Space Technol.* **2015**, *50*, 282–295. [CrossRef]
2. Wang, J.H.; Xu, Z.H.; Wang, W.D. Wall and Ground Movements due to Deep Excavations in Shanghai Soft Soils. *J. Geotech. Geoenviron. Eng.* **2010**, *136*, 985–994. [CrossRef]
3. Zhou, N.; Vermeer, P.A.; Lou, R.; Tang, Y.; Jiang, S. Numerical simulation of deep foundation pit dewatering and optimization of controlling land subsidence. *Eng. Geol.* **2010**, *114*, 251–260. [CrossRef]
4. Tan, Y.; Lu, Y.; Wang, D. Interactive behaviors of four closely spaced mega excavations in soft clays: Case study on an excavation group in Shanghai, China. *Tunn. Undergr. Space Technol.* **2023**, *138*, 105186. [CrossRef]
5. Hu, Y.; Wang, W.; Shen, J. Field monitoring and analysis of stresses for dual-purpose pile wall. *Chin. J. Geotech. Eng.* **2015**, *37*, 197–201. [CrossRef]
6. Wu, C.; Yu, J.; Cao, X.; Shen, W. Study on Design Method of Pile Wall Combination Structure in a Deep Foundation Pit Considering Deformation Induced by Excavation. *Front. Earth Sci.* **2022**, *10*, 837950. [CrossRef]
7. Xiao, J.; Huang, L.; He, Z.; Qu, W.; Li, L.; Jiang, H.; Zhong, Z.; Long, X. Probabilistic models applied to concrete corrosion depth prediction under sulfuric acid environment. *Measurement* **2024**, *234*, 114807. [CrossRef]
8. Cheng, X.S.; Zheng, G.; Diao, Y.; Huang, T.M.; Deng, C.H.; Nie, D.Q.; Lei, Y.W. Experimental study of the progressive collapse mechanism of excavations retained by cantilever piles. *Can. Geotech. J.* **2017**, *54*, 574–587. [CrossRef]
9. Seo, M.; Im, J.-C.; Kim, C.; Yoo, J.-W. A Study on the Applicability of a Retaining Wall Using Batter Piles in Clay. *Can. Geotech. J.* **2016**, *53*, 1195–1212. [CrossRef]
10. Shi, J.; Ding, C.; Ng, C.W.W.; Lu, H.; Chen, L. Effects of overconsolidation ratio on tunnel responses due to overlying basement excavation in clay. *Tunn. Undergr. Space Technol.* **2020**, *97*, 103247. [CrossRef]
11. Wei, G.; Feng, F.; Cui, Y.; Wang, X.; Diao, H.; Wu, Y. Research on the influence of foundation pit excavation on the lateral force and deformation of side shield tunnels based on full-scale experiments. *Tunn. Undergr. Space Technol.* **2023**, *140*, 105236. [CrossRef]
12. Yang, X.; Kulatilake, P.H.S.W.; Jing, H.; Yang, S. Numerical simulation of a jointed rock block mechanical behavior adjacent to an underground excavation and comparison with physical model test results. *Tunn. Undergr. Space Technol.* **2015**, *50*, 129–142. [CrossRef]
13. Sun, X.; Chen, F.; He, M.; Gong, W.; Xu, H.; Lu, H. Physical modeling of floor heave for the deep-buried roadway excavated in ten degree inclined strata using infrared thermal imaging technology. *Tunn. Undergr. Space Technol.* **2017**, *63*, 228–243. [CrossRef]
14. Bilgin, Ö. Numerical studies of anchored sheet pile wall behavior constructed in cut and fill conditions. *Comput. Geotech.* **2010**, *37*, 399–407. [CrossRef]
15. Chen, J.; Xu, Q.; Luo, X.; Tian, A.; Xu, S.; Tang, Q. Safety Evaluation and Energy Consumption Analysis of Deep Foundation Pit Excavation through Numerical Simulation and In-Site Monitoring. *Energies* **2022**, *15*, 7099. [CrossRef]
16. Cui, K.; Feng, J.; Zhu, C. A Study on the Mechanisms of Interaction between Deep Foundation Pits and the Pile Foundations of Adjacent Skewed Arches as well as Methods for Deformation Control. *Complexity* **2018**, *2018*, 6535123. [CrossRef]
17. Wang, J.; Xiang, H.; Yan, J. Numerical Simulation of Steel Sheet Pile Support Structures in Foundation Pit Excavation. *Int. J. Geomech.* **2019**, *19*, 05019002. [CrossRef]
18. Guo, C.; Wang, R.; Lin, P.; Wang, F. Numerical analyses of a prefabricated retaining system for foundation pits in silt soils. *Geotech. Res.* **2020**, *7*, 173–190. [CrossRef]
19. Hsiung, B.-C.B.; Yang, K.-H.; Aila, W.; Hung, C. Three-dimensional effects of a deep excavation on wall deflections in loose to medium dense sands. *Comput. Geotech.* **2016**, *80*, 138–151. [CrossRef]
20. Zhang, J.; Xie, R.; Zhang, H. Mechanical response analysis of the buried pipeline due to adjacent foundation pit excavation. *Tunn. Undergr. Space Technol.* **2018**, *78*, 135–145. [CrossRef]
21. Wang, S.; Li, Q.; Dong, J.; Wang, J.; Wang, M. Comparative investigation on deformation monitoring and numerical simulation of the deepest excavation in Beijing. *Bull. Eng. Geol. Environ.* **2021**, *80*, 1233–1247. [CrossRef]
22. Hsiung, B.-C.B.; Phan, H.K. Exploration of maximum wall deflection and stability for deep excavation in loose to medium-dense sand. *Acta Geotech.* **2024**, *19*, 531–547. [CrossRef]
23. Chai, X.; Rózsás, Á.; Slobbe, A.; Teixeira, A. Probabilistic parameter estimation and reliability assessment of a simulated sheet pile wall system. *Comput. Geotech.* **2022**, *142*, 104567. [CrossRef]

24. Wang, J.; Deng, Y.; Ma, R.; Liu, X.; Guo, Q.; Liu, S.; Shao, Y.; Wu, L.; Zhou, J.; Yang, T.; et al. Model test on partial expansion in stratified subsidence during foundation pit dewatering. *J. Hydrol.* **2018**, *557*, 489–508. [CrossRef]
25. Huang, F.; Shi, X.; Wu, C.; Dong, G.; Liu, X.; Zheng, A. Stability analysis of tunnel under coal seam goaf: Numerical and physical modeling. *Undergr. Space* **2023**, *11*, 246–261. [CrossRef]
26. Wang, R.; Yang, H.; Ni, P.; Zhao, C.; Guo, C.; Ma, H.; Dong, P.; Liang, H.; Tang, M. Model test and numerical simulation of a new prefabricated double-row piles retaining system in silty clay ground. *Undergr. Space* **2023**, *13*, 262–280. [CrossRef]
27. Zhang, S.; Yue, Z.; Lu, X.; Zhang, Q.; Sun, T.; Qi, Y. Model test and numerical simulation of foundation pit constructions using the combined artificial ground freezing method. *Cold Reg. Sci. Technol.* **2023**, *205*, 103700. [CrossRef]
28. Sun, H.; Chen, Y.; Zhang, J.; Kuang, T. Analytical investigation of tunnel deformation caused by circular foundation pit excavation. *Comput. Geotech.* **2019**, *106*, 193–198. [CrossRef]
29. Ng, C.W.W.; Shi, M.J.; Mašín, M.D.; Sun, M.H.; Lei, G.H. Influence of sand density and retaining wall stiffness on the three-dimensional responses of a tunnel to basement excavation. *Can. Geotech. J.* **2015**, *52*. [CrossRef]
30. Zhou, F.; Zhou, P.; Li, J.; Lin, J.; Ge, T.; Deng, S.; Ren, R.; Wang, Z. Deformation characteristics and failure evolution process of the existing metro station under unilateral deep excavation. *Eng. Fail. Anal.* **2022**, *131*, 105870. [CrossRef]
31. *DG_TJ 08-61-2018*; Technical Code for Excavation Engineering. Tongji University Press: Shanghai, China, 2018.
32. Jiang, B.; Xia, W.; Wu, T.; Liang, J. The optimum proportion of hygroscopic properties of modified soil composites based on orthogonal test method. *J. Clean. Prod.* **2021**, *278*, 123828. [CrossRef]
33. Yang, X.; Dong, J.; Yang, J.; Han, X. Similar Material Proportioning Tests and Mechanical Properties Based on Orthogonal Design. *Materials* **2023**, *16*, 6439. [CrossRef]
34. Yuan, B.; Liang, J.; Zhang, B.; Chen, W.; Huang, X.; Huang, Q.; Li, Y.; Yuan, P. Optimized reinforcement of granite residual soil using a cement and alkaline solution: A coupling effect. *J. Rock Mech. Geotech. Eng.* **2024**, *in press*. [CrossRef]
35. Pei, H.; Yin, J.; Zhu, H.; Hong, C.; Fan, Y. In-Situ Monitoring of Displacements and Stability Evaluation of Slope Based On Fiber Bragg Grating Sensing Technology. *Chin. J. Rock Mech. Eng.* **2010**, *29*, 1570–1576.
36. Cheng, K.; Xu, R.; Ying, H.; Gan, X.; Zhang, L.; Liu, S. Observed performance of a 30.2 m deep-large basement excavation in Hangzhou soft clay. *Tunn. Undergr. Space Technol.* **2021**, *111*, 103872. [CrossRef]
37. Dong, Y.P.; Burd, H.J.; Houlsby, G.T. Finite-element analysis of a deep excavation case history. *Géotechnique* **2016**, *66*, 1–15. [CrossRef]
38. Yang, Y.; Chen, C.; Liu, C.; Huang, L.; Chen, W.; Lin, N.; Cui, J.; Xie, W. Performance of a deep excavation and the influence on adjacent piles: A case history in karst region covered by clay and sand. *Undergr. Space* **2023**, *8*, 45–60. [CrossRef]
39. Tan, Y.; Lu, Y.; Wang, D. Deep Excavation of the Gate of the Orient in Suzhou Stiff Clay: Composite Earth-Retaining Systems and Dewatering Plans. *J. Geotech. Geoenviron. Eng.* **2018**, *144*, 05017009. [CrossRef]

Disclaimer/Publisher's Note: The statements, opinions and data contained in all publications are solely those of the individual author(s) and contributor(s) and not of MDPI and/or the editor(s). MDPI and/or the editor(s) disclaim responsibility for any injury to people or property resulting from any ideas, methods, instructions or products referred to in the content.

buildings

MDPI

Article

The Performance of a Circular Excavation Supported by a Prefabricated Recyclable Structure in a Full-Scale Test

Lichao Chen [1], Chengchao Guo [1,2,*], Yanhui Pan [2], Huqing Liang [3] and Mengxiong Tang [3]

1 School of Civil Engineering, Sun Yat-sen University, Zhuhai 519082, China; chenlch8@mail2.sysu.edu.cn
2 BestDr Infrastructure Hospital, Zhumadian 463000, China; pyh511@126.com
3 Guangzhou Municipal Construction Group Co., Ltd., Guangzhou 510275, China
* Correspondence: guochch25@mail.sysu.edu.cn

Abstract: Excavations for underground structures, such as working shafts, underground grain silos, and parking garages, are characterized by uniformity, consistent dimensions, large quantities, and strict timelines. Prefabricated recyclable supporting structures (PRSS) are gaining attention over traditional retaining structures due to their standardized design, efficient construction, and reusability, which suit such excavations better. To validate their performance, full-scale tests are conducted to analyze the deformation and stress characteristics of PRSS. The results show that the average maximum lateral displacement of supporting pile is 0.07% of the excavation depth (H_e), roughly half that of steel plate. Differences in ground surface settlement behind steel plates and the supporting piles are not as significant as those in their lateral displacements. While the displacement of the supporting piles is insufficient to induce soil movement into the active limit state on the non-excavation side, the circular excavation's arching effect reduces the earth pressure on this side of the supporting piles below the active earth pressure limit. Furthermore, the earth pressure acting on the steel plates is lower than that acting on the supporting piles, suggesting the presence of a soil arching effect between two adjacent piles. These findings offer valuable insights for guiding the construction of PRSS.

Keywords: circular excavation; prefabricated supporting structure; full-scale test; earth pressure; deformation

1. Introduction

The rapid urbanization in China has spurred a rising demand for underground infrastructure, such as working shafts [1,2], underground grain and oil storage silos [3], and automated underground parking garages [4,5]. Circular excavations for these underground structures play an important role on their construction, characterized by consistent structural forms, similar cross-sectional dimensions, large quantities, temporary, and tight construction schedules. A retaining system is often used to prevent soil collapse and to protect the surrounding environment during the excavation process.

Existing retaining systems mainly include diaphragm walls [6,7], bored pile walls [8–10], soil nail walls [11,12], soil mixing walls [13–15], sheet pile walls [16,17], and so on. Although these retaining walls are widely used in practical engineering, they do not fit well with the characteristics of the excavations mentioned above. For instance, most of them are made of cement-based material, which require extended curing periods to achieve sufficient strength and pose challenges for demolition after the underground structures are completed, resulting in long-term construction time and a large amount of construction waste. Moreover, although diaphragm walls and bored pile walls exhibit high stiffness and strong deformation resistance, they are less cost-effective when used in the excavation mentioned above, considering their complex construction processes. In contrast, while soil nail walls and soil mixing walls have high construction efficiency and low costs, they exhibit weak deformation resistance. Steel sheet pile walls also have a weakness in their deformation resistance as well as leakage at the joints [18], despite the fact that they can be recycled and reused. In recent years, there has been growing concern about the environmental impact of infrastructure development, including carbon dioxide emissions and waste generation [19–21]. As a result, green construction technologies have emerged as a major focus in civil engineering [22,23]. Therefore, it is necessary to research new eco-friendly types of retaining structures based on the characteristics of the excavation mentioned above.

In comparison to traditional retaining structures, prefabricated recyclable supporting structures have attracted attention due to their standardized design, efficient construction, and reusability, making them more suitable for such excavations [24–26]. Guan et al. [27] and Zhao et al. [2] proposed a prefabricated prestressed structure for small-sized working shafts and validated its feasibility through theoretical calculation, numerical simulation, and on-site loading tests. Pan et al. [26] proposed a new recyclable supporting structure for underground grain silos, which has the advantage of stability, rapid construction, recyclable utilization, and low cost. Its mechanical and deformation characteristics when used in rectangular shafts was studied by Guo et al. [24,25] through numerical simulation and model tests. Zhang et al. [28] investigated its bearing performance and stability. However, these studies primarily focus on rectangular excavations and lack consideration of circular excavations. Circular excavations differ significantly from rectangular ones in terms of their stress and deformation characteristics. It is found that the lateral earth pressure acting on a circular excavation, which accounts for the soil arching effects, is less than that predicted by Rankine theory [29–33]. Additionally, the spatial arch effect in circular excavations increases as the diameter decreases, leading to smaller wall deflection [34,35]. Therefore, it is necessary to investigate the performance of the circular excavation retained by PRSS.

In this paper, full-scale tests are conducted to understand the performance of a circular excavation retained by PRSS. The lateral displacements of supporting structures, groundwater levels, ground surface settlements, lateral earth pressures, and axial forces of walings are measured and analyzed to draw conclusions for the further application of similar structures. In addition, this study provides comprehensive field measurements on PRSS, enhances the existing experimental database on eco-friendly retaining systems, and contributes to reducing the environmental impact of the construction industry.

2. Full-Scale Test

2.1. Site and Geological Conditions

The full-scale test was conducted in Pingyu county, Henan Province, China, situated in the Ru River alluvial plain. According to the geotechnical investigation report, the subsoils within a depth range of 50.0 m from the ground surface are classified as late and middle Pleistocene alluvial deposits, predominantly composed of silty clay. A series of site investigations and laboratory tests were performed to determine the soil properties at different depths, as depicted in Figure 1. The groundwater table is located at a depth of approximately 3.5 m.

2.2. Design Scheme and Instrumentation Layout

As shown in Figure 2, the excavation has a diameter of 10.0 m and a final excavation depth of 9.0 m. The supporting piles are 12.0 m in length and composed of H350 and L125 steel. There were a total of 20 supporting piles in the excavation with an interval of 1.56 m between two adjacent piles. Moreover, four walings were installed at elevation levels of 0.0, 3.0, 6.0, and 9.0 m below the ground surface, respectively. In order to improve the impermeability of PRSS, a two-component expansive polymer was grouted behind steel plates [26,36].

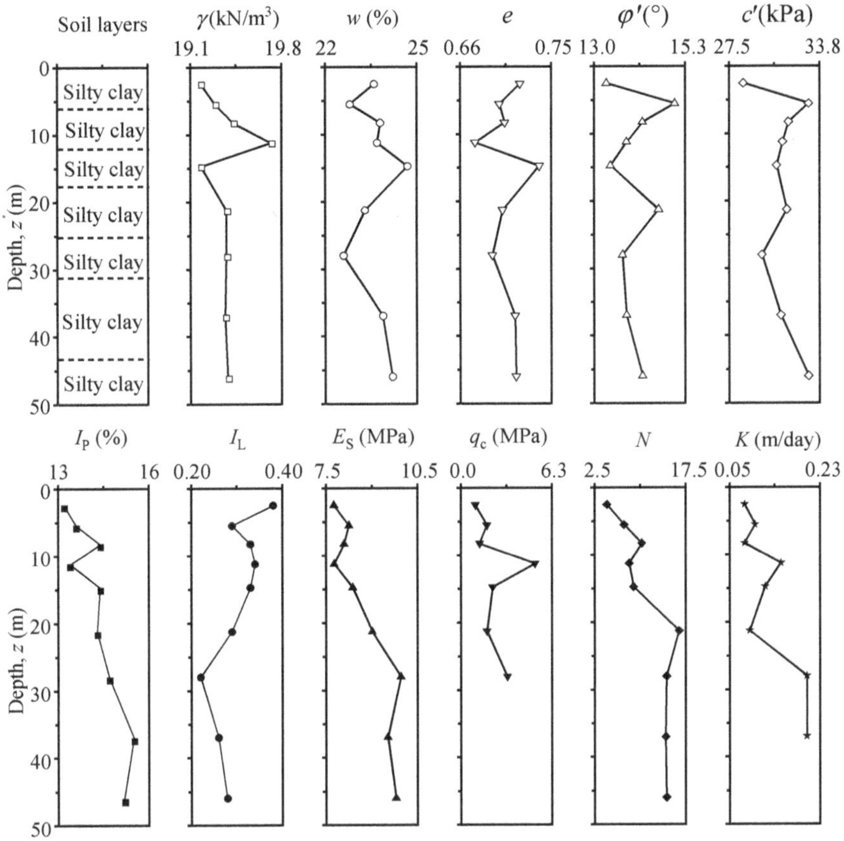

Figure 1. In situ geotechnical profile and soil properties. Note: γ represents unit weight, w represents water content, e represents the void ratio, φ' represents effective internal friction angle, c' represents effective cohesion, I_P represents plasticity index, I_L represents liquidity index, E_s represents compression modulus, q_c represents cone resistance in CPT, N represents the blow count of SPT, and K represents permeability coefficient.

Various instruments were installed to measure the deformations and forces of the PRSS caused by excavation. Inclinometer casings were firmly attached to the supporting piles and the steel plates to monitor their lateral displacements. To monitor the ground surface settlements around the excavation, seven hydrostatic levels were installed behind Pile A and Plate B, respectively. Earth pressure cells were installed on Pile A, Pile B, and Plate A. On the non-excavation side, six soil pressure cells were buried at depths of 2.0 m, 4.0 m, 6.0 m, 8.5 m, 10.0 m, and 11.5 m below the ground surface. On the excavation side, three soil pressure cells were buried at depths of 8.5 m, 10.0 m, and 11.5 m below the ground surface. In order to assess the performance of walings, a pair of sister bar strain gauges were welded on two flanges of each monitored waling to monitor their axial forces. To ensure a dry working face during excavation, four dewatering wells were installed evenly around the perimeter of the excavation at 1.5 m away. Three groundwater observation wells were installed outside the excavation on the west and north sides, respectively.

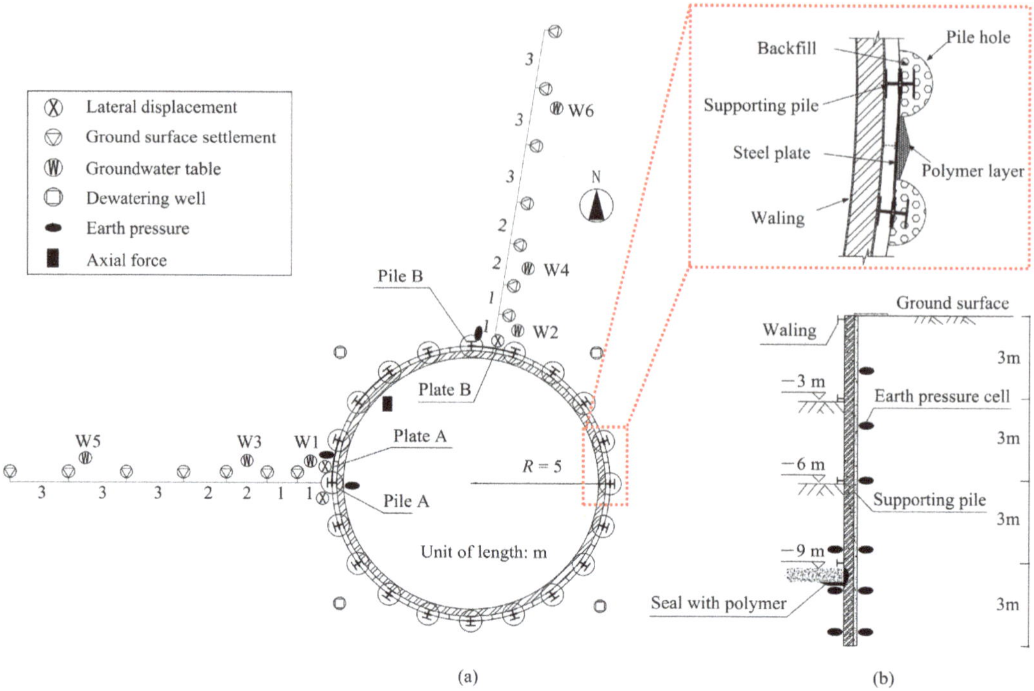

Figure 2. The excavation retained by PRSS and the monitoring scheme: (**a**) plane view; (**b**) cross-section view.

2.3. Construction Sequence

As shown in Figure 3, the walings and supporting piles are prefabricated in the factory using specialized equipment according to design requirements, and then transported to the site for secondary assembly and installation. The waling is designed as a ring, which requires bending processing of the profile steel to ensure they meet design specifications and assembly accuracy. Plasma cutting is used to create connection holes on the walings and supporting piles according to the design. The supporting piles are assembled from H-section and L-section steels.

Figure 3. Prefabrication on the PRSS: (**a**) waling; (**b**) supporting piles.

Figure 4 presents the construction sequence of the PRSS. Firstly, following the boring of the pile holes at the designated location, the supporting piles are lifted into the holes and the holes are then backfilled (Figure 4a). After all the supporting piles are installed, the first-level waling is assembled (Figure 4b). Next, a hydraulic pile press machine is used to drive a custom-made sheet pile into the soil along the grooves of the supporting piles to form a slot, and then a steel plate is inserted into the slot (Figure 4c). Subsequently, excavation is carried out to a depth of 3 m below the ground surface, and the second-level waling is installed (Figure 4d). Expansion polyurethane is then grouted behind the steel plates (Figure 4e). These steps (i.e., excavation, installation of waling, and grouting) are repeated to complete the excavation for the second stage (H_e = 6 m) and the third stage (H_e = 9 m) (Figure 4f).

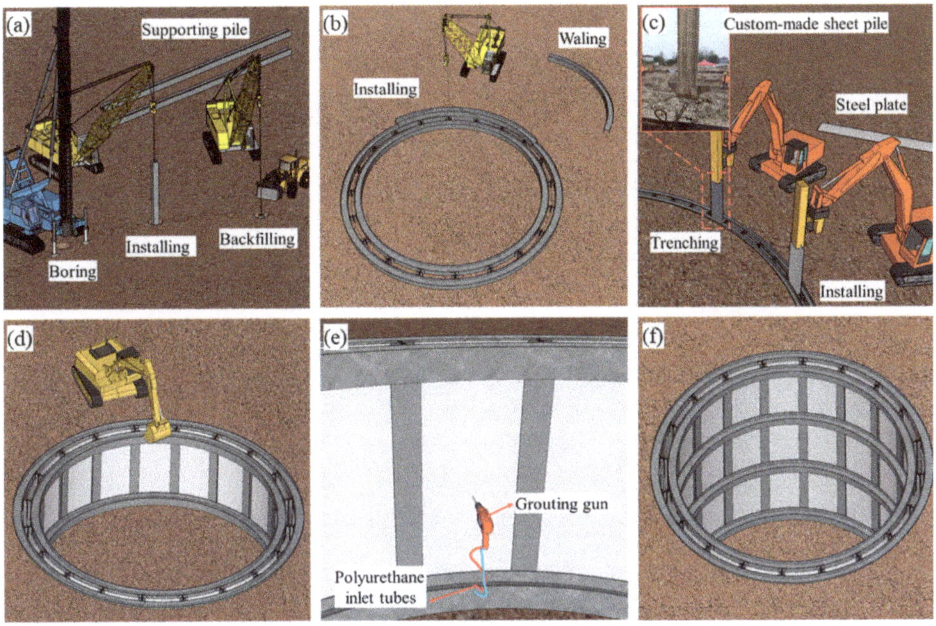

Figure 4. Construction sequence of the PRSS: (**a**) installation of supporting piles; (**b**) installation of first-level waling; (**c**) installation of steel plates; (**d**) excavation of soil and installation of second-level waling; (**e**) grouting behind the steel plates; (**f**) repeating soil excavation, installation of waling, and grouting until reaching the final depth.

3. Results and Discussion

3.1. Lateral Displacement

3.1.1. Supporting Piles

Figure 5 shows the variations of lateral pile displacements with depth at different excavation stages. After the first stage of excavation, supporting piles deform as a cantilever, with an average lateral displacement of 1.5 mm at the top of supporting piles A and B. This is attributed to the loose connection between the supporting piles and the crown beams caused by installation errors, as well as the absence of preload in crown beams, resulting in insufficient stiffness to restrain inward movement. As the excavation progresses to the deeper levels, lateral pile displacements gradually increase, and their maximum values occur at deeper positions because upper pile movements are constrained by new walings or stiffening of existing walings. It should be noted that at the last excavation stage (H_e = 9 m), the maximum lateral displacement of supporting piles reaches 9.1 mm, which is far less than the warning value of 18 mm [37]. Supporting piles A and B have a deep inward type of movement, consistent with findings of Clough and O'Rourke [38] and Ou et al. [39]. However, significant movements occur at supporting pile toes, with an average value of 4.9 mm for the two tested piles. It indicates the lack of resistance from the soil beneath the excavation bottom due to shallow embedment depth of the supporting piles.

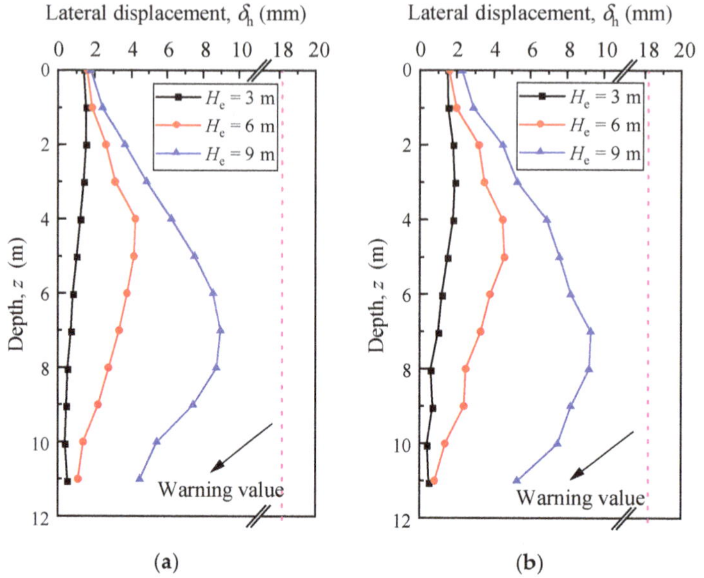

Figure 5. Lateral displacements of supporting piles: (**a**) pile A; (**b**) pile B.

3.1.2. Steel Plates

The lateral displacements of steel plates A and B are illustrated in Figure 6. It is evident from the figure that these displacements increase progressively with excavation depth. Since the steel plates are attached to the supporting piles, they move along with the supporting piles. Consequently, the total displacements of the steel plates equal the sum of the displacements of the supporting piles and the relative displacements of the steel plates with respect to the supporting piles. It can be observed that although the total lateral displacements of the steel plates follow a similar pattern to those of the supporting piles, they are significantly larger. Notably, at the final excavation stage, the average total lateral displacement of steel plates A and B grows to 16.2 mm, approaching the warning value. In terms of the shape of the deformation, the distribution of the relative lateral displacement of the steel plate along the depth is similar to its total lateral displacement. With regard to the magnitude of the deformation, the relative value of the steel plate is approximately half of its total lateral displacement. Additionally, the lateral displacement of the supporting piles is about half of the total lateral displacement of the steel plates.

Figure 7 shows the relationship between maximum lateral displacement and excavation depth, and the corresponding locations of δ_{hm} against H_e. As illustrated in Figure 7a, the δ_{hm}/H_e ratio for supporting pile ranges from 0.05% to 0.10%, with an average value of 0.07% for supporting pile. For steel plates, the δ_{hm}/H_e ratio ranges from 0.10% to 0.18%, averaging 0.14%. Nevertheless, the average δ_{hm}/H_e ratios for both supporting piles and steel plates in this study are close to those reported for diaphragm walls by Liu et al. [40], but are smaller than those for contiguous pile walls studied by Ni et al. [41] and sheet pile walls investigated by Mana et al. [42]. Furthermore, it can be seen that as excavation progresses, the location of δ_{hm} shifts gradually downward. However, for the supporting piles, their maximum lateral displacements occur at a higher level compared to the steel plates, as shown in Figure 7b. The locations of δ_{hm} of the supporting piles are near the excavation bottom, whereas for the steel plates, δ_{hm} locates at a position that is 1.5 m above the excavation bottom. This difference is mainly due to the higher stiffness of the supporting piles compared to that of the steel plates.

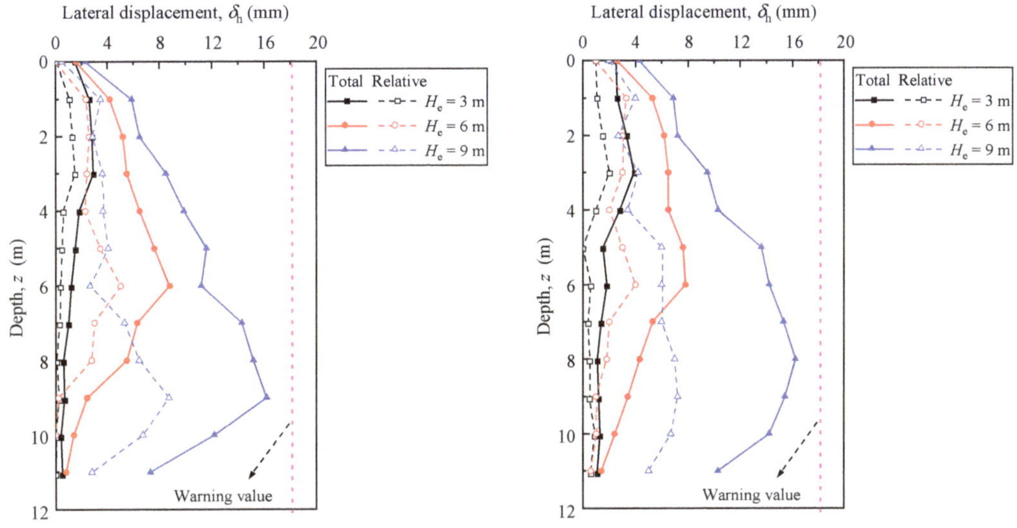

(**a**) (**b**)

Figure 6. Lateral displacements of steel plates: (**a**) plate A; (**b**) plate B.

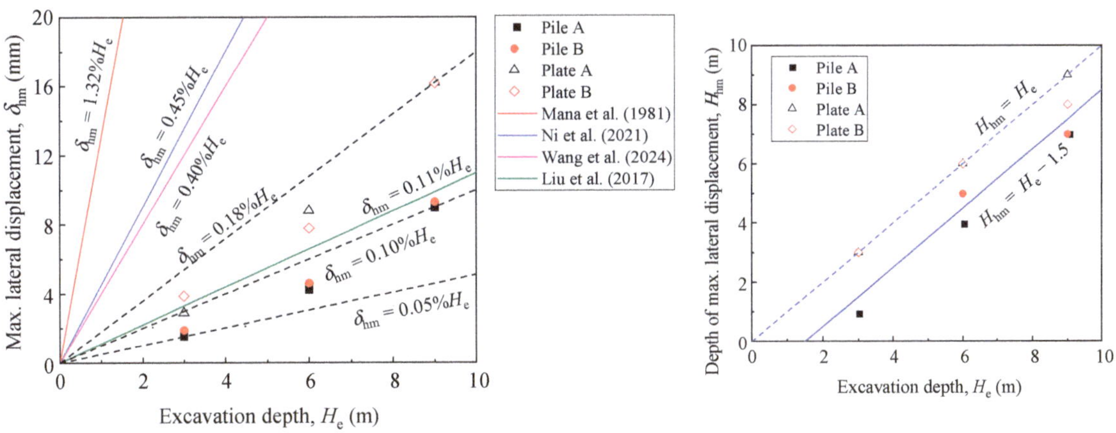

(**a**) (**b**)

Figure 7. Variations of maximum lateral displacement: (**a**) maximum lateral displacement versus excavation depth [36,40–42]; (**b**) maximum lateral displacement depth versus excavation depth.

3.2. Groundwater

Figure 8 shows the variations of the groundwater table during construction. It can be seen that prior to dewatering, the groundwater level was approximately 3.4 m below the ground surface. After dewatering, the groundwater level at 0.5, 3.0, and 11.0 m from the edge of the excavation stabilized at 10.28, 9.27, and 6.18 m below the ground surface, respectively. Site observations revealed that the excavation bottom remained dry throughout the entire excavation process, confirming that the groundwater level consistently stayed below the maximum excavation depth.

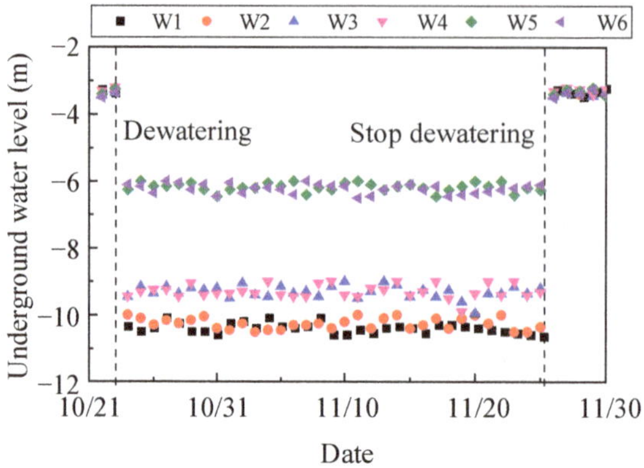

Figure 8. Variations of underground water level.

In order to verify the impermeability of PRSS, dewatering is halted after the completion of the base slab, and changes in the groundwater level outside the excavation and leakage are monitored. According to Figure 8, after dewatering is stopped, the groundwater table outside the excavation returns to approximately 3.4 m below the ground surface. However, as shown in Figure 9, no obvious water leakage is found inside the excavation. This confirms the efficacy of the supporting structure, comprised of steel plates and grouted polymer, in preventing leakage. Since the waterproofing performance of the PRSS has only been preliminarily validated, further improvements in construction and design, as well as additional model tests, are necessary to enhance this supporting system.

Figure 9. Interior of the excavation after stopping dewatering.

3.3. Ground Surface Settlement

Figure 10 shows the variations of ground surface settlements behind the supporting piles and the steel plates at different construction stages. It can be seen that prior to excavation, dewatering outside the excavation leads to ground surface settlement, with an average maximum value of 3.0 mm. As excavation progresses, ground surface settlements show an increasing trend, which is more pronounced at a distance of 12 m (1.3 H_e) from the excavation and diminishes with greater distance. Moreover, the apparent influence range expands with increasing excavation depth. The settlement profile changes from a spandrel type to a concave type, indicating that the location of maximum settlement shifts outward from the excavation edge. The ultimate maximum ground surface settlement behind supporting pile reaches 6.80 mm, representing 0.075% of the excavation depth, occurring at a distance of 2 m (0.22 H_e) from the excavation edge. On the other hand, there are no significant differences in curve shape, settlement value, affected area, or depth-related changes in ground surface settlement behind steel plates compared to supporting piles. However, the maximum ground surface settlement behind steel plates at the final excavation stage is 7.28 mm, accounting for 0.081% of the excavation depth. Furthermore, throughout the entire excavation process, the maximum ground surface settlement is less than the warning value of 25 mm. In contrast to lateral displacements shown in Figures 3 and 4, it is observed that the differences in ground surface settlement behind steel plates and the supporting piles are not as significant as those in their lateral displacements. The primary reason for this phenomenon is that behind the steel plates, soil arching effects are formed between two adjacent piles, which provide support for the soil outside of them. As a result, the displacement of the soil outside the soil arch is significantly smaller than that of the steel plate.

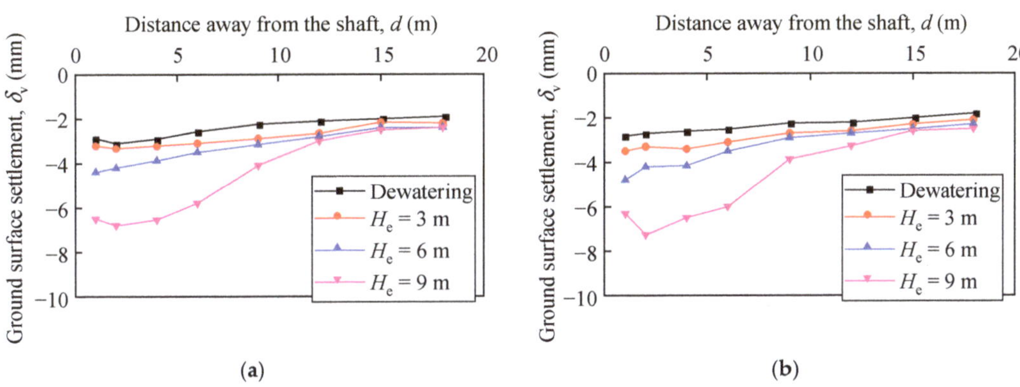

Figure 10. Ground surface settlement outside the excavation: (**a**) behind the supporting pile; (**b**) behind the steel plate.

Figure 11 presents the maximum ground surface settlements (δ_{vm}) against the maximum lateral displacements (δ_{hm}) of supporting piles and steel plates. As mentioned above, the lateral displacements of steel plates are significantly greater than those of supporting piles, while ground surface settlements behind steel plates appear to be close to those behind supporting piles. Therefore, it can be seen that the δ_{vm}/δ_{hm} ratio of steel plates is smaller than that of supporting piles. The δ_{vm}/δ_{hm} ratio of steel plates ranges from 0.4 to 1.2, with an average value of 0.7. For supporting piles, the δ_{vm}/δ_{hm} ratio ranges from 0.7 to 2.2, averaging 1.2. This findings aligns with Goldberg et al. [43] and Moormann [44], who observed that the δ_{vm}/δ_{hm} ratio generally falls in the range of 0.5 to 2.0 for excavations in soft ground supported by various retaining systems. Other studies examining the relationship between the movements of retaining walls and ground surface in deep excavations in soft soils [42,45,46] indicate that the δ_{vm}/δ_{hm} ratios are bounded by the range of 0.5 to 1.0. Ali and Khan [47] reported that the ratio of δ_{vm}/δ_{hm} generally fell into the range between 0.7 and 1.4 for anchored pile walls in soft to stiff clays. These variations may be attributed to differences in soil conditions, types of retaining walls, and excavation dimensions.

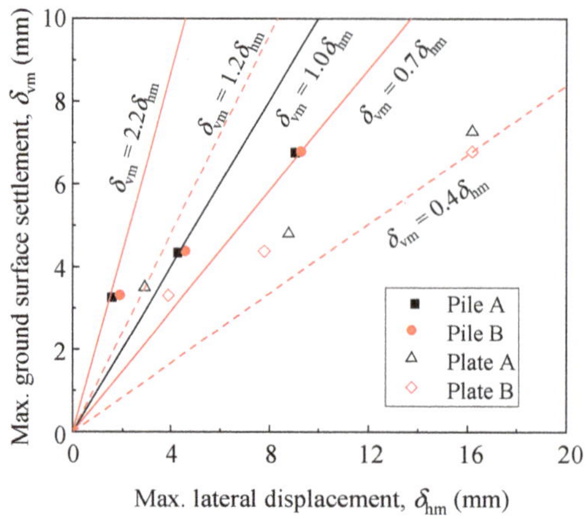

Figure 11. Relationship between maximum ground settlement and maximum lateral displacement.

3.4. Lateral Earth Pressure

3.4.1. On Supporting Piles

Figure 12 presents the distribution of earth pressure on the supporting piles. It can be observed that before excavation (H_e = 0 m), the measured earth pressure on the non-excavation side of the supporting piles is lower than the earth pressure at rest (σ_{h0}). This shows the impact of supporting pile installation, which induces unloading in the surrounding soil [48]. Although the drilled pile holes are backfilled, they are not consolidated as much as undisturbed soil. In addition, the soil around the borehole exhibits a certain self-stability due to the soil arching effect. This effect, if not disturbed before excavation, partially reduces the lateral earth pressure acting on the supporting piles. When the first soil layer is excavated (H_e = 3 m), there is a small inward movement on the upper part of the supporting pile. It is widely acknowledged that when a retaining wall moves away from the soil, the soil behind the wall will gradually transition from at an rest state to an active state, resulting in a decrease in earth pressure acting on the wall [49]. Nevertheless, in Figure 10a,b, there is no obvious trend of decreasing earth pressure. This indicates that in practical applications of PRSS, when the movement is small, the earth pressure acting on supporting piles is influenced not only by displacement but also by pile installation. As the excavation deepens (H_e = 6 m), there is a noticeable decrease in the soil pressure on the supporting piles. At the final excavation stage (H_e = 9 m), the lateral earth pressure acting on the non-excavation side of the supporting pile continues to decrease. For the upper part of the supporting pile at a depth of 6 m, the earth pressure lies between the active (σ_{ha}) and at rest (σ_{h0}) earth pressure. However, for the lower part of the supporting pile at the same depth, the earth pressure is less than the active earth pressure, even though the lateral pile displacement has not reached the threshold for entering the active state. This phenomenon is due to the soil arching effect formed by circular excavation [50–52].

Figure 12c shows the distribution of lateral earth pressure on the excavation side of supporting pile A, along with theoretical at rest (σ_{h0}) and passive (σ_{hp}) earth pressures for comparison. It can be found that at the same excavation stage, the earth pressure increases with depth. According to Figures 5 and 12c, although the lateral displacement of the embedded part of the supporting pile increases significantly with excavation depth, reaching a maximum of 7.5 mm, the earth pressure decreases. Throughout the entire construction process, the passive limit state was never reached and the measured earth pressure consistently remained below the passive earth pressure. However, the earth pressure in the passive zone is still moving towards the passive limit state, and the gap between the measured and passive earth pressure is gradually narrowing. This indicates that the excavated soil inside the excavation has a greater influence on reducing the passive soil pressure in the passive zone than the effect of the displacement of the supporting pile towards the excavation. On the other hand, as the lateral displacement of the embedded portion of the supporting piles increases, the soil in the passive zone is gradually compressed, and the soil arching effect of the circular excavation gradually comes into play. This effect enhances the soil's deformation resistance in the passive zone to some extent.

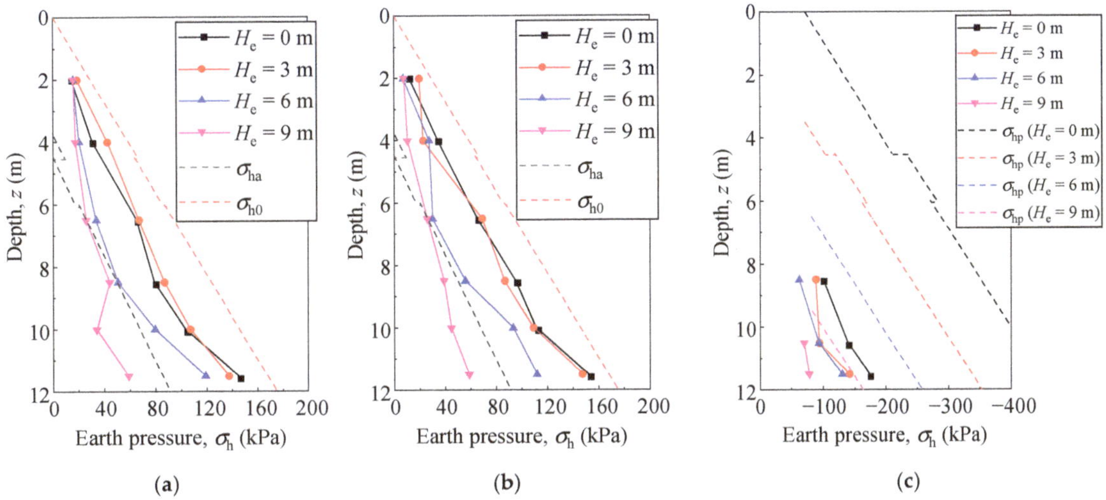

Figure 12. Lateral earth pressure on supporting piles: (**a**) non-excavation side of pile A; (**b**) non-excavation side of pile B; (**c**) excavation side of pile A.

3.4.2. On Steel Plates

Figure 13 shows the earth pressure distribution on the non-excavation side of steel plate A. It can be observed that prior to excavation (H_e = 0 m), the earth pressure on the steel plate is significantly less than the earth pressure at rest. Similar to the supporting piles, this is due to the steel plates between piles being inserted after trenching and then backfilled, causing some disturbance to the original soil. Upon completion of the first layer of excavation (H_e = 3 m), the earth pressure near the bottom of the excavation decreases significantly and is even less than the earth pressure on the supporting piles at the same depth, reflecting the soil arching effect between the piles. However, the change in the earth pressure on the steel plates at greater depths is less pronounced, which is consistent with the lateral displacement trend of the steel plates shown in Figure 6. After completing the second layer of excavation (H_e = 6 m), there is minimal change in earth pressure above 4 m, whereas the earth pressure below 4 m decreases significantly. Upon completion of the third layer of excavation (H_e = 9 m), minimal change is observed above 8.5 m, with a decrease in earth pressure below this depth. This indicates an increasing force imbalance between the two sides of the supporting structure with greater excavation depth. Due to the soil arching effect between the two adjacent supporting piles, the steel plates mainly bear the local earth pressure. Consequently, the earth pressure on the steel plates does not vary significantly with increasing excavation depth but remains relatively constant. Given the limited number of soil pressure measurement points in front of the steel plates, the exact horizontal distribution of the soil pressure at specific depths cannot be precisely determined. However, it is evident that the earth pressure acting on the steel plates is lower compared to that on the supporting piles, highlighting the presence of the soil arching effect between the piles.

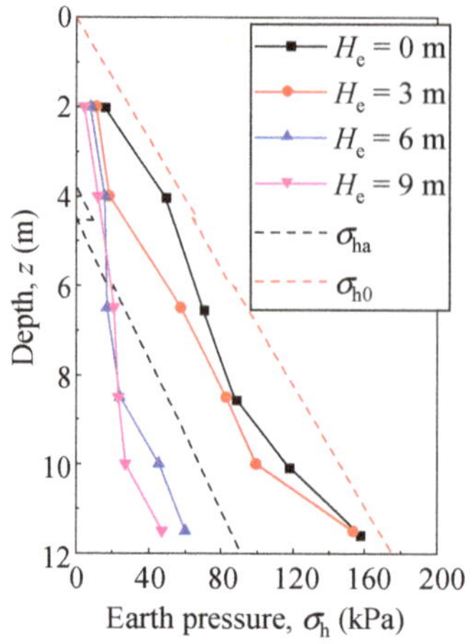

Figure 13. Lateral earth pressure on the non-excavation side of the steel plates.

3.5. Axial Force of Walings

Figure 14 presents variations of axial forces of walings during the excavation process. The axial forces of walings increase gradually with the excavation depth. At the three excavation stages, the axial forces of the first-level waling are 43.73 kN, 67.07 kN, and 86.64 kN, respectively. According to Figure 5, it can be observed that at the third stage (H_e = 9 m), the displacements of the supporting piles at their connections with the first-, second-, and third-level walings are 2.1 mm, 5.1 mm, and 8.1 mm, respectively. Consequently, the third-level waling experiences the highest axial force compared to the upper-level walings, due to the largest displacement increment observed at that location under all conditions.

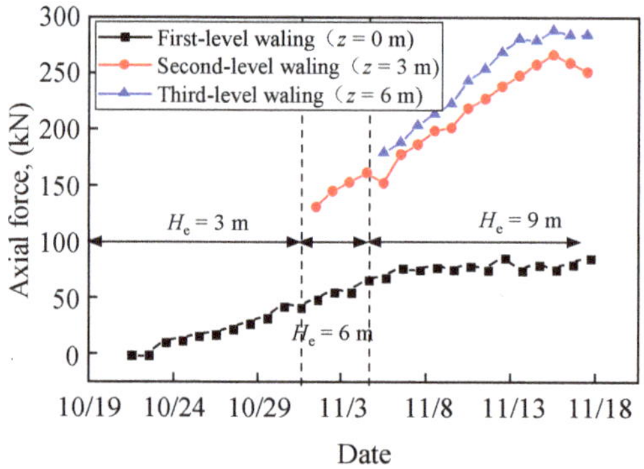

Figure 14. Variations of axial force of the walings.

4. Conclusions

In this paper, full-scale tests were conducted to verify the performance of PRSS in circular excavation, and the deformation and forces of the PRSS were analyzed. The conclusions drawn from this study are summarized below:

(1) The lateral displacements of both the supporting piles and the steel plates exhibit a convex shape; however, the lateral displacements of the steel plates are significantly greater than those of the supporting piles. The maximum lateral displacement values are 0.10% H_e for the supporting piles and 0.18% H_e for the steel plates. The maximum lateral

displacements of the supporting piles and steel plates occur above the excavation bottom, at depths of 0.80 H_e and 0.91 H_e, respectively.

(2) As excavation depths increase, the ground surface settlement profiles behind both the supporting piles and steel plates transition from a spandrel to a concave shape, with maximum settlement values of 0.075% H_e and 0.081% H_e occurring at a distance of 0.22 H_e from the supporting structure, respectively. The differences in ground surface settlements behind the supporting piles and steel plates are not as obvious as their lateral displacements.

(3) The arching effects caused by the circular excavation result in the earth pressure on the supporting piles being less than the active earth pressure. Moreover, due to the soil arching effect between two adjacent piles, the earth pressure acting on the upper part of the steel plates shows little change with increasing excavation depth, while the soil pressure on the lower part decreases noticeably.

(4) As the excavation depth increases, the supporting piles move toward the inside of the excavation, and the walings gradually provide support for the piles, while the axial force on the walings tends to increase.

Author Contributions: Conceptualization, C.G. and Y.P.; methodology, L.C. and Y.P.; formal analysis, L.C.; investigation, L.C.; resources, C.G.; data curation, L.C.; writing—original draft preparation, L.C.; writing—review and editing, C.G., Y.P. and L.C.; visualization, L.C.; supervision, C.G. and Y.P.; project administration, H.L., Y.P. and M.T.; funding acquisition, C.G., M.T. and H.L. All authors have read and agreed to the published version of the manuscript.

Funding: This research was funded by the Science and Technology Program of Guangzhou Municipal Construction Group Co., Ltd., China, grant number 2022-KJ004. The funder had the following involvement with the study: the construction and monitoring of the PRSS.

Data Availability Statement: The data presented in this study are available on request from the corresponding author due to restrictions on privacy.

Conflicts of Interest: Authors Huqing Liang and Mengxiong Tang were employed by the company Guangzhou Municipal Construction Group Co., Ltd. The remaining authors declare that the research was conducted in the absence of any commercial or financial relationships that could be construed as a potential conflict of interest. The authors declare that this study received funding from the Science and Technology Program of Guangzhou Municipal Construction Group Co., Ltd., China. The funder had the following involvement with the study: the construction and monitoring of the PRSS.

References

1. Zhang, C.; Zhang, Y.; Xia, Y.; Fang, H.; Zhao, P.; Wang, C.; Li, B.; Pan, Y.; Zou, Z.; Rabczuk, T.; et al. Risk assessment and optimization of supporting structure for a new recyclable pipe jacking shaft during excavation process. *Process Saf. Environ. Prot.* **2023**, *172*, 211–224. [CrossRef]
2. Zhao, G.; Meng, S.; Guan, C.; Yang, Y. Test Study on the Stress and Deformation Behaviors of a Shaft Supported by a Prefabricated Prestressed Structure. *Appl. Sci.* **2019**, *9*, 629. [CrossRef]
3. Zhang, H.; Wang, H.; Yang, J.; Wang, F. A novel vertical waterproofing joint with trapezoidal steel plate connections for steel–concrete underground silos: Bending test and numerical simulation. *Tunn. Undergr. Space Technol.* **2023**, *137*, 105150. [CrossRef]
4. Xin, J.; Duan, Q.; Jin, K.; Sun, J. A reduced-scale experimental study of dispersion characteristics of hydrogen leakage in an underground parking garage. *Int. J. Hydrogen Energy* **2023**, *48*, 16936–16948. [CrossRef]
5. Rapti, D.; Tinti, F.; Caputo, C.A. Integrated Underground Analyses as a Key for Seasonal Heat Storage and Smart Urban Areas. *Energies* **2024**, *17*, 2533. [CrossRef]
6. Liu, B.; Zhang, D.; Wang, Y.; Wang, N.; Xu, W. Design optimization and observed performance of a super-large foundation pit excavation subjected to unsymmetrical loading in water-rich floodplain: A case study. *Soils Found.* **2023**, *63*, 101329. [CrossRef]
7. Nejjar, K.; Dias, D.; Cuira, F.; Chapron, G.; Le Bissonnais, H. Numerical modelling of a 32 m deep excavation in the suburbs of Paris. *Eng. Struct.* **2022**, *268*, 114727. [CrossRef]
8. Lu, T.; Wu, K.; Liu, S.; Cai, G. Method for estimating three-dimensional effects on braced excavation in clay. *Tunn. Undergr. Space Technol.* **2023**, *141*, 105355. [CrossRef]
9. Mishra, A.; Sawant, V.A. A Detailed Investigation on Contiguous Pile Wall with Homogeneous Backfill. *Geotech. Geol. Eng.* **2023**, *41*, 2065–2089. [CrossRef]

10. Ramadan, M.I.; Meguid, M. Behavior of cantilever secant pile wall supporting excavation in sandy soil considering pile-pile interaction. *Arab. J. Geosci.* **2020**, *13*, 466. [CrossRef]
11. Zolqadr, E.; Yasrobi, S.S.; Norouz Olyaei, M. Analysis of soil nail walls performance-Case study. *Geomech. Geoengin. Int. J.* **2016**, *11*, 1–12. [CrossRef]
12. Liu, L.; Wu, R.; Congress, S.S.C.; Du, Q.; Cai, G.; Li, Z. Design optimization of the soil nail wall-retaining pile-anchor cable supporting system in a large-scale deep foundation pit. *Acta Geotech.* **2021**, *16*, 2251–2274. [CrossRef]
13. Shao, Y.; Macari, E.J.; Weiming, C. Compound deep soil mixing columns for retaining structures in excavations. *J. Geotech. Geoenviron. Eng.* **2005**, *131*, 1370–1377. [CrossRef]
14. Nishanthan, R.; Liyanapathirana, D.S.; Leo, C.J. Deep Cement Mixed Walls with Steel Inclusions for Excavation Support. *Geotech. Geol. Eng.* **2018**, *36*, 3375–3389. [CrossRef]
15. Waichita, S.; Jongpradist, P.; Patawanit, P.; Jamsawang, P.; Arangelovski, G.; Likitlersuang, S. Deformation and failure mechanism of deep cement mixing walls: Experimental study using physical model tests. *Arch. Civ. Mech. Eng.* **2021**, *21*, 127. [CrossRef]
16. Tan, Y.; Paikowsky, S.G. Performance of Sheet Pile Wall in Peat. *J. Geotech. Geoenviron. Eng.* **2008**, *134*, 445–458. [CrossRef]
17. Caputo, G.V.; Conti, R.; Viggiani, G.M.B.; Prüm, C. Improved Method for the Seismic Design of Anchored Steel Sheet Pile Walls. *J. Geotech. Geoenviron. Eng.* **2021**, *147*, 4020154. [CrossRef]
18. Sobala, D.; Rybak, J. Steel Sheet Piles-Applications and Elementary Design Issues. *IOP Conf. Ser. Mater. Sci. Eng.* **2017**, *245*, 22072. [CrossRef]
19. Lu, W.; Lou, J.; Webster, C.; Xue, F.; Bao, Z.; Chi, B. Estimating construction waste generation in the Greater Bay Area, China using machine learning. *Waste Manag.* **2021**, *134*, 78–88. [CrossRef]
20. Yücel, M.; Kayabekir, A.E.; Bekdaş, G.; Nigdeli, S.M.; Kim, S.; Geem, Z.W. Adaptive-Hybrid Harmony Search Algorithm for Multi-Constrained Optimum Eco-Design of Reinforced Concrete Retaining Walls. *Sustainability* **2021**, *13*, 1639. [CrossRef]
21. Seol, Y.; Lee, S.; Lee, J.; Han, J.; Hong, G. Excavation Method Determination of Earth-Retaining Wall for Sustainable Environment and Economy: Life Cycle Assessment Based on Construction Cases in Korea. *Sustainability* **2021**, *13*, 2974. [CrossRef]
22. Wasim, M.; Vaz Serra, P.; Ngo, T.D. Design for manufacturing and assembly for sustainable, quick and cost-effective prefabricated construction—A review. *Int. J. Constr. Manag.* **2022**, *22*, 3014–3022. [CrossRef]
23. Wang, W.; Ding, W.; Yang, X.; Zheng, G.; Xu, Z. Deep excavation engineering and underground engineering—New techniques of high-efficiency and energy-saving, low environmental impact, and sustainable development. *China Civ. Eng. J.* **2020**, *53*, 78–98. [CrossRef]
24. Guo, C.; Ye, J.; Zhao, C.; Wang, F. Mechanical and deformation characteristics of composite assembled supporting structure. *Geotech. Res.* **2020**, *7*, 230–243. [CrossRef]
25. Guo, C.; Wang, R.; Lin, P.; Wang, F. Numerical analyses of a prefabricated retaining system for foundation pits in silt soils. *Geotech. Res.* **2020**, *7*, 173–190. [CrossRef]
26. Pan, Y.; Fang, H.; Li, B.; Wang, F. Stability analysis and full-scale test of a new recyclable supporting structure for underground ecological granaries. *Eng. Struct.* **2019**, *192*, 205–219. [CrossRef]
27. Guan, C.; Yang, Y.; Wang, C. Rapid excavation with a newly developed retaining system: Spiral assembly steel structure. *J. Cent. South Univ.* **2015**, *22*, 2719–2729. [CrossRef]
28. Zhang, C.; Zhai, W.; Xia, Y.; Wang, C.; Zhao, P.; Li, B.; Pan, Y.; Fang, H.; Ummin, O.; He, W.; et al. Analysis of the stability of a novel assembled working shaft support structure during pipe jacking construction: Experiments and numerical simulations. *Eng. Fail. Anal.* **2024**, *162*, 108418. [CrossRef]
29. Kim, K.; Lee, D.; Cho, J.; Jeong, S.; Lee, S. The effect of arching pressure on a vertical circular shaft. *Tunn. Undergr. Space Technol.* **2013**, *37*, 10–21. [CrossRef]
30. Tran, V.D.H.; Meguid, M.A.; Chouinard, L.E. Discrete Element and Experimental Investigations of the Earth Pressure Distribution on Cylindrical Shafts. *Int. J. Geomech.* **2014**, *14*, 80–91. [CrossRef]
31. Tangjarusritaratorn, T.; Miyazaki, Y.; Sawamura, Y.; Kishida, K.; Kimura, M. Numerical investigation on arching effect surrounding deep cylindrical shaft during excavation process. *Undergr. Space* **2022**, *7*, 944–965. [CrossRef]
32. Xiong, G.J.; Wang, J.H.; Chen, J.J. Theory and practical calculation method for axisymmetric active earth pressure based on the characteristics method considering the compatibility condition. *Appl. Math. Model.* **2019**, *68*, 563–582. [CrossRef]
33. Tobar, T.; Meguid, M.A. Experimental Study of the Earth Pressure Distribution on Cylindrical Shafts. *J. Geotech. Geoenviron. Eng.* **2011**, *137*, 1121–1125. [CrossRef]
34. Tan, Y.; Wang, D. Characteristics of A Large-Scale Deep Foundation Pit Excavated by Central-Island Technique in Shanghai Soft Clay. Part II: Top-Down Construction of the Peripheral Rectangular Pit. *J. Geotech. Geoenviron. Eng.* **2013**, *139*, 1894–1910. [CrossRef]
35. Qiao, Y.; Xie, F.; Bai, Z.; Lu, J.; Ding, W. Deformation characteristics of ultra-deep circular shaft in soft soil: A case study. *Undergr. Space* **2024**, *16*, 239–260. [CrossRef]
36. Wang, F.; Chen, L.; Pan, Y.; Guo, C.; Guo, C.; Yue, L.; Chu, X. Deformation characteristics of excavation supported by prefabricated recyclable structures. *Proc. Inst. Civ. Eng. Geotech. Eng.* **2024**, 1–43. [CrossRef]
37. *GB 50497-2019*; Technical Standard for Monitoring of Building Excavation Engineering. Ministry of Housing and Urban-Rural Development: Beijing, China, 2019.

38. Clough, G.W.; O'Rourke, T.D. Construction induced movements of in situ walls. In Proceedings of the ASCE Conference on Design and Performance of Earth Retaining Structures, New York, NY, USA, 18–21 June 1990; pp. 439–479.
39. Ou, C.; Liao, J.; Lin, H. Performance of Diaphragm Wall Constructed Using Top-Down Method. *J. Geotech. Geoenviron. Eng.* **1998**, *124*, 798–808. [CrossRef]
40. Liu, M.; Zhang, D.; Fang, Q.; Hou, Y. Wall and Ground Movements due to Deep Excavation of Tianjin Subway Station. *DEStech Trans. Mater. Sci. Eng.* **2017**, *2017*, 9944. [CrossRef] [PubMed]
41. Ni, X.; Lu, J.; Wang, Y.; Shi, J.; Chen, W.; Tang, L. Field investigation of the influence of basement excavation and dewatering on ground and structure responses. *Tunn. Undergr. Space Technol.* **2021**, *117*, 104121. [CrossRef]
42. Mana, A.I.; Clough, G.W. Prediction of Movements for Braced Cuts in Clay. *J. Geotech. Eng. Div.* **1981**, *107*, 759–777. [CrossRef]
43. Goldberg, D.T.; Jaworski, W.E.; Gordon, M.D. *Lateral Support Systems and Underpinning;* Federal Highway Administration: Washington, DC, USA, 1976.
44. Moormann, C. Analysis of Wall and Ground Movements Due to Deep Excavations in Soft Soil Based on a New Worldwide Database. *Soils Found.* **2004**, *44*, 87–98. [CrossRef]
45. Ou, C.; Hsieh, P.; Chiou, D. Characteristics of ground surface settlement during excavation. *Can. Geotech. J.* **1993**, *30*, 758–767. [CrossRef]
46. Hsieh, P.; Ou, C. Shape of ground surface settlement profiles caused by excavation. *Can. Geotech. J.* **1998**, *35*, 1004–1017. [CrossRef]
47. Ali, J.; Khan, A.Q. Behaviour of anchored pile wall excavations in clays. *Proc. Inst. Civ. Eng. Geotech. Eng.* **2017**, *170*, 493–502. [CrossRef]
48. Richards, D.J.; Clark, J.; Powrie, W. Installation effects of a bored pile wall in overconsolidated clay. *Géotechnique* **2006**, *56*, 411–425. [CrossRef]
49. Terzaghi, K. *Theoretical Soil Mechanics;* John Wiley and Sons, Inc.: New York, NY, USA, 1943; p. 510.
50. Guojun, X.; Jianhua, W. A rigorous characteristic line theory for axisymmetric problems and its application in circular excavations. *Acta Geotech.* **2020**, *15*, 439–453. [CrossRef]
51. Liu, F.Q.; Wang, J.H. A generalized slip line solution to the active earth pressure on circular retaining walls. *Comput. Geotech.* **2008**, *35*, 155–164. [CrossRef]
52. Liu, F.Q.; Wang, J.H.; Zhang, L.L. Axi-symmetric active earth pressure obtained by the slip line method with a general tangential stress coefficient. *Comput. Geotech.* **2009**, *36*, 352–358. [CrossRef]

Disclaimer/Publisher's Note: The statements, opinions and data contained in all publications are solely those of the individual author(s) and contributor(s) and not of MDPI and/or the editor(s). MDPI and/or the editor(s) disclaim responsibility for any injury to people or property resulting from any ideas, methods, instructions or products referred to in the content.

buildings

MDPI

Article

Influence of Different Construction Methods on Lateral Displacement of Diaphragm Walls in Large-Scale Unsupported Deep Excavation

Hao-Wei Chiu [1], Chia-Feng Hsu [2,*], Fu-Huan Tsai [3] and Shong-Loong Chen [1]

[1] Department of Civil Engineering, National Taipei University of Technology, Taipei 10608, Taiwan; herciam.eu@gmail.com (H.-W.C.); f10391@ntut.edu.tw (S.-L.C.)
[2] Department of Civil Engineering, Chienkuo Technology University, Changhua City 500020, Taiwan
[3] Huaku Development Co., Ltd., Taipei 110501, Taiwan; tim30151@gmail.com
* Correspondence: chiafeng1013@gmail.com

Abstract: This study examines the influence of different construction methods on the lateral displacement of diaphragm walls in large-scale, unsupported deep excavations. Using the three-dimensional finite element method (FEM) with PLAXIS 3D 2017 software, the research assesses how varying construction techniques impact wall stability, particularly in proximity to sensitive structures like metro systems. The project uniquely integrates peripheral top-down and central bottom-up approaches to minimize environmental disruption. Key focus areas include the roles of back-pull slabs, zoned excavation, and cross walls in reducing wall deformation. Findings reveal that zoned excavation significantly controls lateral displacement on longer site sides, enhancing adjacent structure safety and overall construction integrity. Back-pull slabs are shown to effectively decrease top wall deformation, thereby increasing structural stiffness. Moreover, despite their considerable length (nearly 60 m), cross walls play a crucial role in controlling lateral deformation along the excavation's length. These insights offer valuable guidance for future projects, especially in regions like Taiwan, where experience with such large-scale, unsupported excavations is limited.

Keywords: large-scale deep excavation; unsupported; FEM; zoned excavation; back-pull slab; cross wall

1. Introduction

This study explores the complexities of and innovative solutions for large-scale, unsupported deep excavations in Taiwan's urban environments. These excavations, essential for the development of high-rise buildings and extensive subterranean structures, are driven by the growing demand for space. Emphasizing the management of excavation projects that exceed 10,000 square meters, the study focuses on the 'Peripheral Top-Down and Central Bottom-Up' approach. Although this method is not widely used in Taiwan, it presents substantial benefits, including a reduced reliance on temporary supports, faster excavation processes, and the use of structurally robust diaphragm walls to maintain foundational stability effectively.

Employing Plaxis 3D 2017 for numerical analysis, the research corroborates its findings with field data, specifically evaluating the method's capacity to control wall deformation. This includes the strategic use of buttress walls, cross walls, and structural slabs in managing diaphragm wall deformation. A key aspect of the study is the detailed examination of back-pull slabs and the use of cross walls extending nearly 60 m, assessing their effectiveness in reducing wall deformation. The investigation, through both field monitoring and numerical simulation, seeks to determine the efficiency of these cross walls in mitigating deformation. The study compiles the various advantages of this construction technique, thereby providing valuable insights for future design practices and serving as a resource for academic research.

In summary, this study combines practical engineering case examples with monitoring data for validation. Currently, such cases are relatively rare in Taiwan. Therefore, it provides an excellent demonstration of design and construction for future large-scale, irregular, unsupported deep excavation projects in Taiwan. By referring to the various numerical scenario analyses and associated construction methods detailed in this case, the goal of achieving optimized design and construction can be accomplished.

Literature Review

- **Peripheral Top-Down and Central Bottom-Up Method**

The 'Peripheral Top-Down and Central Bottom-Up' method, highlighted in Liu's Foundation Pit Engineering Handbook (second edition) [1], represents a sophisticated approach for managing ultra-large-area deep excavations, often covering areas beyond tens of thousands of square meters. This technique involves initially constructing a rigid ring beam slab around the excavation site's perimeter. The process commences with central area excavation, moving downwards to the designated level, and is followed by upward construction that connects back to the peripheral ring slab. Meanwhile, the peripheral sections undergo a simultaneous process of layered excavation and top-down construction. This method is particularly effective for sites with a minimum of two basement levels and a polygonal excavation plan. Critical aspects include ensuring the peripheral ring slab's width for optimal rigidity, angling the surrounding soil for stability, and preserving a significant central space for bottom-up construction once the top-down section is complete.

Lim and Ou's [2] case studies of unsupported excavations in soft soil [2] show that buttress walls and internal structural slabs are often used in large-scale unsupported excavations to serve as a support structure system. This strategy effectively reduces both construction costs and timeframes. The final lateral displacement of the wall relative to the final excavation depth (ratio δ_{hm}/H_e) is observed to range between 0.27 and 0.55. For the longer sides of the excavation site, diaphragm walls exhibit translational movement at the wall's toe, as depicted in Figure 1a, resulting in a cantilevered wall deformation. Conversely, on the shorter sides, the diaphragm walls show curvilinear deformation, with the maximum wall deformation occurring below the final excavation level, as illustrated in Figure 1b.

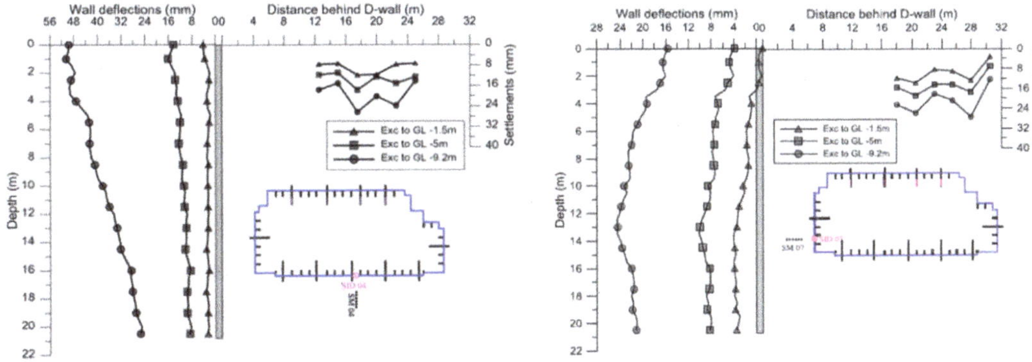

(**a**) (**b**)

Figure 1. Monitoring curves for lateral wall deformation and ground settlement at the site: (**a**) longitudinal; (**b**) transverse (Lim and Ou [2]).

- **Impact of Zoned Excavation**

Ou [3] mentioned that for excavation sites using diaphragm walls as retaining walls, due to the arching effect of the concrete walls, the deformation and surface settlement at and near the corners where the diaphragm walls meet are relatively small. Additionally, the deformation on the shorter sides of the excavation site is less than that on the longer sides. Zoned excavation utilizes this principle to reduce deformation of the retaining walls and surface settlement during excavation.

Jeng and Ho [4] conducted a simulation study focusing on a Taipei site using the Peripheral Top-Down and Central Bottom-Up method, primarily examining the stability and deformation of retaining walls during excavation. Their approach included both three-dimensional and simplified one-dimensional analyses for comparison. Findings indicated that the maximum deformation of diaphragm walls in the three-dimensional analysis closely aligned with actual field measurements, though variations in deformation patterns were noted. Comparisons between full area and zoned excavation revealed that zoned excavation, as per numerical simulations, significantly reduced the lateral displacement of the diaphragm wall bodies.

In a study by Li [5], the excavation process at Shanghai Xi Railway Station metro station was scrutinized. To safeguard the railway and subway systems, a zoned construction approach was proposed, incorporating the erection of several cross walls alongside necessary support structures. This technique partitioned the extensive deep excavation area into smaller, manageable zones. This phased and zoned construction method successfully mitigated conflicts between construction activities and railway operations. The outcomes showed that this approach effectively protected adjacent infrastructures, with maximum lateral wall displacements varying between 0.05% H_e and 0.42% H_e, as depicted in Figure 2. This is within the range of 0.1% H_e and 1.0% H_e proposed by Wang et al. [6], where H_e is the final excavation depth. The study suggested that utilizing the Time–Space Effect (TSE) in phased and zoned construction techniques can aid in the control of deformation in deep excavations.

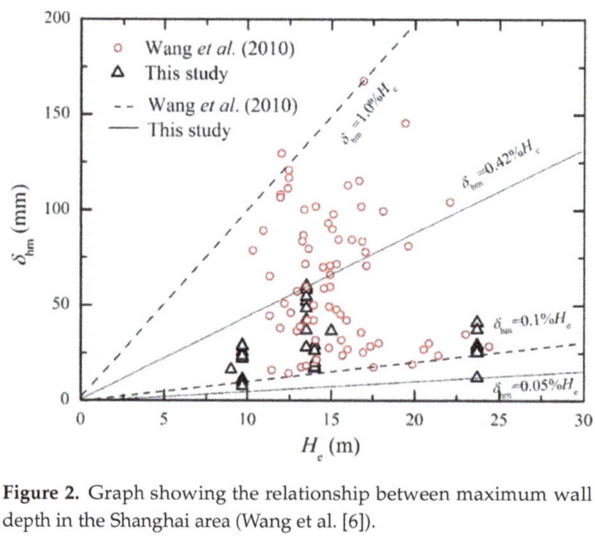

Figure 2. Graph showing the relationship between maximum wall displacement and excavation depth in the Shanghai area (Wang et al. [6]).

- **Lateral Deformation of Retaining Walls Due to Deep Excavation**

Clough and O'Rourke's research [7] delineates three distinct patterns of deformation in retaining walls during excavation projects: cantilever displacement, deep lateral inward displacement, and a combination of both. In deep excavations lacking top supports or countermeasures, the walls typically exhibit a cantilever-like displacement. This behavior is influenced by the depth-to-width ratio and the restraint provided by soil at penetration points. The study also notes that different types of soil, such as hard clay, residual soils, and sandy soils, result in varying degrees of ground surface settlement around these structures. It was found that, on average, the maximum lateral displacement of a wall is approximately 0.2% of the excavation depth (*H*), seldom exceeding 0.5% *H*. Factors such as poor construction quality or insufficient wall depth could lead to excessive displacement. Particularly in softer clay layers, there is a notable correlation between wall displacement and excavation depth, linked to the safety factor against bottom heave. Ou et al. [8] further expand on this topic by discussing how the location of maximum lateral displacement in diaphragm wall bodies changes relative to the excavation depth. While the initial excavation phase predominantly causes cantilever displacement with maximum lateral displacement at the top of the wall, in later phases, this maximum displacement tends to occur closer to the excavation face's depth. According to a finite element method-based parameter analysis of deep excavation wall deformations conducted by Manna and Clough [9], which was compared with field monitoring results of deep excavations, several critical factors can be identified that influence the deformation behavior of deep excavation walls, including the safety factor against heave, the stiffness of the retaining wall and its support systems, the magnitude of support pre-stress, the excavation area's geometric shape, and the duration of the construction period.

Masuda et al. [10] took a broader approach, analyzing 52 excavation cases to identify key factors influencing excavation stability. They concluded that soil stiffness and support stiffness are primary influencers. Additionally, they identified 11 specific factors contributing to retaining wall deformation, including soil type and properties, wall and support stiffness, number and spacing of supports, support pre-stress, excavation method, wall length, ground improvement measures, excavation scale, groundwater conditions, and other concurrent construction activities. Notably, cross walls, which link diaphragm walls either on the north–south or east–west axes, serve as a crucial support system. These walls, typically reinforced below the excavation face and progressively removed during excavation, enhance rigidity, reduce wall deformation, and provide resistance against uplift. Wu et al. [11] focused on the application of cross walls in regions like Japan and Taiwan. Their research, involving 11 real cases with cross walls and 11 models without, found that cross walls significantly reduce wall displacement and ground surface settlement. They developed a simplified regression model for predicting the deformation of diaphragm wall bodies with cross walls. Their research highlights the corner effect in cross walls, akin to that in diaphragm walls, where displacement at the T-junction is minimal. The study also notes that the displacement profile of adjacent cross walls is similar to a pattern illustrated in their diagram, shown in Figure 3.

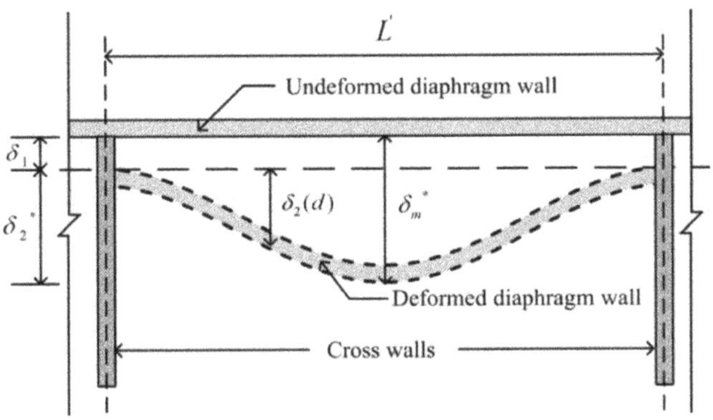

Figure 3. Wall displacement for excavation with cross walls (Wu et al. [11]).

In the figure, δ_m, in general, denotes the maximum wall displacement, while $\delta_m(d)$ denotes the δ_m measured at a distance d away from the corner. This notation of d is needed because the corner effect shows that δ_m decreases with decreasing d. A special case of δ_m is $\delta_m\left(d = \frac{L'}{2}\right)$, denoted by $\delta_m{}^*$, the maximum wall displacement at the middle.

- **Application of Numerical Analysis in Deep Excavation Engineering**

In their study, Do et al. [12] conducted a detailed analysis using the finite element method with PLAXIS software, focusing on a deep excavation project in soft soil that experienced three failures and one instance of excessive deformation. The research brought attention to a unique phenomenon dubbed "toe-kicking" occurring at the bottom end of the diaphragm wall. This phenomenon involves the soil situated behind the wall moving downward and forward into the excavated area. As a result, the embedded portion of the wall is compelled to move and bend toward the excavation area. The study observed that the maximum horizontal displacement happens at the toe of the wall, as depicted in Figure 4. Crucially, the occurrence of this toe-kicking phenomenon is conditional on specific angular relationships: it is not observed when α1 is less than α2, but it does occur when α1 exceeds α2. This finding adds a vital understanding of wall behavior and soil interaction in deep excavation scenarios, particularly in soft soil conditions.

$$\alpha 1 = tan^{-1}\left(\frac{H_e}{D_{He} - D_o}\right) \tag{1}$$

$$\alpha 2 = tan^{-1}\left(\frac{H_p}{D_{He+Hp} - D_{He}}\right) \tag{2}$$

where H_e: excavation depth; H_p: penetration depth; Do: ground horizontal displacement; D_{He}: final excavation depth horizontal displacement; D_{He+Hp}: toe displacement of the wall.

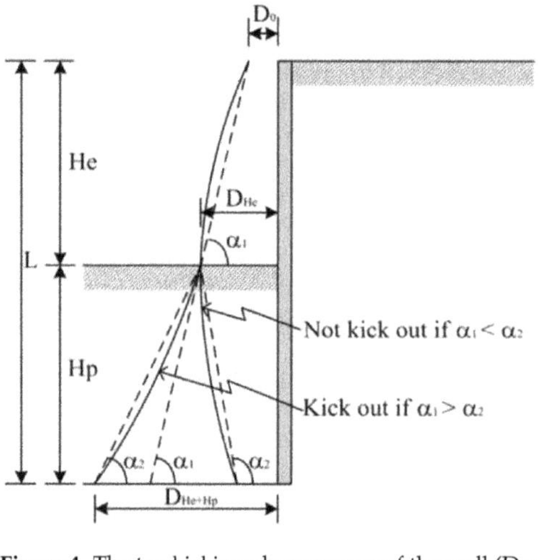

Figure 4. The toe-kicking phenomenon of the wall (Do et al. [12]).

Chheng et al. [13] undertook a comprehensive analysis of a deep excavation project in Bangkok through a three-dimensional finite element approach using PLAXIS 3D. This analysis, fine-tuned with data from both laboratory and field tests, aimed to draw a comparison between the simulated wall deformations and ground settlements in three dimensions against the conventional two-dimensional analyses and actual monitoring data. The study highlighted the limitations of two-dimensional methods, particularly their inability to account for the corner effect. In contrast, the three-dimensional model revealed distinct variations in lateral deformations along the longer and shorter sides of the excavation site, with an average difference of about 10%. A notable observation was the reduced wall movement near the excavation corners. The three-dimensional method's primary advantage lies in its capacity to replicate construction sequences realistically, thereby providing more accurate predictions of wall deformation and ground settlement at different stages of construction.

Similarly, Chen et al. [14] utilized a 3D software tool to investigate the role of exterior struts in minimizing the maximum lateral deformation in the middle of diaphragm walls, particularly in soft ground excavations. The study, corroborated by field data from the Taipei 101 and Neihu basement projects, delved into how the geometry of buttress walls influences their stabilizing effect. While simplifying the problem to two dimensions is common, it often fails to capture the full picture, especially for nearly square excavation sites. The research suggests that for buttress walls, the optimal spacing should be less than double the depth of the excavation, with T-shaped designs proving more effective than I-shaped ones. The study also emphasized the importance of not removing these walls sequentially during excavation to ensure optimal stability of the diaphragm wall.

Hsiung et al. [15] conducted a case study on wall deformation in a major deep excavation project in central Jakarta, Indonesia. This study employed three-dimensional numerical simulations to analyze wall deformation and bending moments, comparing these results with those from two-dimensional simulations. The findings indicated a high level of consistency between the 2D and 3D results. Moreover, the corner areas in the 3D simulation displayed smaller displacements compared to plane strain conditions, aligning more closely with the actual conditions observed on site.

Lim et al. [16] introduced new metrics to assess the impact of wall deformation and ground settlement, as illustrated in Figure 5: the maximum deflection ratio (MDR), maximum settlement ratio (MSR), and deflection area ratio (DAR). These ratios are defined and calculated using Equations (3)–(5). This innovative approach offers a more nuanced understanding of the effects of wall deformation and ground settlement in deep excavation projects.

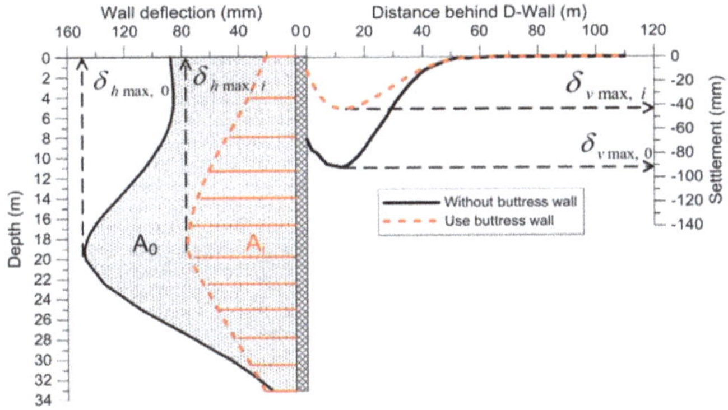

Figure 5. Illustration of parameters used for calculating MDR, MSR, and DAR values (Lim et al. [16]).

In this subsection, the performance of buttress walls is summarized and assessed quantitatively. Therefore, three parameters, namely, the maximum deflection ratio (MDR), the maximum settlement ratio (MSR), and the deflection area ratio (DAR), are proposed. The MDR, MSR, and DAR are defined as Equations (3)–(5):

$$\mathrm{MDR} = \frac{\delta_{hm-0} - \delta_{hm-i}}{\delta_{hm-0}} \times 100\% \tag{3}$$

$$\mathrm{MSR} = \frac{\delta_{vm-0} - \delta_{vm-i}}{\delta_{vm-0}} \times 100\% \tag{4}$$

$$\mathrm{DAR} = \frac{A_0 - A_i}{A_0} \times 100\% \tag{5}$$

where:

- $\delta_{hm\text{-}0}$ = the maximum lateral wall deflection without buttress walls;
- $\delta_{hm\text{-}i}$ = the maximum lateral wall deflection with buttress walls;
- $\delta_{vm\text{-}0}$ = the maximum ground surface settlement without buttress walls;
- $\delta_{vm\text{-}i}$ = the maximum ground surface settlement with buttress walls;
- A_0 = the plane area of lateral wall deflection without buttress walls;
- A_i = the plane area of lateral wall deflection with buttress walls.

In their study of a deep excavation project in Singapore, Chuah [17] explored the use of an unsupported excavation method that incorporated internal buttresses and structural slabs, ideal for large-scale, enclosed excavations. This innovative approach led to a cantilever pattern in wall deformation due to the lack of internal supports. Chuah utilized the finite element method with PLAXIS modeling to analyze this. The study highlighted that, unlike the more traditional 2D analyses, the 3D analysis provided a comprehensive and accurate representation of the deformation patterns. This insight is particularly vital in unsupported conditions, where controlling lateral wall deformation is crucial to prevent significant ground settlement.

- **Utilization of Corners**

According to Ou [3], diaphragm walls exhibit an "arch effect" at corners due to their horizontal stiffness, leading to less deformation and surface settlement near these areas compared to the central sections. This contrasts with steel sheet piles and soldier piles, which, lacking this horizontal stiffness, do not demonstrate the same effect, resulting in similar deformations at corners and central sections. Therefore, in scenarios where a building is close to the excavation site's corners or along its shorter side, using diaphragm walls as retaining structures can be advantageous to capitalize on this corner effect and better protect the surrounding environment.

- **Large-Scale Deep Excavation**

Liu et al. [18] reported on a massive 10.9–15.6 m deep, 70,500 m^2 irregular pit in Shanghai's soft clay, excavated using the bottom-up method and supported by auger cast-in-place pile walls and concrete struts. The key findings include: (1) the prolonged construction and large pit scale resulted in maximum wall deflection and ground subsidence exceeding 0.86% of the final excavation depth (H_e), significantly higher than other Shanghai excavations and surpassing allowable deformation limits; (2) basal heave displayed a dome-shaped pattern with minimal distortion between adjacent columns, but maximum distortion between columns and retaining walls neared the 0.2% limit; (3) significant axial force discrepancies were noted, with some axial forces in diagonal struts being 15% over the design value due to the spatial effects of internal supports; (4) the surrounding pipelines showed three-dimensional deformation and significant settlement, with a maximum of 56 mm.

Liu et al. [19] studied an 18.4 m deep triangular foundation pit in soft clay, covering about 10,000 m^2, at Shanghai's Hongqiao transportation hub. Key observations from the monitoring system during excavation include wall deflections, vertical wall and column movements, and ground subsidence. The pit's unique triangular shape, with acute corners, resulted in deformation characteristics distinct from other metro station or rectangular excavations in Shanghai. These findings are crucial for designing and constructing special-shaped pits in complex environments.

Ren et al. [20] conducted a study on a large 25,720 m^2, 20 m deep foundation pit in Chengdu, China, excavated in 2012–2013. The monitoring process included observing column movements, deflections, internal stress, anchor cable forces, ground settlement, and water levels. Analysis of this excavation, along with ten others in Chengdu, provided key insights: the maximum column deflection generally occurs at the top, vertical movements stabilize after excavation, and anchor cable forces range from 20 to 70 kN. Notably, minimal ground settlement was observed, influenced by the depth of the excavation and its aspect ratio. This research contributes to a better understanding of deformation in large-scale excavations.

Liang et al. [21] conducted a study detailing a large-scale basement excavation adjacent to a metro station, measuring 236.7 m in length, 145.8 m in width, and 23.2–27.3 m in depth, in soft clay. The excavation, divided into six zones by five partition walls, used a zoned-excavation approach to minimize impact on nearby metro and shield tunnels. Extensive monitoring covered lateral wall deflections, surface settlements, and vertical/horizontal displacements of the metro station and tunnels. Analysis revealed diaphragm walls moving towards the excavation, significant ground settlements, tunnel track heave within the station, and substantial shield tunnel settlements. Horizontal movements and elongations in the shield tunnel were also observed due to excavation-induced stress relief.

Ho and Gao [22] noted that their research case was located in Xizhi District, New Taipei City, Taiwan, with an excavation depth of 9.2 m and an area exceeding 2 hectares. The excavation was conducted without support, using diaphragm walls in conjunction with internal and external buttress walls of varying lengths. Additionally, a spanning floor slab around the perimeter was constructed using the top-down approach and also functioned as part of the retaining structure system. For geotechnical analysis, a three-dimensional numerical method (GTS) was employed to simulate and analyze the stability and deformation of the retaining walls. Besides the three-dimensional analysis, the case was also simplified for comparison using a one-dimensional method. Due to some uncertainties in the analysis results, emergency response plans were prepared in advance, allowing for immediate reinforcement if necessary. The actual deformation measured post-excavation was smaller than anticipated. This paper compares and analyzes the results with actual measurements, reviewing areas for improvement in the analysis process to serve as a reference for similar future cases.

This section's cases are summarized in Table 1 and compared large-scale deep excavation cases between Mainland China and Taiwan. Given that the majority of recent literature from Mainland China is based on studies of actual monitoring cases, subsequent discussions will primarily focus on two specific Taiwanese cases as the main points of reference for summarization.

Table 1. Large-scale deep excavation case comparison table.

Researcher	Location	Depth/Area	Numerical Analysis Method	Construction Method (Supported/Unsupported)	Field Monitoring	Key Findings
Liu et al. [18]	Shanghai/China	10.9–15.6 m/ 70,500 m^2	3D Numerical Method	Bottom-Up (Unsupported)	Wall Deflection, Ground Subsidence	Maximum wall deformation and ground subsidence exceed 0.86% of excavation depth
Liu et al. [19]	Shanghai/China	18.4 m/ approx. 10,000 m^2	No	Bottom-Up (Supported)	Wall Deflection, Vertical Wall and Column Movements, Ground Subsidence	Unique deformation characteristics due to the triangular shape of the pit
Ren et al. [20]	Chengdu/China	20 m/25,720 m^2	No	Bottom-Up Central-island (Supported)	Column Movements, Deflections, Internal Stress, Anchor Cable Forces, Ground Settlement, Water Levels	Maximum column deflection usually occurs at the top
Liang et al. [21]	Shanghai/China	23.2–27.3 m/ 31,309 m^2.	Finite Element Method	Zoned Excavation (Unsupported)	Lateral Wall Deflections, Surface Settlements, Metro Station and Tunnels Vertical/Horizontal Displacements	Significant ground settlements and horizontal movements in metro station and tunnels
Ho and Gao [22]	Xizhi District, New Taipei City/ Taiwan	9.2 m/over 2 hectares	GTS 3D Numerical Method	Peripheral Top-Down and Central Bottom-Up (Unsupported)	Wall Stability and Deformation	Actual deformation smaller than anticipated
This Study	Zhonghe District, New Taipei City/ Taiwan	12.25 m/ 13,720 m^2.	Plaxis 3D Finite Element Method	Peripheral Top-Down and Central Bottom-Up (Unsupported)	Lateral Displacement of Diaphragm Walls, Impact of Back-Pull Slabs and Cross Walls	Zoned excavation effectively controls lateral displacement on longer sides of the site

2. Materials and Methods

2.1. Research Methodology and Steps

This study utilizes real engineering case data and the finite element software PLAXIS 3D 2017 for its analysis. It involves a detailed comparison of field monitoring data with the results obtained from the analysis, thereby verifying the appropriateness of the chosen soil and structural parameters. Additionally, the study delves into the effects of zoned excavation, back-pull slabs, and cross walls on the deformation of retaining walls, using practical models for a more accurate representation.

The simulation process begins with the PLAXIS 3D 2017 software, tailored to the specific construction site conditions. This includes incorporating data from drilling reports, soil experimental parameters, and monitoring reports to set accurate parameters for the numerical simulation. The model is then constructed, taking into account the site's geometrical conditions and boundary settings. A key aspect of the simulation is its adherence to the actual construction conditions of the site. The simulation outcomes are then cross-referenced with on-site monitoring data, ensuring the validity of the soil and structural parameters used, as well as the construction steps simulated. These findings establish a baseline for the subsequent section, which focuses on analyzing the effectiveness of techniques like zoned excavation, the application of back-pull slabs, and the utilization of cross walls.

2.2. Case Introduction

This case study examines a construction project in the Zhonghe District of New Taipei City, featuring a notable 17-floor above-ground structure, standing 69.55 m tall, with three basement levels. Occupying a trapezoidal area of approximately 13,720 square meters, the site extends from 90.1 to 181.6 m east–west and 91.5 to 114.4 m north–south. It is encircled by various urban elements, including roads, an elevated MRT bridge, a gas station, metal buildings, and a seven-story building, as illustrated in Figure 6. The construction includes a 0.8 m thick diaphragm wall with depths ranging between 27 and 30 m. To control wall displacement, a "Peripheral Top-Down and Central Bottom-Up" method was employed, integrating cross walls, multiple buttress walls, and 0.4 m thick back-pull slabs. The excavation reached a depth of 12.25 m, supported by a raft foundation.

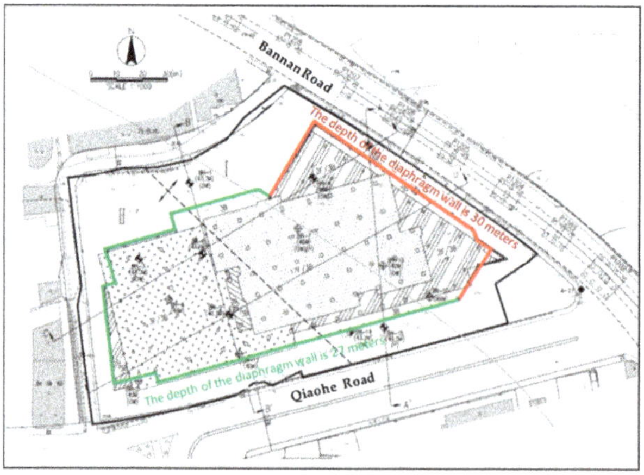

Figure 6. Site location map.

(1) Soil Layer Overview

Extensive data collection from 15 boreholes on-site, combined with laboratory test results and data from neighboring sites, led to the identification of around 13 distinct soil layers within a 50 m depth, as detailed in Appendix A, Figure A1. The borehole profiles, along with the distribution depth, physical properties, and engineering characteristics of each soil layer, are elaborated in Table A1 and Supplementary Figures S1–S5.

(2) Diaphragm Wall Installation

In this engineering case study, the diaphragm walls were categorized into two depth groups: 30 m and 27 m, each with a thickness of 0.7 m. The project aimed to minimize wall deformation and its impact on adjacent structures during excavation. To achieve this, a combination of cross walls, internal buttress walls, and external buttress walls were strategically installed, as illustrated in Figure 7. The installation method for these structures mirrored that of the diaphragm walls.

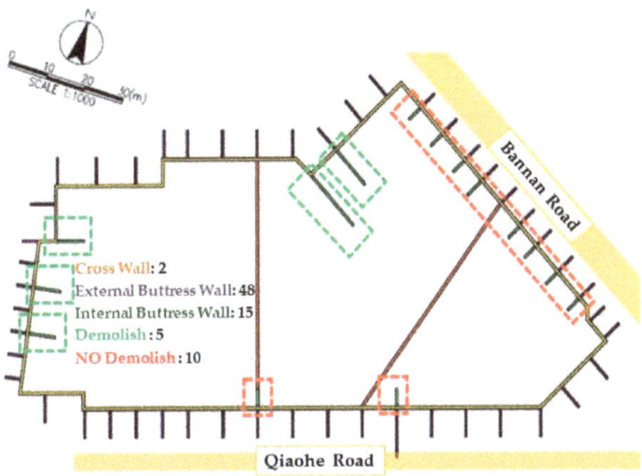

Figure 7. Plan layout of diaphragm wall, internal buttress wall, external buttress wall, and cross wall.

Specifically, two cross walls were erected; one intersected the diaphragm wall perpendicularly, while the other was positioned at a non-perpendicular angle. These walls, which extended approximately 60 m from a depth of GL −2.4 m to −25 m, were gradually removed as excavation progressed. The site also included fifteen internal buttress walls. Notably, eight of these walls near Banan Road and two near Qiaohe Road were not removed and later integrated into the final structure. In contrast, the remaining five internal buttress walls were removed during the excavation process. The depth of these internal and external buttress walls varied: the eight internal walls near the MRT on Banan Road extended from GL −2.4 m to −30 m, while the others ranged from GL −2.4 m to −27 m. An important aspect of this project was the joint connections between the diaphragm walls, cross walls, and both internal and external buttress walls, which were executed using T-shaped and cross-shaped joints to ensure structural integrity and stability.

(3) Monitoring System

Given the site's proximity to neighboring buildings, roads, and the MRT, rigorous construction control was essential. The primary goal of the monitoring system was to ensure the safety of the excavation work and the integrity of nearby structures. Monitoring included diaphragm wall deformation, groundwater level changes, and the settlement and tilting of neighboring buildings and roads. The monitoring equipment comprised internal and external wall inclinometers, strain gauges, water level observation wells, building tiltmeters, settlement observation points, and piezometers. This study particularly focused on inclinometer monitoring data to validate the numerical model's accuracy in comparison to the simulation results.

2.3. Numerical Simulation

PLAXIS, renowned for its proficiency in finite element analysis, particularly in geotechnical engineering, is increasingly favored in both academic and professional realms due to its comprehensive features and user-friendly graphical interface. This research utilizes PLAXIS 3D 2017, and the following outlines the model establishment procedure for simulating deep excavation engineering using this software:

(1) Defining the Overall Model Boundary Profile: The initial step involves calculating the dimensions of the construction site based on the area and excavation depth. For this model, the boundary extends 3 to 5 times the excavation depth from the diaphragm wall. Due to adjacent building loads, a fivefold extension is used in this case. Drilling reports inform the thickness and groundwater level of each soil layer, with additional consideration for irregular layers. The soil body's boundaries are defined with roller supports, allowing only vertical movement and hinge supports to limit both vertical and horizontal displacements. The bottom boundary is set at either a hard layer or a non-influential depth, using hinged nodes, while the side boundaries employ roller supports.
(2) Defining Soil Parameters: Soil parameters for each layer are determined from drilling reports, indoor experiments, and empirical formulas. These parameters are then applied to the already drawn model, defining each soil layer's characteristics.
(3) Simulation of Deep Excavation Site: The structures involved in deep excavation engineering—including diaphragm walls, cross walls, internal and external buttress walls, structural slabs, back-pull slabs, and foundation slabs—are simulated using plate elements. Geometric shapes are drawn and defined as plate elements, followed by applying material parameters to each structure.
(4) Simulation of Neighboring Building Loads: Due to the absence of detailed construction data for neighboring buildings, the simulation uses plate and rigid body elements, with assumptions based on relevant literature and common engineering practices.
(5) Setting Construction Steps and Water Levels: The construction process is set according to the actual steps taken on site. Water levels are adjusted around the excavation face to facilitate dry conditions for construction and ensure quality.
(6) Defining Grid and Grid Density: Once the previous steps are complete, the grid is generated, with densification around the excavation face to enhance the accuracy of the analysis. This study opts for a medium grid density.
(7) Starting the Calculation: With all parameters and configurations set, the final step is to initiate the calculation process in PLAXIS, leading to the analysis of the simulation results.

Based on the aforementioned model establishment procedure, the detailed content is as follows:

(1) Model Setup

This stage involves configuring the numerical model to mirror the actual case study. The model's excavation area and geometric shape are precisely aligned with the original site's size and geometry. The setup includes accurate depth settings for each stage of excavation, diaphragm walls, cross walls, internal and external buttress walls, internal structural slabs, and back-pull slabs, all of which are matched to the depths in the original construction. The model's boundary is defined with a horizontal distance of five times the excavation depth from the diaphragm wall. The vertical depth of the model is set to 50 m, resulting in an overall model size of 313 m × 222 m × 50 m.

(2) Selection of Soil Parameters

The soil in this deep excavation case study is categorized into 13 distinct layers, comprising 5 sandy soil layers, 7 clay soil layers, and a bottommost gravel layer (as detailed in Appendix A, Figure A1 and Table A1). The Mohr–Coulomb soil model is employed for the analysis. For sandy and gravel layers, a drained analysis approach is used, while clay layers are analyzed using an undrained approach. The Mohr–Coulomb model primarily involves four parameters: c, ψ, E, and ν. For drained analysis, the model requires the following: effective modulus of elasticity (E'), effective Poisson's ratio (ν'), effective angle of internal friction (φ'), and effective cohesion (c'). The undrained model used in the study is Undrained C. The Undrained C model analysis requires the following: undrained modulus of elasticity (E_u), undrained Poisson's ratio (ν_u), and undrained shear strength (S_u).

A. Young's Modulus (E)

Due to the disturbance during soil sampling, the Young's Modulus (E) obtained from experiments cannot be directly used for analysis. Therefore, it is often estimated using empirical formulas in practice. Common methods for determining E include using the Standard Penetration Test (*SPT-N*) values for sandy soils and converting them using empirical formulas, and for soft clays, since their strength comes from undrained conditions (S_u), they are converted into E for soft clay using empirical relationships. According to Liao [23] in "Application of Empirical Soil Parameters in Numerical Analysis", the empirical formulas for laboratory values of E proposed by various scholars are used for estimation, as shown in Equations (6) and (7):

For clay, the Young's Modulus (E_u) is used as:

$$E_u = 800\ S_u \tag{6}$$

where the value of Su is obtained from indoor triaxial tests (UU).

For sandy soil, E' is used as:

$$E' = 2800\ N \tag{7}$$

B. Poisson's Ratio (ν)

The recommended range of values for Poisson's ratio for different soils is referenced from Ou [24].

C. Wet Soil Unit Weight (γ_t), Effective Cohesion (c'), and Effective Friction Angle (ψ')

Values for wet soil unit weight, effective cohesion, and effective friction angle can be obtained through drilling reports and indoor experiments.

D. Interface Element Strength

The interaction between soil and structural elements involves a partial transmission of shear stress, modeled using interface elements that are divided into elastic and plastic stages. The elastic stage represents conditions of small strain, while the plastic stage accounts for larger sliding displacements. Interface elements employ a strength reduction factor (R_{inter}) in their calculations. A factor of 1 suggests complete shear stress transmission, but typically in geotechnical engineering, the interaction between soil and structures is weaker than the soil body itself. Hence, a reduction factor (Rinter) of less than 1 is often used, with the PLAXIS user manual recommending a setting of 2/3 for a more realistic approximation.

In summary, the relevant soil parameters are organized as shown in Tables 2 and 3.

Table 2. Undrained material parameters.

Depth (m)	γ (kN/m^3)	S_u (kN/m^2)	E_u (kN/m^2)	ν_u	R_{inter}
0~0.5	18.6	29.4	23,520	0.495	0.67
5.5~7.2	18.3	39.24	31,392	0.495	0.67
11.0~14.1	18.5	49.05	39,240	0.495	0.67
17.4~20.6	18.6	58.86	47,088	0.495	0.67
25.1~29.3	18.8	78.48	62,784	0.495	0.67
31.6~38.6	18.5	98.1	78,480	0.495	0.67
38.6~43.2	20.7	107.91	86,328	0.495	0.67

Reference Source: (E) from Liao [23]; (ν) from Ou [24]; Refer to the recommended values in the Plaxis 3D manual and consider local drilling data for the remaining parameters.

Table 3. Drained material parameters.

Depth (m)	γ (kN/m^3)	φ' (°)	E' (kN/m^2)	v'	R_{inter}
7.2~11.0	19.5	30	21,924	0.2	0.67
14.1~17.4	18.5	32	34,104	0.2	0.67
20.6~25.1	19.4	32	36,540	0.2	0.67
29.3~31.6	18.9	32	38,976	0.2	0.67
43.2~44.8	19.7	33	56,028	0.2	0.67
44.8~50	21.6	40	121,800	0.2	0.67

Reference Source: (E') from Liao [23]; (v') from Ou [24]; Refer to the recommended values in the Plaxis 3D manual and consider local drilling data for the remaining parameters.

(3) Structural Parameters

The diaphragm wall has a thickness of 0.7 m. The construction depth near the MRT side (Banan Road) is from GL −2.4 m to −30 m, with a penetration depth from GL −12.25 m to −30 m. On the non-MRT side, the construction depth is from GL −2.4 m to −27 m, with a penetration depth from GL −12.25 m to −27 m, as shown in Appendix A Figure A1. The internal and external buttress walls have the same construction depth as the diaphragm wall. To reduce deformation of the surrounding structures, 8 internal buttress walls near Banan Road and 2 near Qiaohe Road above the excavation face are not removed and will be considered as part of the structure in the future. All external buttress walls in this case are not removed. The cross wall has a thickness of 0.7 m, with a construction depth from GL −2.4 m to −25 m, and is removed stage by stage above the excavation face as the excavation depth increases. In this project, the joints between the diaphragm wall, cross wall, internal and external buttress walls are either T-shaped or cross-shaped.

The internal structural slab and back-pull slabs have a thickness of 0.4 m. The back-pull slabs is installed at GL −2.4 m and goes around the diaphragm wall of the site, as shown in Figure 8. It overlaps with the external buttress wall and diaphragm wall to enhance the overall stiffness of the diaphragm wall, reducing wall deformation. The top is then backfilled with soil. The internal structural slab is installed near the MRT side, as shown in Figure 8, located at GL −3.55 m. The diaphragm wall and internal buttress wall are set up with reserved reinforcement to connect with the internal structural slab, increasing the stiffness near the MRT side and reducing deformation in that area. The foundation slab has a thickness of 0.8 m and is placed at GL −12.25 m. Due to the large excavation area of this case, zoned excavation is used during construction. In areas where excavation is completed first, the foundation slab structure is implemented to enhance the stiffness of that area and reduce wall deformation.

The moduli of elasticity for the diaphragm wall, external buttress wall, non-removable internal buttress wall, and cross wall are calculated using the concrete's modulus of elasticity $E_c = 15,000 \times \sqrt{f_c'}$ (kgf/cm^2), where f_c' is the 28-day compressive strength of concrete. For this study, f_c' is taken as 280 (kgf/cm^2) for calculation. Parts of the internal buttress wall that need to be removed stage by stage will use low-strength backfill concrete, calculated with f_c' of 140 (kgf/cm^2). Cross walls above the excavation face are removed as the excavation depth increases. To reduce costs, low-strength concrete with f_c' 140 (kgf/cm^2) is used in the parent elements, and sandy soil is backfilled in the child elements, as shown in Figure 9. In numerical simulations, the removed part of the cross wall is simplified to low-strength concrete with f_c' = 140 (kgf/cm^2). Table 4 lists the selected parameters for the diaphragm wall, internal buttress wall, external buttress wall, cross wall, back-pull slabs, internal structural slab, and foundation slab. Plaxis 3D uses a three-dimensional analysis approach, employing wall elements to simulate the diaphragm wall, buttress walls, and cross wall. It is assumed that the diaphragm wall, buttress walls, and cross wall are homogeneous and fully integrated, without considering poor construction conditions.

Therefore, a linear state is used, and G12 = G13 = G23 = $E_c/2(1 + \nu_{12})$, where G is the shear modulus. Table 4 shows the parameters for the diaphragm wall, buttress walls, and cross wall.

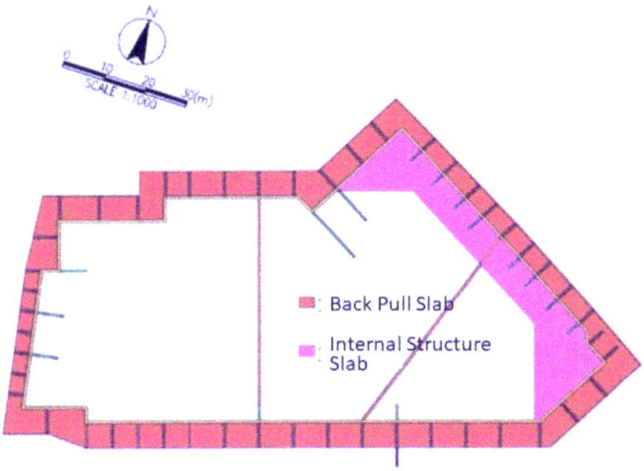

Figure 8. Location diagram of back-pull slab and internal structural slab.

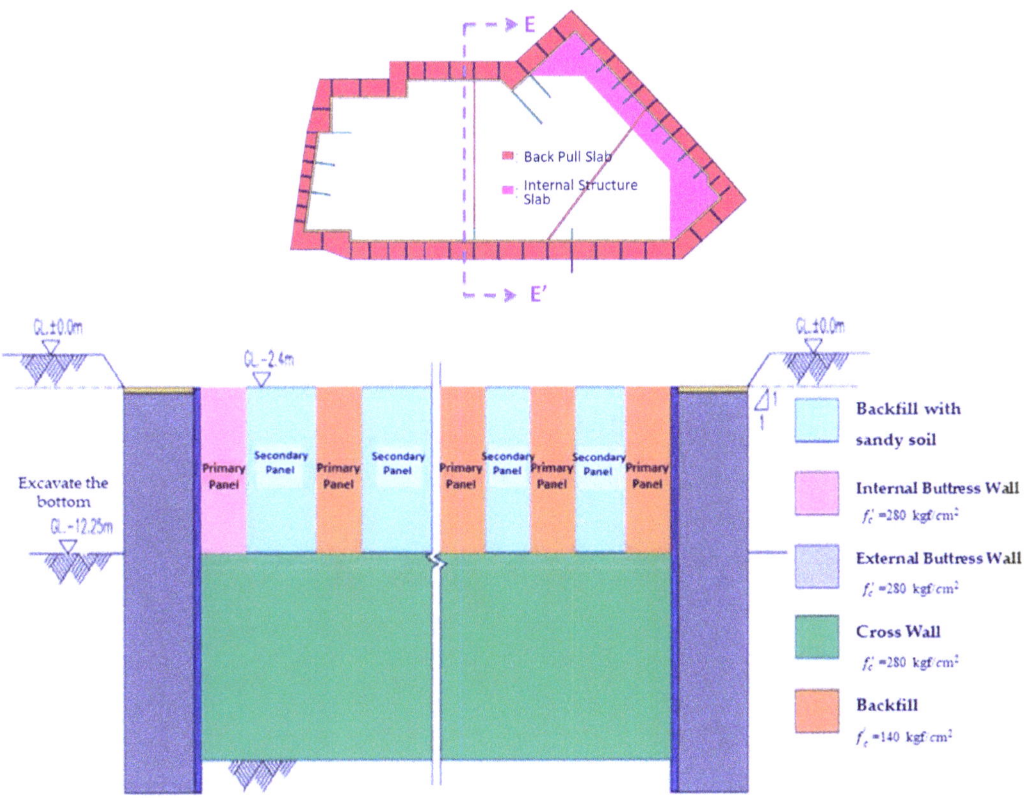

Figure 9. Cross-section diagram of excavation and cross wall.

Table 4. Structural parameters of diaphragm wall, buttress wall, cross wall, back-pull slabs, external structural slab, and foundation slab.

Type	Thickness (m)	Depth (m)	E (kN/m^2)	Poisson's Ratio υ	f_c' (kg/cm^2)
Diaphragm walls	0.7	2.4~27 2.4~30	2.42×10^7	0.15	280
Buttress walls (Layered Elimination)	0.7	2.4~12.25	1.74×10^7	0.15	140
Buttress walls	0.7/0.6	2.4~27 2.4~30	2.42×10^7	0.15	280
Cross walls (Layered Elimination)	0.7	2.4~12.25	1.74×10^7	0.15	140
Cross walls	0.7	12.25~27	2.42×10^7	0.15	280
1FL Back-Pull Slabs	0.4	2.0~2.4	2.42×10^7	0.15	280
B1F Internal Structure Slabs	0.4	3.55~3.95	3.02×10^7	0.15	420
Foundation Slab Back-pull Slabs	0.8	11.45~12.25	2.42×10^7	0.15	280

Data Source: Provided by the on-site construction team of Huida Construction for this case.

(4) Road and Nearby Structural Load Parameter Settings

In this case study, there are structures and roads surrounding the site. To closely simulate the actual conditions, the analysis considers the impact of structural load and road load on the excavation of the site. In previous studies, the simulation of structural and road loads was typically applied as surface loads on the ground, simplifying the analysis. The road and neighboring building loads remain active on the surface, representing conditions that already existed before the start of the project.

Based on common engineering experience, it is assumed that the load of one floor above ground is 1 t/m^2; the road load is also assumed to be 1 t/m^2. For the concrete of the MRT pile cap structure on the road, the compressive strength f_c' is 280 (kgf/cm^2), the thickness is 3 m, and the unit weight γ is 23.54 (kN/m^3). The structural load values of neighboring buildings (1F to 3F and 7F) are, respectively, 9.81, 19.62, 29.43, and 68.67 KN/m^2. The structural model of the analysis is shown in Figure 10.

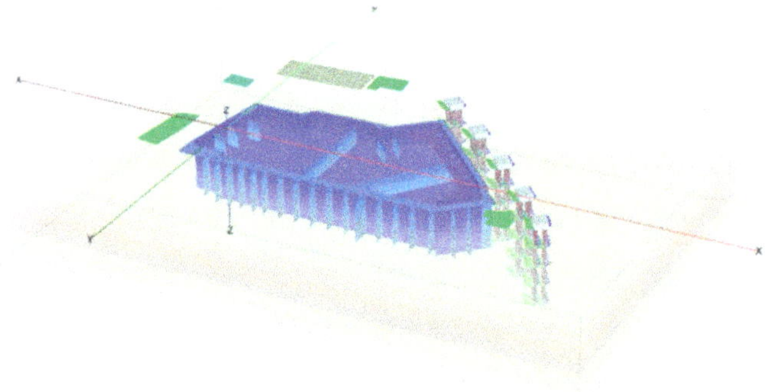

Figure 10. Structural model diagram.

(5) Zoned Excavation

Due to the large site area of 13,720 m^2 in this case study, it is necessary to conduct zoned excavation. Based on the construction conditions of the site, the daily excavation volume is approximately 1500 m^3. After excavating each zone, foundation slab construction must be carried out, which typically takes about 7 days per zone. According to the site construction plan, the entire site is divided into 6 areas for zoned excavation, as shown in Figure 11. The excavation areas of each zone are detailed in Table 5.

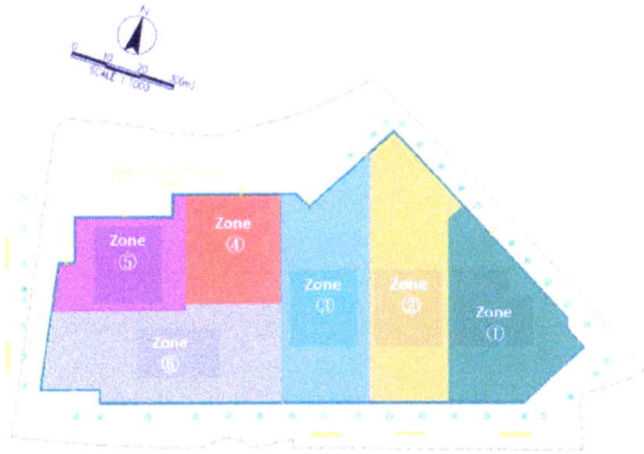

Figure 11. Zoned excavation area diagram.

Table 5. Area of zoned excavation.

Excavation Zone	Area	Location Photo
Zone ①	1307	Figure S6
Zone ②	1506	Figure S7
Zone ③	1500	Figure S8
Zone ④	822	Figure S9
Zone ⑤	884	Figure S10
Zone ⑥	1664	Figure S11
Total Area	7683	

2.4. Construction Steps

Finite element method analysis is used to simulate the construction steps, divided into 13 stages, to investigate the impact of the excavation process on lateral displacement changes. The aim is to find the optimal analysis method, aligning the simulation more closely with actual conditions. Due to the implementation of internal structural slabs in certain areas of this case, the construction steps are divided into the process shown in Figure 12.

During the analysis simulation, the construction process is divided into 13 steps:

(1) Load neighboring buildings and roads, and calculate initial stress.
(2) Excavate the entire site to (GL −2.4 m).
(3) Install diaphragm walls, cross walls, internal buttress walls, and external buttress walls, and reset displacements to zero.
(4) Construct external structural slab (GL −2.4 m), backfill certain areas.
(5) Excavate areas with internal structural slabs to (GL −5.5 m), and lower the groundwater level to (GL −5.5 m).
(6) Install internal structural slabs (GL −3.55 m).
(7) Excavate the first zone to (GL −12.25 m), and lower the groundwater level to (GL −12.25 m).
(8) Excavate the second zone to (GL −12.25 m), and install foundation slab for the first zone (GL −12.25 m).
(9) Excavate the third zone to (GL −12.25 m), and install foundation slab for the second zone (GL −12.25 m).
(10) Excavate the fourth zone to (GL −12.25 m), and install foundation slab for the third zone (GL −12.25 m).

(11) Excavate the fifth zone to (GL −12.25 m), and install foundation slab for the fourth zone (GL −12.25 m).
(12) Excavate the sixth zone to (GL −12.25 m), and install foundation slab for the fifth zone (GL −12.25 m).
(13) Install foundation slab for the sixth zone (GL −12.25 m). Undrain (B) parameters.

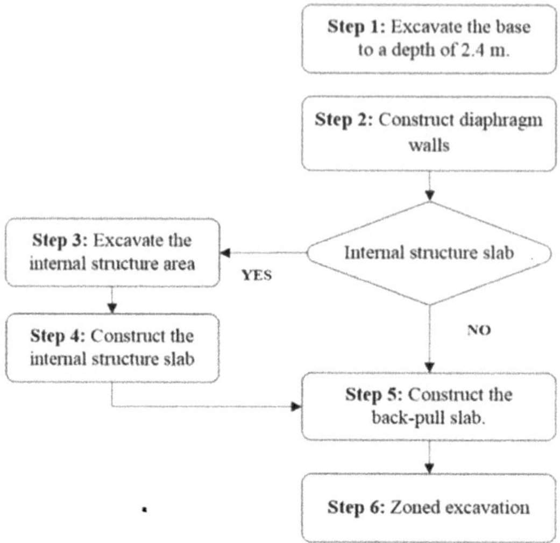

Figure 12. Process flow diagram of construction with and without internal structural slab.

2.5. Analysis Results and Case Verification

This section will discuss the comparison of the results from the finite element method software PLAXIS 3D for excavation engineering simulation analysis with the actual site monitoring data to verify the accuracy of the simulation. In this case, the site was monitored with 10 inclinometers to measure the lateral displacement of the diaphragm wall. The distribution map of the inclinometer monitoring locations is shown in Figure 13. This study focuses on verifying five inclinometers around the site, namely SIS02, SIS05, SIS06, SIS07, and SIS09. Figure 14a compares numerical analysis and monitoring data at the site's eastern monitoring point SIS02. At surface level GL −2.4 m, influenced by back-pull slabs, the analysis closely matches monitoring results. At excavation level GL −12.25 m, the monitored lateral displacement of the wall is 0.79 cm, while the simulation shows a slightly higher value of 0.4 cm. A 0.26 cm discrepancy is noted at the bottom of the diaphragm wall (GL −30 m) between internal and external inclinometers, indicating wall displacement. The simulation accounts for the back-pull slabs, non-removable internal and external buttress walls, and internal structural slab, which increase the wall's moment of inertia and stiffness, resulting in a cantilever deformation pattern.

Figure 14b shows that at the southern monitoring point SIS05, the numerical analysis matches the monitoring data, with cantilever-style deformation similar to that described by Wu et al. [25]. The lack of internal supports in this case study leads to slightly greater simulated lateral wall deformations compared to the monitored values, with the largest discrepancy being about 0.8 cm. The comparison at monitoring points SIS06 and SIS07 (Figure 14c,d) indicates similar trends in lateral wall displacement, with SIS06 showing less deformation due to its location between cross walls. The maximum displacement difference here is around 1.7 cm. At the northern monitoring point SIS10 (Figure 14e), the maximum error in lateral wall displacement is 0.3 cm above the excavation level, with a 0.73 cm difference observed at the wall bottom. The simulation shows increased wall stiffness near the corner due to back-pull slabs, leading to reduced displacement. The maximum wall displacement in the simulation occurs at the excavation level, with a 0.3 cm difference from the monitored value.

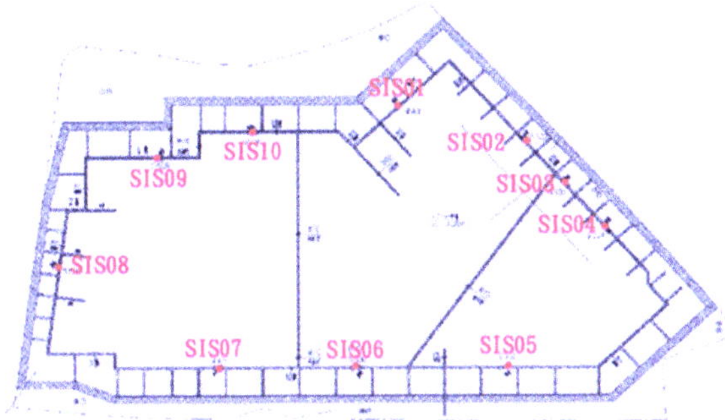

Figure 13. Distribution map of inclinometer monitoring locations (where "SISXX" refers to monitoring point number XX).

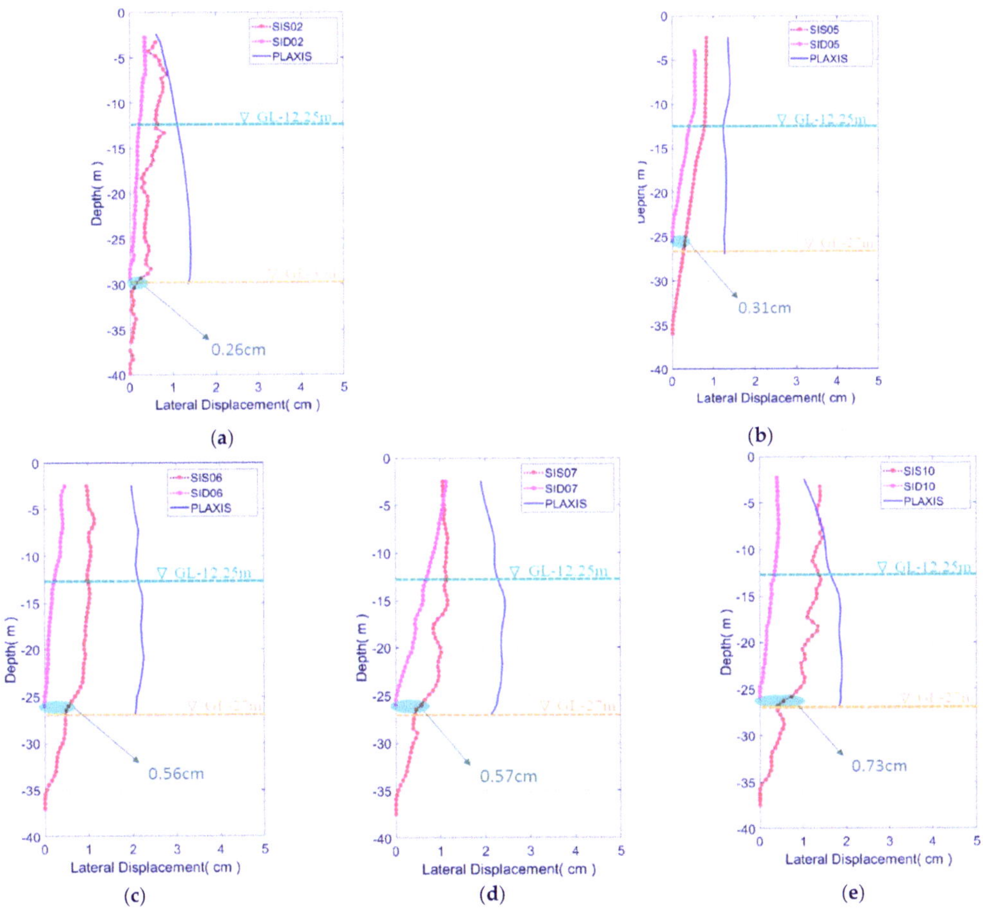

Figure 14. Comparison of numerical analysis results with monitoring values at each monitoring point: (**a**) SIS02; (**b**) SIS05; (**c**) SIS06; (**d**) SIS07; (**e**) SIS09.

The maximum lateral displacement (δ_{hm}) of the wall is compared. In this case, the data collected fall within the range of 0.2% to 0.5% of H_e (H_e being the final excavation depth), as outlined by Ou et al. [8] in their collection of research cases in the Taipei area. The maximum displacement of the wall in this study is on the lower end of the range proposed by Ou, as shown in Figure 15. This is likely due to the effects of zoned excavation and the implementation of the back-pull slabs, consequently reducing the overall lateral displacement of the wall.

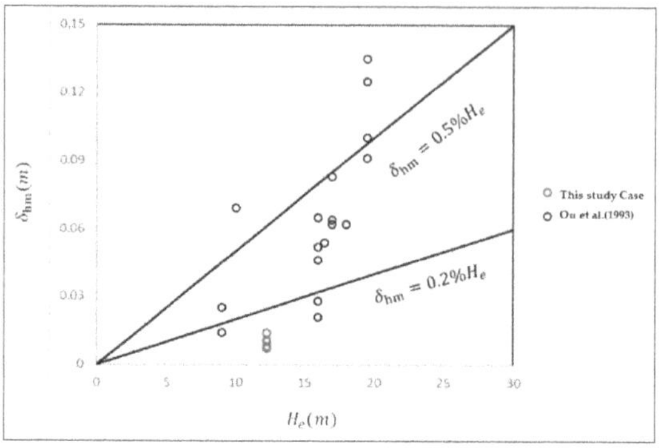

Figure 15. Relationship between maximum wall displacement and excavation depth [8].

From the comparison results, it is observed that the overall deformation trends of the diaphragm wall on the south side of the site in the analysis simulation are closer to the monitoring results. The wall deformation at the surface level GL −2.4 m in the simulation is generally lower. It is believed that the back-pull slabs can effectively suppress the lateral displacement of the diaphragm wall, primarily because the back-pull slab is implemented at the surface, enhancing the overall stiffness of the diaphragm wall at that location. To better understand whether the implementation of the back-pull slabs indeed effectively suppresses wall deformation, further discussion will follow in Section 3. Due to the large excavation area of this site, and the considerable length and width of the site without internal supports, a slight inward translation of the diaphragm wall body was observed in the numerical simulation. According to Ou [3], in unsupported excavation cases in soft soil, a toe translation phenomenon is also observed on the longitudinal side of the excavation. To better align with the actual construction process on-site, the simulation of construction procedures also adopts the same zoned excavation model as used on-site. Whether this directly affects the overall lateral displacement of the wall will be discussed in the next section.

2.6. Limitations and Impacts of the Plaxis 3D Model

PLAXIS, a finite element software for geotechnical and soil–structure interaction analysis, presents several limitations despite its robustness and versatility. These constraints are mainly in the following areas:

- Element Number: PLAXIS has a maximum limit on elements, which could restrict modeling in complex or large-scale projects, and more elements lead to increased computation time and resource usage.
- Material Properties: The software may not always accurately model the properties of heterogeneous soils or complex materials, and handling non-linear materials can present challenges.
- Boundary Conditions: Although PLAXIS offers various options for boundary conditions, it might fall short in specialized or extreme cases, and replicating real-world conditions accurately can be difficult.

- Loading Conditions: Simulating certain dynamic or cyclic loads, as well as extreme load scenarios, might exceed PLAXIS's capabilities.
- Analysis Types: The software has limitations in long-term consolidation and creep analysis, as well as simulating certain environmental effects.
- Software and Hardware Interaction: The performance of PLAXIS is hardware-dependent, and there may be constraints in parallel processing efficiency for large or complex models.

Engineers and researchers using PLAXIS should be aware of these limitations and consider supplementing their work with other tools, tests, or models to achieve comprehensive and accurate outcomes. It is also important to regularly consult PLAXIS's latest manuals and technical documents for updated information on its capabilities and limitations.

3. Case Study

This section aims to explore three major aspects related to the case study: the impacts of zoned excavation, back-pull slabs, and cross walls on the lateral displacement of diaphragm walls. Individual benefit analyses will be conducted to examine the effectiveness of these methods. The goal is to determine if they can effectively suppress the lateral deformation of diaphragm walls, thereby indirectly reducing the impact on neighboring structures and ensuring the safety of the site to prevent engineering disasters. The research in this section is based on the model, soil layers, and retaining structure parameters established in the previous sections.

This study uses an evaluation metric for wall displacement, adopting the maximum deflection ratio (MDR), as proposed by Lim et al. [16], to assess the impact of wall deformation, as shown in Equation (1).

3.1. Impact of Excavation

This section explores the impact of changing the excavation area on the lateral deformation of diaphragm walls. Using the model from the original engineering case, which involved excavation in six zones, this study investigates the effects of simplified excavation methods: three-zone excavation and layer-by-layer excavation. The analysis focuses on the longitudinal displacement of the diaphragm walls.

(1) Impact of Zoned Excavation

In the original case, the entire site was divided into six zones for excavation, as shown in Figure 11. To examine the effect of zoned excavation on diaphragm walls, the excavation zones were simplified by dividing the site into three zones based on the placement of cross walls, as illustrated in Figure 16. The respective areas of excavation zones 1 to 3 are 1487, 3089, and 3107 m^2, with the total area being 7683 m^2. The study focuses on two cross-sections: the ground surface at GL −2.4 m and the excavation face at GL −12.25 m. Lateral displacement of the diaphragm walls for these two sections is depicted in Figure 17.

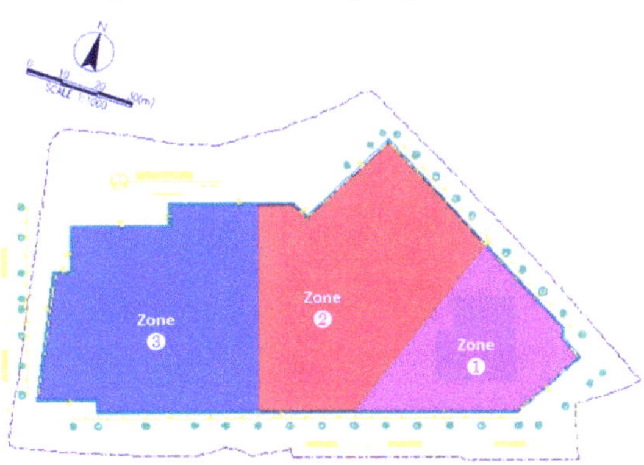

Figure 16. Simplified zoned excavation Areas.

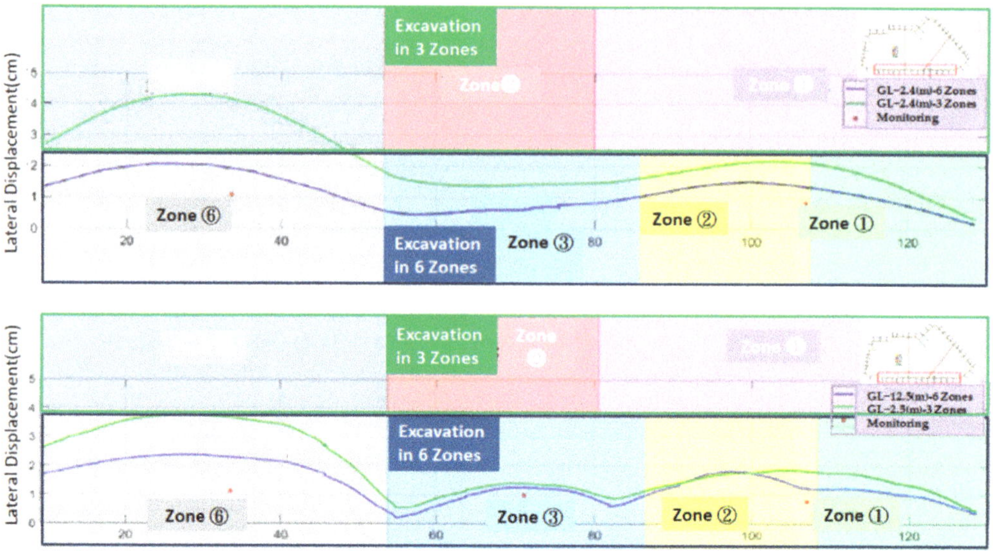

Figure 17. Lateral displacement diagrams of the diaphragm wall at GL −2.4 m and −12.25 m.

Based on the effect of corner arching, the maximum lateral displacement of the diaphragm wall occurs in the longitudinal direction. In this case, the longest segment of the diaphragm wall is located near the bridge and road, with a length of 118.30 m. Therefore, the analysis and comparison are focused on the wall in this area.

For two different zoned excavations with excavation sequences as shown in Figures 11 and 16, both progress from right to left in the diagrams. The impact of different zoned excavations on the maximum deformation of the diaphragm wall is summarized in Table 6. It is observed that dividing the zoned excavation into six zones results in a smaller maximum wall displacement than that of three-zone excavation. In both types of zoned excavation, when the length of the first excavation zone is 21.39 m, the lateral displacement of the wall at GL −2.4 m is 1.33 cm. When the length increases to 46 m, the lateral displacement of the wall increases to 2.09 cm, a significant increase of 0.75 cm. The six-zone excavation simulation, located at the GL −12.25 m excavation face, shows that the first and second excavation zones have a mitigating effect on wall deformation, as clearly seen in Figure 17. The third excavation zone of the six-zone excavation has the same length as the second excavation zone of the three-zone excavation (27 m). The lateral displacement of the wall is 1.51 cm and 1.63 cm, respectively, with no significant increase in wall lateral displacement. The slight increase in displacement is mainly influenced by the previous excavation zone. Finally, comparing the sixth excavation zone of the six-zone excavation with the third excavation zone of the three-zone excavation, both have excavation ranges of 44.95 m. The maximum lateral displacement of the wall at GL −2.4 m is 2.06 cm and 4.28 cm, respectively, showing that this zone significantly suppresses wall deformation, with a difference of 2.22 cm. In summary, zoned excavation, in the absence of internal support, has a significant effect on suppressing lateral deformation of the diaphragm wall, enhancing overall construction safety, and indirectly reducing its impact on surrounding structures.

(2) The impact of excavation by layers:

The original case study divided the site into six areas for excavation to explore the impact of sectional excavation on the lateral displacement of the diaphragm wall. For analysis, the excavation area was simplified, using layer-by-layer excavation. This study utilizes the model from the original engineering case, which involved sectional excavation in six areas. In this study, layer-by-layer excavation was employed, divided into four stages. The study focuses on two cross-sections: the ground surface after excavation at GL −2.4 m and the excavation face at GL −12.25 m. The lateral displacement of the diaphragm wall is illustrated in Figure 18 (top and middle diagrams).

Table 6. Maximum wall displacement in zoned excavation at GL −2.4 m.

Zone	Max. Displacement in 6 Zones (cm)	Max. Displacement in 3 Zones (cm)	Max. Displacement in 6 Zones (cm)	Max. Displacement in 3 Zones (cm)
	Local @ GL −2.4 m		Local @ GL −12.25 m	
Zone 1	1.33	2.09	1.22	1.87
Zone 2	1.51	1.63	1.84	1.23
Zone 3	1.08	4.28	1.12	3.77
Zone 6	2.06	-	2.37	-

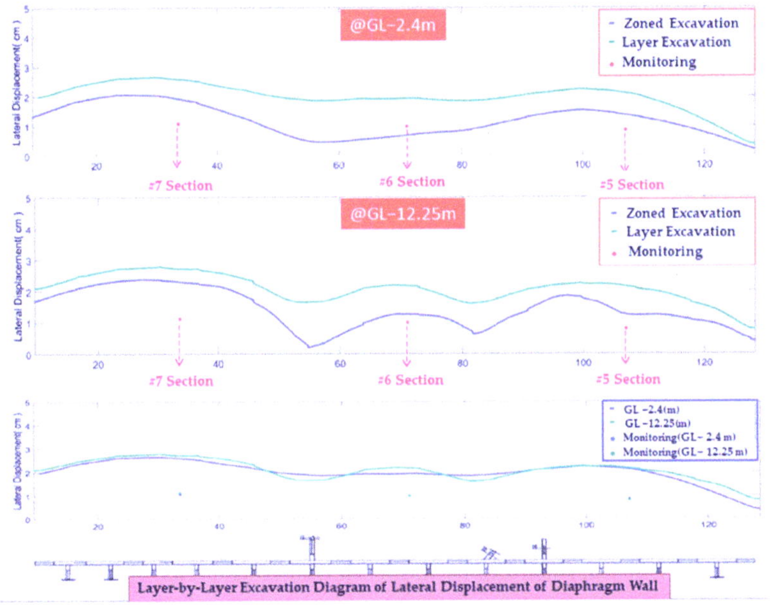

Figure 18. Comparative diagram of lateral displacement of diaphragm wall in zoned excavation and layer-by-layer excavation.

By comparing the lateral displacement of the diaphragm wall in sectional excavation with layer-by-layer excavation from the previous section, analysis is conducted on three cross-sections. At the ground surface GL −2.4 m, the lateral displacement values of the diaphragm wall in all three cross-sections are greater in layer-by-layer excavation compared to sectional excavation, as shown in Table 7. The differences at the three cross-sections are 0.67 cm, 1.25 cm, and 0.74 cm, respectively. Similarly, at the final excavation face GL −12.25 m, the lateral displacement values in layer-by-layer excavation exceed those in sectional excavation, with differences of 0.41 cm, 0.91 cm, and 0.9 cm at the three cross-sections. In Figure 18 (bottom diagram), it is evident that the excavation values at the ground surface and the deformation trends at the final excavation face are very similar in layer-by-layer excavation.

Table 7. Comparison of displacements in sectional excavation and layer-by-layer excavation.

Zone	Zoned Exc.	Layer Exc.	Zoned Exc.	Layer Exc.
	Displacement (cm)			
	Local @ GL −2.4 m		Local @ GL −12.25 m	
#5 Section	1.91	2.58	2.29	2.7
#6 Section	0.67	1.92	1.25	2.16
#7 Section	1.33	2.07	1.22	2.12

From the above data, a conclusion can be drawn: in large-scale deep excavations, dividing the site into several independent small excavation areas can effectively suppress the lateral displacement of the main diaphragm wall. This construction method can effectively protect surrounding facilities. In this study's case, the maximum displacement of the wall is located in the lower half of the range proposed by Ou [8]. The maximum lateral displacement of the wall, measured as δ_{hm}, is within 0.2% of H_e.

3.2. *Effectiveness of Back-Pull Slabs*

This section explores the impact of back-pull slabs on reducing the lateral deformation of continuous walls. The study uses the model from the original engineering case and analyzes two scenarios: with and without back-pull slabs. The analysis focuses on the site's lengthwise side, where the lateral displacement of the continuous wall is greatest. Comparisons are made at cross-sections 5, 6, and 7, where monitoring instruments are located. The maximum deflection ratio (MDR) is used to assess wall deformation, as shown in Equation (1).

The study, detailed in Table 8 and Figures 19–21 shows that back-pull slabs significantly reduce the lateral displacement of diaphragm walls, especially at cross-sections 6 and 7. With the slab, displacement at cross-section 6 is 0.63 cm compared to 1.81 cm without, and at cross-section 7, it is 1.91 cm versus 3.77 cm. The effect is weaker at cross-section 5 due to interactions between corner effects, sectional excavation, and cross walls. The slabs, installed at GL −2.4 m, increase wall stiffness and effectively control deformation. Without the slabs, walls experience greater inward displacement due to the lack of internal supports. This indicates that in unsupported conditions, back-pull slabs are effective in reducing lateral wall displacement.

Table 8. Wall displacement with and without back-pull slab.

Section	Wall Displacement (cm) with Back-Pull Slab	Wall Displacement (cm) without Back-Pull Slab	MDR (%)
#5 Section	1.33	1.46	8.9
#6 Section	0.63	1.81	65.2
#7 Section	1.91	3.77	49.3

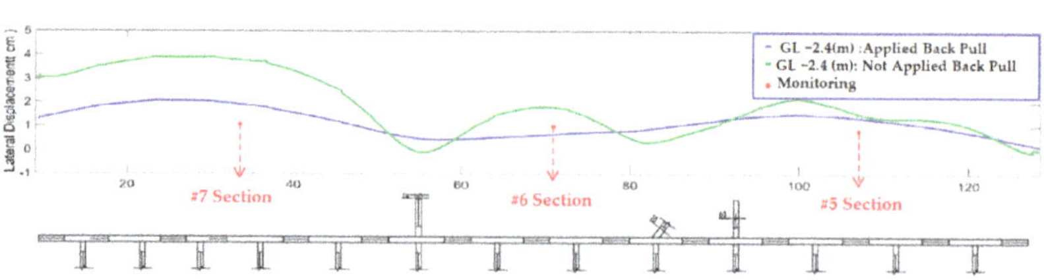

Figure 19. Graph of wall displacement with and without back-pull slab.

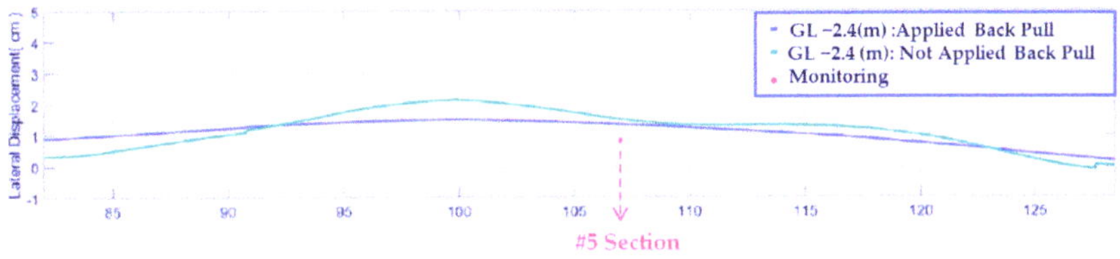

Figure 20. Graph of wall displacement at cross-section #5 with and without back-pull slab.

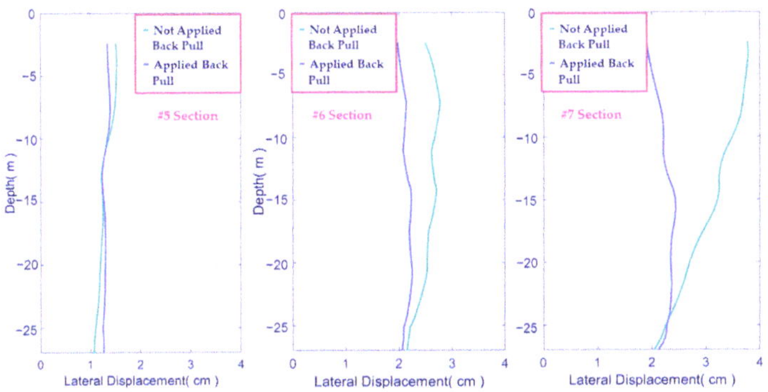

Figure 21. Lateral displacement graph of the wall at cross-sections #5 to #7.

3.3. *The Effectiveness of Cross Wall Construction*

This section evaluates the impact of cross walls on reducing lateral deformation of diaphragm walls in a large excavation site, as depicted in Figure 8. The study focuses on a 118 m segment along the bridge and road, assessing areas with cross walls both perpendicular and non-perpendicular to the diaphragm wall.

Figure 22 highlights the effect of cross walls on diaphragm wall stability. Figure 23 illustrates this in three areas (A, B, C), showing reduced wall displacement at cross wall locations. In Area A, the maximum deformation at GL −12.25 m reduces from 2.37 cm to 0.19 cm near the cross wall; at GL −2.4 m, it decreases from 2.05 cm to 0.49 cm. In Area B, deformation at GL −12.25 m drops from 1.25 cm to as low as 0.19 cm, and at the surface, from 0.89 cm to 0.45 cm. Area C shows similar trends, with maximum displacement decreasing from 1.84 cm to 0.63 cm and 0.38 cm at various points.

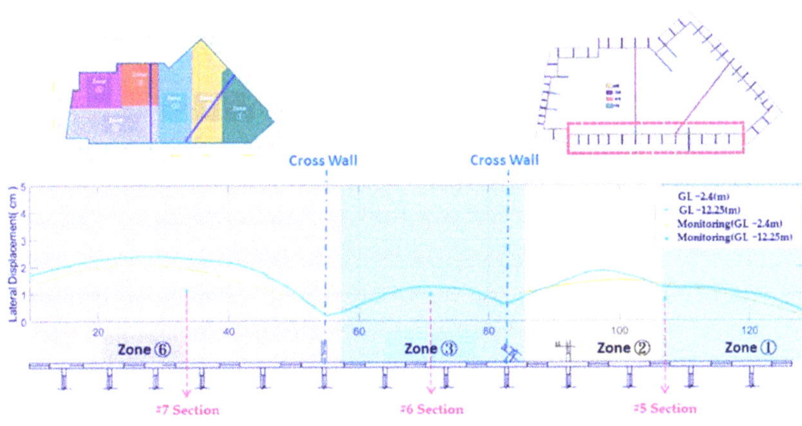

Figure 22. Lateral displacement graph of the diaphragm wall.

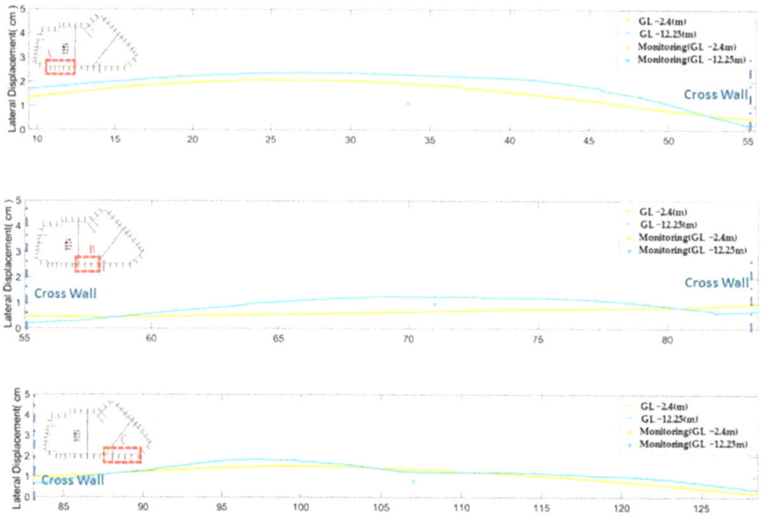

Figure 23. Lateral deformation graphs of the diaphragm wall in Areas A, B, and C.

The study confirms that cross walls, especially when perpendicular to diaphragm walls, significantly limit lateral displacement. This is quantified in Table 9, where perpendicular cross walls demonstrate greater deformation control. The analysis suggests that future designs should ideally align cross walls perpendicular to diaphragm walls for optimal deformation control.

Table 9. Benefits of perpendicular and non-perpendicular cross walls.

Level	Perpendicular Cross Wall Displacement (cm)	Non-Perpendicular Cross Wall Displacement (cm)
Surface Level (GL −2.4 m)	0.49	0.89
Final Excavation Level (GL −12.25 m)	0.19	0.63

4. Conclusions

This study utilizes PLAXIS 3D 2017 as a research tool, simulating a real-world excavation project for validation. By approximating the construction site conditions with the site's geometry, soil and structural parameters, and construction steps, and comparing these simulations with on-site monitoring results, the study verifies the model's credibility. Based on this model, further analyses were conducted to assess the impact of excavation methods, back-pull slabs, and cross walls on diaphragm wall displacement, aiming to explore effective ways to control deformation in large-scale deep excavations. The numerical analysis simulations are intended to provide references for both academia and the engineering field. The following are the conclusions and recommendations from this study.

4.1. Summary

1. Impact of Excavation Methods on Lateral Displacement of Diaphragm Walls in Large Unsupported Deep Excavations:
    - Sectional excavation significantly reduces wall deformation. Designing smaller excavation areas in accordance with on-site construction can markedly lower lateral deformation.
    - Layer-by-layer excavation in this method shows less significant reduction in lateral wall deformation.

- Numerical simulations and monitoring data indicate that this construction method results in a cantilever-type wall deformation.

2. Lateral Displacement Phenomena of Diaphragm Walls:
    - The construction method in this study causes displacement at the toe of the diaphragm wall, as observed through external and internal inclinometers.
    - Different inclinometer installation depths may alter the assumed position of the neutral point. Both numerical analysis and monitoring show toe displacement, so there remains some discrepancy in accurately estimating lateral displacement through either method.
3. Influence of Back-Pull Slabs on Lateral Wall Displacement:
    - In the construction method involving peripheral top-down and central bottom-up approaches, the study finds that back-pull slabs, when feasible in large site areas, significantly enhance the stiffness and connection between the main diaphragm wall and external buttress walls, effectively reducing lateral displacement. In this case, it could reduce up to 65.19% of the lateral deformation.
4. Influence of Cross Walls on Lateral Wall Displacement:
    - Due to the large excavation area and elongated site width, this study analyzes the effect of two cross walls on wall displacement. Results show that even with a width of 60 m, cross walls can effectively control lateral displacement of the main diaphragm wall.
    - One of the cross walls, due to the site's shape, is not perpendicular to the diaphragm wall and shows less effective deformation control compared to perpendicular walls.
    - Horizontal steel stress meters in cross walls and inclinometer installation positions do not directly prove the mechanical behavior of cross walls; their effectiveness is primarily observed through numerical simulation.

This study significantly impacts large unsupported deep excavations, offering key insights into design and construction. It demonstrates that sectional excavation effectively reduces wall deformation, providing a critical reference for future projects. The research also highlights the importance of understanding lateral displacement in diaphragm walls, as observed through inclinometer data, enhancing predictive accuracy and design safety. Additionally, it reveals the effectiveness of back-pull slabs in increasing wall stiffness and connectivity, reducing lateral displacement. The analysis of cross walls' influence, particularly in wide excavations, further informs deformation control strategies. Overall, this study provides essential guidance for the design and execution of similar constructions, emphasizing innovative methods like back-pull slabs and cross walls for effective deformation management.

Based on the findings of this study, there are several significant applications in practical engineering contexts. The demonstrated effectiveness of sectional excavation in reducing wall deformation provides a valuable strategy for managing lateral displacement in large unsupported deep excavations. This approach can be particularly beneficial in urban areas where space constraints and the proximity of existing structures demand precise control of excavation impacts.

The study suggests that layer-by-layer excavation is a tailored solution for certain projects and that numerical simulations are key in designing against wall deformation. It also finds that back-pull slabs improve structural stability in large excavations, providing a cost-effective alternative to traditional supports. Furthermore, cross walls prove effective for lateral support across various excavation widths and shapes. Collectively, these insights inform better practices for deep excavation, enhancing safety and efficiency.

4.2. Recommendations

1. The biggest challenge in this study was model construction. The case is extremely complex, including diaphragm walls, cross walls, internal and external buttress walls, back-pull slabs, and internal structural slabs, combined with the site's irregular geometry, multiple turns, and sloping edges. This complexity often led to meshing failures at interfaces, requiring adjustments in overall mesh density and fine-tuning of conflicting structural coordinates to avoid small area mesh generation and model failures.
2. In Taiwan, engineering cases using this excavation method are currently very rare, which limits the scope of research findings. In the future, collecting and analyzing more cases related to this method could help in substantiating the research with additional relevant data.
3. For future monitoring device setup, placing inclinometers at the junctions of cross walls and diaphragm walls could allow for directly observing the impact of cross walls on diaphragm walls, facilitating validation with numerical simulations. Steel stress meters in cross walls should be installed on both sides of the wall to explore the force behavior of cross walls.
4. The analysis could be further enhanced by using a wider variety of soil models. Comparing a hardening soil model or hardening soil with a small strain model could identify the most conservative design approach, thereby increasing overall safety.

Supplementary Materials: The following supporting information can be downloaded at: https://www.mdpi.com/article/10.3390/buildings14010023/s1, Figure S1: Location Map of Borehole Profiles on the Site; Figure S2: Columnar Section Diagram of Borehole A A′; Figure S3: Columnar Section Diagram of Borehole B B′; Figure S4: Columnar Section Diagram of Borehole C C′; Figure S5: Columnar Section Diagram of Borehole D D′; Figure S6: Construction Status Map of Zone 1; Figure S7: Construction Status Map of Zone 2; Figure S8: Construction Status Map of Zone 3; Figure S9: Construction Status Map of Zone 4; Figure S10: Construction Status Map of Zone 5; Figure S11: Construction Status Map of Zone 6.

Author Contributions: Conceptualization, H.-W.C., C.-F.H., F.-H.T. and S.-L.C.; methodology, H.-W.C., F.-H.T. and S.-L.C.; formal analysis, H.-W.C. and F.-H.T.; investigation, F.-H.T.; writing—original draft preparation, F.-H.T. and C.-F.H.; writing—review and editing, C.-F.H. All authors have read and agreed to the published version of the manuscript.

Funding: This research was funded by the Ministry of Science and Technology, Taiwan, under contract No. NSTC 112-2221-E-027-047.

Data Availability Statement: The data presented in this study are available in supplementary material.

Conflicts of Interest: Author Fu-Huan Tsai was employed by the company Huaku Development Co., Ltd. The remaining authors declare that the research was conducted in the absence of any commercial or financial relationships that could be construed as a potential conflict of interest.

Appendix A

Table A1. Simplified soil layer parameter chart (provided by drilling report).

Soil Layer	Average Depth	γ_t (kN/m^3)	N	S_u (kN/m^2)	e	C'_{ref} (kN/m^2)	φ'(°)
CL1	0~−5.5	18.6	6	29.4	0.92	0.2	28
CL/ML1	−5.5~−7.2	18.3	3	39.24	1.00	0.2	27
SM1	−7.2~−11.0	19.5	9	–	0.65	0.2	30
CL/ML2	−11.0~−14.1	18.5	7	49.05	0.87	0.2	28
SM2	−14.1~−17.4	19.1	14	–	0.71	0.2	32
CL2	−17.4~−20.6	18.6	7	58.86	0.85	0.2	28
SM3	−20.6~−25.1	19.4	15	–	0.60	0.2	32

Table A1. *Cont.*

Soil Layer	Average Depth	γ_t (kN/m^3)	N	S_u (kN/m^2)	e	C'_{ref} (kN/m^2)	φ'(°)
CL3	−25.1~−29.3	18.8	7	78.48	0.79	0.2	29
SM4	−29.3~−31.6	18.9	16	–	0.76	0.2	32
CL-ML	−31.6~−38.6	18.5	10	98.10	0.90	0.2	30
CL4	−38.6~−43.2	20.7	22	107.91	0.55	0.2	32
SM5	−43.2~−44.8	19.7	23	–	0.60	0.2	33
GW/GP	−44.8 down	21.6	50	–	–	0.2	40

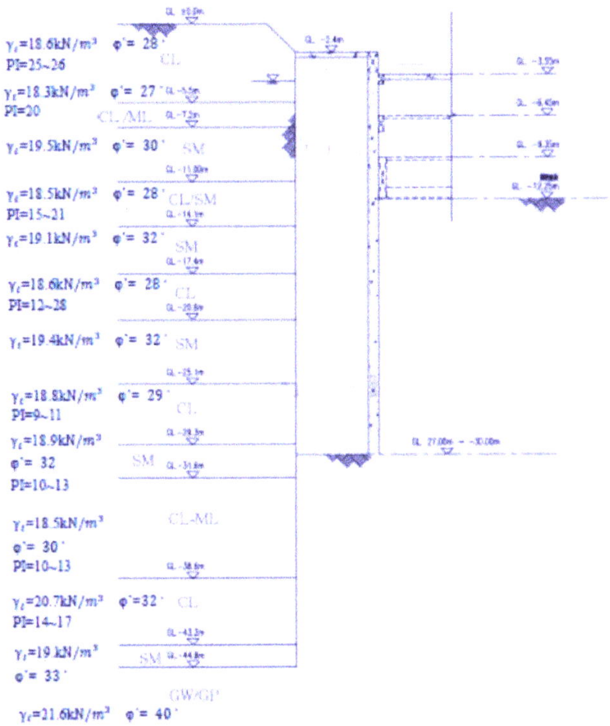

Figure A1. Excavation and underground cross-section diagram.

References

1. Weidong, L.G.W. *Foundation Pit Engineering Handbook*, 2nd ed.; China Architecture & Building Press: Beijing, China, 2009.
2. Lim, A.; Ou, C.-Y. Case Record of a Strut-free Excavation with Buttress Walls in Soft Soil. In Proceedings of the 2nd International Symposium on Asia Urban GeoEngineering; Springer Singapore: Singapore; 2018.
3. Ou, C.Y. *Advanced Deep Excavation Engineering Analysis and Design*; Tech Books Incorporated: Taipei, Taiwan, 2017.
4. Jeng, C.-J.; Ho, S.-K. A Case Study of 3-D Numerical Analysis for Nonstrut Construction Method in Huge Area Foundation Excavation. *J. Art Des. Huafan Univ.* **2016**, *11*, 75–87.
5. Li, M.-G.; Zhang, Z.-J.; Chen, J.-J.; Wang, J.-H.; Xu, A.-J. Zoned and staged construction of an underground complex in Shanghai soft clay. *Tunn. Undergr. Space Technol.* **2017**, *67*, 187–200. [CrossRef]
6. Wang, J.H.; Xu, Z.H.; Wang, W.D. Wall and Ground Movements due to Deep Excavations in Shanghai Soft Soils. *J. Geotech. Geoenviron. Eng.* **2010**, *136*, 985–994. [CrossRef]
7. Clough, G.W.; O'Rourke, T.D. *Construction-Induced Movements of In Situ Wall, Design and Performance of Earth Retaining Structure*; ASCE: Reston, VA, USA, 1990.
8. Ou, C.-Y.; Hsieh, P.-G.; Chiou, D.-C. Characteristics of ground surface settlement during excavation. *Can. Geotech. J.* **1993**, *30*, 758–767. [CrossRef]

9. Mana, A.I.; Clough, G.W. Prediction of movements for braced cuts in clay. *J. Geotech. Eng. Div.* **1981**, *107*, 759–777. [CrossRef]
10. Masuda, T.; Einstein, H.H.; Mitachi, T. Prediction of lateral deflection of diaphragm wall in deep excavations. *Doboku Gakkai Ronbunshu* **1994**, *1994*, 19–29. [CrossRef] [PubMed]
11. Wu, S.-H.; Ching, J.; Ou, C.-Y. Predicting wall displacements for excavations with cross walls in soft clay. *J. Geotech. Geoenviron. Eng.* **2013**, *139*, 914–927. [CrossRef]
12. Do, T.-N.; Ou, C.-Y.; Lim, A. Evaluation of factors of safety against basal heave for deep excavations in soft clay using the finite-element method. *J. Geotech. Geoenviron. Eng.* **2013**, *139*, 2125–2135. [CrossRef]
13. Chheng, C.; Likitlersuang, S. Underground excavation behaviour in Bangkok using three-dimensional finite element method. *Comput. Geotech.* **2018**, *95*, 68–81. [CrossRef]
14. Chen, S.-L.; Ho, C.-T.; Li, C.-D.; Gui, M.-W. Efficiency of buttress walls in deep excavations. *J. Geoeng.* **2011**, *6*, 145–156.
15. Hsiung, B.-C.; Yang, K.-H.; Aila, W.; Ge, L. Evaluation of the wall deflections of a deep excavation in Central Jakarta using three-dimensional modeling. *Tunn. Undergr. Space Technol.* **2018**, *72*, 84–96. [CrossRef]
16. Lim, A.; Hsieh, P.-G.; Ou, C.-Y. Evaluation of buttress wall shapes to limit movements induced by deep excavation. *Comput. Geotech.* **2016**, *78*, 155–170. [CrossRef]
17. Chuah, S.; Harry, S.T. Numerical study on a new strut-free counterfort embedded wall in Singapore. In Proceedings of the Earth Retention Conference 3, Bellevue, Washington, 1–4 August 2010.
18. Liu, G.; Guo, J.; Li, M.; Qin, T.; Huang, P. Measured behaviors of an oversized irregular basement excavation and its surrounding responses in thick soft clay. *Arab. J. Geosci.* **2019**, *13*, 3. [CrossRef]
19. Liu, Y.; Xiang, B.; Fu, M. Observed Performance of a Large-Scale Deep Triangular Excavation in Shanghai Soft Clays. *Geotech. Geol. Eng.* **2019**, *37*, 2791–2809. [CrossRef]
20. Ren, D.; Kang, C.; Liu, H.; Li, Y.; Wang, J. Characteristics of a Large-Scale Deep Foundation Pit Excavated by the Central-Island Technique in Chengdu Soft Clay. *KSCE J. Civ. Eng.* **2022**, *26*, 2610–2623. [CrossRef]
21. Liang, R.; Wu, J.; Sun, L.; Shen, W.; Wu, W. Performances of adjacent metro structures due to zoned excavation of a large-scale basement in soft ground. *Tunn. Undergr. Space Technol.* **2021**, *117*, 104123. [CrossRef]
22. Ho, S.K.; Gao, F. Case Study of Large-Scale Unsupported Excavation. *Sino-Geotech.* **2011**, *128*, 89–98.
23. Liao, N.-H. Application of Empirical Soil Parameters in Numerical Analysis. Master's Thesis, National Cheng Kung University, Tainan, Taiwan, 2003.
24. Ou, C.-Y. *Deep Excavation Engineering: Analysis, Design Theories and Practices*; Science and Technology Books: Taipei, Taiwan, 2002.
25. Wu, P.-Z.; Wang, M.-J.; Peng, Y.-R. Investigation of Continuous Wall Deformation Behavior. In Proceedings of the 7th Geotechnical Engineering Research Symposium, Berlin, Germany, 8–11 September 1997.

Disclaimer/Publisher's Note: The statements, opinions and data contained in all publications are solely those of the individual author(s) and contributor(s) and not of MDPI and/or the editor(s). MDPI and/or the editor(s) disclaim responsibility for any injury to people or property resulting from any ideas, methods, instructions or products referred to in the content.

buildings

MDPI

Article

Study on the Impact of Deep Foundation Excavation of Reclaimed Land on the Deformation of Adjacent Subway Tunnels

Fenghai Ma [1,2], Senlin Li [1] and Qiongyi Wang [2,*]

[1] College of Architecture and Engineering, Dalian University, Dalian 116622, China; mfh@dlu.edu.cn (F.M.)
[2] School of Mechanics and Engineering, Liaoning Technical University, Fuxin 123000, China
* Correspondence: wangqiongyi1358@163.com

Abstract: The objective of this research is to investigate the characteristics of the deformation response in adjacent subway tunnels caused by deep foundation excavation of reclaimed land. Focusing on a deep foundation excavation project situated in proximity to Line 11 of the subway in Shenzhen, this study employs theoretical analysis, numerical simulation, and on-site measurements to thoroughly investigate the deformation issues induced by the unloading of the excavation. The research results are as follows: using the energy method to calculate the uneven deformation of adjacent subway tunnels caused by the excavation can overcome the limitations of traditional algorithms, which treat the subway tunnel as a uniformly elastic foundation beam, resulting in more reasonable calculation results. Increasing the self-stiffness $(EI)_{eq}$ of the tunnel can effectively reduce the maximum displacement (w_{max}) of the tunnel, and as $(EI)_{eq}$ increases, its "weakening effect" on w_{max} gradually diminishes. Underground continuous walls can effectively control tunnel deformation, with tunnel displacement decreasing as the thickness and concrete strength of the continuous walls increase. "Long excavation" deep foundation excavations can impact the displacement and uplift range of the tunnel, with the maximum tunnel displacement showing a nonlinear decrease with increasing excavation depth. Tunnel displacement decreases as geotechnical parameters (elastic modulus E, internal friction angle φ, and cohesion C) increase, with the elastic modulus being the most sensitive parameter. The research findings can be applied to tunnel construction, maintenance, and safety evaluations, providing valuable references for similar engineering projects in the future.

Keywords: reclaimed land; deep foundation excavation; adjacent subway tunnel; theoretical analysis; numerical simulation; on-site measurements

1. Introduction

With the expansion of urban construction and rapid growth in high-rise buildings and subway networks, deep foundation excavation projects have become significant endeavors. Among these projects, the impact of foundation pit construction on adjacent subway tunnels has become a focal point for urban underground engineering construction management. Excavation of foundation pits alters the stress and deformation fields of the soil, potentially leading to stress and displacement exceeding allowable values in subway structures, thus affecting the normal operation of the subway [1–6]. Therefore, accurately assessing and predicting the impact of foundation pit excavation on nearby subway tunnels is an urgent issue that needs to be addressed.

Theoretical analysis is an important method for studying the mechanical characteristics of deep foundation excavation processes. Wang et al. [7] studied the influence of clayey soil adhesion on the clogging of tunnel boring machines through experiments and proposed a new semi-empirical diagram, which found that the maximum adhesion occurred in the region of strong clogging, indicating the feasibility of using adhesion stress as an index to evaluate the clogging of tunnel boring machines. Fu et al. [8] derived analytical solutions for the quarter-space grouting of rigid support structures based on the original-original principle and cavity expansion theory, and validated tunnel horizontal displacements calculated using Paschenak foundation beam theory with engineering examples from Tianjin and Shenzhen. Zhang et al. [9] studied the distribution characteristics of soil pressure in limited spaces through the expansion project of a small space subway station in Shenzhen, and deeply investigated the mechanical properties of pile-supported structures using theoretical analysis and indoor tests. Zhang et al. [10] divided the influence area of sidewalls by considering the deformation of retaining structures and the spatial effect of foundation pit excavation, analyzed and calculated the additional stress generated, and further introduced a shield tunnel rotation and dislocation cooperative deformation model to study the longitudinal deformation of adjacent tunnels under multiple factors during foundation pit construction. Huang et al. [11] theoretically analyzed the inadequacy of current force input conditions and the reasonability of displacement input conditions, distinguished between two Winkler foundation moduli under different input conditions, and proposed to use passive and active Winkler foundation moduli for displacement and force input conditions, respectively, to obtain a more reasonable tunnel response to upper foundation pit excavation. Wei et al. [12] introduced a rotational and staggered deformation-coordinated shield tunnel model, derived a formula for calculating the longitudinal deformation of the lower shield tunnel caused by foundation pit excavation based on the image source method, and validated the calculation method by comparing B-W and B-P methods with three actual engineering monitoring data, and analyzed the four influencing factors, namely, maximum uplift of the pit bottom, excavation width, excavation depth, and tunnel burial depth.

In addition to theoretical analysis, numerical simulation and on-site measurements play important roles in studying the mechanical characteristics of deep foundation excavation projects. Niu et al. [13] conducted in-depth research on the deformation of adjacent subway tunnels caused by foundation pit excavation in the influence zone of a fault zone, using theoretical analysis, numerical simulation, and on-site measurements. Zhang et al. [10] analyzed the impact mechanism of foundation pit excavation on adjacent shield tunnels based on the theoretical change of lateral forces in adjacent shield tunnels caused by foundation pit excavation, and further calculated the additional stress generated in the soil and the longitudinal deformation of the tunnel. Wang et al. [14] established a Winkler foundation–Timoshenko beam model for longitudinal deformation analysis of adjacent station (working well) shield tunnels under foundation pit excavation unloading, and strictly derived the differential solution of the analytical model based on the force method principle. Wei et al. [15] used a small-strain hardening model of soil and a two-dimensional finite element numerical method to conduct numerical simulations on the deformation of adjacent existing shield tunnel segments caused by the excavation of a foundation pit in soft soil and analyzed the variation in tunnel segment vertical and horizontal displacements with different influencing factors using a single-factor analysis method and a multi-factor analysis method. Chen et al. [16] proposed a theoretical calculation method for lateral force in shield tunnels considering the influence of retaining structure deformation, verified the reliability of the theoretical calculation method through numerical simulation and centrifuge tests, and explored the correlation mechanism between tunnel lateral force, deformation, and internal force based on actual engineering site monitoring data. Liu et al. [17] used numerical methods to elucidate the impact of excavation of a foundation pit adjacent to an existing tunnel on the existing tunnel structure and compared the measured data with finite element data to verify the effectiveness of the adopted three-dimensional model. Feng et al. [18] simplified the tunnel into a Timoshenko beam that considers both longitudinal stiffness and shear deformation, used a three-parameter Kerr foundation model for the foundation, and obtained the analytical solution for longitudinal deformation of the tunnel using finite difference method while considering the boundary conditions at both ends of the tunnel. The rationality of this method was verified by comparing it with finite element data, existing foundation model theory, and measured engineering data. Wu et al. [19] studied the influence of deep foundation excavation of soft soil layers on the lower subway tunnel, conducted deformation monitoring of the lower subway tunnel during the deep foundation excavation of a bridge abutment foundation pit, and analyzed the influence of different construction conditions on the displacement and deformation of the subway tunnel through the analysis of accumulated settlement and horizontal displacement data. Zhao et al. [20] established a three-dimensional finite element model based on a typical case of foundation pit and tunnel, using various methods to determine the geomechanical properties such as standard penetration tests and heavy dynamic penetration. Zheng et al. [21] used a large-scale foundation pit project adjacent to a nearby subway structure as an example to analyze the deformation characteristics of subway structures during foundation pit construction and conducted experiments and practical applications to actively control the horizontal deformation of soil and tunnel by grouting. On this basis, they compared and evaluated several tunnel deformation control methods through numerical simulation and conducted optimization research on grouting schemes. Wei et al. [22] established a three-dimensional finite element model for foundation pit excavation above an existing shield tunnel arranged with anti-floating anchor rods, considered the length, diameter, angle, spacing, and number of pipe sections of the anti-floating anchor rods, and analyzed the influence of the arrangement scheme of the anti-floating anchor rods on the anti-floating effect of the shield tunnel, while considering the excavation conditions of the foundation pit and the relative position relationship with the shield tunnel, and analyzed the effect of the anti-floating anchor rods under different conditions. Wu et al. [23] conducted numerical simulations on the process of constructing a gantry anti-floating framework and the influence of large-scale foundation pit excavation on the lower shield tunnel, explored the interaction mechanism and anti-floating effect of the gantry anti-floating framework and the strata, and finally proposed optimized design suggestions for the anti-floating framework structure. Liu et al. [24] collected 42 engineering cases of foundation pit excavation affecting adjacent side tunnels, analyzed the influence of important factors on vertical and horizontal displacements of the tunnels, determined the discriminant conditions for uplift and subsidence of the tunnel, gave a zoning of vertical displacement of the tunnel outside the pit, and proposed a comprehensive predictive indicator for horizontal displacement of the tunnel and gave empirical predictive formulas for tunnel horizontal displacement under three different strata conditions. Based on the research results of different scholars, it can be observed that when studying the deformation characteristics of adjacent subway tunnels caused by foundation pit excavation, there are certain differences in the research results due to factors such as research methods and regional conditions. Therefore, it is necessary to use different research methods to conduct in-depth studies on specific engineering cases.

Based on the research analysis of the literature mentioned above, there is currently a wealth of studies on foundation pit engineering projects under complex geological conditions both domestically and internationally. However, research on foundation pit engineering specifically focused on reclaimed land is relatively scarce. Therefore, it is necessary to conduct research on the impact of deep foundation pit excavation of reclaimed land on the deformation of nearby subway tunnels. This study is based on a deep foundation pit project located near a subway tunnel in Shenzhen. The study employs theoretical analysis, numerical simulation, and on-site measurements to investigate the uneven deformation caused by foundation pit excavation in the adjacent subway tunnel. The energy method is used to calculate the uneven deformation of the adjacent subway tunnel caused by foundation pit excavation, resulting in more reasonable calculation results. The study also examines the influence of underground structure parameters, foundation pit excavation parameters, and geological parameters on tunnel deformation, and analyzes the sensitivity of geological parameters. The research findings have certain theoretical value and engineering significance, providing references for the design of deformation control schemes for adjacent subway tunnels in similar projects.

2. Project Overview

2.1. Project Introduction

This paper is based on a project located near the Deep Bay Metro Station in the reclamation area of Shenzhen City. The surrounding area of the foundation pit covers approximately 10,670 m^2 and is within the protected zone (50 m) of Shenzhen Metro Line 9 and Line 11. The foundation pit is located about 5 m away from the closest structure of Metro Line 9 station and approximately 10.5 m away from the outer boundary of Metro Line 11 tunnel structure. The elevation of the nearby tunnel top is approximately −17.4 m. The outer diameter of the tunnel is 6.7 m, and the thickness of the concrete lining segments is 350 mm with a concrete strength of C35. The plan view of the foundation pit–tunnel layout and aerial photograph of the site are shown in Figures 1–3. The parameters of the supporting structure are presented in Table 1.

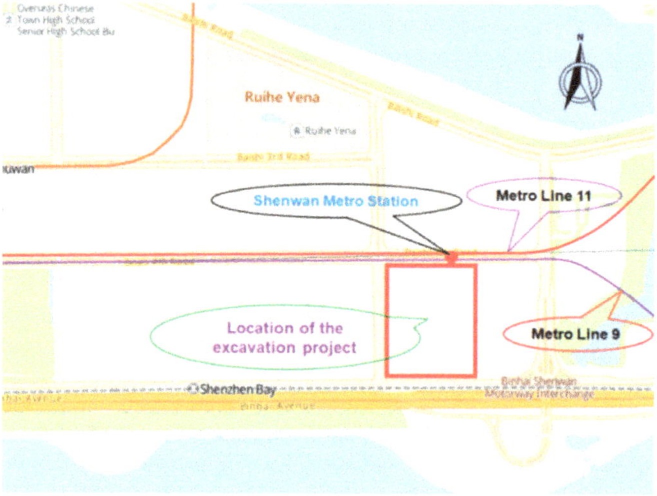

Figure 1. Plan view of foundation pit and tunnel location.

Figure 2. Aerial photograph of the site.

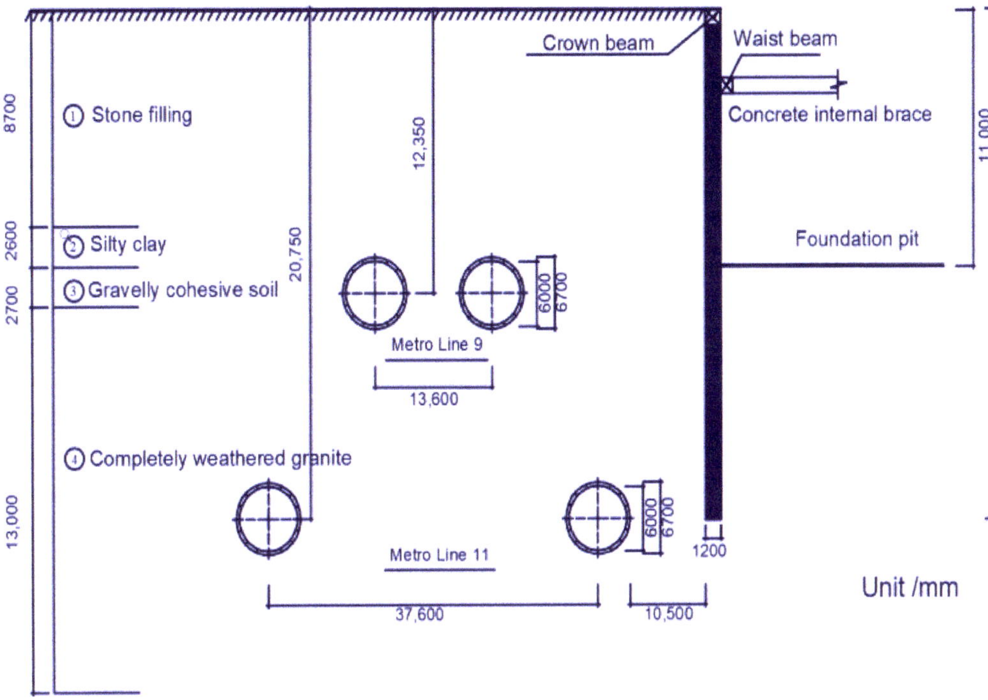

Figure 3. Foundation pit–tunnel cross-section.

Table 1. The cross-sectional dimensions and physical–mechanical properties of the support structure.

Name	Material	Section Dimensions/(mm)	Unit Weight/(kg/m^3)	Elastic Modulus/(GPa)	Internal Friction Angle/(°)	Poisson's Ratio
Retaining pile (Diaphragm wall)	C30	Thickness 880	2500	30	25	0.22
Column pile	C30	Ø1200	2500	30	25	0.22
Crown beam	C30	1000 × 1200	2500	30	25	0.22
Support beam	C30	1000 × 1200	2500	30	25	0.22
Waist beam	C30	1000 × 1200	2500	30	25	0.22

2.2. Geotechnical Conditions of the Project

The simplified stratigraphy of the site from top to bottom is as follows: ① stone filling layer; ② silty clay layer; ③ gravelly cohesive soil layer; ④ completely weathered granite layer; ⑤ highly weathered granite layer; ⑥ moderately weathered granite layer; ⑦ slightly weathered granite layer, which remains unexposed. The relevant parameters of the strata were obtained from triaxial consolidation drainage tests, triaxial loading–unloading tests, standard curing tests, and resonance column tests. In consideration of the foundation pit excavation as a stress-release process, the rebound modulus of the soil is three times the deformation modulus. The main geotechnical and physical parameters of the site are shown in Table 2.

Table 2. Main physical and mechanical parameters of rock–soil.

Stratigraphic (Genetic)	Natural Weight/(kN/m^3)	Elasticity Modulus/(MPa)	Cohesion/(kPa)	Internal Friction Angle/(°)	Permeability Coefficient/(m/Day)
① Stone filling	20	20	–	30	8.0
② Silty clay	17.0	4	9	4	0.005
③ Gravelly cohesive soil	19.0	18	21	23	0.1
④ Completely weathered granite	20.0	50	31	28	0.2
⑤ Highly weathered granite	22.0	80	35	31	0.5
⑥ Moderately weathered granite	22.0	80	35	31	0.5
⑦ Slightly weathered granite	22.0	500	–	–	0.05

3. Theoretical Calculation and Analysis of Tunnel Displacement

The subway tunnel exhibits inhomogeneous deformation, whereas the traditional algorithm, which assumes it to be a mean-averaged elastic foundation beam, exhibits homogeneous deformation. This is an idealized assumption that leads to errors in the calculation results. Considering the inadequacy of treating the subway tunnel as an average elastic foundation beam in traditional algorithms, the energy method (Zhou et al. [25]) is adopted for a more rational calculation of the uneven deformation of the adjacent subway tunnel caused by foundation pit excavation. Assuming that the tunnel structure is composed of a series of shear springs connected to an elastic foundation short beam, the deformation of the tunnel lining due to the excavation of the foundation pit is considered; thus, a longitudinal deformation calculation model of the tunnel is established, as shown in Figure 4. The force analysis is performed on the nth ring with a ring width of ξ.

$$F_i(x) = d\sigma_i(x) = k_1 du_i(x) - k_2[u_i((n+1)\xi) - u_i(n\xi)] - k_2[u_i(n\xi) - u_i((n-1)\xi)] \quad (1)$$

where $F_i(x)$ represents the additional load acting on the tunnel; d is the outer diameter of the tunnel; $\sigma_i(x)$ is the additional stress acting on the tunnel (the specific solution can be found in a study by Niu et al. [13]); i represents horizontal and vertical directions; k_1 and k_2 are foundation bed coefficients and shear stiffness between rings, respectively; and u represents the tunnel displacement.

The work carried out by the additional load on the tunnel, W_1, can be expressed as:

$$W_1 = \int_{-T\xi}^{T\xi} u_i(x)F_i(x)dx \quad (2)$$

The work performed by the shear force between tunnel rings, W_2, is given by:

$$W_2 = -\sum_{n=-T}^{T-1} \frac{1}{2}k_2[u_i((n+1)\xi) - u_i(n\xi)]^2 \quad (3)$$

The work carried out by the resistance of the strata, W_3, is expressed as:

$$W_3 = \int_{-T\xi}^{T\xi} \frac{1}{2}k_1 du_i^2(x)dx \quad (4)$$

By combining Equations (2)–(4), the total potential energy, E_i, of the adjacent tunnel caused by the excavation of the foundation pit can be obtained as:

$$E_i = W_1 + W_2 + W_3 = \int_{-T\xi}^{T\xi} u_i(x)F_i(x)dx - \sum_{n=-T}^{T-1} \frac{1}{2}k_2[u_i((n+1)\xi) - u_i(n\xi)]^2 + \int_{-T\xi}^{T\xi} \frac{1}{2}k_1 du_i^2(x)dx \quad (5)$$

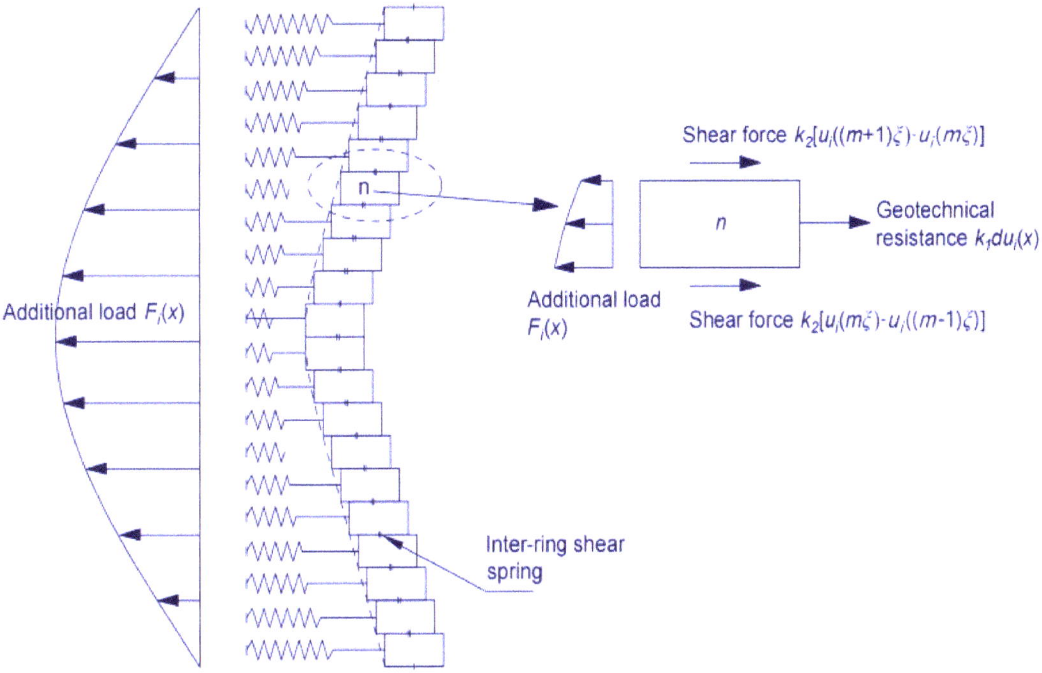

Figure 4. Tunnel deformation calculation model.

By using the Ritz method ($\partial W/\partial R = 0$), the vertical and horizontal displacements, $u_i(x)$, of the adjacent subway tunnel caused by foundation pit excavation can be determined.

4. Finite Element Numerical Simulation Analysis

4.1. Three-Dimensional Finite Element Analysis Model

MIDAS GTS NX three-dimensional finite element software (version 2019) was used to simulate the design conditions and study the deformation effects of the surrounding environment during the excavation and support process of the foundation pit. According to the Saint-Venant's principle, the influence range of the foundation pit excavation on the surrounding environment is approximately three to five times the excavation depth. Therefore, the three-dimensional numerical model's dimensions were chosen as 400 m × 450 m × 40.5 m. The three-dimensional finite element model is shown in Figure 5.

The waist beams and support elements of the foundation pit support were represented using one-dimensional beam elements, with their cross-sectional shapes and dimensions identical to the actual structures. The foundation pit support piles were replaced by equivalent stiffness and represented using two-dimensional plate elements. The support anchors were represented using one-dimensional implanted truss elements. The station lining structure and tunnels were represented using two-dimensional plate elements, with their cross-sectional dimensions matching the actual structures. In the geotechnical finite element analysis, one-dimensional elastic rod elements were used to simulate the internal supports in the foundation pit, while plain plate elements were used to simulate the foundation pit retaining walls, floor supports, tunnel linings, and channel structures. All soil layers were assumed to be elastic-plastic materials, and the Mohr–Coulomb criterion was used as the failure criterion, with the constitutive model selected as the Modified Mohr–Coulomb model. A mixed mesh generator was used to generate the mesh for the rock–soil materials, resulting in a total of 33,020 nodes and 190,960 elements. The material parameters are detailed in Table 2.

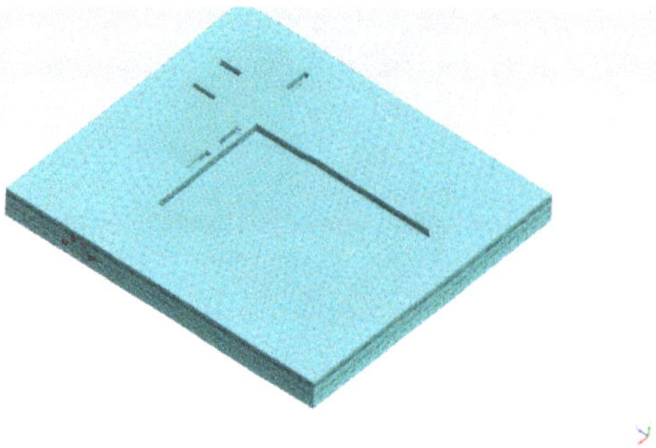

(a)

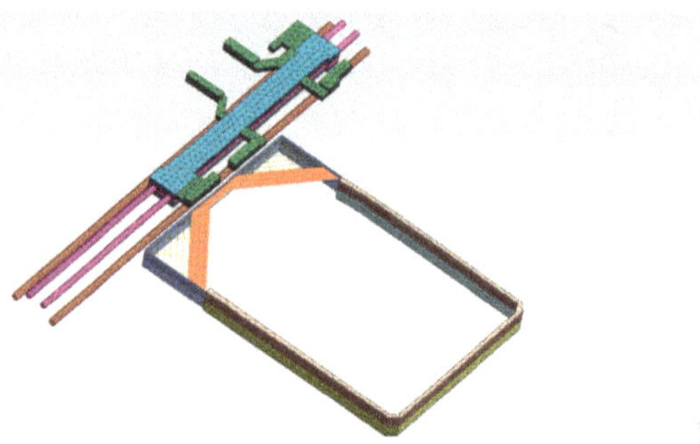

(b)

Figure 5. Three-dimensional finite element analysis model. (**a**) Three-dimensional finite element model; (**b**) relationship between tunnel supporting structure and tunnel.

4.2. Boundary Conditions and Loads

In the calculation model, the vertical boundary conditions were constrained by horizontal hinge supports, and the base of the model was constrained in the vertical direction. The boundary conditions were kept constant during each calculation construction stage. The initial vertical stress field at the site was set as the self-weight of the soil, and the horizontal stress σ_x was determined according to the following equation:

$$\sigma_x = K_0 \sigma_z \tag{6}$$

where K_0 represents the coefficient of static earth pressure, with values of 0.35 for completely weathered rock, 0.18 for highly weathered rock, and 0.1 for moderately weathered rock, calculated using the following equation for other soils:

$$K_0 = 1 - \sin \varphi \tag{7}$$

4.3. Construction Conditions Setting

The construction condition setting is shown in Table 3.

Table 3. Setting of construction conditions.

Operating Conditions	Construction Procedures
1	Site initial stress analysis, implementing network displacement reset
2	Simulation of the site after the completion of the subway station and tunnel, displacements reset to zero.
3	Construction of the foundation pit retaining structure, excavation of the foundation pit to the bottom.
4	Construction of the underground basement floor and the negative one-floor slab, removal of the supports.

4.4. Calculation Results and Analysis

The horizontal and vertical displacement numerical results were extracted from the mid-axis of the right bottom surface of the tunnel structure and plotted in Figure 6. During the excavation of the foundation pit, the horizontal displacement of the mid-axis of the tunnel sidewall is shown in Figure 6a. As the excavation depth of the foundation pit increases, the horizontal displacement gradually increases. Especially within the range of the foundation pit excavation (−90 m to 90 m), the tunnel is significantly affected, and there is clear correspondence between the overall lateral displacement deformation region of the tunnel and the range of the foundation pit excavation. Within the foundation pit excavation range, the tunnel exhibits larger horizontal displacements, noticeably exceeding the horizontal displacements generated by the tunnel sections outside the excavation range. As the distance from the foundation pit excavation range increases, the horizontal displacement of the tunnel gradually decreases. The horizontal displacement curve shows that the model is symmetrical, with a maximum horizontal displacement of 1.4 mm, indicating small deformations.

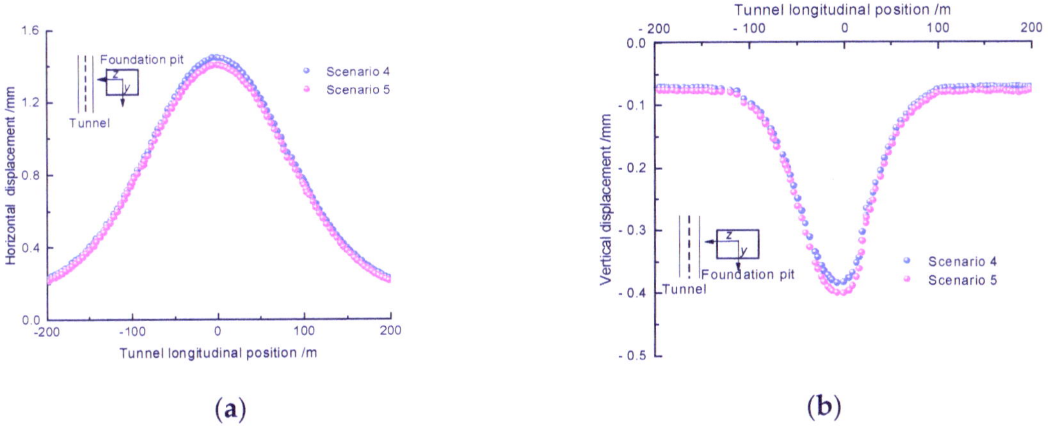

Figure 6. Displacement and deformation curve of up-line tunnel. (**a**) Horizontal displacement; (**b**) vertical displacement.

The variation in vertical displacement along the mid-axis of the right bottom surface of the tunnel structure is shown in Figure 6b. It can be observed that throughout the excavation process, the settlement of the bottom of the tunnel on both sides increases with the increasing depth of the foundation pit excavation. Similarly, within the foundation pit excavation range, the settlement of the tunnel significantly increases. Within the foundation pit excavation range, the settlement is greater than the settlement generated by the tunnel sections outside the excavation range. As the distance from the excavation range increases, the settlement of the tunnel gradually decreases.

4.5. Comparison and Analysis of Monitoring Results

The comparison and analysis of theoretical, simulated, and monitored horizontal and vertical displacements of the tunnel are shown in Figure 7. The influence law of pit excavation on horizontal displacement and vertical displacement of adjacent subway tunnels basically conforms to the normal distribution law. Considering factors such as the overall excavation of the foundation pit, delayed support reinforcement, and surface loading, the monitored horizontal displacement of the tunnel is noticeably larger than the theoretical and simulated values. Similarly, the numerical simulation of the foundation pit excavation did not consider the coupled effect of seepage, resulting in larger vertical displacement values in the actual tunnel monitoring. Overall, the numerical results of the three methods show consistent trends, validating the feasibility of numerical calculations and the reliability of simulation results.

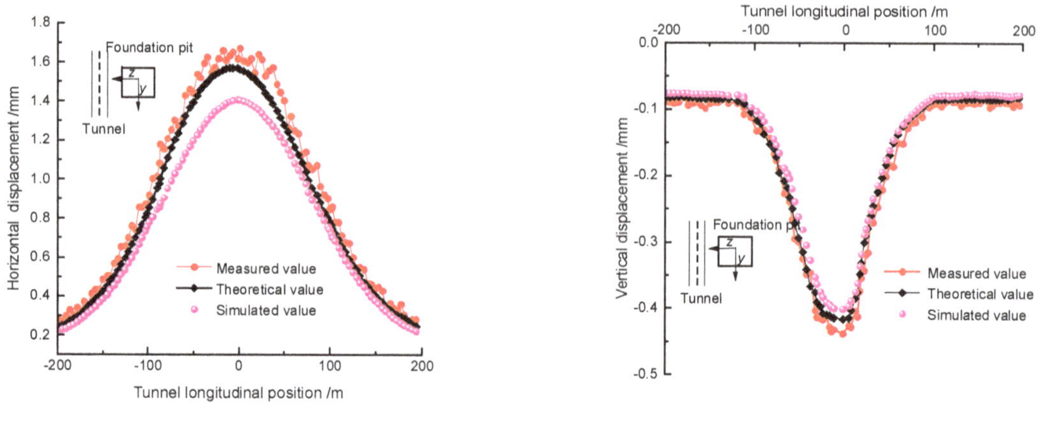

(a) (b)

Figure 7. Comparative analysis of monitoring results. (**a**) Horizontal displacement; (**b**) vertical displacement.

5. Influence Analysis of Structural Parameters

5.1. Impact of Tunnel Equivalent Flexural Stiffness

As shown in Figure 8, the relationship curve between the tunnel's longitudinal equivalent flexural stiffness $(EI)_{eq}$ and the maximum displacement (w_{max}) under different equivalent flexural stiffness values is presented. As the equivalent flexural stiffness increases to 100, the reduction in w_{max} under the four working conditions is 47.7%, 46.7%, 45.0%, and 40.2%, respectively. It can be observed that increasing the tunnel's own stiffness effectively reduces the maximum displacement (w_{max}). Simultaneously, with the gradual increase in the equivalent flexural stiffness $(EI)_{eq}$, its weakening effect on w_{max} gradually diminishes. Furthermore, from Figure 8, it is evident that w_{max} follows a non-linear decreasing trend of "rapid decrease followed by a slow decrease" with the increasing equivalent flexural stiffness $(EI)_{eq}$. This indicates that appropriately increasing $(EI)_{eq}$ can significantly improve w_{max}. However, excessively increasing $(EI)_{eq}$ may weaken the improvement effect on w_{max}. Therefore, it is necessary to pay attention to not excessively increasing $(EI)_{eq}$ while improving the stiffness to avoid compromising the improvement effect on w_{max}.

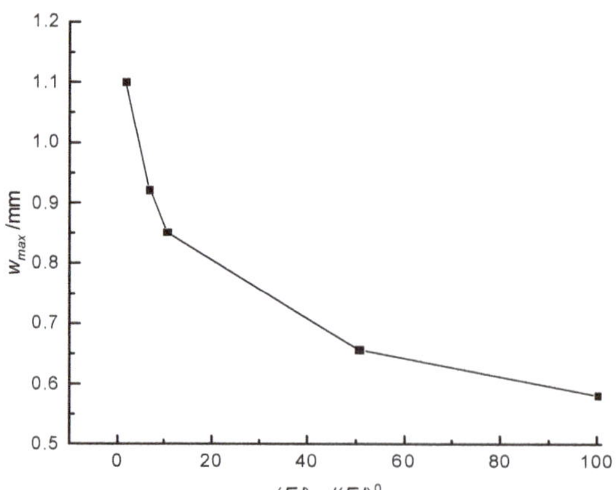

Figure 8. Maximum displacement curve of tunnel under different equivalent bending stiffness.

5.2. Impact of Underground Continuous Wall Parameters

5.2.1. Influence of Wall Thickness

The underground continuous wall plays an indispensable role in deep foundation excavation support projects as it has the ability to withstand substantial soil pressures. The deformation of the tunnel under different wall thicknesses is shown in Figure 9. An increase in the wall thickness of the underground continuous wall can reduce the displacement of the surrounding soil near the foundation pit. Consequently, the smaller the thickness of the underground continuous wall, the more severe the impact on the tunnel. Meanwhile, in Figure 9, the variation patterns of horizontal and vertical displacements of the tunnel are similar. With an increasing wall thickness, the deformation extent of the tunnel gradually decreases, and the deformation rate also reduces. By increasing the thickness of the underground continuous wall, the displacement of the surrounding soil near the foundation pit can be effectively reduced, and the tunnel's deformation can be minimized. This further proves the crucial role of the underground continuous wall in deep foundation excavation projects.

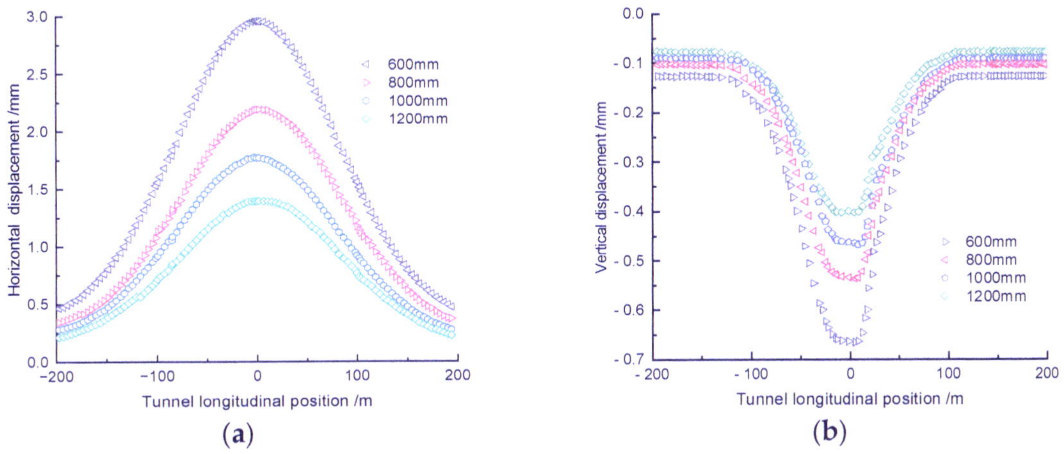

Figure 9. Influence of ground wall thickness on tunnel deformation. (**a**) Horizontal displacement; (**b**) vertical displacement.

5.2.2. Influence of Wall Material

Currently, concrete is commonly used as the material for the underground continuous wall in excavation projects. Different types of concrete mainly differ in their elastic modulus. The elastic moduli of different concretes are shown in Table 4.

Table 4. Elastic modulus of concrete.

Concrete	C20	C25	C30	C35
Elastic modulus/(GPa)	25	28	30	31.5

The response of the tunnel when using different types of concrete for the underground continuous wall is shown in Figure 10. The elastic modulus of the underground continuous wall reflects its deformation capacity. The higher the elastic modulus, the greater the deformation capacity of the underground continuous wall, which can reduce soil displacement and decrease tunnel deformation. In Figure 10, the variations in horizontal and vertical displacements of the tunnel structure are relatively small. With an increase in the elastic modulus of the underground continuous wall, the displacements decrease, and the curves show a pattern of higher values in the middle and lower values on the sides. The elastic modulus of the underground continuous wall significantly affects the reduction of soil displacement and the decrease in tunnel deformation. Selecting concrete materials with higher elastic moduli can enhance the deformation capacity of the underground continuous wall, thereby improving the stability and safety of the project.

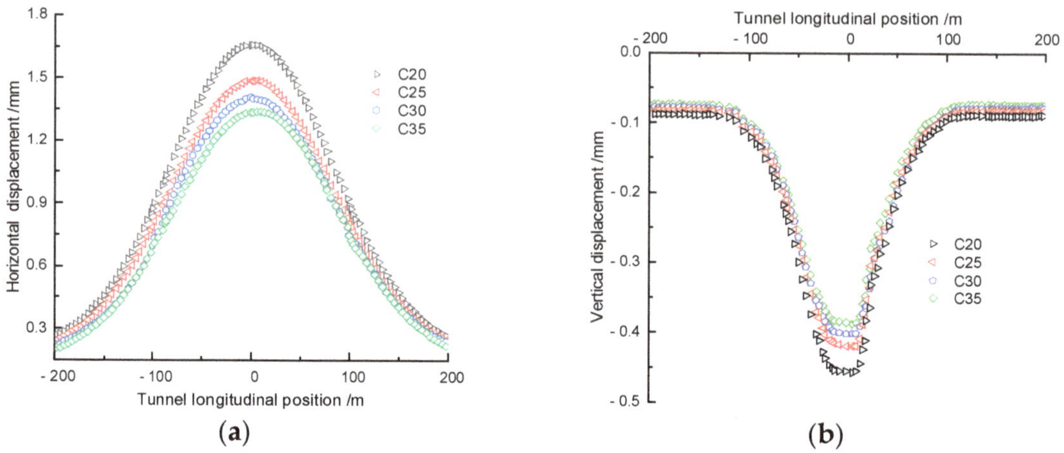

Figure 10. Influence of ground wall material on tunnel deformation. (**a**) Horizontal displacement; (**b**) vertical displacement.

6. Influence of Foundation Pit Excavation Parameters

6.1. Impact of Excavation Depth

With the soil layer, tunnel, and foundation pit retaining structure parameters kept constant, the effect of different foundation pit excavation depths (11 m, 23 m, 28 m, 32 m, and 37 m) on the deformation of adjacent tunnel structures was analyzed. The calculated results are shown in Figure 11. From Figure 11, it can be observed that foundation pit excavation leads to deformation of the adjacent metro tunnel structure, and as the excavation depth of the foundation pit increases, the deformation effect becomes more evident. When the foundation pit depth is 11 m, the maximum vertical and horizontal displacements of the tunnel structure (near the foundation pit side) are 0.4 mm and 1.4 mm, respectively. When the foundation pit depth is 23 m, the maximum vertical and horizontal displacements of the tunnel structure are 2 mm and 3.4 mm, respectively. When the foundation pit depth is 37 m, the maximum vertical and horizontal displacements of the left tunnel are 3.19 mm and 6.37 mm, respectively. As the foundation pit excavation depth increases, the deformation of the tunnel will exceed the deformation control value. Therefore, in practical engineering, attention should be paid to adopting additional reinforcement measures to avoid affecting the safety of the operating metro.

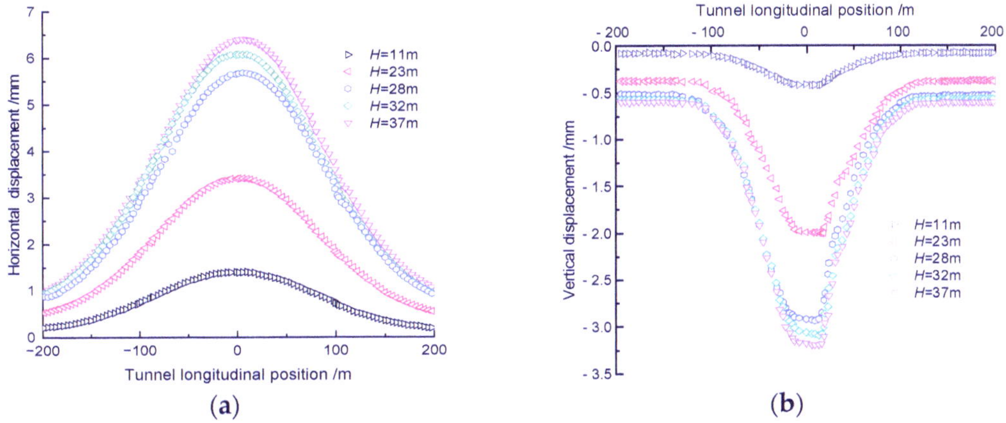

Figure 11. Tunnel structure deformation at different pit depths. (**a**) Horizontal displacement; (**b**) vertical displacement.

6.2. Impact of Excavation Length

For "long excavation" scenarios of the foundation pit (i.e., the foundation pit width B remains constant while the excavation length L increases), as L increases, the maximum displacement $w_{\max}$ of the tunnel increases significantly, and the area affected by the foundation pit excavation on the tunnel also increases notably. Long excavation not only affects the magnitude of tunnel displacement but also changes the uplift range of the tunnel.

It is worth noting that when L is approximately nine times the foundation pit excavation depth (L = 9d), $w_{\max}$ reaches its maximum value, and thereafter, $w_{\max}$ does not continue to increase with further increases in L. However, with increasing L, the uplift range of the tunnel still increases. Figure 12 shows the curve of $w_{\max}$ versus L/d for long excavation scenarios. It can be observed that the curve follows a non-linear increasing pattern of "rapid increase followed by a slow increase". When L is increased to $9d$, $w_{\max}$ increases rapidly. However, for a given foundation pit excavation area ($L \times B$), further increasing L does not significantly affect the tunnel displacement, and the displacement of the tunnel tends to stabilize. Therefore, in the design and construction process, a reasonable evaluation of the long excavation of the foundation pit should be conducted, and corresponding support measures should be taken to control the tunnel displacement and ensure tunnel stability.

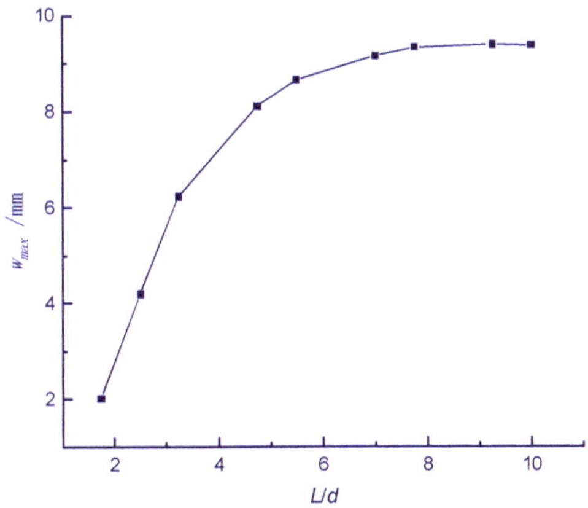

Figure 12. Maximum displacement curve of tunnel under different excavation lengths of foundation pit.

7. Influence Analysis of Geotechnical Parameters

7.1. Impact of Elastic Modulus

In the context of geotechnical survey design in Shenzhen, the values of soil parameters were found to exhibit a degree of variation. Values of 15%, 30% and 45% were identified as representative of this variation. Based on the original design parameters, considering the differences in elastic modulus of different soil layers, various scenarios were designed and calculated by increasing the elastic modulus of each soil layer by 15%, 30%, and 45%, respectively, while keeping other parameters constant. The variations in tunnel displacement were studied.

The relationship between tunnel displacement and the soil elastic modulus is shown in Figure 13. With the continuous increase in the elastic modulus, the lateral deformation and settlement of the tunnel further decrease. Additionally, as the elastic modulus increases to 45%, the reduction rate of tunnel displacement also increases. The distribution pattern of tunnel deformation remains consistent, with the maximum values of horizontal displacement and settlement located in the central region of the excavation. The changes in displacement are most significant within the excavation range (−90 m~90 m). Therefore, considering the differences in elastic modulus of different soil layers has a noticeable impact on tunnel displacement. As the soil's elastic modulus increases, the lateral deformation and settlement of the tunnel decrease. These results have important guiding significance for displacement control in engineering design and construction processes.

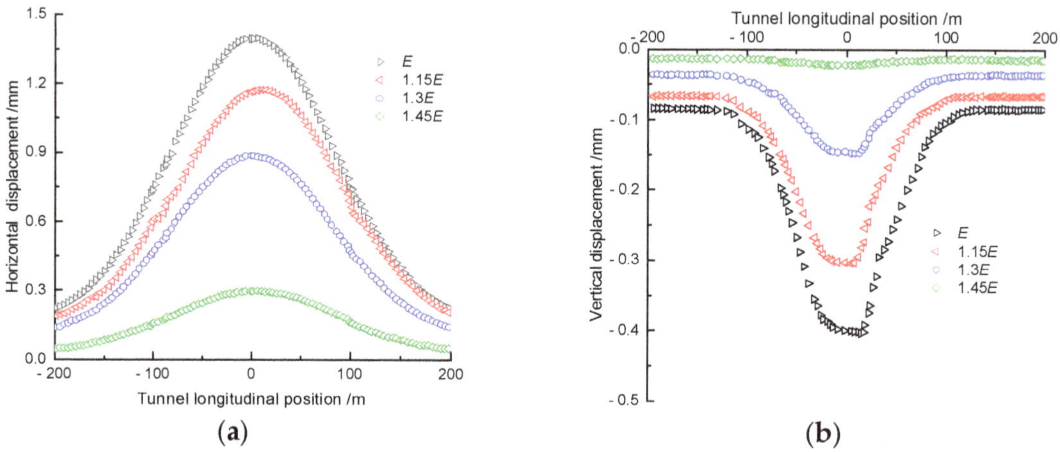

Figure 13. Influence of increasing elastic modulus on tunnel deformation. (**a**) Horizontal displacement; (**b**) vertical displacement.

7.2. Impact of Internal Friction Angle

Similar to the elastic modulus analysis, based on the original design parameters, the internal friction angle of the soil was increased by 15%, 30%, and 45%, respectively, to investigate the tunnel deformation upon completion of excavation. The relationship between tunnel displacement and soil internal friction angle is shown in Figure 14. With an increase in the internal friction angle, the tunnel deformation gradually decreases. This is because the internal friction angle reflects the frictional properties between soil particles, and an increase in the internal friction angle leads to an increase in the frictional resistance, thereby reducing the soil's deformation.

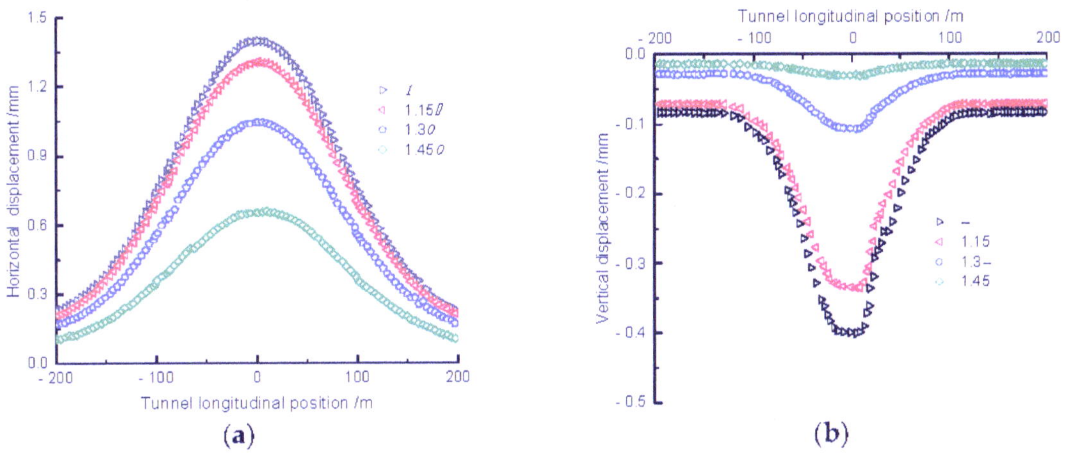

Figure 14. Influence of increasing internal friction angle on tunnel deformation. (**a**) Horizontal displacement; (**b**) vertical displacement.

Numerically, with an increase in the soil's internal friction angle, the lateral displacement gradually decreases. The variation in the internal friction angle also significantly affects tunnel displacement. The values of horizontal displacement and vertical displacement of the tunnel decrease as the internal friction angle increases. Additionally, as the internal friction angle increases to 45%, the reduction rate of tunnel displacement also increases. The distribution pattern of tunnel deformation remains consistent, with the maximum values of horizontal displacement and settlement located in the central region of the excavation (−90 m~90 m). Moreover, the increase in vertical displacement is greater than horizontal displacement, indicating that the changes in the internal friction angle have a larger impact on tunnel settlement, with higher sensitivity.

Therefore, increasing the soil's internal friction angle can reduce tunnel deformation. With an increase in the internal friction angle, the frictional resistance between soil particles increases, resulting in reduced soil deformation and, consequently, reduced tunnel deformation. These results are of significant importance for displacement control in engineering design and construction processes.

7.3. Impact of Cohesion

On the basis of the original design parameters, the cohesion of the soil was increased by 15%, 30%, and 45%, respectively, to investigate the structural deformation of the tunnel upon completion of excavation. The relationship between tunnel displacement and soil cohesion is shown in Figure 15. With an increase in the cohesion of the soil, the tunnel's displacement and deformation gradually decrease. This is because the increase in cohesion enhances the physical and chemical forces between soil particles, thus reducing the soil's deformation.

Numerically, with an increase in the soil's cohesion, the lateral displacement gradually decreases. The changes in cohesion also have a significant impact on tunnel deformation, with the values of horizontal and vertical displacements decreasing as the cohesion increases. Additionally, as the cohesion increases to 45%, the reduction rate of tunnel displacement also increases. The distribution pattern of tunnel deformation remains consistent, with the maximum values of horizontal displacement and settlement located in the central region of the excavation, while the largest variations in displacement occur within the excavation range.

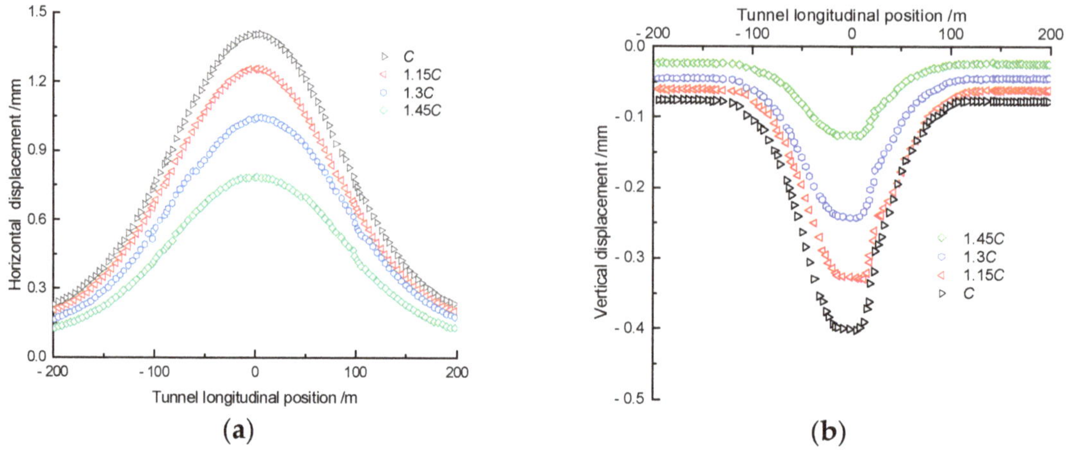

Figure 15. Influence of increasing cohesion on tunnel deformation. (**a**) Horizontal displacement; (**b**) vertical displacement.

7.4. Sensitivity Analysis of Parameters

As shown in Figure 16, by analyzing the impact curves of parameter variations on the maximum displacement growth rate of the tunnel, it can be observed that the increase in displacement does not exhibit a simple linear growth but rather a process of gradually accelerating growth. Among these parameter variations, the increase in elastic modulus has the most sensitive influence on tunnel displacement, with both its growth rate and acceleration exceeding those of the internal friction angle and cohesion. The effect of the internal friction angle on displacement follows next. Considering that the lateral deformation of the tunnel is usually greater than the vertical deformation, the elastic modulus is chosen as the most sensitive parameter affecting the overall tunnel displacement, followed by the internal friction angle, while the variation in cohesion has the least impact on displacement.

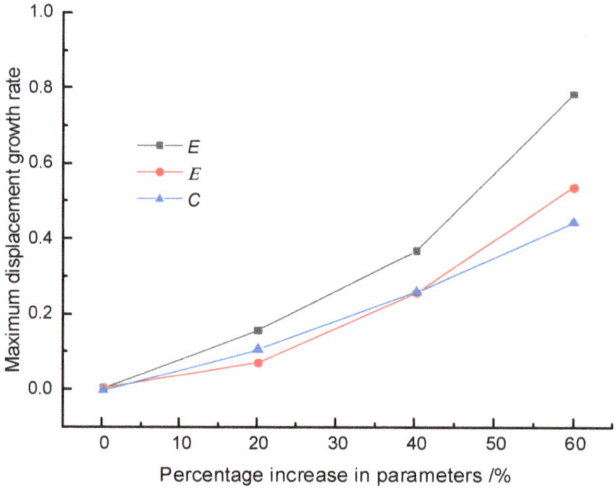

Figure 16. Comparison of sensitivity of each parameter to station and tunnel deformation.

These results are of significant importance for displacement control in engineering design and construction processes. In practical engineering, particular attention should be given to the influence of elastic modulus on displacement, and corresponding measures should be taken to control the changes in elastic modulus during design and construction. At the same time, the changes in the internal friction angle also require careful consideration and adjustments. Although the effect of cohesion variation on displacement is relatively small, it still needs to be comprehensively considered during the design process. By comprehensively analyzing the influence of these parameter variations on displacement, it is possible to better control and manage displacement issues in engineering projects.

8. Conclusions

This study investigated the rare case of the nearby metro tunnel in the construction of the Zhongxing Headquarters Seaside Deep Foundation Pit Project through theoretical analysis and numerical simulation. The following conclusions were drawn:

(1) To overcome the shortcomings of treating the tunnel structure as a uniformly elastic foundation beam in traditional algorithms, this study treated the metro structure as an elastic foundation short beam connected by shear springs, which made the calculation of the uneven deformation caused by the foundation pit excavation near the metro tunnel more reasonable.

(2) Increasing the tunnel's own stiffness $(EI)_{eq}$ can effectively reduce w_{max}, and the weakening effect of $(EI)_{eq}$ on w_{max} gradually diminishes with the increase in stiffness. The tunnel displacement decreases with the increase in the thickness of the underground continuous wall and the concrete strength. The "long excavation" of the foundation pit not only affects the magnitude of tunnel displacement but also changes the uplift range of the tunnel, posing a greater hazard to the tunnel, and the greater the depth of foundation pit excavation, the larger the deformation of the tunnel.

(3) Tunnel displacement decreases with an increase in the soil's cohesion, elastic modulus, and internal friction angle. Among them, the elastic modulus is the most sensitive parameter affecting the overall tunnel displacement, followed by the internal friction angle, while the variation in cohesion has the least impact on displacement.

Author Contributions: Conceptualization, S.L. and F.M.; methodology, S.L.; software, Q.W. and F.M.; formal analysis, Q.W. and F.M.; investigation, Q.W.; writing—original draft, S.L.; funding acquisition, F.M. All authors have read and agreed to the published version of the manuscript.

Funding: National Natural Science Foundation of China (51474045).

Data Availability Statement: The original contributions presented in the study are included in the article, further inquiries can be directed to the corresponding authors.

Conflicts of Interest: The authors declare no conflicts of interest.

References

1. Dong, X.; Mei, L.; Yang, S.Y.; He, L. Deformation Response Research of the Existing Subway Tunnel Impacted by Adjacent Foundation Pit Excavation. *Adv. Mater. Sci. Eng.* **2021**, *2021*, 5121084. [CrossRef]
2. Zhang, J.C. Analysis of the Impact of Deep Foundation Pit Excavation on the Safety of Adjacent Metro Structures: Review of "Analysis of Typical Accident Cases of Metro Structural Safety Protection". *Ind. Constr.* **2021**, *51*, 231.
3. Qiao, J.G.; Peng, R.; Li, J.W.; Guo, F. Study on the Impact of Foundation Pit Excavation on Tunnel Safety Based on Modified Mohr-Coulomb Model. *China Saf. Sci. Technol.* **2022**, *18*, 177–183.
4. Duan, Z.H. Analysis of the Safety Impact of Deep Large Foundation Pit Excavation on Existing Subway Interval Tunnel. *Build. Struct.* **2020**, *50*, 747–752.
5. Bian, R.; He, L.; Yan, Y.Z.; Wu, B. Study on the Deformation of Foundation Pit Retaining Structure and its Influence on Adjacent Existing Tunnels. *Build. Struct.* **2021**, *51*, 1532–1537.
6. Zhang, M.; Yan, L.; Pan, C.W.; Liu, W.B.; Xie, M. Impact of Deep Foundation Pit Construction on Safety of Adjacent Metro Station. *J. Civ. Eng. Manag.* **2020**, *37*, 99–104+121.
7. Wang, G.; Zhang, X.W.; Liu, X.Y.; Gao, H.D.; An, R.; Yin, S. Experimental assessment of clayey layers for clogging potential in TBM tunnel driving and the influence of compositional factors. *Tunn. Undergr. Space Technol.* **2023**, *137*, 105113. [CrossRef]
8. Fu, Y.B.; Wang, B.L.; Wu, H.; Chen, X.S.; Sun, X.H.; Bian, Y.W.; Shen, X. Theoretical analysis on horizontal rectification of tunnel near deep foundation pit by grouting. *Tunn. Undergr. Space Technol.* **2023**, *133*, 104977. [CrossRef]

9. Zhang, Q.; Hou, L.L.; Hu, J.; Wang, J.; Lin, P.; He, P.; Song, S.G. Study on the mechanical behavior of a foundation pit retaining structure adjacent to the pile foundation of a subway station. *Environ. Earth Sci.* **2021**, *80*, 704. [CrossRef]
10. Zhang, X.H.; Wei, G.; Jiang, C.W. The study for longitudinal deformation of adjacent shield tunnel due to foundation pit excavation with consideration of the retaining structure deformation. *Symmetry* **2020**, *12*, 2103. [CrossRef]
11. Huang, M.S.; Li, H.; Yu, J.; Zhang, C.R.; Ni, Y.P. Analysis method of longitudinal deformation of adjacent subway tunnel caused by upper foundation pit excavation. *Chin. J. Geotech. Eng.* **2023**, *45*, 1–9.
12. Wei, G.; Guo, B.L.; Wang, Z.; Liu, J.Y. Calculation of deformation in underlaid shield tunnel induced by foundation pit unloading considering excavation width. *J. Rock Mech. Eng.* **2023**, *42*, 1019–1030.
13. Niu, Y.G.; Wang, Q.Y.; Ma, F.H. Study on the Influence of Foundation Pit Excavation on the Deformation of Adjacent Subway Tunnel in the Affected Area of Fault Zones. *Sustainability* **2023**, *15*, 9462. [CrossRef]
14. Wang, Z.X.; Shi, C.H.; Gong, C.J.; Cao, C.Y.; Liu, J.W.; Peng, Z. Analytical Calculation Method for the Influence of Adjacent Station (Shaft) Foundation Pit Excavation on Downward Shield Tunnel. *Rock Soil Mech.* **2022**, *43*, 2176–2190.
15. Wei, Z.K.; Chen, J.; Chen, B.; Huang, Y.H. Study on the Deformation Effect of Soft Soil Foundation Pit Excavation on Adjacent Existing Tunnels. *People's Yangtze River* **2022**, *53*, 198–206.
16. Chen, R.P.; Liu, M.C.; Meng, F.Y.; Li, Z.C.; Wu, H.L.; Cheng, H.Z. Study on Lateral Force and Deformation of Adjacent Shield Tunnel Structure Caused by Foundation Pit Excavation. *Chin. J. Geotech. Eng.* **2023**, *45*, 24–32.
17. Liu, J.W.; Xue, B.S.; Wang, H.B.; Zhang, X.M.; Zhang, Y.X. Numerical Study on the Behavior of an Existing Tunnel during Excavating Adjacent Deep Foundation Pit. *Sustainability* **2023**, *15*, 9740. [CrossRef]
18. Feng, G.H.; Xu, C.J.; Zheng, M.W.; Xue, W.J.; Yang, K.F.; Guan, L.X. Study on the Longitudinal Deformation of Downward Existing Tunnel Induced by Foundation Pit Excavation Considering Shear Deformation. *China Railw. Soc.* **2022**, *44*, 132–141.
19. Wu, L.L.; Wang, W.; Zhong, Y.; Ren, Z.Y.; Wang, Y.; Guo, D. Field measurement analysis of soft soil deep foundation pit excavation on the underlying subway tunnel. *J. Guilin Univ. Technol.* **2023**, *43*, 1–7.
20. Zhao, X.; Li, Z.W.; Dai, G.L.; Wang, H.X.; Yin, Z.W.; Cao, S.N. Numerical Study on the Effect of Large Deep Foundation Excavation on Underlying Complex Intersecting Tunnels. *Appl. Sci.* **2022**, *12*, 4530. [CrossRef]
21. Zheng, G.; Pan, J.; Cheng, X.S.; Bai, R.B.; Du, Y.M.; Diao, Y. Study on Passive and Grouting Active Control of Longitudinal Deformation of Downward Existing Tunnel Induced by Foundation Pit Excavation. *Chin. J. Geotech. Eng.* **2019**, *41*, 1181–1190.
22. Wei, G.; Zhao, Y.; Sun, Q. Prediction of the effect of foundation pit excavation on shield tunnel with lower anti-floating anchor rod. *J. Undergr. Space Eng.* **2022**, *18*, 603–610.
23. Wu, H.N.; Feng, D.L.; Liu, Y.; Lan, G.Z.; Chen, R.P. Control of downward tunnel deformation during foundation pit excavation based on gantry anti-floating frame. *J. Shanghai Jiaotong Univ.* **2022**, *56*, 1227–1237.
24. Liu, B.; Zhang, D.W.; Li, J.C. Prediction formula and engineering application of side tunnel deformation caused by foundation pit excavation based on multiple case statistics. *Rock Soil Mech.* **2022**, *43*, 501–512.
25. Zhou, S.H.; He, C.; Xiao, J.H. Energy calculation method for deformation of adjacent metro shield tunnel caused by foundation pit excavation under staggered platform effect. *J. China Railw. Sci.* **2016**, *37*, 53–60.

Disclaimer/Publisher's Note: The statements, opinions and data contained in all publications are solely those of the individual author(s) and contributor(s) and not of MDPI and/or the editor(s). MDPI and/or the editor(s) disclaim responsibility for any injury to people or property resulting from any ideas, methods, instructions or products referred to in the content.

buildings

MDPI

Article

Three-Dimensional Numerical Analysis on the Influence of Buttress Wall Removal Timing on the Lateral Deformation of Diaphragm Walls during Deep Excavation

Chia-Feng Hsu [1,*], Chung-Fu Kuan [2] and Shong-Loong Chen [3]

1 Department of Civil Engineering, ChienKuo Technology University, Changhua City 500020, Taiwan
2 ChienKuo Construction Co., Ltd., Taipei 106045, Taiwan; chungfu_kuan@ckgroup.com.tw
3 Department of Civil Engineering, National Taipei University of Technology, Taipei 10608, Taiwan; f10391@ntut.edu.tw
* Correspondence: chiafeng1013@gmail.com; Tel.: +886-958-786-134

Abstract: Urban areas in Taiwan are densely populated with limited land. Excavation often takes place near existing buildings, necessitating protective measures for adjacent properties. Among these measures, cross walls or buttress walls are commonly employed, especially in weak foundation soils, which have seen many successful applications. These mechanisms mainly contribute to reducing lateral deformation of the diaphragm walls, ground subsidence, and excavation face uplift. However, the behavior of these walls is essentially three-dimensional, and common engineering analyses employing one-dimensional elastoplastic beam analysis cannot adequately simulate their mechanical behavior. This study utilized the PLAXIS 3D 2018 software to analyze real-life cases of buttress walls and cross walls. Then, the results of the numerical models were validated against actual field measurements, and the outcomes were satisfactory, and within the regulatory allowable values. The primary objective of this study was to find the influence of different buttress wall removal timings on the lateral deformation of the diaphragm wall. The findings suggest that the gradual removal of buttress walls will effectively control the lateral deformation of diaphragm walls during the excavation of deeper floors like mezzanines.

Keywords: diaphragm wall; deep excavation; weak foundation soils; buttress walls; removal timing

1. Introduction

Land utilization remains a pertinent challenge in Taiwan's metropolitan regions. To maximize space, existing buildings are either renovated or replaced by new structures that expand vertically and increasingly, underground. Consequently, foundations are being dug deeper, expanding the scope of deep excavation projects. The Taipei Basin predominantly comprises thick layers of soft clay, known for its high water content and weak consistency. To ensure stability within the excavation area and safety for surrounding buildings, diaphragm walls are typically designed to penetrate significantly below the excavation face. The present design strategy often involves the use of buttress walls or cross walls to reduce lateral wall displacement and protect neighboring properties. This study focused on a large-scale deep excavation project on Liaoning Street, Chunghsiao Section, Taipei City, comparing numerical results with on-site measurements. The influence of different internal bracing removal timings on the lateral displacement of diaphragm walls was also explored.

1.1. Literature Review

In densely populated urban regions like Taipei, Shanghai, and Singapore, the necessity for underground space is inevitable. Over the past decades, the trend of utilizing subterranean spaces in urban developments, such as basements for parking and commercial uses or subway systems, has been evident [1]. Deep excavation is typically employed for basement or cut-and-cover tunnel construction. In some instances, these excavations occur near existing subway tunnels or adjacent buildings [2–5]. Thus, the design and construction of deep pits should be meticulously executed to prevent excessive wall deflection, ground subsidence, or even wall collapses, as seen in cases like the Nicoll Highway failure [6]. Recent incidents, such as the substantial subsidence at the Dazhi Jitai construction site in Taipei in 2023, caused by inadequate ground improvement, ignorance of anomalous monitoring system alerts, and diaphragm wall damages, underline the profound impact of deep excavation construction on neighboring structures [7].

The literature review is organized into three main themes as shown in Tables 1–3.

Table 1. Literature Compilation Table on the Lateral Displacement of Diaphragm Wall.

Author	Abstract
Manna and Clough [8]	They utilized the finite element method to analyze factors influencing the deformation of excavation retaining walls. When comparing their findings with field observations from deep excavations, they identified several key influencing factors: the safety factor against uplift, stiffness of the retaining wall and support system, pre-stress magnitude of supports, plan geometry of the excavation area, and project duration.
Ou et al. [9]	They examined ten cases of deep excavation in the Taipei Basin where either pre-bored piles or diaphragm walls were employed. Their findings indicated that lateral displacements increased with excavation depth. The maximum lateral displacement (*δh, max*) ranged between 0.2% and 0.5% of the final excavation depth (*He*). Notably, in soft clays, the maximum displacement tends towards the upper limit, while in sandy grounds it leans towards the lower limit.
Clough and O'Rourke [10]	The patterns of retaining wall deformation due to excavation with internal struts and tie-backs can be categorized into cantilever displacement, deep inward displacement, and a combination of the two. Their analysis across different soil types—stiff clay, residual soil, and sand—revealed that the average maximum surface settlement was about 0.15 times the excavation depth. Cases with lateral displacement exceeding 0.5% of the excavation depth were attributed to construction malpractice or insufficient wall penetration.
Masuda et al. [11]	They categorized influential factors affecting excavation performance into two main groups based on an analysis of 52 excavation cases: soil stiffness and support stiffness. Furthermore, they summarized 11 reasons affecting retaining wall deformation, such as soil type, soil properties, wall stiffness, support quantity and spacing, pre-stress on supports, excavation method, wall length, ground improvement presence, scale of excavation, groundwater conditions, and other construction activities.
Wu et al. [12]	They consolidated data from Taipei's metro construction and past excavation projects in the city. Their results showed a relationship between the maximum lateral wall displacement (δ_{max}) and excavation depth (D) as δ_{max} = (0.07% to 0.2%) D. The depth at which this maximum displacement occurs, Za, was related to the excavation depth as Za = (0.8 to 1.2) D, averaging around the excavation depth.
Wang et al. [13]	They analyzed deep excavation cases in Kaohsiung's clay layer. Their findings suggested a relationship between the maximum lateral wall displacement ($\delta_{h,max}$) and the excavation depth (H) as $\delta_{h,max}$ = (0.1% to 0.4%) H.
Surarak et al. [14]	They back-calculated the Eu/Su ratio using the MC model, providing a reliable prediction for the lateral movement of retaining walls. However, this model was less successful in predicting surface settlements. Utilizing soil parameters from lab and field tests for both SSM (Small Strain Model) and HSM (Hardening Soil Model) analyses led to improved agreement with observed lateral wall movements and surface settlements.
Hsieh et al. [2]	They proposed a simplified method for predicting wall displacement and designing diaphragm walls to ensure safety standards are met. They used a three-dimensional numerical analysis to study the factors influencing wall displacement during excavation. These factors encompass excavation geometry, diaphragm wall thickness, wall spacing, wall penetration depth, diaphragm wall stiffness, strut axial stiffness, diaphragm wall axial stiffness, and undrained shear strength of clay.

Table 2. Literature Compilation Table on the Characteristics of Subsidence Induced by Deep Excavation.

Author	Abstract
Peck [15]	He compiled observational data on ground subsidence from excavation cases in areas such as Chicago and Oslo, proposing a relationship between ground subsidence (δ_v) and distance from the retaining wall (d).
Ou and colleagues [9,16,17]	They delineated that subsidence from deep excavation generally exhibits two patterns: triangular trough and concave trough.
Wang et al. [13]	They analyzed deep excavation cases in Kaohsiung, concluding that the relationship between maximum ground subsidence and excavation depth is approximately $\delta_{v,max}$ = (0.04–0.25%)*He*, where *He* is excavation depth and $\delta_{v,max}$ = (0.21–1.10) $\delta_{v,max}$ for maximum wall displacement.
Avanti [18]	He observed that the finite element analysis of ground subsidence in the Jakarta Metro project shows an arch-shoulder type. Larger R_{inter} values (as a parameter to represent interfaces between soil and structure) lead to a concave trough type of ground subsidence, which accurately predicts building damage, as illustrated in Figure 1a. If the subsidence is of the concave type, taking the TNEC (Taipei National Enterprise Center) case as an example (Ou [19]), it can be discerned from the ground subsidence contour that Building A has tilted and begun to sink during excavation, as shown in Figure 1b.

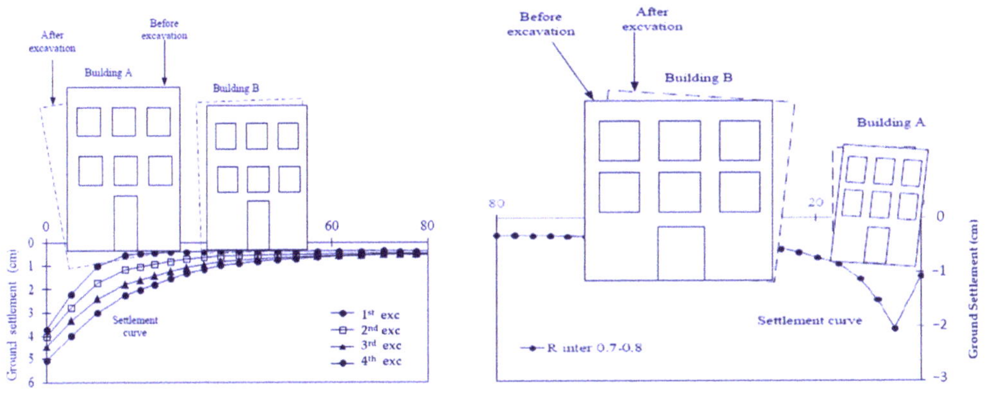

(**a**) (**b**)

Figure 1. (**a**) Predicted Arch-Shoulder Surface Settlement after Deep Excavation in Jakarta Metro Project; (**b**) Predicted Building Tilt Post Deep Excavation in the TNEC Case Study.

Table 3. Literature Compilation Table on the Application of Numerical Analysis in Deep Excavation Engineering.

Author	Abstract
Liao [20], Wang [21] and Hsieh [22]	They utilized numerical analysis to study the behavior of wall deformation and ground subsidence in Taipei's deep excavation projects. The results indicated a significant underestimation of maximum ground subsidence ($\delta_{v,max}$), which often appeared further from the retaining wall than the actual observations. This led to overestimations in secondary affected areas, and unconverged values at the end of these regions. The primary reason for these discrepancies was that the existing analysis could not smoothly simulate the small strain behavior of the soil.
Hsieh [23] and others	They pointed out that three-dimensional numerical analysis simulating excavations in Taipei with central walls and struts compared well with monitoring data.
Chen et al. [24]	They used a 3D software tool to study the effect of exterior struts in reducing the maximum lateral deformation in the middle of the Diaphragm Wall.
Khoiri and Ou [25]	They explored wall displacement and ground subsidence for the Kaohsiung Metro System O6 station using both MC (Mohr-Coulomb Model) and HS (Hardening Soil Model).

Table 3. *Cont.*

Author	Abstract
Ou et al. [26]	They showed that installing a central wall can reduce the maximum wall deformation by 75% and maximum ground subsidence by 82%. They discussed the corner effects of retaining walls using a 3D numerical method.
Ye [27]	They used PLAXIS 3D Foundation to simulate a real project, suggesting that central walls and struts can significantly enhance the stiffness of the Diaphragm Wall.
Feng [28]	They employed the "Plaxis 3D" software to study the effects of different spacings, thicknesses, and penetration depths for struts and central walls, emphasizing their effectiveness in reducing lateral wall movement and ground subsidence.

1.2. Introduction to Buttress Walls, Cross Walls, and Wall Piles

With the increasing depth of excavations in recent constructions, retaining walls are subjected to immense hydrostatic pressures, posing significant construction risks. Insufficient rigidity of diaphragm walls, inadequate lateral support, construction flaws leading to prestress losses, or excessive base size can result in considerable lateral displacement of these walls. When excavations are conducted in soft clay layers, the low elastic modulus and strength of the soil can cause substantial displacement of the diaphragm wall and excessive ground settlement, leading to potential damage to adjacent properties. Nevertheless, recent research and case studies have demonstrated that the implementation of buttress Walls and in-ground walls can effectively enhance the rigidity of diaphragm walls, reduce their displacement, and ground settlement, and provide resistance against uplift forces.

1. Internal Buttress Walls: Positioned on the excavation side of the diaphragm wall, these walls primarily serve to enhance the wall's rigidity, curbing deformations due to excavation-induced decompression. They are progressively removed as excavation proceeds and can be regarded as a temporary support system.
2. External Buttress Walls: Located on the non-excavated side of the diaphragm wall, external Buttress Walls differ from their internal counterparts. They are seen as an integral part of the diaphragm wall structure and are usually reinforced. Their primary function is to increase the rigidity of the diaphragm wall. Unlike internal walls, they are not removed during excavation phases.
3. Cross Walls: These walls are set up to connect the Diaphragm Wall on either the north-south or east-west sides and can be seen as a support system. They offer significant rigidity to the diaphragm walls, effectively limiting wall deformation, differential foundation settlement, and providing resistance against uplift. They are generally reinforced below the excavation face and are removed progressively with excavation.
4. Wall Piles: A subset of foundation piles, wall piles play a vital role in preventing uplift, enhancing foundational load-bearing capacity, and minimizing differential settlement.

2. Materials and Methods

2.1. Research Methodology and Steps

The behavior of retaining walls during foundational excavation and its subsequent impact on the surrounding ground surface has been the subject of numerous studies. The most commonly utilized one-dimensional retaining wall analysis software in the industry is the RIDO program developed by Robert Fages Logiciels and the TORSA program devised by the Geotechnical Technology Research Development Foundation. Moreover, two-dimensional and even three-dimensional numerical analysis software are employed for simulation reviews. This research relied on the excavation data from the "Taipei Liaoning Street Changchun Section New Construction Project" and employed the finite element method via the PLAXIS 3D 2018 software for simulation. The results from the simulation were then juxtaposed against onsite monitoring data to validate the accuracy of soil and structural parameters, as well as the chosen soil constitutive laws. Prior studies have shown that the timing of removal for Buttress Walls and cross walls can influence wall displacement and ground settlement. Hence, this study investigated the efficiency of displacement reduction in retaining walls by varying the timing of strut and in-ground wall removals. The procedural layout for this research is illustrated in Figure 2.

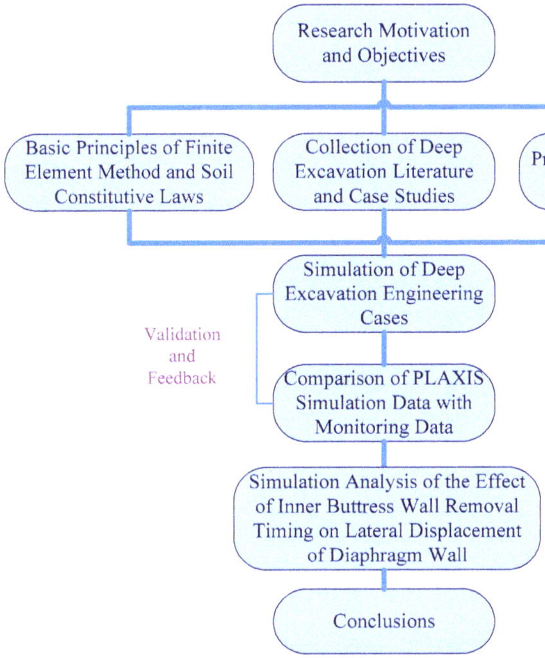

Figure 2. Research Process Diagram.

2.2. *Research Content*

- Validation of PLAXIS Numerical Analysis Simulation and Monitoring Data

An introduction to the PLAXIS numerical analysis software, which encompasses the theory of the finite element method, the constitution laws of soil materials, an overview of different soil models within Plaxis 3D, and the usage of related parameters.

- Simulation Analysis of the Timing for Internal Buttress Wall Removal

Analysis of the timing for the removal of buttress Walls and cross walls. By altering the removal timings, the study aimed to discern its influence on wall displacement.

2.3. *Numerical Simulation*

(1) Software Overview

PLAXIS is a finite element analysis software, due to its ability to easily tackle a plethora of geotechnical engineering problems and its intuitive graphical user interface, its adoption has progressively increased in both academic and professional sectors. The version adopted for this research is PLAIXS 3D 2018.

Plaxis 3D is adept at simulating the deformation and stability of complex geotechnical engineering tasks, such as large-scale foundation excavation and its mutual influence with surrounding environments, shield tunnel construction's interaction with neighboring structures, pile foundations, etc. Its features include:

- Simulation capabilities encompassing: elements, soil bodies, various elements and soil body interfaces, wall, plate, beam structures, pile foundations, braces and anchors, tunnels, and groundwater seepage analysis.

- Analytical capability for deformation, consolidation, graded loading, stability, and seepage, as well as variations in shear, bending moment, and axial stress for various structures, hydrosols, and temporary support elements. It can also consider the impact of low-frequency seismic loads.
- In terms of soil material stress-strain constitutive models, it offers:
  1. Linear-elasticity model.
  2. Mohr-Coulomb model.
  3. Soft Soil Creep Model.
  4. Hardening Soil Model.
  5. Hardening Soil Model with Small-strain.

(2) Hardening Soil Model

- Principles

In this study, we predominantly used the Hardening Soil model (HS) from the five soil material constitutive laws embedded within the Plaxis 3D software. The HS model is an advanced soil model capable of simulating behaviors of both soft and hard soils. This model adopts the Mohr-Coulomb failure criterion and employs plasticity theory, making its stress-strain behavior more aligned with the real-world soil responses than the conventional Mohr-Coulomb. As illustrated in Figure 3a, while an idealized soil model exhibits a linear-elastic perfect plastic behavior, the actual stress-strain relationship of soil is nonlinear. The HS model offers a closer approximation to this real-world stress-strain relationship. Furthermore, the HS model incorporates the shear dilation of the soil and features a yield cap, facilitating a more realistic simulation of soil behaviors.

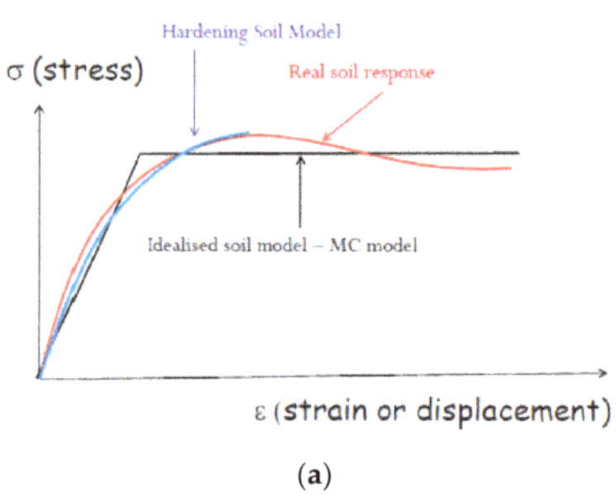

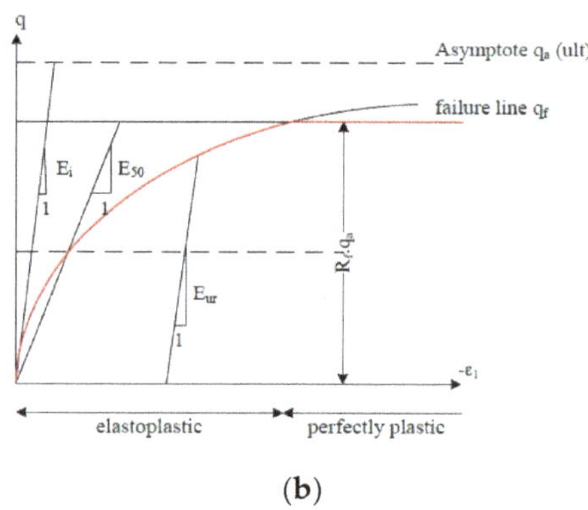

Figure 3. (**a**) The HS model more closely resembles real-world soil behavior; (**b**). Stress-strain relationship in the soil hardening model.

The HS model can simulate the nonlinear stress-strain relationship of soils, meaning that the stiffness at each stress level (Stress-Level) varies during the loading process. As depicted in Figure 3b, when observed through drained triaxial tests, the stress-strain exhibits a hyperbolic relationship:

$$\text{axial strain} : -\varepsilon_1 = \frac{1}{2E_{50}} \times \frac{q}{1 - \frac{q}{q_a}} \quad (1)$$

Here, q denotes axial stress, q_a is the shear strength asymptote, and q_f is the stress value corresponding to the asymptote of the hyperbola (stress-strain relationship). E_i represents the initial modulus of elasticity.

When q < qf, $R_f = \frac{q_f}{q_a}$, and Equation (1) can be rewritten as $E_i = \frac{2E_{50}}{2-R_f}$. The stiffness modulus of the confining stress under the main load can be found using Equation (1):

$$E_{50} = E_{50}^{ref}\left(\frac{c{\cdot}cos\varphi - \sigma_3{\prime}{\cdot}sin\varphi}{c{\cdot}cos\varphi + P^{ref}{\cdot}sin\varphi}\right)^m \quad (2)$$

Here, P_{ref} signifies the reference pressure (set internally at 100 kN/m^2), E_{50}^{ref} is the secant modulus of elasticity at 50% limit strength under the reference pressure, $\sigma_3{}'$ is the lateral effective stress, φ is the internal friction angle, and m is the exponent for the modulus of elasticity concerning stress.

Not only can the HS Model simulate nonlinear stress-strain relationships, but by utilizing plasticity theory, it can also compute the plastic strain amount at each stress stage. Additionally, plastic strains commence from the initial stages of loading. Moreover, its yielding criterion not only encompasses the yield surface caused by axial differential stress but also contemplates the yield surface induced by confining stress.

- Input Parameters for the HS Model

1. General Parameters

Saturated unit weight γ_{sat} and unsaturated unit weight γ_{unsat} (kN/m^3), along with the porosity ratio e.

2. Stiffness Parameters

E_{50}^{ref}: Triaxial drained secant modulus under reference stress.
E_{oed}^{ref}: One-dimensional oedometric secant modulus under reference stress.
E_{ur}^{ref}: Modulus of elasticity for loading-unloading under reference stress.
m: Power for the relation between elastic modulus and stress.

3. Strength Parameters

C_{ref}: Effective cohesion
φ: Effective friction angle (deg)
Ψ: Dilation angle (deg)

4. Advanced Strength Parameters

ν_{ur}: Poisson's ratio for unloading-reloading. By default in the PLAXIS 3D manual, it is 0.2.
p^{ref}: Reference stress. According to the PLAXIS 3D manual, the default is 100 stress units.
R_f: Failure ratio. By default in the PLAXIS 3D manual, it is set to 0.9.
$\sigma_{tension}$: Tensile strength. The default value is 0 (kN/m^2).

2.4. Parameter Settings for Soil Layer Drainage Conditions

- Drained Conditions

Primarily simulates drainage or long-term loading behavior. Both the stiffness and strength parameters input are based on effective values.

- Undrained (*A*) Parameters

This is mainly for simulating short-term behaviors under undrained or loading conditions. A high bulk modulus of water is applied to render the soil incompressible. The undrained (*A*) computation is based on undrained effective stresses, and both stiffness and strength parameters input are effective values. Pore water pressure (including excess pore water pressure) is computed, with accuracy dependent on the chosen model and parameters.

- Undrained (*B*) Parameters

Similar to the undrained (*A*) condition, this simulates short-term behaviors under undrained or loading situations. The soil is rendered incompressible using a high bulk modulus of water. The undrained (*B*) computation is conducted based on undrained effective stresses, where stiffness parameters are based on effective values while strength parameters utilize undrained shear strength. Pore water pressure (including excess pore water pressure) is computed, but results might be significantly imprecise.

- Undrained (C) Parameters

This is for modeling short-term behaviors under undrained or loading conditions. Both stiffness and strength parameters are based on total stress values. The stiffness is simulated using the elastic modulus (E_u) and the undrained Poisson's ratio (ν_u). The strength is modeled using the undrained shear strength (S_u) and friction angle (ϕ_u). Note, that if an undrained Poisson's ratio (ν_u) of 0.5 is chosen, the software cannot compute it; thus, an input value of 0.495 should be used. This mode does not compute excess pore water pressure.

- Non-Porous Condition

No pore water pressure is generated.

2.5. Model Creation Process

The procedure for establishing a model in PLAXIS 3D for deep excavation projects is as follows. (For a detailed explanation, please refer to the software user manual).

1. Define the overall boundary profile of the model.
2. Specify the soil parameters.
3. Structure simulation.
4. Adjacent building load simulation.
5. Define construction steps and water level settings.
6. Specify the mesh and mesh density.

2.6. Model Limitations

PLAXIS is a finite element software used for geotechnical and soil-structure interaction analysis. While it is highly robust and versatile, like any software, it has its limitations. Here are some of the constraints related to the maximum allowable limits when using PLAXIS for modeling:

- Element Number Limitations:
    - Complexity: PLAXIS has a limit on the number of elements that can be modeled. For highly complex structures or extensive geotechnical scenarios, this could be a limiting factor.
    - Computational Efficiency: More elements mean longer computation times and more intensive use of computer resources.
- Material Properties:
    - Accuracy: There are limits in modeling the exact material properties, especially for heterogeneous soil conditions or complex materials.
    - Non-linearity: Although PLAXIS can handle non-linear materials, there can be complexities and limitations in extreme scenarios.
- Boundary Conditions:
    - Flexibility: While PLAXIS offers a range of options for setting boundary conditions, in some specialized or extreme cases, the options might not be adequate.
    - Real-World Representations: Representing real-world boundary conditions accurately can sometimes be challenging.
- Loading Conditions:
    - Dynamic Loads: There might be limitations in simulating certain types of dynamic or cyclic loading conditions.

  - Extreme Loads: While PLAXIS is capable of simulating a variety of load types, extreme loading scenarios can sometimes push beyond its capabilities.
- Analysis Types:
  - Consolidation Analysis: PLAXIS has certain constraints in dealing with very long-term consolidation and creep analysis.
  - Environmental Factors: The simulation of some specific environmental influences might be constrained.
- Software and Hardware Interaction:
  - Hardware Limitations: The performance of PLAXIS is also dependent on the hardware on which it is run. Hence, limitations in hardware can impact the efficiency and speed of analysis.
  - Parallel Processing: There could be limitations in the efficiency of parallel processing for very large models or complex analyses.

In using PLAXIS, engineers and researchers need to be mindful of these limitations and consider augmenting their analyses with additional tools, testing, or modeling approaches where necessary to ensure comprehensive and accurate results. Always refer to the latest manuals and technical documents of PLAXIS for the most updated and specific details on its capabilities and constraints.

3. Numerical Simulation and Monitoring Data Validation

In this chapter, the results from the numerical analysis of deep excavation projects using PLAXIS 3D 2018 were compared with actual monitored values. Real-world project data, including drilling reports, soil experiment parameters, site geometry, support structure configurations, construction processes, etc., were taken into consideration. The closer the simulation was to real-world conditions, the better. Later, the simulation results were juxtaposed with on-site monitoring data to verify if assumptions about soil parameters, structural parameters, model geometric conditions, construction, dewatering steps, and neighboring building loads are consistent with actual conditions. This will serve as the basis for future analyses on the timing of bracing and diaphragm wall removal.

3.1. Case Overview

The real-world project selected for this study is located in Liaoning Street, Zhongshan District, Taipei City. The site, roughly rectangular, covers an area of 9081 m^2. The original site was a flat parking lot. To the east, it is closely bordered by two-story, three-story, and six-story neighboring buildings. On the west side, the site faces Liaoning Street, which is 11 m wide, with a park located across the street. The northern edge of the site also borders an 11-m wide Liaoning Street (Lane 185), with another park situated across this lane. To the south, the site is bounded by an 8-m wide Liaoning Street (Lane 155), beyond which lie a three-story building and two larger buildings of 14 and 11 stories with 3 and 1 basement levels respectively. The plan was to construct a building with twenty floors above ground and six underground. The main excavation area measured approximately 98.0 m to the south, 80.0 m to the north, and 52.4 m for both the east and west sides, resulting in an excavation area of around 1900 square meters. The site layout is shown in Figure 4. According to structural design data, the excavation depth at this site was 25.7 m. The diaphragm wall, used as a retaining measure, was 54 m deep on the south side, and 52 m on the other three sides, with a thickness of 1.2 m throughout. The top-down construction method was adopted. To mitigate potential wall displacement and surface subsidence for the safety of neighboring properties, three diaphragm walls oriented north-south, each 60 cm thick, were placed within the excavation area. Additionally, on both the north and south sides, there were eight bracings, each 10 m long and 60 cm thick. Both the east and west sides had four bracings, each 12 m long and 60 cm thick. Horizontal bracing was positioned at a depth of 21.8 m.

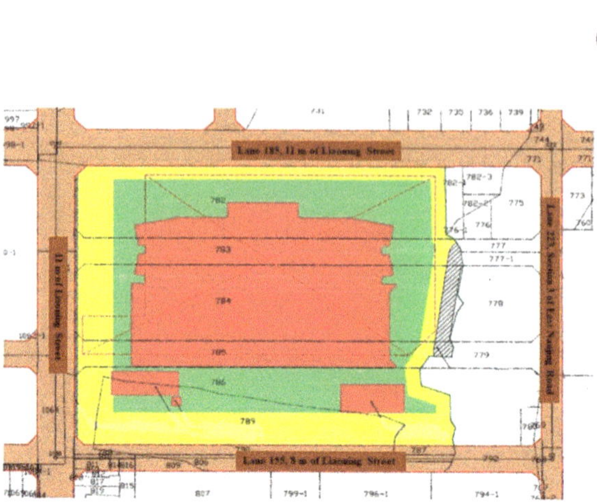

Figure 4. Site Location Map.

(1) Stratigraphy Overview

Based on the on-site drilling investigation report, the drilling depth of this study reached 62.0 m, broadly divided into seven strata, alternating between clay and sandy layers. The groundwater level during drilling was found to be approximately GL-2.8 m. A concise soil layer description table from the drilling report is provided in Table 4 (For additional details, please refer to Supplementary Table S1, with a sectional view of the soil layers presented in Supplementary Figure S1.

Table 4. Concise Soil Layer Description Table.

Layer	Soil Layer Description	Layer Bottom Distribution Depth (GL.−m)	Average Distribution Elevation (GL.−m)
1	Backfill Layer (SF)	0.1~0.9 (0.5)	0.0~0.5
2	Silty Clay Layer (CL)	1.8~3.7 (3.0)	0.5~3.0
3	Silty Sand Layer (SM)	9.2~10.6 (9.9)	3.0~9.9
4	Silty Clay Layer (CL)	19.8~27.4 (23.1)	9.9~23.1
		40.0~41.5 (40.7)	23.1~40.7
5	Silty Sand Layer (SM)	40.9~45.0 (42.2)	40.7~42.2
6	Silty Clay Layer (CL)	51.9~54.3 (52.7)	42.2~54.7
7	Gravel Layer (GW/GP)	- (Drilling Depth)	>52.7

(2) Diaphragm Wall, Buttress Wall, and Cross Wall Overview

For this study, the excavation depth was 25.7 m and a 1.2-m thick diaphragm wall served as the retaining measure. Both the east and west sides had four bracings, each 12 m long and 60 cm thick, as depicted in Figure 5. These temporary installations, the buttress and diaphragm walls, were designed to control excessive displacement of the diaphragm wall and were removed sequentially during the excavation process. For deeper excavation levels, especially for the raft foundation level, horizontal bracing was positioned at a depth of 21.8 m to prevent excessive wall displacement.

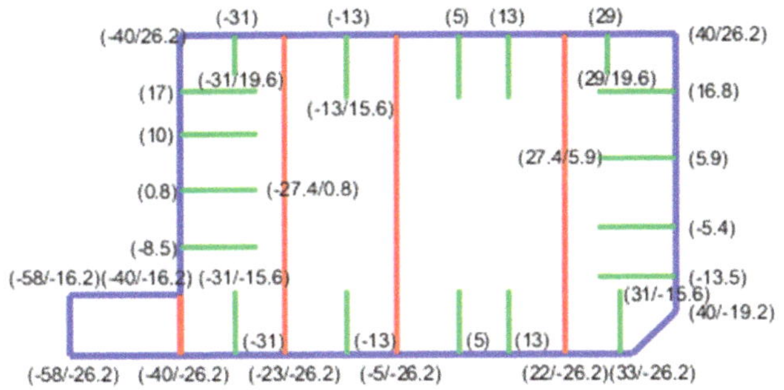

Figure 5. Diagram illustrating the colored positions of each wall. (where, the blue line represents diaphragm wall, the green line represents buttress wall, and the red line represents cross wall).

(3) Overview of the Top-down Construction Method

For this case, with an excavation depth of 25.7 m, the top-down construction method, also known as the reversed construction method, was employed. Before excavation, a diaphragm wall is erected around the excavation area of the site. Coupled with the foundation pile construction, steel columns (also known as internal columns) are suspended to bear the load. Once these structures have cured, excavation and floor slab construction proceed sequentially from the surface downwards. As each level's excavation is completed, the slab for that level is constructed. Upon reaching the required strength, excavation and slab construction for the subsequent level begin, continuing in this manner until the desired depth is achieved. Concurrently, the above-ground structure is built as per design load specifications. This method differs from the conventional open-cut with shoring, minimizing issues like wall displacement and surface subsidence during deep excavation. It is particularly suited for urban construction or underground projects like subways and metros. The floor slabs in the underground structure also serve as internal shoring but are stiffer than traditional steel shoring, thereby enhancing safety during excavation operations. However, during the final excavation to the lowest level, the raft foundation and the lowest floor are excavated at once. To prevent excessive wall displacement due to the removal of the floor slab's supporting effect, horizontal bracing is installed at a depth of 21.8 m. This method's illustration is presented in Figures 6 and 7.

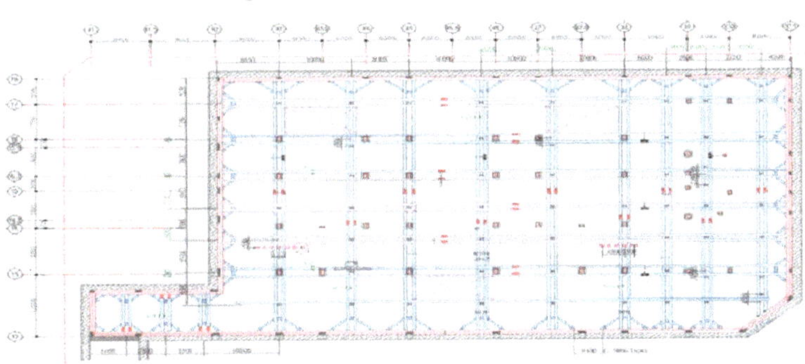

Figure 6. Plan Layout of Excavation Base Support.

(4) Introduction to the Monitoring System

Although this engineering case has established methodologies and measures for stable excavation, there are local variations within the site that cannot be fully understood with limited borehole data. Moreover, the analytical theory applied is still based on accumulated empirical formulas. Concurrently, unforeseeable factors during the construction process could potentially impact the work. Hence, to ensure the safety of the basement excavation, adjacent buildings, surrounding roads, and public facilities, it is imperative to install a safety monitoring system both inside and outside the site during construction. The data obtained during construction should be organized, interpreted, and analyzed to verify the safety of each excavation stage and to predict potential situations in each construction phase. If necessary, mitigation measures should be taken in advance to prevent disasters. The layout of the monitoring system for this case study is shown in Figure 8.

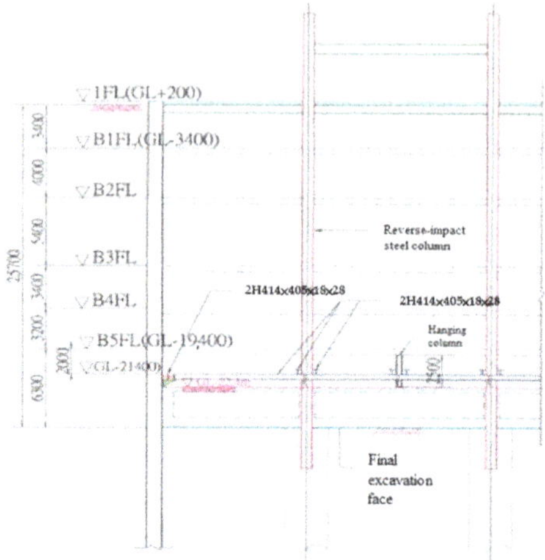

Figure 7. Cross-sectional Schematic of Excavation Base Support at depth of 21.8 m.

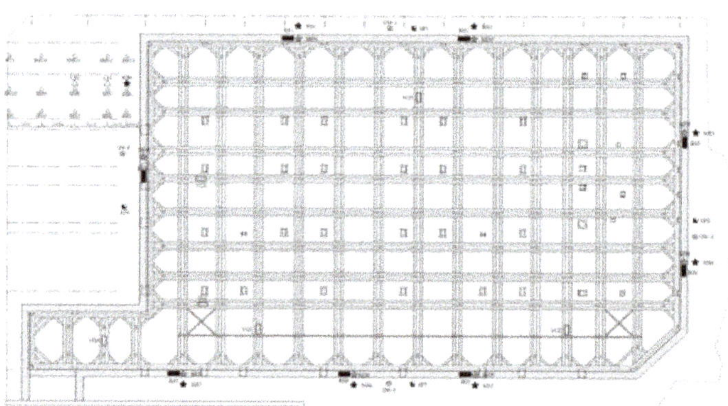

Item	Monitoring item	Symbol	No	Quantity	Instrument name
1	Lateral displacement of diaphragm wall	☆	SID	8	Inner Wall Inclination Meter
2	Lateral displacement of geological strata	★	SIS	8	Outer Wall Inclination Meter
3	Steel reinforcement stress	▬	RS	8	Reinforcement Stress Gauge
4	Support stress	▭	VG	6	Strut Strain Gauge
5	Groundwater level	⊗	OW	4	Water Level Observation Well
6	Groundwater pressure	◩	EP	4	Electronic Hydraulic Pressure Gauge*4
7	Top-down steel column settlement point	—	RHE	20	Settlement Scale
8	Surface settlement	○	SM/SB	-	Ground Settlement Monitoring Point
9	Building inclination	-	-	-	Building Inclinometer
10	Strata settlement	△	LS	2	Multi-Point Extensometer

Figure 8. Layout of the Monitoring System Plan.

Additionally, the monitoring frequency often varies due to factors such as project scale, site geological conditions, chosen construction methods, construction sequence, construction techniques, and construction duration. Therefore, it is essential to adjust based on on-site conditions, providing ample data for analytical interpretation (serving the function of hazard early warning), and establishing the principle for determining monitoring frequency.

3.2. Case Study Simulation

(1) Model Geometry

This research employed the three-dimensional finite element method software PLAXIS 3D 2018 for numerical simulation. The soil element mesh consisted of tetrahedral elements with 10 nodes. The excavation site chosen was somewhat rectangular in shape, with a small chamfer on the southeast side and a protruding rectangle on the southwest. This model adopted a full-section analysis. According to the study by Heish and Ou [29], the ground subsidence caused by deep excavation, whether triangular or trough-shaped, primarily affects the area within a distance of twice the excavation depth from the diaphragm wall. Moreover, the primary influence zone is approximately equivalent to the secondary influence zone (where the surface subsidence area comprises the primary influence zone (PIZ) and the secondary influence zone (SIZ)). The former exhibits a larger subsidence magnitude and a wider affected area, characterized by a steeper subsidence curve. In contrast, the latter has a smaller subsidence magnitude and a more limited affected area, featuring a gentler subsidence curve. Consequently, during excavation, surface subsidence is primarily governed by the PIZ, with an approximate range extending up to twice the excavation depth). Therefore, during analysis, selecting four times the excavation depth adequately captured the subsidence range affected by excavation. Since this research took into account the effects of neighboring buildings and road loads during simulation, a distance of five times the excavation depth from the diaphragm wall was set as the model's horizontal boundary (in the X and Y directions). The model extended vertically into the cobblestone layer, with a mesh depth (in the Z direction) of 62.0 m. In summary, the overall model boundary dimensions were 260 m × 240 m × 62 m. The mesh boundary conditions were set with hinge supports at the base boundary and roller supports at the peripheral boundaries. The model geometry is depicted in Figure 9.

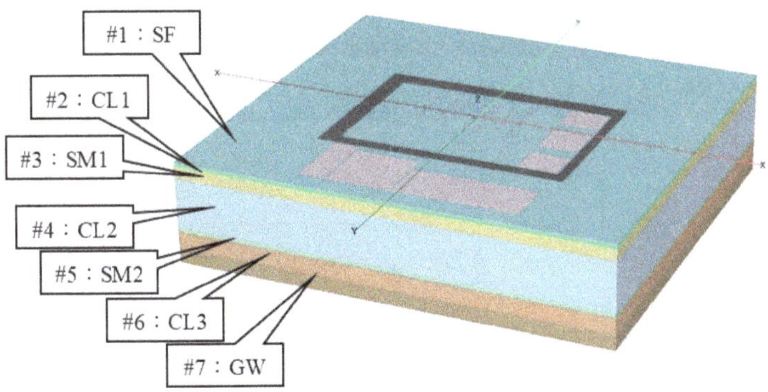

Figure 9. Geometry and Stratum Construction of the PLAXIS 3D Model.

(2) Initial Soil Stress

Based on the study by Heish [2], the silt clay in Taipei, under regular compaction, at-rest earth pressure coefficient closely approximates Equation (3). Therefore, when considering clay, Equation (4) can also be adopted. While accounting for the over-consolidation behavior of clay, Equation (4) is first used to calculate k_0, and then, as suggested by Ladd [30], Equation (5) is employed to estimate $k_{0,OC}$:

$$k_0 = 1 - sin\phi' \tag{3}$$

$$k_{0,NC} = 1 - sin\phi' \tag{4}$$

$$k_{0,OC} = k_0(OCR)^{\alpha} \tag{5}$$

where, $k_{0,OC}$ = static earth pressure coefficient of over-consolidated soil, α = empirical coefficient (for PI = 20, α = 0.4; for PI = 40, α = 0.32)

(3) Soil Parameter Selection

This study's case was situated in the Taipei Basin and comprised seven soil layers. The first layer was a backfill layer, while the third and fifth layers were sandy soil layers. The second, fourth, and sixth layers were clay layers, with the bottommost seventh layer being a gravel layer. The soil layer setting depth is aligned with a geological drilling depth of 62 m. The numerical model's soil layer setup is illustrated in Figure 9.

For both sandy and clayey soils, this research employed the Hardening Soil Model for analysis. Fundamental parameters required for the analysis with the Hardening Soil Model include γ_t, e, c, ϕ, ν, stiffness parameters E_{50}^{ref}, E_{oed}^{ref}, E_{ur}^{ref}, permeability coefficients, drainage properties, and more. The following provides a sequential introduction to the bases of their selections (see Tables 5–7).

Table 5. Soil Parameter Input for the Layers in This Study.

Modulus	Clay	Sand
Soil Unloading Elastic Modulus	$E_{ur}^{ref} = \frac{3(1+e)p^{ref}(1-2\nu_{ur})}{Cs/2.303}$	$E_{ur}^{ref} = 3E_{50}^{ref}$
Soil Shear Elastic Modulus	$E_{50}^{ref} = 1/3E_{ur}^{ref}$	$E_{50} = E_{50}^{ref}\left\{\frac{c\prime COS\varnothing' - \sigma_3\prime \sin\varnothing\prime}{c\prime COS\varnothing' + p^{ref}\sin\varnothing\prime}\right\}$
Soil Bulk Elastic Modulus	$E_{oed}^{ref} = 0.8E_{50}^{ref}$	$E_{oed}^{ref} = E_{50}^{ref}$

Table 6. Poisson's Ratios for Different Soils (Ou [31]).

Soil Type	Poisson's Ratio ν
Saturated Clay (Undrained)	0.5
Unsaturated Clay (Undrained)	0.35–0.4
Silty Sandy Clay	0.3–0.4
Silty Sand	0.3–0.4
Sand and Gravel Soil	0.15–0.35
Silt Soil	0.3–0.35
Rock	0.1–0.4 (varies by type)
Concrete	0.15

(4) Setting of Neighboring Building and Road Load Parameters

The case study under consideration was surrounded by neighboring buildings and roads. To accurately simulate the actual on-site conditions, it was essential to consider the loads from neighboring buildings and road surfaces when assessing the impact of excavation on the site. Therefore, prior to the excavation simulation, the loads from the surrounding buildings and roads were applied directly to the ground surface using a surface load approach. This method simplified the analysis and accelerated the computation speed. The parameters related to neighboring structures are presented in Table 8, and the 3D model configuration of the neighboring buildings is illustrated in Figure 10.

Table 7. Soil Material Parameter Table.

Item No.	Soil Layer	Depth (m)	γ_{unsat} (kN/m^3)	γ_{sat} (kN/m^3)	c' (kN/m^2)	φ' (°)	$E_{50}{}^{ref}$ (kN/m^2)	$E_{oed}{}^{ref}$ (kN/m^2)	$E_{ur}{}^{ref}$ (kN/m^2)	ν_{ur}	ψ	m
#1	SF	−0.5	19.1	19.6	0.5	30	20,000	20,000	60,000	0.20	0	0.5
#2	CL1	−3.0	19.0	19.3	0.7	29	9000	7200	27,000	0.20	0	1.0
#3	SM1	−10.0	19.5	20.1	0.7	30	28,000	28,000	84,000	0.20	0	0.5
#4	CL2	−40.0	18.8	19.0	1.1	30	17,000	13,600	51,000	0.20	0	1.0
#5	SM2	−43.0	19.0	19.5	1.0	31	34,500	34,500	103,500	0.20	1.0	0.5
#6	CL3	−53.0	19.2	19.5	1.5	31	19,000	15,200	57,000	0.20	1.0	1.0
#7	GW	−62.0	21.5	21.7	1.0	35	70,000	70,000	210,000	0.20	5.0	0.5

Soil Parameter	Soil Parameter Selection Description
Drainage Properties	According to the PLAXIS manual, a model's drainage behavior can be categorized into drained, undrained A, undrained B, undrained C, and impervious scenarios. Given that sandy soil quickly expels excess pore water pressure due to its larger voids, effective stress analysis with drainage behavior is utilized. Clay sections are modeled with undrained behavior, and for analysis consistency with sandy soil, the undrained A scenario is chosen. Undrained A uses effective stiffness and strength parameters to simulate undrained behavior.
Wet Soil Unit Weight γ_t, Effective Cohesion c', Effective Friction Angle ϕ'	Values for wet soil unit weight, effective cohesion, and effective friction angle can be derived from drilling reports and laboratory tests. The PLAXIS manual suggests inputting a slightly larger value when effective cohesion is 0 to speed up analysis.
Void Ratio *e*, Specific Gravity ***Gs***	The soil parameter table in this study did not provide specific gravity and void ratio data. Following recommendations from Ou's "Deep Excavation Engineering Analysis, Design Theory, and Practice", an assumption was made that each soil layer has a specific gravity of 2.7 [31]. The void ratio was then inferred using the saturated unit weight and specific gravity.
Soil Elastic Modulus ***Es***	Attention is crucial when analyzing sandy soil in deep excavation projects. Based on research by Ou et al. [26], the effective internal friction angle reflects the friction between soil particles, related to the roughness, shape, and compression of the soil particles. The elastic modulus is associated with the physical properties and inter-particle forces. In practice, empirical formulas often estimate these. In this research, the soil layers utilized the Hardening Soil Model, with the soil parameters summarized in Table 5. According to Khoiri and Ou [25], the recommended elastic modulus Es for sandy soil using the HS model ranges from Es = 2000–4000 N. This study adopts Es = 3000 N.
Permeability Coefficient ***K***	Based on Ou [31], various soil types' permeability coefficients have been collated. This study adopted the permeability coefficients for Taipei silty clay K = (0.5 − 2.0) × 10^{-7} cm/s and Taipei silty sand K = (0.5 − 6.0) × 10^{-4} cm/s.
Poisson's Ratio ν	Ou [31] provided ranges for the Poisson's ratio of various soils, as shown in Table 6.
Other Relevant Parameters	The parameters obtained from empirical formulas, drilling reports, and laboratory experiments are listed in Table 7.

Note: The PLAXIS manual recommends a soil parameter $c' \geq 0.2$ kN/m^2 to prevent convergence issues during analysis.

Table 8. Parameters of Neighboring Structures.

Item	Material Type	Thickness t (m)	Distribution Depth d (m)	Unit Weight r (kN/m^3)	Modulus of Elasticity E (kN/m^2)	Poisson's Ratio ν′
2F Neighbor	Plate	1.0	0	20.0	2.1 × 10^7	0.15
3F Neighbor	Plate	1.3	0	30.0	2.1 × 10^7	0.15
6F Neighbor	Plate	2.1	0	50.0	2.1 × 10^7	0.15
11F Neighbor	Plate	3.1	0	90.0	2.1 × 10^7	0.15
14F Neighbor	Plate	3.8	0	100.0	2.1 × 10^7	0.15
Road				5.0	9.81	0.15

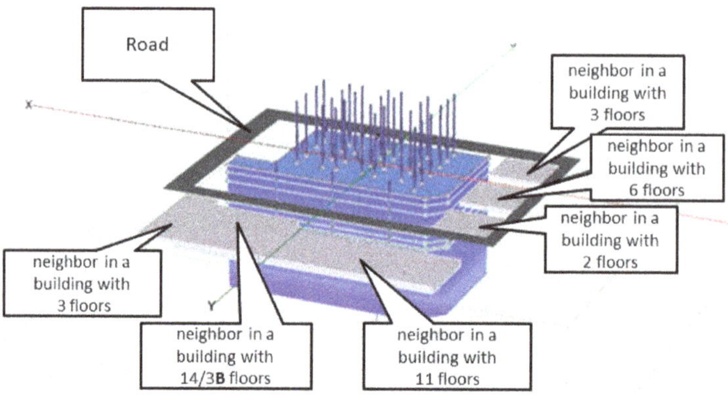

Figure 10. Configuration of neighboring buildings in the PLAXIS 3D model.

The estimated load per square meter for each floor of the neighboring buildings was derived from Taiwan's Building Technical Regulations for Building Structures. The static load was approximated as 550 kgf/m^2 (D ≈ 2400 kgf/m^3 × 0.2 + 70 kgf/m^2), and the live load for residences was roughly L ≈ 200 kgf/m^2. Based on the safety factors stipulated in the Reinforced Concrete Design Code, the estimated load for each floor of the neighboring buildings was calculated as 980 kgf/m^2 (1.2 × 550 kgf/m^2 + 1.6 × 200 kgf/m^2).

For the roads and sidewalks surrounding the construction site, considering the static load of subgrade grading, reinforced concrete, and asphalt concrete, as well as the live load from vehicular traffic and incorporating load safety factors, the approximate load of the nearby roads was estimated to be about 1 tf/m^2 (9.81 kN/m^3).

(5) Parameters for Diaphragm Wall, Buttress Wall, Cross Wall, Floor Slabs, and Foundation Slab

The earth-retaining structure in this study was represented by a diaphragm wall with a thickness of 1.2 m. Its penetration depth was 54 m on the south side and 52 m on the other three sides. Within the excavation zone, three Cross walls oriented from south to north, each with a thickness of 60 cm, were established. Additionally, there were eight buttress walls on both the north and south sides, each with a length of 10 m and a thickness of 60 cm. Four buttress walls, 12 m in length and 60 cm thick, were situated on both the east and west sides. The thickness of the floor slabs was 0.45 m, while the foundation slab was 1.2 m thick. The vertical distribution was designated as 1FL at −0.3 m, B1FL at −3.4 m, B2FL at −7.4 m, B3FL at −12.8 m, B4FL at −16.2 m, B5FL at −19.4 m, B6FL at −22.6 m, and FS at −25.7 m. The detailed configurations can be seen in Figure 11a,b and Figure 12b.

The modulus of elasticity for the diaphragm wall, buttress wall, Cross wall, floor slabs, and foundation slab employs the concrete's elasticity modulus $E_c = 150{,}000 \times \sqrt{f_c'} \times 9.8\ (\mathrm{kN}/cm^2)$. Here, f_c' signifies the 28-day compressive strength of the concrete. For the diaphragm wall, buttress wall, and Cross wall, f_c' is taken as 280 (kgf/cm^2), while for the floor and foundation slabs, it is taken as 350 (kgf/cm^2). When considering the stiffness of the aforementioned structures, it is noted that when these structures undergo substantial bending moments, their stiffness may vary due to concrete cracking. According to research by Ou [31], the wall stiffness should be multiplied by a reduction coefficient ranging from 0.6 to 0.8 for design purposes. Additionally, according to Ou [32] in "Advanced Deep Excavation Engineering Analysis and Design", it was suggested that under the same conditions, the influence of construction factors on floor slabs was relatively minor compared to the support, resulting in a smaller reduction in axial stiffness for the floor slabs than for the support. Therefore, in this study, integrating the content of the aforementioned literature, a reduction coefficient of 0.6 was used for the diaphragm wall, buttress wall, and Cross wall, whereas a coefficient of 0.7 was adopted for the floor and foundation slabs. The selected parameters for these structures are presented in Table 9. In the PLAXIS 3D analysis, plate elements were employed to simulate the diaphragm wall, buttress wall, and Cross wall. It was assumed that these walls are homogeneous and perfectly bonded, without considering the possibility of construction defects.

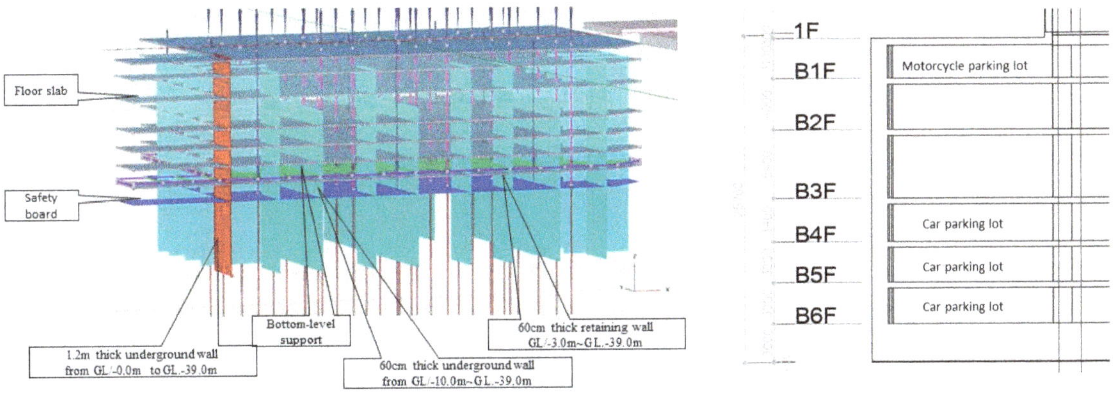

(a) (b)

Figure 11. (**a**) PLAXIS 3D configuration of Cross wall, buttress wall, and pile wall; (**b**) Elevation layout of floor and foundation slabs.

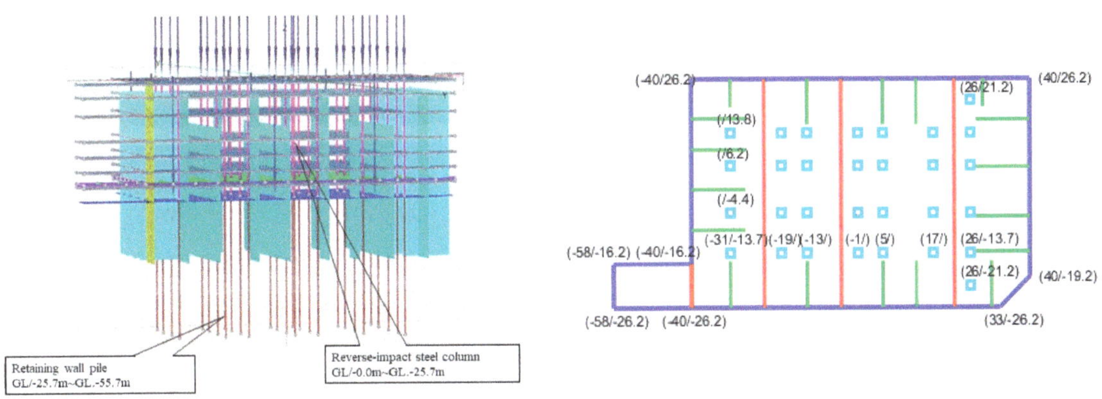

(a) (b)

Figure 12. (**a**) PLAXIS 3D configuration of Pile Wall and Top-Down Steel Columns; (**b**) Plan layout showing positions of the Pile Wall and Top-Down Steel Columns. (where, the blue line represents diaphragm wall, the green line represents buttress wall, and the red line represents cross wall).

(6) Pile Wall Parameter Configuration

In this project, the foundation employed pile walls with a thickness of 1.2 m. Their lengths range from 2.5 m to 5.0 m, with a depth of 30 m (from GL.−25.7 to GL.−55.7). To streamline the analysis, the foundation piles in this study were designed with dimensions of 1.2 m × 3.5 m × 30 m. For the PLAXIS 3D analysis, embedded beam elements were utilized to simulate the pile walls. The material parameters for these are provided in Supplementary Table S2.

Table 9. Parameters for the Diaphragm Wall, Buttress Wall, and Cross Wall.

Item	Thickness (m)	Unit Weight γ (kN/m^3)	Young's Modulus E (kN/m^2)	Poisson's ratio υ	f_c' (kg/cm^2)	Remark (Ou [32])
Diaphragm wall (DW)	1.2	24	1.48×10^7	0.17	280	Reduction coefficient of 0.6
Buttress walls, Cross walls (BW&CW)	0.6	24	1.48×10^7	0.17	280	Reduction coefficient of 0.6
Floors 1F to B6F Slabs	0.45	24	1.93×10^7	0.17	350	Reduction coefficient of 0.7
Foundation Slab (FS)	1.2	24	1.93×10^7	0.17	350	Reduction coefficient of 0.7

(7) Top-Down Construction Steel Column Parameter Configuration

The construction process employed in this project is the top-down method, which included 30 inverted steel support columns. Each column had dimensions of 1.2 m × 1.2 m, and their designed length was 28.7 m (from GL.0 to GL.−25.7, with an additional 3 m embedded into the pile wall). The columns were filled with self-compacting concrete with a compressive strength $f_c' = 560$ kg/cm^2. In the PLAXIS 3D analysis, beam elements were used to represent the top-down steel columns. The material parameters for these columns are detailed in Supplementary Table S3. The positioning of these columns can be seen in Figure 12.

(8) Bracing Parameter Configuration

According to the structural design data, horizontal steel bracing was erected at GL.−21.8 m. This bracing adopted a 2H414 × 405 × 18 × 28 steel profile with intervals less than or equal to 6 m. The surrounding structure utilized the same 2H414 × 405 × 18 × 28 steel profile, while diagonal bracing employed the H350 × 350 × 12 × 19 steel profile. Each bracing unit was pre-stressed with a load of 100 t. Based on the study by Ou [31], the stiffness of the bracing is appropriately reduced, with general reduction coefficients ranging between 0.5 to 0.7. In this research, the reduction coefficient for bracing stiffness was set at 0.7.

For the PLAXIS 3D analysis, top-down steel columns utilize beam elements, and the encirclement and bracing were simulated using node-to-node anchor elements. When simulating with beam elements, it was assumed that the steel profiles were homogenous, thus a linear condition was selected. The material parameters related to this are detailed in Supplementary Table S4. The positioning of these elements is illustrated in Supplementary Figure S2 and Figure 13.

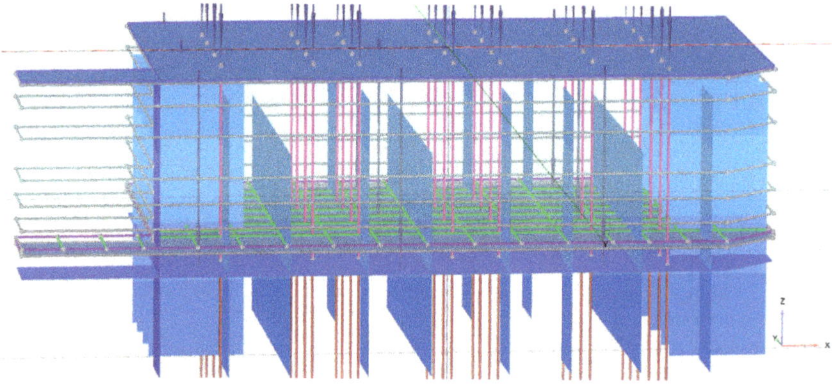

Figure 13. PLAXIS 3D Ground Level Horizontal Bracing Configuration.

3.3. Construction Steps

The construction procedure was simulated using Plaxis 3D 2018, broken down into ten distinct construction phases as illustrated in Figure 14. The purpose was to study the impact of excavation on lateral displacement changes, subsequently pinpointing the most accurate analytical method that aligns the simulation with real-world scenarios.

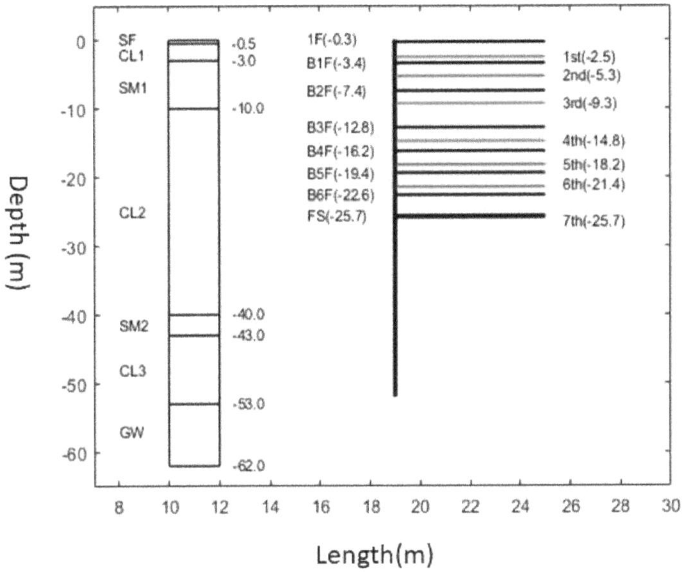

Figure 14. Excavation Construction Cross-Section.

The simulated construction phases encompass the following ten steps:

(1) Initial phase: Establishment of initial ground stresses.
(2) Phase 1:
   a. Load application from adjacent buildings and roads.
   b. Construction of diaphragm walls, cross walls, buttress walls, wall piles, and top-down steel columns, followed by zeroing displacements.
(3) Phase 2:
   a. First excavation to GL.−2.5 m, concurrently demolishing the metro exit cross wall to GL.−2.5 m.
   b. Dewatering corresponding with the excavation depth.
   c. Construction of 1F floor slab at GL.−0.3 m.
(4) Phase 3:
   a. Second excavation to GL.−5.3 m, with simultaneous removal of the buttress walls and metro exit cross wall to that depth.
   b. Corresponding dewatering.
   c. B1F floor slab establishment at GL.−3.4 m.
(5) Phase 4:
   a. Third excavation to GL.−9.3 m, with parallel demolition of the buttress walls and metro exit cross wall.
   b. Dewatering as per excavation.
   c. B2F floor construction at GL.−7.4 m.
(6) Phase 5:
   a. Fourth excavation to GL.−14.8 m, with concurrent demolition activities.

b. Relevant dewatering.
c. B3F floor establishment at GL.−12.8 m.

(7) Phase 6:
a. Fifth excavation to GL.−18.2 m, with simultaneous demolition tasks.
b. Dewatering in line with excavation.
c. B4F floor construction at GL.−16.2 m.

(8) Phase 7:
a. Sixth excavation to GL.−21.4 m, alongside demolition activities.
b. Associated dewatering.
c. B5F floor establishment at GL.−19.4 m.

(9) Phase 8:
a. Seventh excavation to GL.−22.6 m, in parallel with demolition.
b. Corresponding dewatering.
c. Horizontal bracing installation at GL.−21.8 m.

(10) Phase 9:
a. Eighth excavation reaching GL.−25.7 m, alongside the usual demolition tasks.
b. Dewatering to match excavation.
c. Foundation slab (FS) construction at GL.−25.7 m and B6 floor slab at GL.−22.6 m.
d. Removal of the horizontal bracing at GL.−21.8 m.

3.4. Analysis Results and Model Validation

This section elucidated the analysis results from the deep excavation simulation using the finite element software PLAXIS 3D2018 and their comparison with actual on-site measurements to verify the accuracy of the simulation.

1. Lateral Displacement Analysis and Comparison:

The excavation site had eight inclinometers within the wall, each measuring the wall displacement at various excavation stages. The study made a comparison for each. From the analysis, the trend in the deformation of the diaphragm wall was consistent. By the last stage (eighth stage) of excavation, the location and magnitude of maximum displacement in the simulation were approximately in line with the measurements. The detailed numerical comparisons are shown in Table 10.

Table 10. Comparison of Analysis Results with On-Site Maximum Displacement Measurements.

Inclinometer ID	HS Model (Plaxis 3D 2018)		Actual Monitoring Data		Remark
	Depth (Location)	Displacement Value	Depth (Location)	Displacement Value	
SID-1	−21.35 M	−5.06 mm	−22.00 M	−5.33 mm	North side
SID-2	−21.42 M	−5.68 mm	−23.00 M	−8.18 mm	North side
SID-3	−21.80 M	−7.06 mm	−18.00 M	−6.74 mm	East side
SID-4	−20.90 M	−3.51 mm	−21.00 M	−3.6 mm	East side
SID-5	−21.46 M	−5.52 mm	−25.00 M	−5.86 mm	South side
SID-6	−24.27 M	−7.76 mm	−25.00 M	−10.32 mm	South side
SID-7	−21.24 M	−3.23 mm	−21.00 M	−4.30 mm	South side
SID-8	−21.46 M	−6.73 mm	−25.00 M	−6.79 mm	West side

Displacement Negative Values on the Excavation Side. Location Negative Values Below Ground Level.

2. Settlement Analysis and Comparison:

Based on the road location, the excavation site was examined at three locations on the north side and two on the west, totaling five positions. The author has organized the numerical analysis and actual monitoring values in Table 11. Observations from the table indicate that, regardless of whether it is on the north or west side, the maximum subsidence occurs in the eighth phase of excavation. Moreover, the results are all less than the 20 mm neighboring building settlement allowance specified by Taiwanese regulations [33].

Table 11. Stage-wise Excavation Settlement (PLAXIS Analyzed Values and Measured Values).

Point	Source	Phase 1 Excavation	Phase 2 Excavation	Phase 3 Excavation	Phase 4 Excavation	Phase 5 Excavation	Phase 6 Excavation	Phase 7 Excavation	Phase 8 Excavation
SM21~SM23	Analyzed Values	−13.52	−13.54	−13.42	−13.05	−13.13	−13.37	−13.91	−14.69
(North)	Measured Values	−8.8	−8.8	−12.6	−13.00	−13.00	−13.50	−13.90	−19.00
SM29~SM31	Analyzed Values	−13.76	−13.88	−13.90	−13.58	−13.65	−13.88	−14.49	−15.33
(North)	Measured Values	−8.0	−8.50	−9.90	−10.90	−11.30	−12.60	−13.60	−18.50
SM37~SM39	Analyzed Values	−13.11	−13.18	−13.14	−12.84	−12.92	−13.16	−13.72	−14.53
(North)	Measured Values	−8.7	−9.50	−10.80	−12.60	−12.00	−13.50	−14.30	−18.10
SM11~SM13	Analyzed Values	−12.16	−12.27	−12.43	−12.5	−12.66	−12.87	−13.27	−13.74
(West)	Measured Values	−5.10	−6.40	−7.50	−8.10	−9.20	−9.80	−11.10	−15.10
SM3~SM5	Analyzed Values	−13.17	−13.10	−12.97	−12.76	−12.89	−13.15	−13.58	−14.21
(West)	Measured Values	−7.30	−9.20	−10.90	−12.40	−13.30	−13.60	−14.60	−15.90

By comparing the results, it is evident that the site is rectangular in shape. The numerical simulation results for the shorter east-west sides of the site were closer to the actual measurements. For the longer north side, due to the even distribution of the strut and the diaphragm wall, the simulation results were also closely aligned with the measurements. However, for the south side, the uneven placement of the strut and the diaphragm wall, combined with the multiple corners, resulted in a more significant disparity between the simulation and the measurements. As per the research by Ou [19], the mechanism by which the strut suppresses wall displacement primarily arises from the friction between the strut and the soil, as well as the bearing capacity of the strut's end. This leads to a more complex mechanical behavior in the suppression of wall displacement by the strut, making simulations more challenging with reduced accuracy. Furthermore, because the counter-driven construction has a better capability of suppressing wall displacement, with deformations mostly below 10 mm, the measurements often deviate due to human error, making comparisons with simulation analyses less precise. Thus, trends in displacement are typically recognized. Comparisons of the numerical results for wall displacement and the trend in measurements are illustrated in Supplementary Figures S8–S15. Similarly, comparisons for settlement are shown in Supplementary Figures S3–S7.

4. Simulation Analysis of Timing for Removing Buttress Walls

Urban areas are densely populated with limited space. With the advent of subway systems and the burgeoning economic activities, many business activities have gradually moved underground, such as shopping streets and banquet halls. As a result, the height of a single underground floor has shifted from the traditional 3.2~3.4 m (typical of parking lots) to a more substantial 5~6 m to meet commercial demands. Therefore, controlling the deformation of the diaphragm wall has become increasingly crucial.

The primary function of the struts is to restrain the deformation of the diaphragm wall. While they are typically removed in stages based on excavation depth, this study proposed delaying the removal of the struts on the southern side of the B3F (with a height of 5.4 M) until after the completion of B5FL, to simulate and analyze deformation differences (see Figure 14).

4.1. Simulation Steps

The simulation includes the following ten construction phases, as illustrated in Supplementary Materials (Figures S16–S19):

(1) Initial Phase: Establishment of the initial geotechnical conditions.
(2) Phase 1:
- Establishment of loadings from adjacent buildings and roads.
- Construction of Diaphragm Wall, cross walls, buttress walls, wall piles, and top-down steel columns, followed by resetting displacements to zero.

(3) Phase 2:
- First excavation to GL.−2.5 m while simultaneously demolishing the subway exit cross walls down to GL.−2.5 m.
- Dewatering in coordination with the excavation to GL.−2.5 m.
- Construction of the 1F floor slab at GL.−0.3 m.

(4) Phase 3:
- Second excavation to GL.−5.3 m, accompanied by the demolition of the buttress walls and subway exit cross walls down to GL.−5.3 m.
- Dewatering in sync with the excavation to GL.−5.3 m.
- Construction of the B1F floor slab at GL.−3.4 m.

(5) Phase 4:
- Third excavation to GL.−9.3 m, concurrently demolishing the buttress walls and subway exit buttress walls down to GL.−9.3 m.
- Dewatering in tandem with the excavation to GL.−9.3 m.
- Construction of the B2F floor slab at GL.−7.4 m.

(6) Phase 5:
- Fourth excavation to GL.−14.8 m, while simultaneously removing the buttress walls, cross walls, and subway exit buttress walls down to GL.−14.8 m, with a gradual demolition of the southern buttress walls.
- Dewatering in alignment with the excavation to GL.−14.8 m.
- Construction of the B3F floor slab at GL.−12.8 m.

(7) Phase 6:
- Fifth excavation to GL.−18.2 m, coinciding with the demolition of the buttress walls, cross walls, and subway exit buttress walls down to GL.−18.2 m.
- Dewatering alongside the excavation to GL.−18.2 m.
- Construction of the B4F floor slab at GL.−16.2 m.

(8) Phase 7:
- Sixth excavation to GL.−21.4 m, in conjunction with the removal of the retaining walls, cross walls, and subway exit buttress walls down to GL.−21.4 m.
- Dewatering congruent with the excavation to GL.−21.4 m.
- Construction of the B5F floor slab at GL.−19.4 m.

(9) Phase 8:
- Seventh excavation to GL.−22.6 m, simultaneous to the demolition of the retaining walls, cross walls, and subway exit buttress walls down to GL.−22.6 m.
- Dewatering in parallel with the excavation to GL.−22.6 m.
- Installation of horizontal bracing at GL.−21.8 m.

(10) Phase 9:
- Eighth excavation to GL.−25.7 m, accompanied by the removal of the retaining walls, cross walls, and subway exit buttress walls down to GL.−25.7 m, plus the demolition of the previously gradually removed southern buttress walls at B3F.
- Dewatering coordinated with the excavation to GL.−25.7 m.
- Construction of the foundation slab (FS) at GL.−25.7 m; Construction of the B6 floor slab at GL.−22.6 m.
- Dismantling of the horizontal bracing at GL.−21.8 m.

4.2. Analysis Results

The Displacement Reduction Ratio (DRR) was adopted as the metric for evaluating wall deformation. This ratio assesses the impact of different configurations on the lateral displacement of the diaphragm wall, as expressed in Equation (6):

$$\mathrm{DRR} = \frac{\delta_{hm,0} - \delta_{hm}}{\delta_{hm,0}} \times 100\% \qquad (6)$$

where: $\delta_{hm,0}$: represents the maximum lateral displacement when the strut is removed with the excavation, δ_{hm}: is the maximum lateral displacement when the strut's removal is delayed based on functional requirements.

1. On the direct southern side, analysis from HS-5 (Figure A1) indicates:
   - Maximum lateral displacement with immediate strut removal was 5.52 mm.
   - Delaying B3F strut removal until after B5F completion resulted in a maximum lateral displacement of 5.30 mm.
   - This leads to a difference of 0.22 mm and a DRR of 3.98%. One reason might be the proximity to the margin and the reinforcing effect of the vertical diaphragm wall.
2. For the southeast side, as deduced from HS-6 (Figure A2):
   - Maximum lateral displacement with immediate strut removal was 7.76 mm.
   - With the delayed removal approach, it reduced to 3.63 mm.
   - The displacement difference amounted to 4.13 mm, with a DRR of 53.22%. This could be due to its central location on the diaphragm wall.
3. On the southwest side, data from HS-7 (Figure A3) reveals:
   - The maximum lateral displacement with immediate strut removal was 3.23 mm.
   - The delayed approach yielded a displacement of 3.15 mm.
   - This equates to a 0.08 mm difference and a DRR of 2.47%. This might be attributed to its proximity to the edge, leading to reduced deformation due to the reinforcing effect of the vertical diaphragm wall.

5. Conclusions

5.1. Conclusions

1. Conclusions derived from numerical analysis simulation and validation include:
   - The site in question is rectangular in shape. Simulation results on the shorter east-west sides closely match observed data. On the northern longer side, due to evenly spaced retaining and diaphragm walls, the simulated results also align closely with observations. However, for the southern longer side, which features irregularly placed retaining and diaphragm walls with multiple corner transitions, there is a notable difference between the simulation and observed data.
   - According to the research by Ou [19], the mechanism by which retaining walls control wall displacement primarily stems from the friction between the retaining wall and the soil, as well as the bearing capacity at the end of the retaining wall. This leads to a complex mechanical behavior, making simulations more challenging and less accurate.

- Additionally, top-down construction methods tend to offer better control over wall displacement. However, monitoring often entails human errors, resulting in imprecisions between simulations and observed values. As a result, trends rather than actual measurements in displacement are often relied upon for verification.
- The study indicates that the trend in lateral displacement is consistent with numerical analysis results. For surface subsidence, the observed values exceeded simulation predictions due to repeated resurfacing of the adjacent AC road. When factoring in the effects of repeated road resurfacing and observational errors, the analytical results can be considered consistent with the actual situation.

2. Conclusions regarding the timing of retaining wall removal include:
   - On the direct southern side, analysis from HS-6 shows a Displacement Reduction Rate (DRR) of 53.22%. This can be attributed to its position between the Diaphragm Wall, which inherently exhibits larger displacements and thus has more pronounced reduction effects.
   - On the southeast side, HS-5 analysis indicates a DRR of 3.98%. Being closer to the edge, and influenced by the reinforcing effects of the vertical Diaphragm Wall, the inherent displacements are smaller, leading to less significant reduction effects.
   - On the southwest side, analysis from HS-7 reveals a DRR of 2.47%. Similar to the southeast side, being closer to the edge and under the influence of the vertical Diaphragm Wall, the inherent displacements are minimal, resulting in less pronounced reduction effects.

5.2. Recommendations

1. Simulations indicate that adjusting the removal timing of retaining walls based on functional needs and linking them to floor slabs to form a T-shaped structure can effectively control wall displacement. For deeper basement levels, where temporary support is not available, this method can be considered to minimize displacement.
2. This study solely analyzed the staggered removal of the B3F southern retaining wall. Future studies could consider not removing all retaining walls and analyzing the relationship with wall thickness. If the results are significant, adjustments to the retaining wall's thickness could be made, potentially reducing construction costs or incorporating it as a permanent structural component.
3. Typically, in top-down projects during the final excavation phase, due to a longer unsupported length from deeper excavations, horizontal or diagonal bracing is required to control lateral wall displacement. However, this procedure is time-consuming and poses higher operational risks, increasing construction hazards. Future studies might explore the effects of not removing retaining and diaphragm walls during the final excavation phase on wall displacement control.

Supplementary Materials: The following supporting information can be downloaded at: https://www.mdpi.com/article/10.3390/buildings13112678/s1, Figure S1: Borehole Plan Layout and Cross-Section Line; Figure S2: Ground Level Horizontal Bracing Layout; Figure S3: Comparison of Ground Subsidence Analysis and Monitoring Results for Each Excavation Phase on the Northeast Side; Figure S4: Comparison of Ground Subsidence Analysis and Monitoring Results for Each Excavation Phase on the North Side; Figure S5: Comparison of Ground Subsidence Analysis and Monitoring Results for Each Excavation Phase on the Northwest Side; Figure S6: Comparison of Ground Subsidence Analysis and Monitoring Results for Each Excavation Phase on the Southwest Side; Figure S7: Comparison of Ground Subsidence Analysis and Monitoring Results for Each Excavation Phase on the Northwest Side; Figure S8: Comparison of Numerical Analysis Results with Monitoring Values from Monitoring Point SID 1; Figure S9: C omparison of Numerical Analysis Results with Monitoring Values from Monitoring Point SID 2; Figure S10: Comparison of Numerical Analysis Results with Monitoring Values from Monitoring Point SID 3; Figure S11: Comparison of Numerical Analysis Results with Monitoring Values from Monitoring Point SID 4; Figure S12: Comparison of Numerical Analysis Results with Monitoring Values from Monitoring Point SID-5; Figure S13: Comparison of Numerical

Analysis Results with Monitoring Va lues from Monitoring Point SID 6; Figure S14: C omparison of Numerical Analysis Results with Monitoring Values from Monitoring Point SID 7; Figure S15: C omparison of Numerical Analysis Results with Monitoring Values from Monitoring Point SID 8; Figure S16: Schematic Diagram of the Timing Steps for Removing Buttress Walls (Step 1 to Step 2); Figure S17: Schematic Diagram of the Timing Steps for Removing Buttress Walls (Step 3 to Step 4); Figure S18. Schematic Diagram of the Timing Steps for Removing Buttress Walls (Step 5 to Step 6); Figure S19 Schematic Diagram of the Timing Steps for Removing Buttress Walls (Step 7 to Step 9); Figure S20: Plaxis 3D Simulation Diagram for Buttress Wall Removal Timing; Figure S21: Plaxis 3D Analysis Location Diagram; Table S1: Simplified Soil Layer Parameter Table; Table S2: Material parameters for the Pile Wall (Embedded Beam); Table S3: Material parameters for the Top-Down Steel Column (Beam); Table S4: Node-to-Node Anchor Material Parameters.

Author Contributions: Conceptualization, C.-F.H., C.-F.K. and S.-L.C.; Methodology, C.-F.K. and S.-L.C.; formal analysis, C.-F.K.; investigation, C.-F.K.; writing—original draft preparation, C.-F.K. and C.-F.H.; writing—review and editing, C. F.H. All authors have read and agreed to the published version of the manuscript.

Funding: This research received no external funding.

Data Availability Statement: The data used in this article are public.

Conflicts of Interest: The authors declare no conflict of interest.

Appendix A

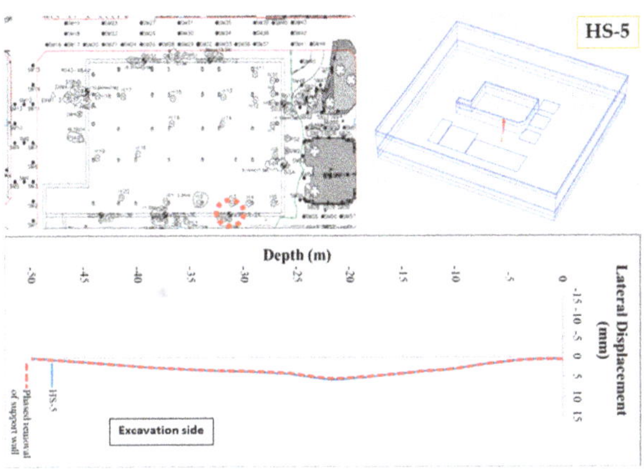

Maximum Displacement of the Diaphragm Wall		
Location	Depth	Displacement
HS-5	−21.46 M	5.52 mm
Gradual Removal of Retaining Wall	−21.46 M	5.30 mm
Difference		0.22 mm
B3F Diaphragm wall Displacement		
Location	**Depth**	**Displacement**
HS-5	−11.8 M	3.28 mm
Gradual Removal of Retaining Wall	−11.8 M	3.25 mm
Difference		0.03 mm

Note: Positive displacement values represent the excavation side. Negative depth values indicate below ground level.

Figure A1. Comparison of Displacement between Gradual Removal and Simultaneous Removal of Buttress Walls on the Southeast Side.

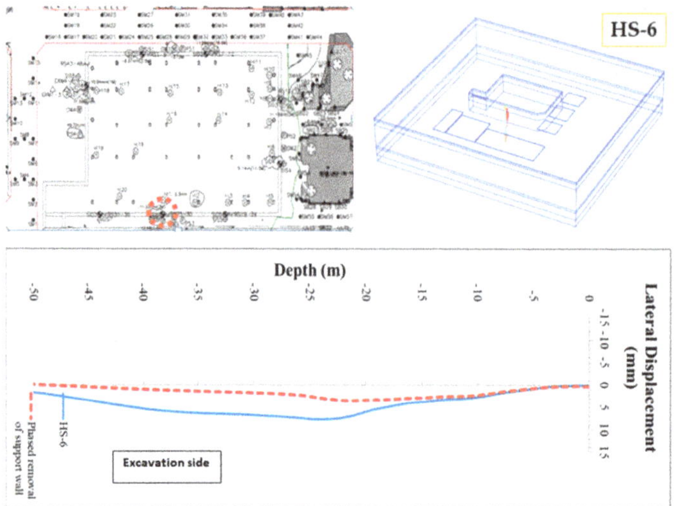

Maximum Displacement of the Diaphragm wall		
Location	Location	Location
HS-6	−24.27 M	7.76 mm
Gradual Removal of Retaining Wall	−21.43 M	3.63 mm
Difference		4.13 mm
B3F Diaphragm wall Displacement		
Location	**Location**	**Location**
HS-6	−11.8 M	3.19 mm
Gradual Removal of Retaining Wall	−11.8 M	2.64 mm
Difference		0.55 mm

Note: Positive displacement values represent the excavation side. Negative depth values indicate below ground level.

Figure A2. Comparison of Displacement between Gradual Removal and Simultaneous Removal of Buttress Walls on the South Side.

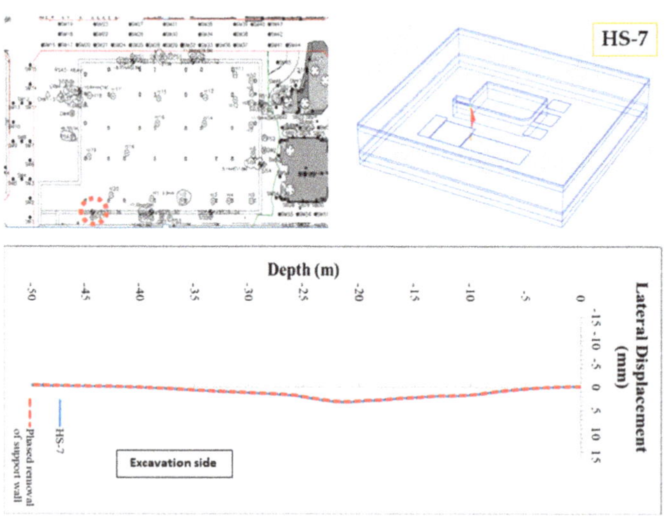

Figure A3. *Cont.*

Maximum Displacement of the Diaphragm wall		
Location	**Location**	**Location**
HS-7	−21.24 M	3.23 mm
Gradual Removal of Retaining Wall	−21.24 M	3.15 mm
Difference		0.08 mm
B3F Diaphragm wall Displacement		
Location	**Location**	**Location**
HS-7	−11.8 M	1.89 mm
Gradual Removal of Retaining Wall	−11.8 M	1.89 mm
Difference		0.00 mm
Note: Positive displacement values represent the excavation side. Negative depth values indicate below ground level.		

Figure A3. Comparison of Displacement between Gradual Removal and Simultaneous Removal of Buttress Walls on the Southwest Side.

References

1. Jian, Z.; Olivia, K. An introduction to connectivity concept and an example of physical connectivity evaluation for underground space. *Tunn. Undergr. Space Technol.* **2016**, *55*, 205–213. [CrossRef]
2. Hsieh, P.-G.; Ou, C.-Y.; Lin, Y.-K.; Lu, F.-C. Lessons learned in design of an excavation with the installation of buttress walls. *J. GeoEng.* **2015**, *10*, 63–73.
3. Shi, J.; Liu, G.; Huang, P.; Ng, C.W.W. Interaction between a large-scale triangular excavation and adjacent structures in Shanghai soft clay. *Tunn. Undergr. Space Technol.* **2015**, *50*, 282–295. [CrossRef]
4. Hong, Y.; Ng, C.W.W.; Liu, G.B.; Liu, T. Three-dimensional deformation behaviour of a multi-propped excavation at a "greenfield" site at Shanghai soft clay. *Tunn. Undergr. Space Technol.* **2015**, *45*, 249–259. [CrossRef]
5. Chen, R.; Meng, F.; Li, Z.; Ye, Y.; Ye, J. Investigation of response of metro tunnels due to adjacent large excavation and protective measures in soft soils. *Tunn. Undergr. Space Technol.* **2016**, *58*, 224–235. [CrossRef]
6. Whittle, A.; Davies, R. Nicoll Highway collapse: Evaluation of geotechnical factors affecting design of excavation support system. In Proceedings of the International Conference on Deep Excavations, Singapore, 28–30 June 2006; p. 30.
7. Abdi, A.S.; Ou, C.-Y. Numerical Study of the Effect of Ground Improvement on Basal Heave Stability for Deep Excavations in Normally Consolidated Clays. *J. Geotech. Geoenviron. Eng.* **2023**, *149*, 04023042. [CrossRef]
8. Mana, A.I.; Clough, G.W. Prediction of movements for braced cuts in clay. *J. Geotech. Eng. Div.* **1981**, *107*, 759–777. [CrossRef]
9. Ou, C.-Y.; Hsieh, P.-G.; Chiou, D.-C. Characteristics of ground surface settlement during excavation. *Can. Geotech. J.* **1993**, *30*, 758–767. [CrossRef]
10. Clough, G.W.; O'Rourke, T.D. *Construction-Induced Movements of In Situ Wall, Design and Performance of Earth Retaining Structure*; ASCE: Reston, VA, USA, 1990; pp. 439–470.
11. Masuda, T.; Einstein, H.H.; Mitachi, T. Prediction of lateral deflection of diaphragm wall in deep excavations. *Doboku Gakkai Ronbunshu* **1994**, *1994*, 19–29. [CrossRef] [PubMed]
12. Wu, P.-Z.; Wang, M.-J.; Peng, Y.-R. Investigation of Continuous Wall Deformation Behavior. In Proceedings of the 7th Geotechnical Engineering Research Symposium, Taiwan, 7–11 September 1997; pp. 601–608.
13. Wang, J.-Z.; Chan, X.-S.; Lei, Y.-M. Analysis of Soil Parameters in the Kaohsiung Metro Red Line Section. *J. Chin. Inst. Civ. Hydraul. Eng.* **2003**, *30*, 101–104.
14. Surarak, C. *Geotechnical Aspects of the Bangkok MRT Blue Line Project*; Griffith University: Nathan, Australia, 2011.
15. Peck, B. Deep excavation and tunnelling in soft ground, State of the art volume. In Proceedings of the 7th ICSMFE, Mexico City, Mexico, 29 August 1969; pp. 225–290.
16. Ou, C.-Y.; Chiou, D.-C.; Wu, T.-S. Three-dimensional finite element analysis of deep excavations. *J. Geotech. Eng.* **1996**, *122*, 337–345. [CrossRef]
17. Ou, C.; Liao, J.; Lin, H. Performance of a topdown basement construction. *J. Geotech. Geoenviron. Eng.* **1998**, *9*, 798–808. [CrossRef]
18. Avanti, A. *Numerical Analyses of Jakarta MRT Deep Excavation Project*; National Taiwan University of Science and Technology: Taipei, Taiwan, 2013.
19. Ou, C.-Y. Finite element analysis of deep excavation problems. *J. Geoengin.* **2016**, *11*, 1–12.
20. Liao, R.-T. *Performances of a Top Down Deep Excavation*; National Taiwan University of Science and Technology: Taipei, Taiwan, 1996.
21. Wang, J.-Z. *Undrained Creep Behavior of Soft Clay Induced by Deep Excavation*; National Taiwan University of Science and Technology: Taipei, Taiwan, 1997.
22. Hsieh, P.-G. *Prediction of Ground Movements Caused by Deep Excavation in Clay*; National Taiwan University of Science and Technology: Taipei, Taiwan, 1999.
23. Hsieh, P.-G.; Lin, Y.-L.; Ou, C.-Y. Three-Dimensional Numerical Analysis and Performance of Deep Excavation with Cross Walls and Buttress Walls. *J. Chin. Inst. Civ. Hydraul. Eng.* **2010**, *22*, 11–22. [CrossRef]

24. Chen, S.-L.; Ho, C.-T.; Li, C.-D.; Gui, M.-W. Efficiency of buttress walls in deep excavations. *J. GeoEng.* **2011**, *6*, 145–156.
25. Khoiri, M.; Ou, C.-Y. Evaluation of deformation parameter for deep excavation in sand through case histories. *Comput. Geotech.* **2013**, *47*, 57–67. [CrossRef]
26. Ou, C.-Y.; Hsieh, P.-G.; Lin, Y.-L. A parametric study of wall deflections in deep excavations with the installation of cross walls. *Comput. Geotech.* **2013**, *50*, 55–65. [CrossRef]
27. Ye, C.-Y. *Efficiency of Buttress Walls and Cross Walls in Deep Excavation*; National Taipei University of Technology: Taipei, Taiwan, 2012.
28. Fong, J.-W. *Numerical Analysis of Buttress Walls and Cross Walls in Deep Excavation*; National Taipei University of Technology: Taipei, Taiwan, 2015.
29. Hsieh, P.-G.; Ou, C.-Y. Shape of ground surface settlement profiles caused by excavation. *Can. Geotech. J.* **1998**, *35*, 1004–1017. [CrossRef]
30. Ladd, R.S. Specimen Preparation and Cyclic Stability of Sands. *J. Geotech. Eng. Div.* **1977**, *103*, 535–547. [CrossRef]
31. Ou, C.-Y. *Deep Excavation Engineering: Analysis, Design Theories and Practices*; Science and Technology Books: Taipei, Taiwan, 2002.
32. Ou, C.Y. *Advanced Deep Excavation Engineering Analysis and Design*; Tech Books Incorporated: Taipei, Taiwan, 2017.
33. Civil Engineering Association of Taipei City. *Practical Manual for Excavation and Earth Retaining Support Engineering Design*; Civil Engineering Association of Taipei City: Taipei, Taiwan, 2002.

Disclaimer/Publisher's Note: The statements, opinions and data contained in all publications are solely those of the individual author(s) and contributor(s) and not of MDPI and/or the editor(s). MDPI and/or the editor(s) disclaim responsibility for any injury to people or property resulting from any ideas, methods, instructions or products referred to in the content.

buildings

MDPI

Article

A Study on Seismic Dynamic Response of Pile-Supported Tunnels in Deep Backfill Area of Soil–Rock Mixture Based on a Model Test

Hanlin Li, Xiaoguang Jin *, Guodong Sun and Jie He

School of Civil Engineering, Chongqing University, Chongqing 400038, China; 20211601018@cqu.edu.cn (H.L.); 20191613025t@cqu.edu.cn (G.S.); 20201601069@cqu.edu.cn (J.H.)
* Correspondence: jinxiaoguang@cqu.edu.cn

Abstract: Aiming to address the problem of construction and environmental risks in tunnel construction through the soil–rock mixture backfill area, this paper carried out a seismic dynamic response model test of a pile-supported tunnel based on practical projects. Firstly, the stress–strain curves and failure characteristics of the soil–rock mixture in the study area were obtained through triaxial tests, and based on this, similar materials for the model test were developed. Then, a vibration table model test was devised to investigate the seismic dynamic response of the pile–tunnel structure. The findings revealed the following: when subjected to seismic waves, the soil–rock mixture stratum displayed a "skin effect" in its acceleration response, indicating that closer proximity to the surface led to heightened horizontal acceleration responses; the horizontal peak acceleration of the grouting mixture stratum in the vertical direction exhibited a "Zigzag" pattern; the peak values of strain response and bending moment in the tunnel lining cross-section exhibited an "X" shape and inverted "V" shape, respectively. The bending moment at the pile crown increased alongside the peak value of the input seismic wave acceleration. The maximum surface settlement in the model ranged from 0.5 to 1 cm, with the tunnel–pile structure effectively mitigating surface settlement.

Keywords: soil–rock mixture backfill area; pile-supported tunnel system; seismic dynamic response; physical model test

1. Introduction

Earthquakes, as the world's largest natural disaster, frequently result in varying degrees of damage to tunnels, such as cracking, collapsing, and displacement of the lining [1]. The shaking-table model test is the most direct method to study the seismic response of structure and the mechanism of earthquake damage [2,3]. The selected seismic wave is output to the shaker table to generate excitation to the shaker model structure through the control room, so that the seismic process can be reproduced relatively realistically. The seismic dynamic response of orthogonal tunnels has been a prominent focus of research in the past. Pai Lifang et al. [4] carried out shaking-table tests based on typical cases of cross-tunnel underpass landslides to reveal the regional spatial dynamic response characteristics of lining structures. Based on the interchange tunnel project, Lai Tianwen et al. [5] designed a large-scale shaking-table model test to study the seismic dynamic response characteristics of the orthogonal tunnel. At present, researchers are beginning to focus on the seismic dynamic response of parallel tunnels. Sun, TC et al. [6] carried out a model test of the portals of two parallel tunnels to learn about the dynamic response of tunnel liner and the interaction between surrounding rock and liner in earthquakes. Tao Yang et al. [7] investigated the dynamic response and the failure characteristics of a parallel overlapped tunnel under seismic forces by employing shaking-table tests. However, a pile-supported tunnel structure is a common form of tunnel construction, on which there is little shaking-table model test research.

The research of similar materials in tunnel shaking-table tests is also a focus, including those in tunnel lining and surrounding rock. Wu Honggang et al. [8] adopted hard PVC as the main material and applied a mixture of gypsum, quartz sand, and water as a 5 mm thick shell to complete the production of the lining model; D. K. Singh et al. [9] used different materials to simulate damaged and undamaged tunnels and studied the seismic performance of damaged tunnels in aftershocks. Hao Zhou [10] studied the seismic dynamic response of a large-section tunnel in compacted clay by filling a container with clay soil; Cho Mya Darli [11] carried out a series of shaking-table tests on the integrated corridor tunnel in the mixed stratum of sand and clay, and the results showed that the dynamic response of the tunnel was significantly different in the clay and sandy stratum. To study the dynamic response of special soil under tunnel shaking-table tests, most of the current studies applied site soil without considering the similarity relationship [12].

In the process of infrastructure construction in mountainous cities, a large number of soil–rock mixed backfills were formed due to site leveling; therefore, some lines inevitably pass through these soil–rock mixed backfill areas. Some other scholars have carried out research on the characteristics of soil–rock mixture (SRM). Simoni et al. [13] conducted triaxial tests on soil–rock mixtures to analyze the stress–strain relationship, and the results showed that such materials had obvious shear expansibility; Huang Wenjian et al. [14] and Cao Wengui et al. [15] used the orthogonal experimental design method to analyze the influence of various factors such as rock content on the physical and mechanical properties of SRM, among which rock content has the greatest influence on the shear strength; Xia Jiaguo et al. [16,17] studied the effects of rock content and confining pressure on the mechanical properties of SRM using a large triaxial shear instrument, the results showed that the stress–shear displacement curve had a "jump" phenomenon, and the shear index parameters first increased and then decreased with the increase in rock content, and always increased with the increase in confining pressure. The preceding research on soil–rock mixture characteristics enables the development of a SRM similar material that meets the similarity relationships and certain failure characteristics in the model test.

In this paper, based on the subway tunnel project in Chongqing, a small clear distance pile-supported tunnel model was made according to the similarity theory. Furthermore, SRM similar materials reflecting certain failure characteristics were developed based on triaxial tests. The study focuses on the seismic dynamic response of tunnel lining structure, and to determine the unfavorable parts of the tunnel–pile structure, which can provide reference for the design, construction, and maintenance of similar projects.

2. Project Overview

2.1. Research Site

The research site is located in the deep soil–rock mixed backfill area of Chongqing Rail Transit Line 10, with an overlying thickness ranging from 11.5 to 56.0 m. The most dangerous section is a three-line tunnel, and the minimum spacing is only 3 m. Given the poor tunnel conditions, a series of precautionary measures were implemented to ensure project safety, including: ① preliminary curtain grouting; ② installation of tunnel piles to bolster the tunnel infrastructure. The profile view is illustrated in Figure 1, while a representative cross-section is delineated in Figure 2.

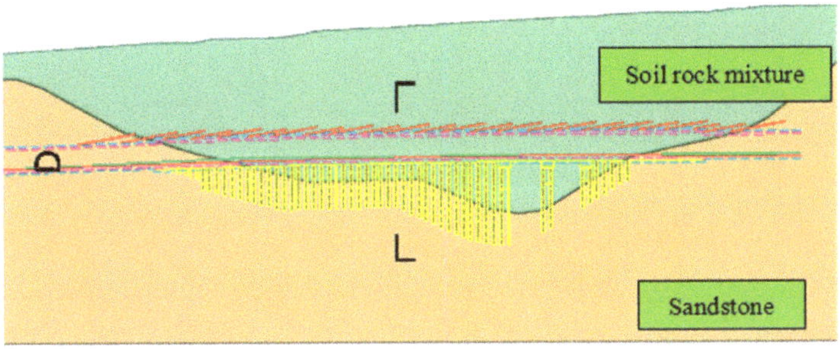

Figure 1. Profile view of the right line.

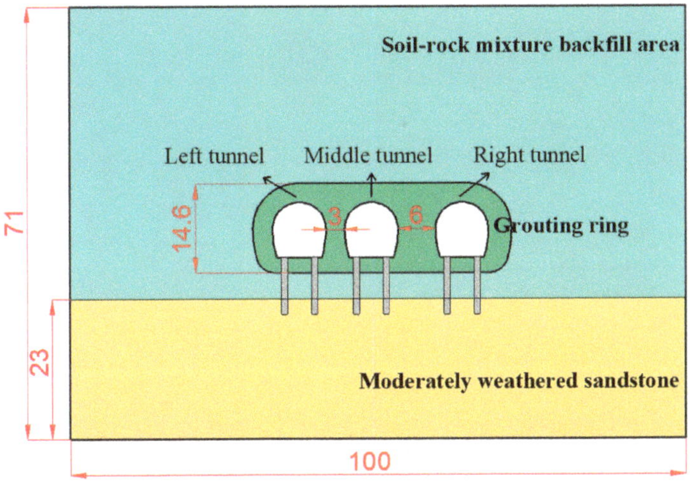

Figure 2. Cross-section of tunnel (unit: m).

2.2. Properties of Soil–Rock Mixture

First, the grading of the soil–rock mixture in the research site was measured, and its original gradation curve is depicted in Figure 3. Then, triaxial tests were conducted on the soil–rock mixture sample, and the stress–displacement curve was obtained as shown in Figure 4. It is worth noting that when there was no obvious mutation during the stress–displacement curve, the stress at 15% strain could be deemed as the failure value of the sample. The morphology of the post-failure samples is shown in Figure 5, showing a "drum shape". Utilizing nonlinear programming, the cohesion (c) was determined as 46.81 kPa, and the internal friction angle (φ) as 36.1° for the soil–rock mixture. The Mohr–Coulomb envelope is portrayed in Figure 6. According to geological investigation data and triaxial tests, the main mechanical properties of the prototype materials involved in this model test (soil–rock mixture, grouting mixture, sandstone, lining, and pile) are shown in Table 1.

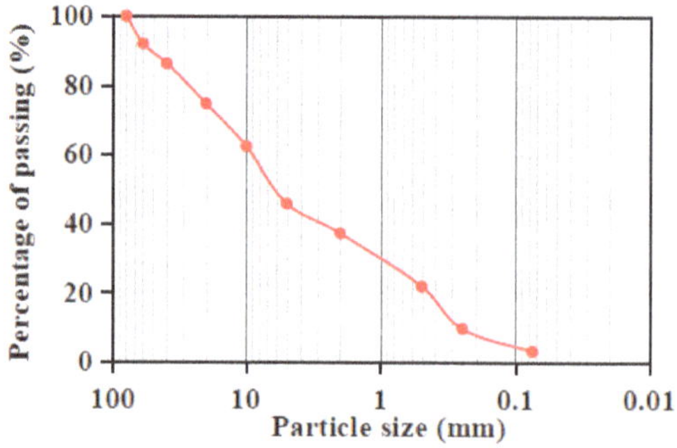

Figure 3. Gradation curve of SRM sample.

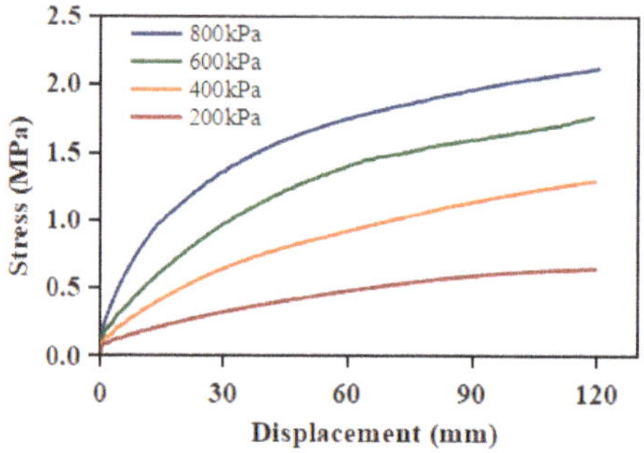

Figure 4. Stress–displacement curve of SRM sample.

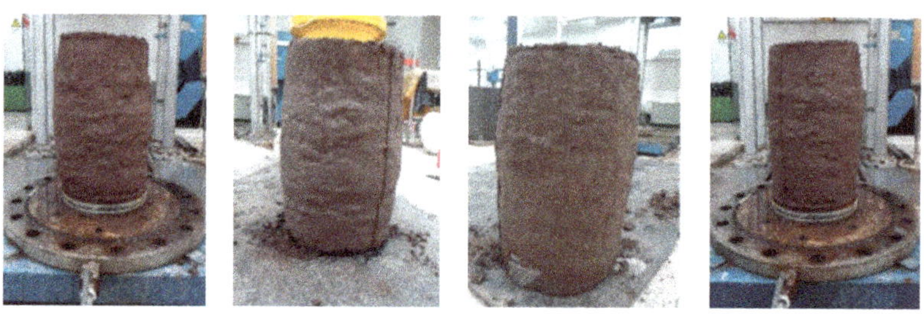

(**a**) 200 kPa (**b**) 400 kPa (**c**) 600 kPa (**d**) 800 kPa

Figure 5. Failure mode of soil–rock mixture sample under different confining stress.

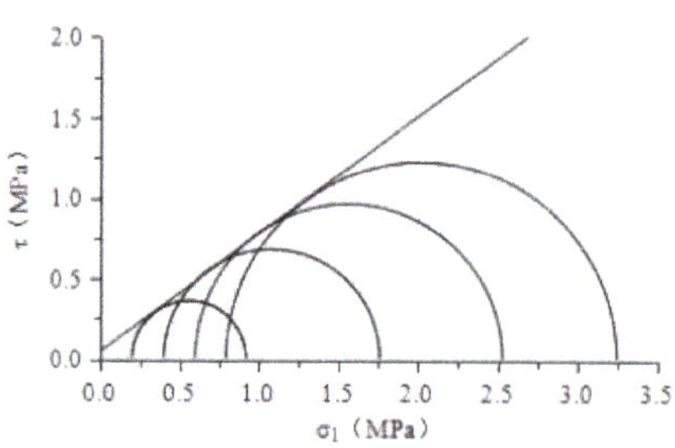

Figure 6. Molar envelope of soil–rock mixture sample.

Table 1. Mechanical properties of prototype materials.

Materials	c (kPa)	φ (°)	E (MPa)	Density (g/cm^3)	Poisson Ratio	Rock Content (%)	Uniaxial Compressive Strength (MPa)
SRM	46.8	36.1	64	1.8	0.3	40–60	
Grouting	400–800	35.8	140–280	2.0	0.28		
Sandstone	4000	41	5000	2.4	0.25		36.8
Lining			32,500	2.5	0.3		26.8
Pile			32,500	2.5	0.3		26.8

3. Design of the Model Test

3.1. Similar Materials

In the model test, the similarity between the model and the prototype should be ensured not only in the geometric scale, but also in the mechanical parameters. However, it is difficult to fully satisfy the similarity relationship between the model structure and the prototype structure, so priority was assigned to selecting key mechanical parameters impacting test outcomes and upholding their similarity in this experiment. Beyond the fundamental physical dimensions of length (l) and force (F), the design of the shaking-table model test must also incorporate the temporal factor (t) to maintain similarity ratios effectively.

In the dynamic model test, the inertia force of the structure is usually the primary load, and the relationship between the inertia force, damping force, and restoring force should be considered first in the study of dynamic problems. The dynamic equation is expressed as follows:

$$m(\ddot{x}(t) + \ddot{x}_g(t)) + c\dot{x}(t) + kx(t) = 0 \tag{1}$$

According to the requirements of equation analysis, the similar relationship of each physical quantity of the dynamic equation satisfies the equation:

$$S_m(S_{\ddot{x}} + S_{\ddot{x}_g}) + S_c S_{\dot{x}} + S_k S_x = 0 \tag{2}$$

Based on the similarity π theorem, applying dimensional analysis to the dynamic Equation (1), we obtain an expression with the fundamental similarity constants of elastic modulus E, density ρ, length l, and acceleration a:

$$S_\rho S_l^3(S_\alpha + S_\alpha) + S_E\sqrt{\frac{S_l^3}{S_\alpha}}\sqrt{S_l S_\alpha} + S_E S_l^2 = 0 \tag{3}$$

$$\frac{S_E}{S_\rho S_\alpha S_l} = 1 \tag{4}$$

Based on the actual engineering and the dimensions of the model test table, the geometric similarity ratio was determined to be 1/80, considering geometric dimensions length l, density ρ, and acceleration a as the fundamental physical quantities. Subsequent parameters were analyzed using dimensional analysis to form similarity relationships, as shown in Table 2.

Table 2. Target parameters of similar materials.

Materials	c (kPa)	φ (°)	E (MPa)	Density (g/cm^3)	Poisson Ratio	Rock Content (%)	Uniaxial Compressive Strength (MPa)
Similarity relationship	$S_c = S_E$	$S_\varphi = 1$	$S_E = S_l S_\rho S_a$	$S_\gamma = S_\rho S_g$	$S_\varphi = 1$		Sucs = S_E
Similarity ratio	1/80	1	1/80	1	1		1/80
SRM	0.58	36.1	0.8	1.8	0.3	40–60	
Grouting	5–10	35.8	1.75–3.5	2.0	0.28		
Sandstone	50	41	62.5	2.4	0.25		0.46
Lining			406	2.5	0.3		0.33
Pile			406	2.5	0.3		0.33

(1) Soil–rock mixture and grouting mixture

In this paper, coarse quartz sand within 1~2 mm particle size and fine quartz sand within 0.5~1 mm particle size were used to simulate the coarse-grain phase in the prototype of soil–rock mixture, and fine sand within 0.1~0.2 mm particle size was used to simulate the fine-grain phase, and clay was selected as the cementing material. To maintain a rock content of 40% to 60% in similarity material, the mass ratio of coarse quartz sand and fine quartz sand needed to be controlled within the 40% to 60% range. Furthermore, the method of increasing the proportion of cementing material (clay) was used to simulate the grouting mixture. Through the above methods, three distinct samples were set, as shown in Table 3, and the configuration process is shown in Figure 7.

Table 3. Mix proportion design of similar materials.

Group	Main Ratio	Others	
	Coarse Quartz Sand (1~2 mm)–Fine Quartz Sand (1~2 mm)–Fine Sand–Clay	Water Content	Rock Content
Sample 1	2:2:5:1	7%	40%
Sample 2	2:2:4:2	10%	40%
Sample 3	2:3:3.5:1.5	10%	50%

(**a**) Stir

(**b**) Standing

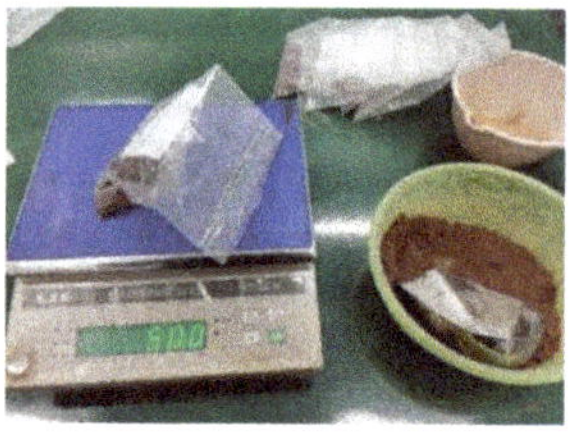

(**c**) Package

Figure 7. Material preparation process.

Since the influence of groundwater was not considered in the actual project, unconsolidated and undrained triaxial tests were conducted to measure the mechanical parameters of similar materials. The stress–strain curves of the three samples subjected to confining stress of 50, 100, and 150 kPa are shown in Figure 8. By employing a nonlinear programming approach, the cohesion values (c) were determined as 0.8 kPa, 28 kPa, and 8 kPa for samples 1, 2, and 3, correspondingly, while the internal friction angles (ϕ) were calculated as 33.5°, 31°, and 38.5° for samples 1, 2, and 3. The corresponding Mohr–Coulomb envelopes are shown in Figure 9.

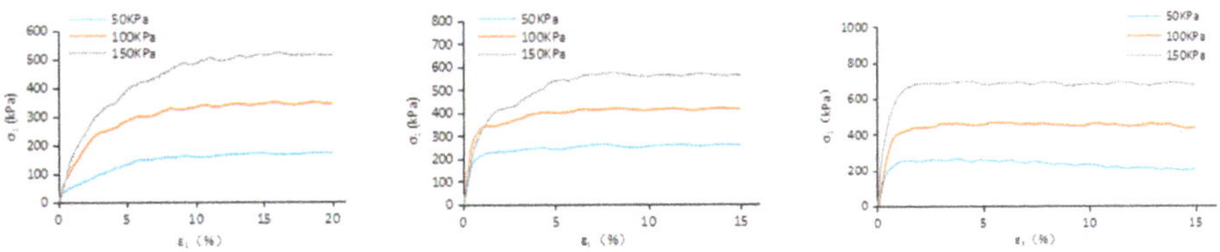

(**a**) Sample 1 (**b**) Sample 2 (**c**) Sample 3

Figure 8. Stress–strain curves of the samples.

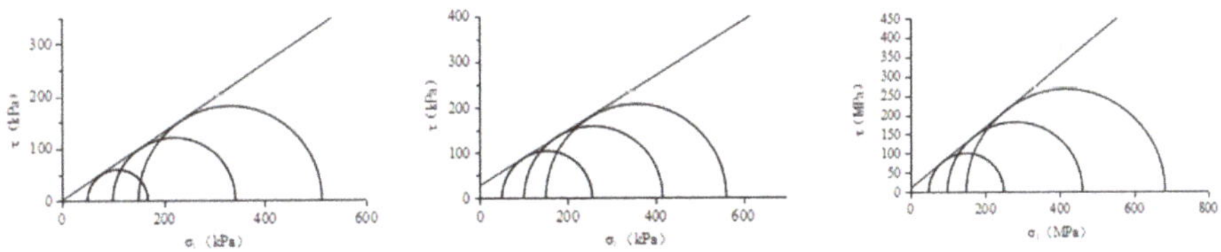

(**a**) Sample 1 (**b**) Sample 2 (**c**) Sample 3

Figure 9. Mohr–Coulomb envelopes of the samples.

Sample 1 was finally selected as the similar material of the soil–rock mixture, and sample 3 was selected as the similar material of grouting mixture. The failure mode of the sample is shown in Figure 10. It can be seen that the deformation of sample 1 under the triaxial compression test resembled a "drum shape", showing the same failure mode as that of the soil–rock mixture prototype. When subjected to a confining pressure of 50 kPa, sample 3 displayed an obvious 45-degree shear plane, and the failure mode was similar to the brittle failure of rock. When subjected to a confining pressure of 150 kPa, sample 3 did not exhibit obvious failure point, which was similar to the failure mode of soil. On the whole, sample 3 presented a failure mode similar to that of grouting mixture obtained by Hu [18].

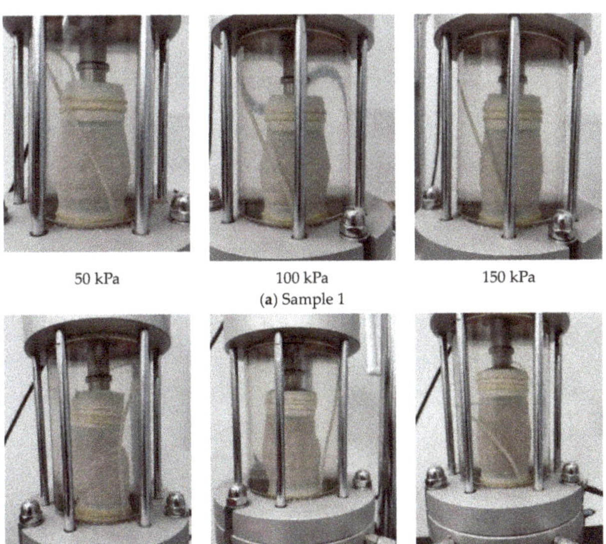

50 kPa 100 kPa 150 kPa
(**a**) Sample 1

50 kPa 100 kPa 150 kPa
(**b**) Sample 3

Figure 10. Failure mode of sample 1 and sample 3.

(2) Lining and pile foundation

In this experiment, plexiglass was used to make the tunnel–pile structure. Flexural stiffness was adopted as the governing criterion for determining the thickness of the tunnel lining and the diameter of the tunnel pile. According to the principle of similarity ratio, the ratio between the flexural stiffness K of the actual tunnel lining and the flexural stiffness K_0 of the model test is $(1/80)^5$. Calculations revealed a tunnel lining thickness h_0 of 4 mm and a pile diameter D_0 of 6.2 mm. The completed model is shown in Figure 11.

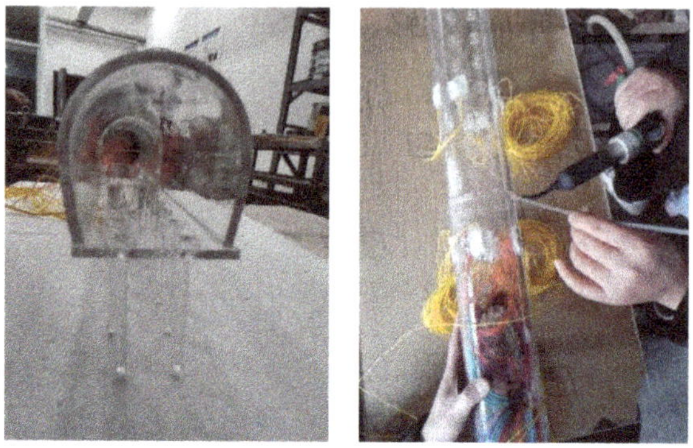

Figure 11. Tunnel–pile structure model.

3.2. Test Scheme

(1) Model box design

A layered shear model box with a width of 0.85 m, a length of 0.95 m, and a height of 0.6 m was selected for the test. The model box was composed of multi-layer rectangular frames, each of which can move relative to the others in a certain amplitude, thus minimizing seismic wave reflection and offering a more accurate simulation of soil shear properties. Limited by the size of the shaking table, the original cross-section was simplified for the model test, as shown in Figure 12. The left and middle tunnels were selected as the research objects in the simplified section, and the tunnel spacing was adjusted to 52 mm in model box, retaining the characteristics of a tunnel with a small distance.

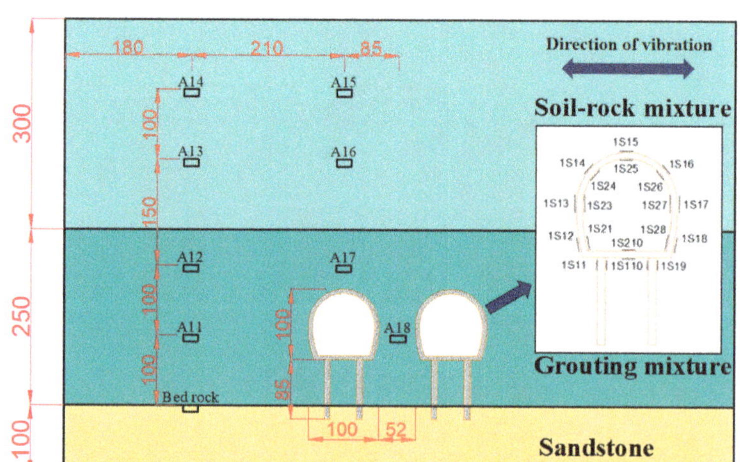

Figure 12. Simplified cross-section of the tunnel and layout of monitoring sensors (unit: mm).

(2) Sensor arrangement

In order to study the acceleration response characteristic of surrounding rock and tunnel–pile structure system, accelerometers and strain gauges were used in this experiment. The sensor layout of the model test is shown in Figure 12.

(3) Seismic wave selection

In this shaking-table model test, seismic waves were applied in the horizontal direction. Kobe wave, Wolong wave, and sine wave were selected as input signals. The original seismic wave exhibited significant displacements, sometimes exceeding 1 m, whereas the maximum displacement capacity of the shaking table was limited to 0.1 m. Hence, baseline adjustments were performed on the original seismic wave to ensure its displacement remained within the 0.1 m threshold. The time similarity constant was 1/10; consequently, the peak vibration time of Kobe and Wolong waves were compressed to 4 s, and that of the sine wave was compressed to 14 s.

The model filling process Is shown in Figure 13.

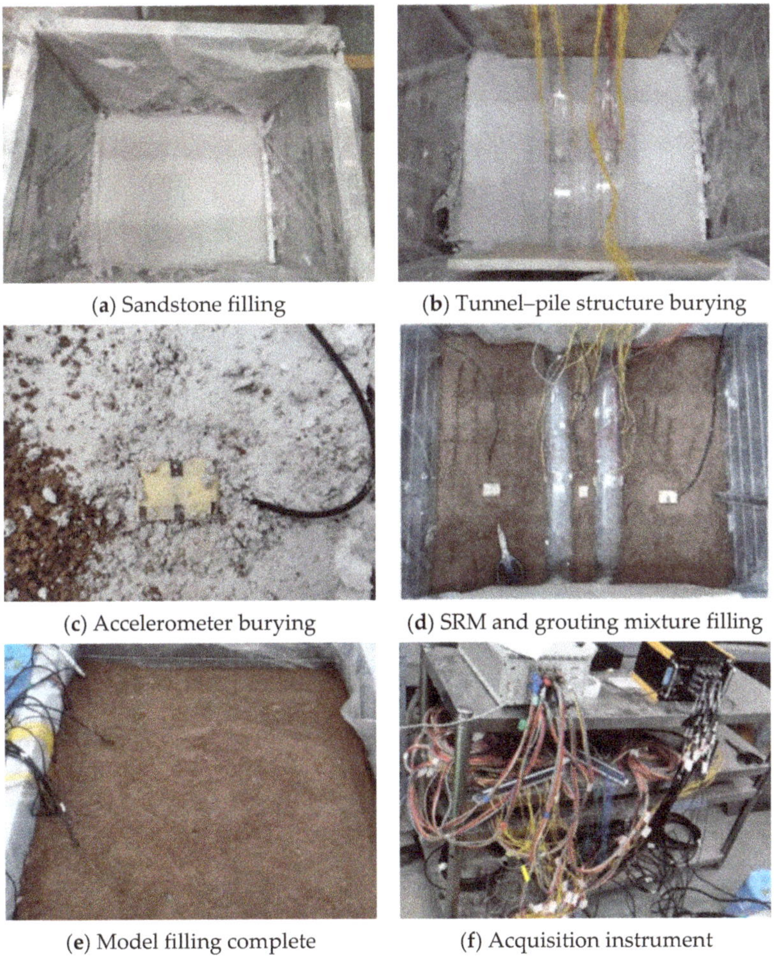

(**a**) Sandstone filling (**b**) Tunnel–pile structure burying

(**c**) Accelerometer burying (**d**) SRM and grouting mixture filling

(**e**) Model filling complete (**f**) Acquisition instrument

Figure 13. Model filling process.

4. Analysis of Test Results

4.1. Acceleration Response Characteristics of Surrounding Rock

After the model filling was completed, the boundary effect of the model was verified first. Kobe wave and Wolong wave with 1.2 g peak acceleration were the input. A comparison of acceleration–time history curves of measuring points A14 and A15 under the action of Kobe wave and Wolong wave is shown in Figure 14. The height of the curves was almost consistent, indicating that the boundary effect produced by the layered shear box in this test was small.

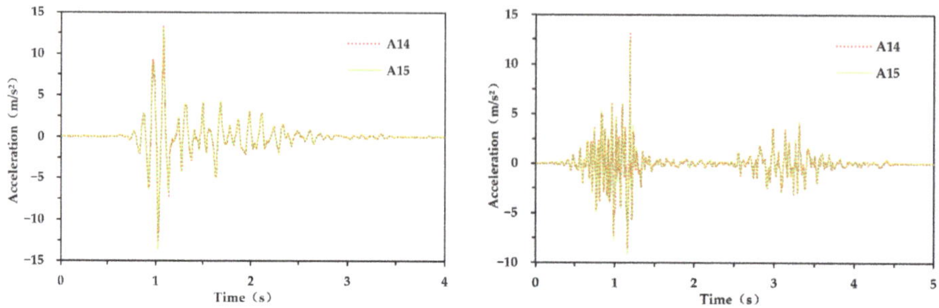

(**a**) Kobe wave (**b**) Wolong wave

Figure 14. Comparison of acceleration–time history curves of measuring points A14 and A15 under the action of Kobe wave and Wolong wave.

(1) Stratum acceleration response

As illustrated in Figure 12, five measuring points were arranged along the stratum height, sequentially identified from bottom to top as bedrock, A11, A12, A13, and A14. Figure 15 shows the curves of peak acceleration variation along the height of the stratum under different seismic waves. It can be seen that the peak acceleration within the stratum exhibited a consistent rise with increasing height from the base, distinctly displaying an amplification effect.

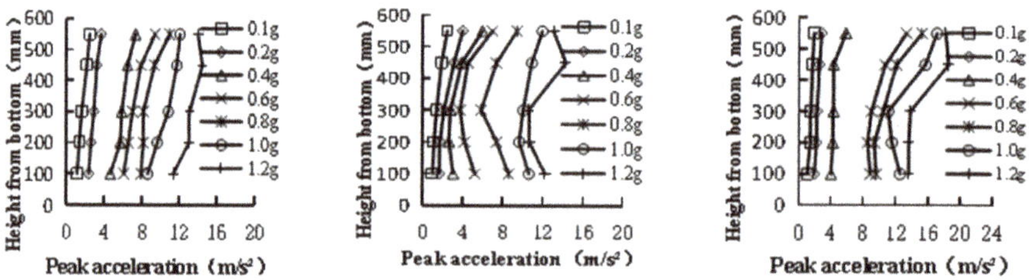

(**a**) Kobe wave (**b**) Wolong wave (**c**) Sine wave

Figure 15. Curves of peak acceleration variation along the height of the stratum under different seismic waves.

Under the action of Kobe wave, input seismic wave acceleration peak values ranging between 0.1 g and 1.0 g triggered a gradual increase in acceleration peak values along the stratum height, displaying a progressively intensifying trend suggestive of a pronounced "skin effect". However, as the input seismic wave acceleration peak value exceeded 1.0 g, the acceleration peak value decreased at the top of the SRM stratum (measuring points A14).

Conversely, under the action of Wolong wave and sine wave, the peak value of stratum acceleration at measuring points A11 and A12 showed a decreasing trend that strengthened with higher input acceleration peak values. This attenuation is credited to the mitigation of stratum acceleration by the tunnel–pile structure when subjected to the Wolong wave and sine wave. However, when the peak acceleration of the input seismic wave exceeds 1.0 g, its peak value showed a "Z" shape along the stratum height, indicating a dual discernible shift in acceleration peak values with increasing distance from the base height.

To comprehensively study the acceleration amplification effect and variation law of the stratum, the peak acceleration ratio of individual points in relation to the peak acceleration at the surface of the platform was defined as the acceleration amplification coefficient for each point. Variation curves of acceleration amplification coefficient along vertical direction (bedrock, A11~A14) under different seismic waves are shown in Figure 16. It can be seen that within the soil–rock mixture layer (350~650 mm), when the input acceleration value remained below 0.2 g, the stratum's acceleration amplification coefficients increased monotonically along the vertical axis, indicating that the stratum response was in a linear stage at this time. As the peak input acceleration reached 0.4 g, a slight reduction in acceleration amplification coefficients occurred, yet the overall trend maintained a monotonic increase, indicating minimal soil damage. When the input acceleration peak value exceeded 1.0 g, the stratum acceleration amplification coefficients decreased significantly to less than one, indicating that the soil–rock mixture had entered a nonlinear reaction stage at this time. Potential plastic deformation within the stratum likely contributed to elevated damping and enhanced dissipation of seismic energy.

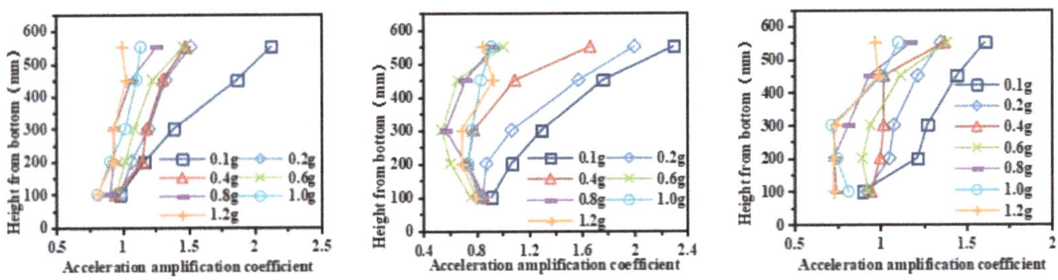

(**a**) Kobe wave (**b**) Wolong wave (**c**) Sine wave

Figure 16. Variation curves of acceleration amplification coefficient along height direction under different seismic waves.

(2) Influence of pile-supported tunnel structure on stratum acceleration

It can also be seen from Figure 16 that in the grouting stratum (100~350 mm), under the action of Wolong wave and sine wave, the acceleration amplification coefficient exhibited a transition from an increasing trend to a decreasing trend with rising input peak acceleration levels. This phenomenon is likely attributed to the emergence of the energy dissipation mechanism within the tunnel–pile system as the input peak acceleration intensified. This mechanism notably mitigated the stratum's acceleration response in this range. In addition, the cumulative damage caused by step-by-step seismic wave loading was also a part of the reason.

Under the action of different seismic waves, a comparison of peak acceleration values between points situated within the free field stratum (A12–A14) and points positioned above the tunnel axis (A15–A17) at equivalent elevations is presented in Figure 17. Across all three seismic wave scenarios, the peak acceleration of the measurement points above the tunnel axis was smaller than that of the free field stratum at the same height. When the input peak acceleration remained below 0.4 g, the difference between the two points was small; however, as the input peak acceleration exceeded 0.4 g, this difference showed a progressively increasing trend.

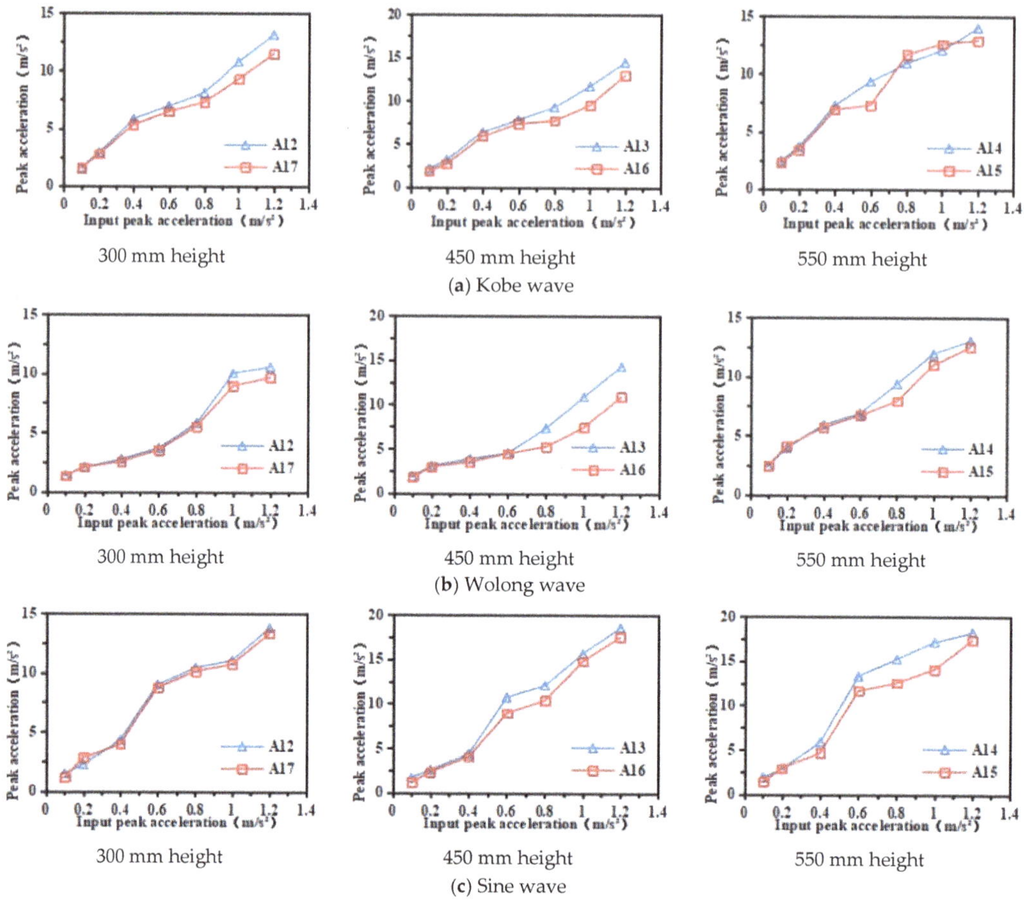

Figure 17. Comparison diagram of peak acceleration at different measuring points at the same height under the action of different seismic waves.

The relative difference percentage was defined as the ratio of the difference between the peak accelerations of two points at the same height to the peak acceleration of the free field stratum, as shown in Table 4. As can be seen from the table, the influence of the tunnel–pile system on stratum acceleration could reach up to 14.5%, and the greatest impact was located above the tunnel (at the height of 450 mm). There are two reasons why the tunnel–pile system had a weakening effect on stratum acceleration: first, the tunnel–pile system exerts a discernible energy dissipation effect on seismic waves; second, the settlement of the stratum above the tunnel was much smaller than that of the free field stratum, indicating the soil in the latter was comparatively looser and consequently led to higher acceleration values.

Table 4. Relative difference percentage of peak acceleration at different points at the same height.

Measuring Points	Height (mm)	Relative Difference Percentage of Peak Acceleration		
		Kobe Wave	Wolong Wave	Sine Wave
A14 A15	550	5.18%	4.07%	13.74%
A13 A16	450	12.61%	14.43%	13.08%
A12 A17	300	8.57%	5.81%	3.43%

4.2. Acceleration Response Characteristics of Pile-Supported Tunnel Structure

(1) Strain acceleration response of tunnel lining

The strain radar diagram of the tunnel lining under the action of Kobe wave, Wolong wave, and sine wave at different peak accelerations is shown in Figure 18. Analysis of the strain radar diagram reveals distinct patterns: On the outer edge of the lining, the highest strain peak occurs at the arch foot, followed by the arch shoulder, with comparatively lower values observed at the arch crown, arch waist, and inverted arch. The distribution of strain response peak value of the tunnel lining cross-section exhibited an "X" shape. On the inner edge of the lining, the peak value of strain at the left arch foot of the left tunnel lining was significantly greater than that at the right arch foot, indicating that the strain dynamic response of the inner edge of the lining was less pronounced than that of the outer edge due to the interaction of the two tunnels.

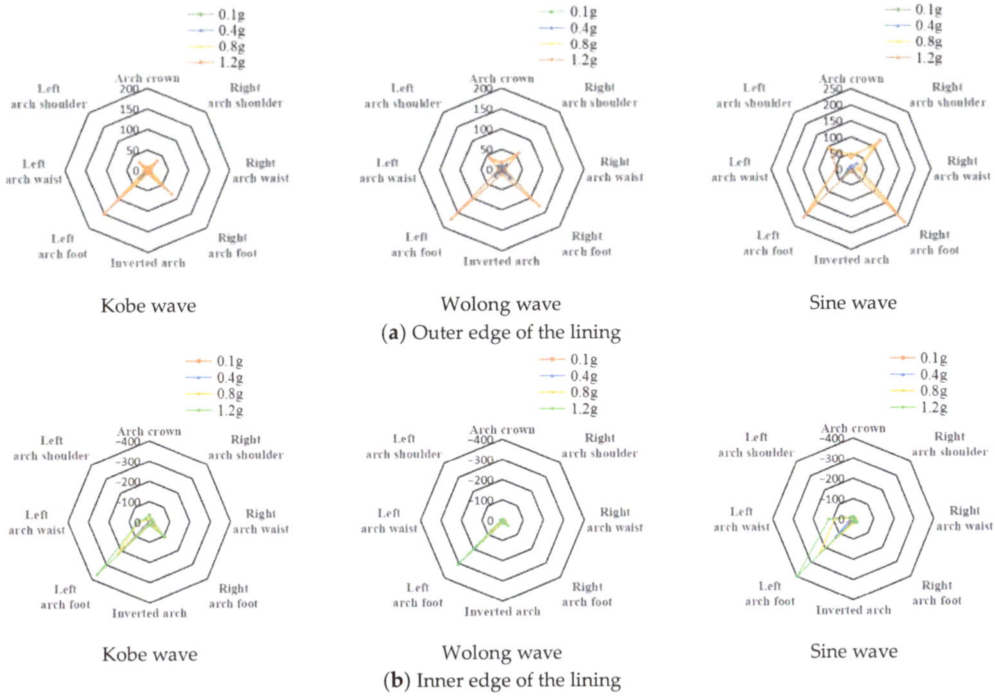

Figure 18. Strain distribution on left tunnel cross-section under different seismic waves (unit: με).

(2) Bending moment acceleration response of tunnel lining

In order to study the bending moment value at each measuring point of the tunnel lining, the strain at the inner and outer edges of the lining was converted to the bending moment value according to Formula (5):

$$M = \frac{1}{2}(\sigma_1 - \sigma_2)W = \frac{1}{2}EW(\varepsilon_1 - \varepsilon_2) = \frac{1}{12}Ebh^2(\varepsilon_1 - \varepsilon_2) \quad (5)$$

where σ_1 and σ_2 are the stress on the inner and outer edges of the measuring point, ε_1 and ε_2 are the strain at the inner and outer edges of the measuring point, E is the elastic modulus of lining material, W is the resistance in cross-section, and b is the section width (1 m).

Under the action of three kinds of seismic waves, the maximum bending moment of each measuring point of the lining cross-section along the axis direction of the tunnel is shown in Figure 19. The bending moment at the arch foot was the largest, followed by the left and right arch shoulder, and the bending moment at the inverted arch and arch waist was relatively small. The distribution of bending moment of the lining cross-section exhibited an inverted "V" shape on the whole. The bending moment increase effect at the right arch foot was smaller than that at the left arch foot, so its maximum bending moment value was 1/2–1/4 of the left arch foot.

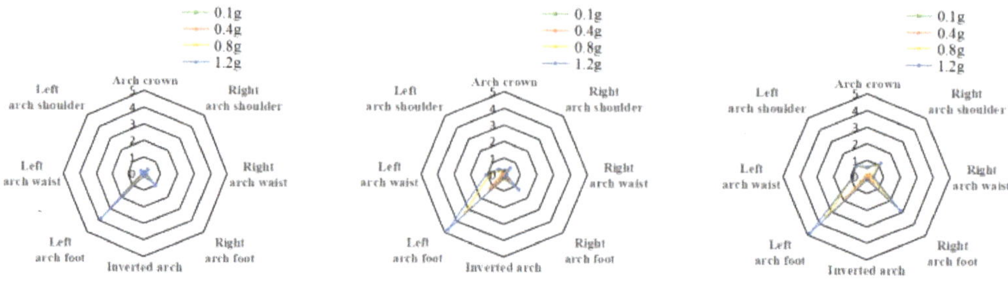

(**a**) Kobe wave (**b**) Wolong wave (**c**) Sine wave

Figure 19. Bending moment distribution of the left tunnel cross-section under different seismic waves (unit: N·m).

It can be seen from Figure 20 that with the increase in the input peak acceleration, the bending moment value at each position of the lining generally keeps increasing. When the input peak acceleration remained below 0.6 g, the bending moments of the lining retained a consistent increase trend except at the arch foot. As the input peak acceleration exceeds 0.6 g, the increase rate of bending moment slowed down at almost each position. As the input peak acceleration exceeds 0.8 g, the bending moment value of all lining positions did not increase except in the left arch foot.

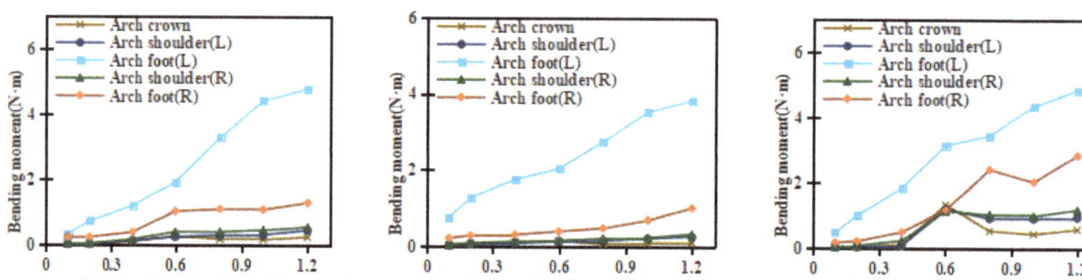

(**a**) Kobe wave (**b**) Wolong wave (**c**) Sine wave

Figure 20. Curves of the tunnel lining bending moment with input peak acceleration.

From Formula (5), the bending moment value at each position of the lining is positively correlated with the difference between the inner and outer strains. Moreover, the strain of the outer edge is positive, and the strain of the inner edge is negative, accentuating that bending moment distribution corresponds closely to higher absolute strain values. In this model test, the strain value at the left arch foot of inner edge is the largest, and the strain value at the right arch foot of outer edge is the second largest but much larger than that at other positions. Therefore, the bending moment of the lining cross-section is mainly affected by the left and right arch foot and exhibit an inverted "V" shape.

(3) Bending moment acceleration response of pile crown

Based on the strain data on the left and right sides of the pile crown, the bending moment changes of the pile crown under different seismic waves were plotted, as shown in Figure 21. Under the action of the three seismic waves, the bending moment of the left pile crown was greater than that of the right pile crown. Under the action of Kobe wave and sine wave, when the peak acceleration of input seismic wave was between 0.1–0.8 g, the bending moment of the pile crown increased gradually. As the peak acceleration of the input seismic wave exceeded 0.8 g, the bending moment of the pile crown approached stability or began to decline. The reason is that when the input peak acceleration exceeded 0.8 g, the grouting stratum besides pile entered large plastic deformation state; the peak dynamic earth pressure tended to be stable, so the dynamic bending moment also tended to be stable. When subjected to Kobe and sine waves, the decrease in the bending moment at the pile crown can be attributed to the energy dissipation after the grouting stratum reached the critical threshold of plastic deformation, causing a sudden decrease in dynamic earth pressure from the maximum value.

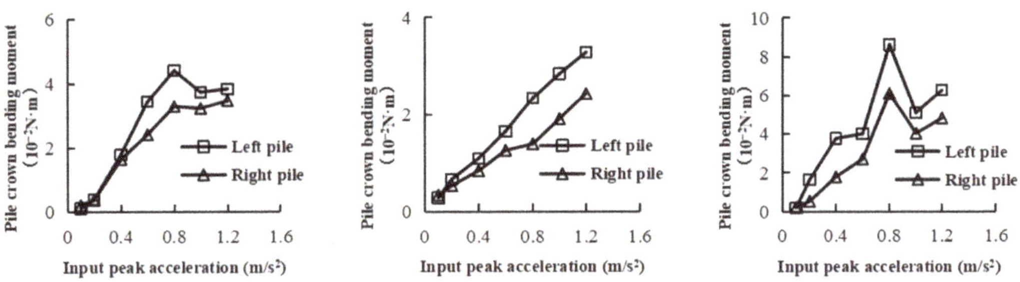

(**a**) Kobe wave (**b**) Wolong wave (**c**) Sine wave

Figure 21. Variation diagram of pile crown bending moment under different seismic waves.

4.3. Surface Settlement

After the shaking-table test was completed, the surface settlement was observed and measured. As shown in Figures 22 and 23, there was no significant settlement of the stratum above the tunnel. However, the free field stratum on both sides had obvious settlement. By measurement, the settlement amount on the left side reached 1 cm, and the right side reached 1.5 cm, indicating that the surface settlement of the soil–rock mixture was obvious under the action of the earthquake. However, the existence of a tunnel–pile structure mitigated the surface settlement obviously, and the uneven surface settlement would cause the cracking and destruction of above-ground roads and structures in practical engineering.

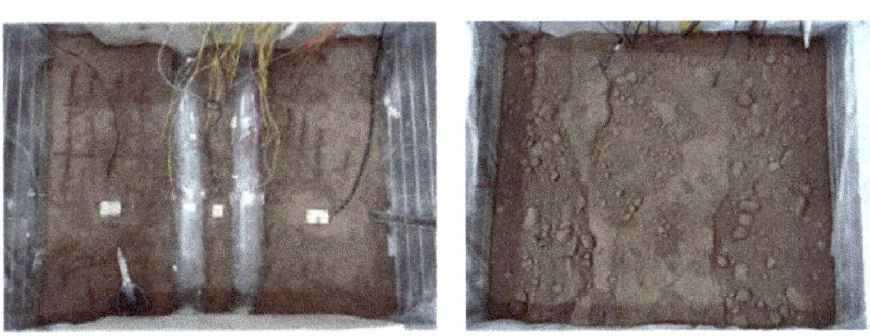

Figure 22. Surface settlement of the model.

Figure 23. Maximum surface settlement of the model.

5. Conclusions

In this paper, based on a subway tunnel project in Chongqing, the mechanical parameters of the soil–rock mixture at the original site of the project were measured based on a large-scale indoor triaxial test, and then the development methods and mechanical parameters of similar materials such as soil–rock mixture, grouting mixture, lining, and pile satisfying the similarity ratio were studied. Finally, the analysis was conducted from the stratum acceleration and its amplification factor, the distribution of strain and bending moment of the tunnel–pile structure, and the surface settlement. The main conclusions are as follows:

(1) Under the action of seismic waves, the acceleration response of soil–rock mixture stratum exhibited obvious "skin effect", indicating that closer proximity to the surface led to heightened horizontal acceleration responses; however, with the increase in the peak value of input seismic waves, the plastic deformation occurred in the upper stratum of the soil–rock mixture; the grouting stratum was obviously affected by the tunnel–pile structure, and the horizontal peak acceleration of the stratum in the vertical direction changes in a "Z" shape.

(2) The tunnel–pile structure had a weakening effect on the horizontal acceleration of both the stratum in its height range and the stratum above the tunnel.

(3) Under the action of seismic waves, the peak strain values at the arch shoulder and arch foot of the tunnel were larger, and the peak strain response of the tunnel lining cross-section generally exhibited an "X" shape. In the direction of the tunnel axis, the bending moment at the arch foot was the largest, followed by the left and right arch shoulder, and the bending moment at the inverted arch and arch waist were smaller. The bending moment of the lining cross-section exhibited an inverted "V" shape.

(4) The bending moment of the pile crown generally increased with the increase in the peak acceleration of the input seismic wave. When the peak acceleration of the input seismic wave exceeded 0.8 g, the increasing trend of the pile crown bending moment decreased.

(5) The maximum surface settlement in the model test was 0.5~1 cm, and the tunnel–pile structure effectively reduced the surface settlement but lead to uneven surface settlement.

Author Contributions: Conceptualization: X.J.; methodology, H.L. and G.S.; data processing, J.H.; writing—original draft preparation, H.L.; writing—review and editing, X.J.; project administration, X.J. All authors have read and agreed to the published version of the manuscript.

Funding: This research was funded by the Science and Technology Project of the Fifth Engineering Co., Ltd., China Railway 21st First Bureau Group (21J5GS-CQGD-JF-2020-003).

Data Availability Statement: The data presented in this study are available in the article.

Conflicts of Interest: The authors declare that this study received funding from the Science and Technology Project of the Fifth Engineering Co., Ltd., China Railway 21st First Bureau Group. The funder had the following involvement with the study: provides background information of the project (including geological data and design data).

References

1. Lai, J.; He, S.; Qiu, J.; Chen, J.; Wang, L.; Wang, K. Characteristics of seismic disasters and aseismic measures of tunnels in Wenchuan earthquake. *Environ. Earth Sci.* **2017**, *76*, 94. [CrossRef]
2. Zhang, F. Analysis of Dynamic Response of Subway Underground Structure Under Seismic and Second Development of Software. Master's Thesis, Huazhong University of Science & Technology, Wuhan, China, 2014. [CrossRef]
3. Thoms, R.; Kuesel, F. Earthquake design criteria for subways. *J. Struct. Div. Proc. Am. Soc. Civ. Eng.* **1969**, *95*, 1225–1328.
4. Pai, L.; Wu, H. Experimental study on dynamic response of tunnel lining structure orthogonal under-crossing a landslide under earthquake. *J. Rock Mech. Geotech. Eng.* **2022**, *41*, 979–994.
5. Lai, T.; Lei, H.; Wu, H. Vibration Acceleration Responses of Crossing Tunnels of Highway and Railway under Different Directions of Seismic Loads. *China Railw. Sci.* **2021**, *42*, 95–104.
6. Sun, T.; Yue, Z.; Gao, B.; Li, Q.; Zhang, Y. Model test study on the dynamic response of the portal section of two parallel tunnels in a seismically active area. *Water Energy Int.* **2011**, *68*, 74. [CrossRef]
7. Yang, T.; Rao, Y.; Wu, H.; Zhang, J.; Lei, H.; Ding, H. Dynamic Response of Parallel Overlapped Tunnel under Seismic Loading by Shaking Table Tests. *Shock Vib.* **2021**, *2021*, 2535762. [CrossRef]
8. Wu, H.; Pai, L.; Pang, W.; He, C.; Zhang, X.; Li, D. Vibration table tests on the acceleration response of a three-dimensional cross tunnel with small net spacing and small angle. *J. Vib. Shock* **2021**, *40*, 298–306. [CrossRef]
9. Singh, D.K.; Mandal, A.; Karumanchi, S.R.; Murmu, A.; Sivakumar, N. Seismic behaviour of damaged tunnel during aftershock. *Eng. Fail. Anal.* **2018**, *93*, 44–54. [CrossRef]
10. Zhou, H.; Wang, X.; He, C.; Huang, C.; Aenlle, M.L. Seismic Response of a Tunnel Embedded in Compacted Clay through Large-Scale Shake Table Testing. *Shock Vib.* **2018**, *2018*, 5968431. [CrossRef]
11. Mya, D.C.; Tang, A.; Huang, D.; Zhang, J. Large scale shaking table model test and analysis on seismic response of utility tunnel in non-homogeneous soil. *Earthq. Eng. Eng. Vib.* **2021**, *20*, 505–515.
12. Li, Y.; Ma, X.; Feng, N.; Shang, J.; Xu, X. Experimental Study on Seismic Response of Ji'nan Silty Clay and the Tunnel in It. *Chin. J. Undergr. Space Eng.* **2019**, *15*, 1652–1660.
13. Simoni, A.; Houlsby, G.T. The Direct Shear Strength and Dilatancy of Sand–gravel Mixtures. *Geotechn. Geol. Eng.* **2006**, *24*, 523–549. [CrossRef]
14. Huang, W. Large-Scale Triaxial Tests on Soil-Rock Aggregate Mixture Based on Orthogonal Design. Master's Thesis, Hunan University, Changsha, China, 2015.
15. Cao, W.G.; Huang, W.J.; Wang, J.Y.; Zhai, Y.C. Large-scale triaxial tests Study om Deformation and Intensity Characteristic of Soil-rock Aggregate Mixture. *J. Hunan Univ.* **2016**, *43*, 142–148.
16. Xia, J.; Hu, R.; Qi, S.; Gao, W.; Sui, H. Large-scale triaxial shear testing of soil rock mixtures containing oversized particles. *J. Rock Mech. Geotech. Eng.* **2017**, *36*, 2031–2039.
17. Xia, J.; Hu, R.; Gao, W. Research on Mechanical Characteristic of the Soil Rock Mixtures by Large-scale triaxial shear Tests. *J. Eng. Geol.* **2016**, *24*, 1211–1218.
18. Hu, S.; Jin, X.; Sun, G.; Li, H.; Nie, Z. Triaxial test and PFC-FLAC coupling simulation study on material parameters and deformation characteristics of soil-rock mixture. *J. Rock Mech. Geotech. Eng.* **2021**, *40*, 3344–3356. [CrossRef]

Disclaimer/Publisher's Note: The statements, opinions and data contained in all publications are solely those of the individual author(s) and contributor(s) and not of MDPI and/or the editor(s). MDPI and/or the editor(s) disclaim responsibility for any injury to people or property resulting from any ideas, methods, instructions or products referred to in the content.

buildings

MDPI

Article

Hybrid Effect of Basalt and Polyacrylonitrile Fibers on Physico-Mechanical Properties of Tailing Mortar

Jianbo Xu [1], Zhijie Pi [2,*], Shibing Huang [2,3,*], Yanzhang Liu [2,3], Ziwen Chen [2] and Yongqi Shen [2]

[1] China First Metallurgical Group Co., Ltd., Wuhan 430080, China; xujb@cfmcc.com
[2] School of Resources and Environmental Engineering, Wuhan University of Science and Technology, Wuhan 430081, China; liuyanzhang@wust.edu.cn (Y.L.); 13571901339@163.com (Z.C.); shenyongqi1999@163.com (Y.S.)
[3] Hubei Key Laboratory for Efficient Utilization and Agglomeration of Metallurgic Mineral Resources, Wuhan 430081, China
* Correspondence: pizhijie1534010242@163.com (Z.P.); huangshibing@wust.edu.cn (S.H.)

Abstract: In this study, 50% iron ore tailings (IOTs) were used to prepare the cemented mortar at low economic costs and with great environmental benefits. Basalt fiber (BF) and polyacrylonitrile fiber (PANF) were added to the tailing mortar to improve the comprehensive performance of tailing mortars, including BF (0~0.5%), PANF (0~0.05%) and the combination of them. The results show that the addition of BF and PANF can significantly improve the ultrasonic velocity, uniaxial compressive strength (UCS), split-tensile strength (STS), flexural strength (FS) and toughness of the tailing mortar. A novel finding is that the enhancement of hybrid fibers is much better than single fiber, and the best hybrid fiber combination is B0.25P0.05 (0.25 wt% BF and 0.05 wt% PANF), because this combination not only causes the most considerable increase in strength but also possesses great cost-effectiveness. Compared to the B0P0 group without fibers, the maximum increments of B0.25P0.05 in UCS, STS and FS are 45.74%, 52.33% and 15.65%, respectively. It is evidenced that the improvement in STS is the largest because the fibers have good cracking resistance and bridging effect in the tailing mortar. The scanning electron microscope (SEM) further confirms that too many hybrid fibers will agglomerate and produce more voids, which is harmful to the development of the internal structure. Beyond B0.25P0.05, the hydration products are also reduced due to the decrease in nucleation sites, observed by combining X-ray diffraction (XRD) tests. Therefore, it is suggested that the hybrid fibers containing 0.25% BF and 0.05% PANF should be used in this tailing mortar.

Keywords: sulfur-containing tailings; basalt fiber; polyacrylonitrile fiber; physico-mechanical properties; fiber-reinforced tailing mortar

1. Introduction

Concrete and mortar are the most widely used construction materials in civil engineering, and approximately 25 billion tons are consumed per year [1]. River sand accounts for about 25% of the concrete. In China, the annual consumption of river sand exceeds 20 billion tons. In addition, the overexploitation or even depletion of river sand has occurred in some areas. This phenomenon has also caused a series of serious ecological destruction and geological disasters [2]. Therefore, it is urgent to seek fine aggregates that can replace river sand, which is also required for the sustainable production of concrete, environmental protection and ecological balance. Iron ore tailings (IOTs) are the waste residue after separating iron ore concentrate [3]. Currently, the storage of IOTs in China has exceeded 5 billion tons and is growing rapidly at an annual rate of more than 300 million tons. Due to the low utilization ratio, a large amount of IOTs can only be piled up in tailing ponds, which not only encroach on the land and pollute the surrounding environment, but also require a large amount of maintenance funds [4–6]. To protect the environment and increase the economic benefits, the secondary use of IOTs in the construction industry has attracted the attention of many scholars [7,8]. Andrews et al. [9] found that the compressive strength and splitting tensile strength of concrete were greatly improved with a 25% IOT substitution. Zhang et al. [10] and Zhao et al. [11] also observed that the maximum compressive strength of concrete could be derived with a 40% replacement of river sand by IOTs. Huang et al. [12] found that the strength of mortar by adding 50% IOTs did not decline compared with the mortar with complete standard sand. Therefore, IOTs can be used as fine aggregate in cemented materials.

However, concrete with a high IOT replacement ratio may also suffer from a low mechanical strength, poor cracking resistance and weak ductility [13,14]. Therefore, fibers are introduced to further improve the mechanical strength and overcome the brittle failure problem [15–18]. In general, the utilization of single fiber only enhances the properties of cemented materials in a certain aspect. For example, High et al. [19] and Zhang et al. [20] reported that basalt fiber (BF) significantly enhanced the tensile strength and flexural strength, but the improvement in compressive strength was insignificant. In addition, Chinchillas et al. [21] claimed that polyacrylonitrile fiber (PANF) could improve the flexural strength much more than the compressive strength. Conversely, the synergistic effect of hybrid fibers may contribute to improving the overall performance through a multi-scale reinforcement mechanism [22–24]. Walton et al. [25] found that the addition of hybrid inorganic and organic fibers in concrete could increase its impact and tensile properties. Wang et al. [26] illustrated that the compressive, tensile and flexural strengths of concrete were greatly improved by the incorporation of hybrid basalt and polypropylene fibers. Currently, the recommended hybrid fiber is the combination of rigid and flexible fibers. Rigid fibers with a high elastic modulus improve the strength of cemented materials, while flexible fibers with a low elastic modulus effectively inhibit the generation and development of cracks [27]. Besides, to reduce the production cost and achieve a high performance of cemented materials, the hybrid fibers with a good price and high quality are needed urgently. Basalt fibers (BFs) are made from basalt ores, which have the advantages of a low production cost, high tensile strength, high modulus of elasticity and good corrosion resistance [28–30]. In addition, BF has better compatibility with cemented materials due to their similar chemical composition [31,32]. Therefore, BF has been widely used in the engineering construction industry [33,34]. Polyacrylonitrile fiber (PANF) is a kind of green, flexible and low-cost material with excellent tensile and crack resistance [35,36]. PANF is a reinforcing material for concrete because it has a positive effect on the improvement in tensile strength, durability and toughness [37,38]. Zeng et al. [39] reported that BF and PANF fiber-reinforced concrete had better mechanical properties and toughness than single fiber-reinforced concrete. However, previous studies have mainly focused on a single kind of fiber, and the combination effect of hybrid BF and PANF fibers in mortars is not fully understood, particularly in tailing mortars. In addition, the best hybrid fiber content in the mortar is significantly influenced by the aggregate, such as tailings. It is essential to investigate the best combination and optimal additive content of hybrid BF and PANF fibers by considering the overall physico-mechanical performance and production costs of the tailing mortar.

The main objective of this study is to investigate the effect of BF, PANF and hybrid fibers on the physico-mechanical properties of fresh and hardened tailing mortars. The specific aims are as follows: (1) to investigate the effect of single and hybrid fibers on the common physico-mechanical properties of tailing mortar; (2) to evaluate the synergetic effect and cost of fiber-reinforced tailing mortar; (3) to interpret the reinforced mechanism of BF and PANF on the tailing mortar by using SEM and XRD tests; (4) tp quantify the most economical and high-performance additive content of hybrid fiber for the tailing mortar with 50% replacement of standard sands.

2. Materials and Methods

2.1. Materials

2.1.1. Cement

The cement is P·O 42.5 produced by Huaxin Cement Co., Ltd. (Ezhou, China), according to Chinese standard GB 175-2007 [40]. The properties and composition are shown in Table 1.

Table 1. Main properties and composition of the cement.

Properties	Value	Element	Cement/wt%
Specific surface area (m^2/kg)	351	CaO	57.18
Soundness	qualified	SiO_2	21.97
Initial setting time/min	230	Al_2O_3	6.55
Final setting time/min	300	SO_3	4.68
3-day compressive strength/Mpa	28.6	Fe_2O_3	2.89
3-day flexural strength/Mpa	5.4	MgO	2.14
Loss on ignition/%	3.73	K_2O	0.958
		P_2O_5	0.509

2.1.2. Aggregates

The aggregates utilized in this study were standard sand (SS) and IOTs. The SS was produced by Xiamen ISO Standard Sand Co., Ltd. (Xiamen, China) (GB/T 17671-2021 [41]) and the main chemical composition was quartz (>98%). The IOTs were selected from a tailing reservoir in Ezhou City, Hubei Province. The chemical composition of the IOTs was measured by XRF and XRD (Table 2, Figure 1), which showed that the main minerals were quartz, sulfide, calcite, etc. Obviously, they are a type of high-sulfur tailings. After drying at 105 °C for 24 h, the fineness modulus of SS and IOTs was determined at 2.38 and 1.54 by the sieving tests, respectively (Figure 2). According to the GB/T 14684-2022 [42], the IOT is too fine to meet the construction requirements. Therefore, further experiments were conducted by mixing IOTs with SS to effectively utilize the IOTs. Earlier works have demonstrated the viability of mixed sand that contains 50% SS and 50% IOTs [12]. The gradation curve of mixed sand was produced through the sieving test, as shown in Figure 2. The fineness modulus of mixed sand is 1.96. Table 3 displays the physical properties of SS, IOTs and mixed sand. The particle gradation and void fraction of mixed sand were improved.

Table 2. Chemical properties of IOTs.

Elements	SiO_2 (%)	CaO (%)	SO_3 (%)	Fe_2O_3 (%)	Al_2O_3 (%)	MgO (%)	K_2O (%)	Na_2O (%)
IOTs/wt%	24.70	19.10	15.06	11.56	6.88	3.86	2.96	1.63

Table 3. Physical properties of SS, IOTs and mixed sand.

Type	Loose Bulk Density (kg/m^3)	Compact Bulk Density (kg/m^3)	Apparent Density (kg/m^3)	Loose Void Ratio (%)	Compacted Void Ratio (%)
SS	1654.47	1735.17	2656.07	37.7	34.7
IOTs	1737.57	1963.13	3104.05	44.0	36.8
Mix sand	1689.84	1857.20	2874.96	41.2	35.4

2.1.3. Fibers

Table 4 lists the morphological and physical characteristics of fibers. Because the size of themortar specimen and aggregates is small, the suitable lengths of BF and PANF are 6 mm and 12 mm, respectively. Figures 3 and 4 depict the macromorphology and micromorphology of fiber filaments. The BF is a rigid fiber with smooth surface, while

PANF is a flexible fiber with a relatively rough surface. Before mixing, a hair dryer was used to disperse the fibers in a closed container to avoid fiber aggregation.

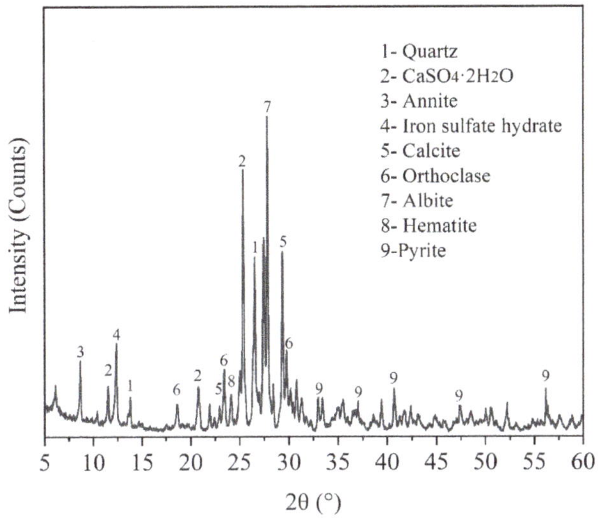

Figure 1. XRD pattern of IOTs.

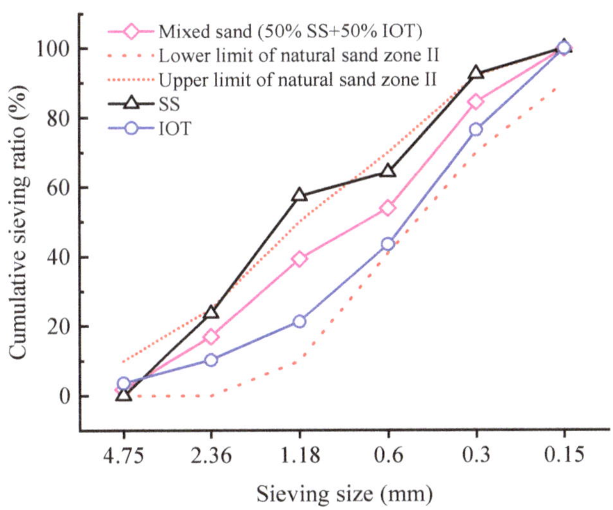

Figure 2. Gradation curve of the SS, IOTs and mixed sand.

Table 4. Physical properties of fibers.

Fiber Type	Length /mm	Diameter /μm	Density/(g/cm³)	Tensile Strength/MPa	Elastic Modulus/GPa	Elongation/%
PANF	12	13	1.18	530	7.2	22
BF	6	17	2.7	1650	63	2.8

2.1.4. Water and Super-Plasticizer

Water was taken from local tap water in Wuhan city. Polycarboxylic super-plasticizer (SP) was used to improve the fluidity and strength of the mortar.

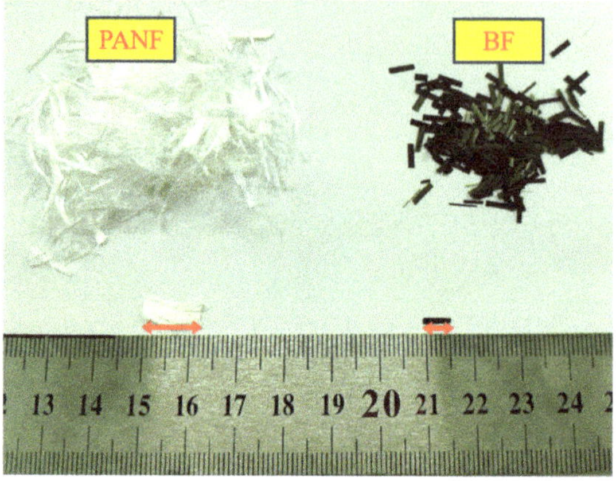

Figure 3. Macromorphology of fiber filaments.

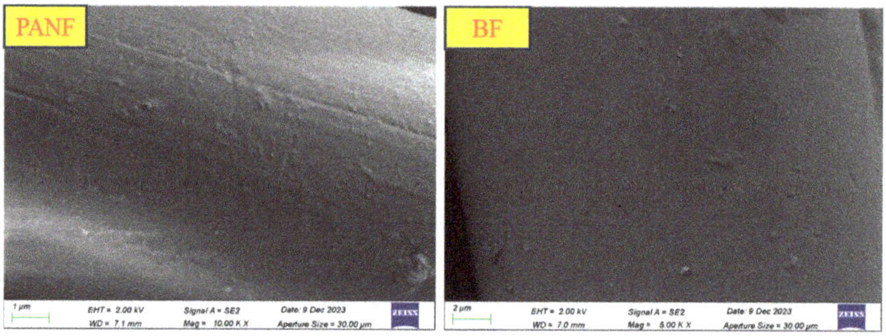

Figure 4. Micromorphology of fiber filaments.

2.2. Specimen Preparation

To explore the effect of hybrid fibers by combining BF and PANF on the tailing mortar, 12 different mixture proportions were designed in this study. The agglomeration of excessive fibers significantly influences the fluidity of mortar; therefore, the BF content was 0.25% and 0.5%, and the PANF content was set as 0.01%, 0.03% and 0.05%, respectively. The cement/SS/IOTs/water/SP = 400:600:600:168:2. Therefore, all of the tailing mortar with 12 different contents of fibers can be divided into 5 kinds of single fiber-reinforced mortar, 6 kinds of hybrid fiber-reinforced mortar and 1 kind of mortar without fibers.

The mixing process was carried out according to Chinese standard JTG 3420-2020 [43]. First, the dry SS and IOTs were mixed in a mixer for 1 min. Second, cement and fibers were added in sequence and mixed for 1 min and 2 min, respectively. Third, the water was added with the simultaneous addition of SP to stir for 5 min. The stirring process included slow stirring at 135 r/min for 2 min followed by fast stirring at 270 r/min for another 3 min. Therefore, the fresh hybrid fiber-reinforced tailing mortar (HRTM) could be derived. The fluidity and density tests were conducted first on the fresh mortar. Then, it was poured into the molds with dimensions of $70.7 \times 70.7 \times 70.7$ mm^3 and $40 \times 40 \times 160$ mm^3. To remove the entrapped air from the matrix, an external shaking table was used to vibrate matrix for 10 s. All specimens were demolded after being cured for 24 h (20 °C, 90 RH%). Thereafter, the specimens were cured in a standard curing box for 28 d (20 ± 2 °C,

≥95 RH%). The detailed mixing and curing process is shown in Figure 5. The specimens of 70.7 × 70.7 × 70.7 mm^3 were used for the uniaxial compression test and splitting tensile test, and the 40 × 40 × 160 mm^3 specimens were used to conduct the flexural test and ultrasonic wave velocity test.

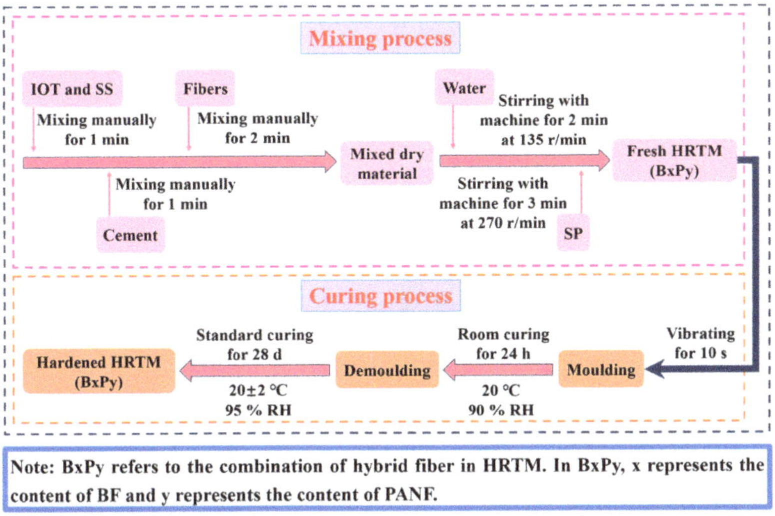

Figure 5. The detailed mixing and curing process.

2.3. Testing Procedure

2.3.1. Fluidity Test

Fluidity and density tests were performed according to Chinese standard JTG 3420-2020 [43]. The consistency was used to quantify the fluidity of fresh mortar. Mortar consistency is the penetration depth of the standard cone into the mortar by self-weight after 10 s. Therefore, a mortar consistometer (Tongli, SZ-145) was used to test the consistency in order to investigate the influence of fiber content on the fluidity of fresh mortar. In addition, the densities of hardened tailing mortars after curing for 28 d were also tested.

2.3.2. Ultrasonic Velocity Test

The primary wave (P-wave) and secondary wave (S-wave) velocities of specimens against curing ages were measured by a non-metallic acoustic wave detector (RSM-SY6C). The velocities were calculated as follows.

$$V = \frac{S}{t} \times 10^6 \tag{1}$$

V is the wave velocity (m/s), S is the interval between transducers (m), and t is the time (μs).

According to ASTM C597-09 (2009) [44], the dynamic modulus of elasticity was calculated as follows:

$$E_d = \frac{(1+\mu)(1-2\mu) \times {v_p}^2 \times \rho}{(1-\mu)} \times 10^{-9} \tag{2}$$

E_d is the modulus of dynamic elasticity (GPa), v_p is the P-wave velocity (m/s), ρ is density (kg/m^3) and μ is the Poisson ratio.

The Poisson ratio can be expressed as follows:

$$\mu = \frac{v_p^2 - 2v_s^2}{2v_p^2 - 2v_s^2} \tag{3}$$

v_s is the S-wave velocity (m/s).

2.3.3. Mechanical Test

The uniaxial compressive strength (UCS), split tensile strength (STS) and flexural strength (FS) of tailing mortar were tested referring to the Chinese standard JTG 3420-2020. The mechanical strengths were measured by using a hydraulic servo-controlled mechanical machine. To reduce errors caused by the molding process and test conditions, three specimens were used for the same condition. The loading rates were 1.5 kN/s, 0.08 Mpa/s and 50 N/s for the uniaxial compressive, split tensile and flexural strengths, respectively. The mechanical test process is displayed in Figure 6.

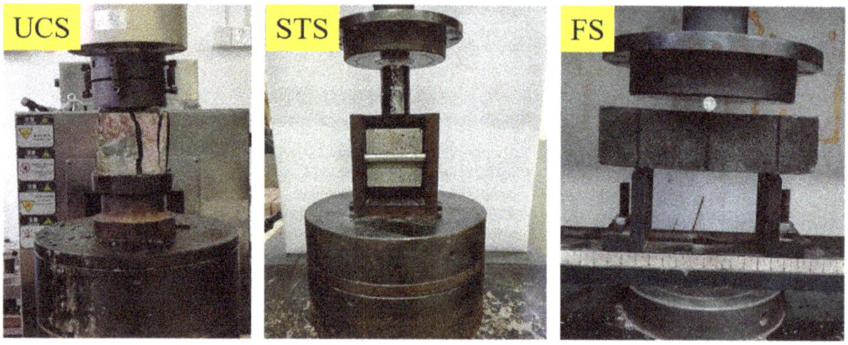

Figure 6. Schematic diagram of the mechanical tests.

2.3.4. Porosity Test

The damaged specimens (40 × 40 × 160 mm^3) were cut into three smaller specimens with dimensions of 40 × 40 × 40 mm^3 for the porosity test. These specimens were first dried at 105 °C for 24 h to reach a constant weight m_d. Thereafter, the completely dried specimens were placed in a vacuum saturation device (Huida, ZK-270; Cangzhou, China) for 48 h at −0.1 MPa. The saturated mass was recorded as m_s. Therefore, the porosity can be calculated as follows:

$$n = \frac{m_s - m_d}{\rho_\omega V_s} \times 100\% \tag{4}$$

n is the porosity (%), m_s is the mass of the saturated specimen (g), m_d is the mass of the completely dried specimen (g), ρ_ω is the density of water (1 g/cm^3) and V_s is the volume of the specimen (cm^3).

2.3.5. Microstructural Analysis

The crushed specimens from mechanical tests were selected for microstructure analysis by using scanning electron microscopy (SEM) and XRD. These crushed specimens were first kept in ethanol to prevent the hydration process and preserve their original pore characteristics. Thereafter, they were dried in an oven at 65 °C until they reached a constant weight. The drying temperature was controlled at 65 °C to avoid the destruction of hydration products [45].

To observe the micromorphology of hydration products and the interaction between the fibers and the matrix, several specimens with different fiber contents were selected for SEM (ZEISS GeminiSEM 300) tests. These test pieces were smaller than 10 mm × 10 mm × 20 mm. When carrying out tests, the pieces were coated with a gold film to improve the conduction and hence the image quality. The acceleration voltage was 3 kV and the scale bar of the images was 10 μm.

To analyze the crystalline phase in hydration products, the XRD tests were carried out by a diffractometer (Bruker D8 Advance, copper target). The 2-theta range was from 5° to 60° at a scanning rate of 2°/min. Small pieces were cut from the core of the crushed specimens and ground into powders for XRD tests.

3. Results and Discussion

3.1. Fluidity and Density

Table 5 shows the fluidity of mortar specimens against fiber types and contents. The fluidity of tailing mortar displays a decreasing trend compared to the control group. The minimum fluidity occurs in the B0.50P0.05 group, decreasing by 33.11%. The addition of BF and PANF increases the friction of tailings to promote the interlocking effect [46–48]. In addition, fibers also absorb some water to reduce the water film thickness (WFT) on the mortar matrix and thus further reduce the fluidity of tailing mortar [49]. For single BF, it does not have a significant effect on fluidity. This may be attributed to the relatively smooth surface and short length of BF, which cannot easily form the network structure. Conversely, due to the rougher surface, larger length (12 mm) and greater water absorption capacity, the adverse effect of PANF on the fluidity is greater than BF. In addition, the addition of hybrid fibers further significantly reduces the fluidity of the mortar compared to the single fibers. The fiber combination (BF: 6 mm; PANF: 12 mm) more easily forms a network structure inside the mortar specimens. Additionally, with increasing fiber content, the specific surface area of the mortar increases and more cement paste is wrapped around the fiber surface [50].

Table 5. The fluidity and density results of the mortar specimens.

Notation	Consistency /cm	Decrement/%	Fresh Density /(kg/m^3)	Increment/%	Hardened Density /(kg/m^3)	Increment/%
B0P0	10.33	0.00	1873.1	0.00	2016.2	0.00
B0P0.01	8.9	13.84	1904.5	1.68	2046.3	1.49
B0P0.03	8.23	20.33	1934.8	3.29	2055.8	1.96
B0P0.05	8.07	21.88	1950.2	4.12	2064.6	2.40
B0.25P0	10.28	0.48	1884.2	0.59	2032.5	0.81
B0.50P0	10.2	1.26	1893.3	1.08	2044.1	1.38
B0.25P0.01	8.62	16.55	1977.1	5.55	2087.4	3.53
B0.25P0.03	7.98	22.75	2048.1	9.34	2099.2	4.12
B0.25P0.05	7.13	30.98	2075.8	10.82	2148.5	6.56
B0.50P0.01	8.52	17.52	2013.2	7.48	2105.2	4.41
B0.50P0.03	7.48	27.59	2054.8	9.70	2140.8	6.18
B0.50P0.05	6.91	33.11	1970.5	5.20	2081.2	3.22

The densities of the fresh and hardened mortar specimens are also presented in Table 5. The fresh and hardened densities show a similar variation trend. When one type of fiber is added, the density increases. The BF improves the density of the tailing mortar because it has a greater density than the matrix [51]. However, it can be observed that when 0.25 wt% or 0.5 wt% BF is added, the fresh densities of B0.25P0 and B0.5P0 do not show any significant increase. First, there is an intrinsic link between the consistency of the mortar and the fresh density. The fluidity of B0.25P0 and B0.5P0 is very close (10.28 cm, 10.20 cm), which will not cause any difference in the fresh density. In addition, the basalt fiber contents in mortar are in smaller quantities and thus contribute less to the fresh density. Li et al. [52] also found that the wet pack densities of mortar with 0~0.5% BF raise slightly with increasing BF contents. PANF is longer and more likely to form a network structure to compact the tailing mortar and further increase the density. In addition, the flexible PANF has a smaller diameter and better dispersion in the tailing mortar, which can easily fill into and pass through the pores between the fine aggregates to increase the density (Figure 7). This finding is consistent with previous research [53]. Therefore, when the hybrid fibers are added to the tailing mortar, the density increases significantly. However, excessive added fibers can agglomerate together and introduce some air into the mortar, resulting in an increase in the porosity and a reduction in the density [54].

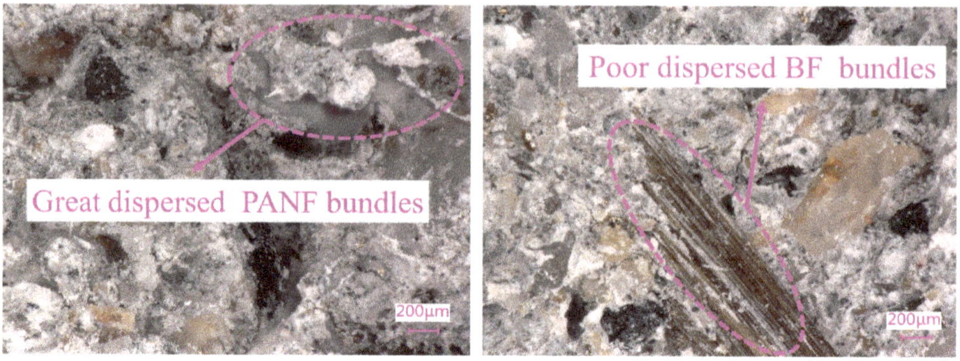

Figure 7. The dispersion of single PANF and BF bundles in the mortar specimens.

3.2. Porosity

Figure 8 displays the variation in porosity against the BF and PANF contents. It is evident that the addition of fibers significantly decreases the porosity compared with the control mixture. The fibers have a bridging effect that fills the pores and prevents the growth of cracks, thus reducing the porosity. When single BF fibers are added at 0.25% and 0.5%, the porosity decreases by 2.23% and 4.70%, respectively. Additionally, by increasing single PANF to 0.01%, 0.03% and 0.05%, the porosity decreases by 0.73%, 2.05% and 2.85%, respectively. This shows that the improvement in relatively low fiber content is not significant due to the weak bridging effect. Compared to the single type of fibers, hybrid fibers cause a greater improvement in porosity (Figure 8). Additionally, the B0.25P0.05 group of specimens have minimum porosity at 18.54%, which is 25.55% lower than the control group. With increasing fiber contents, the bridging effect is gradually developed to form the dense three-dimensional network structure. However, when the hybrid fiber content exceeds the critical value, poor dispersion occurs and more pores are produced, which causes an increase in the porosity [55]. The porosity of the B0.50P0.05 group is only 5.10% lower than the control mix.

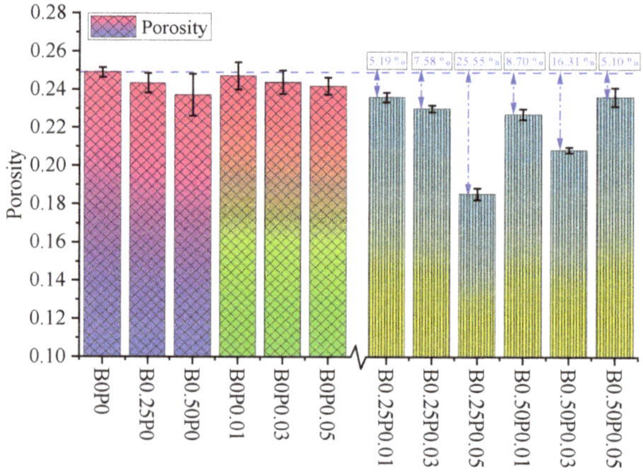

Figure 8. The porosity of the mortar specimens.

3.3. Ultrasonic Velocity

Figures 9–11 show the variation in P-wave velocity, S-wave velocity and the dynamic modulus of elasticity against the curing age and fiber contents. When increasing the curing age, they have the same growth trend due to the continuous hydration reaction of the cement matrix and production of hydration products [56]. The addition of BF and PANF significantly increases the ultrasonic velocity compared to the control mix. When single BF or PANF are added, the ultrasonic velocity increases with increasing fiber content. In terms of hybrid fibers, when the BF content is 0.25%, they increase quickly with increasing PANF content. However, at a 0.50% BF content, they first increase but then decrease with the increasing content of PANF. For hybrid fiber specimens, the B0.25P0.05 group and B0.50P0.05 group possess the maximum and minimum ultrasonic velocities, respectively. For example, the P-wave velocity, S-wave velocity and dynamic elastic modulus of the B0.25P0.05 group at 28 d are 7.39%, 5.75% and 20.52% higher than the B0P0 group. Additionally, the P-wave velocity, S-wave velocity and dynamic elastic modulus of the B0.50P0.05 group at 28 d are only 2.87%, 2.52% and 8.68% higher than the B0P0 group, respectively. Overall, the increment in the ultrasonic velocity by adding hybrid fibers is better than a single fiber. The correlation between the ultrasonic velocity and porosity is also significant (Figure 12). In addition, although the tailing mortar and the concrete are different kinds of cemented materials, they have very similar physico-mechanical properties. Due to the lack of sufficient research on the tailing mortar, a similar relationship between the ultrasonic velocity and porosity for concrete can be also found in Benaicha [56] and Ridengaoqier [57], which further verifies the reliability of the present results.

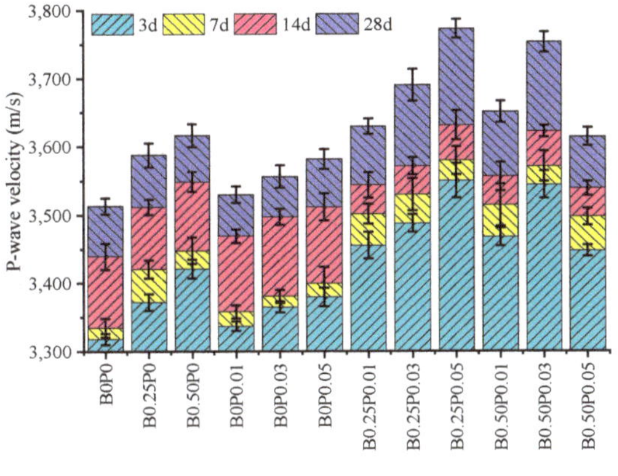

Figure 9. The P-wave velocity against the added content of fibers.

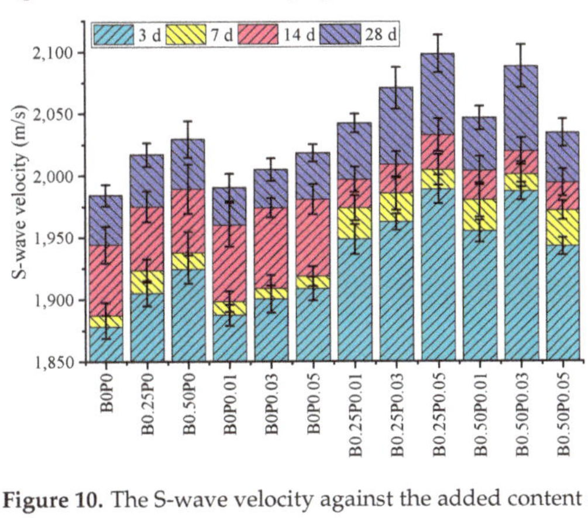

Figure 10. The S-wave velocity against the added content of fibers.

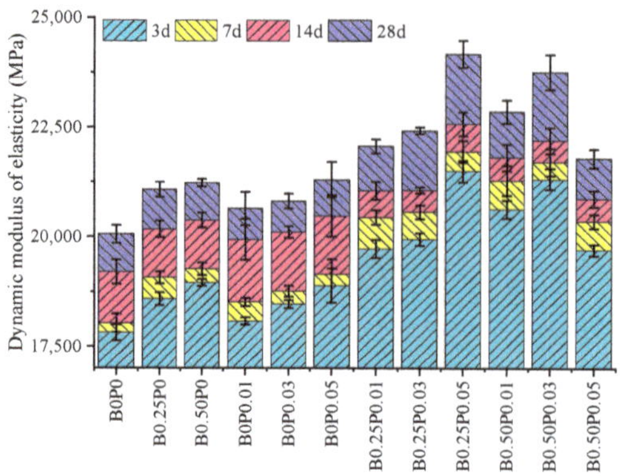

Figure 11. The dynamic modulus of elasticity against the added content of fibers.

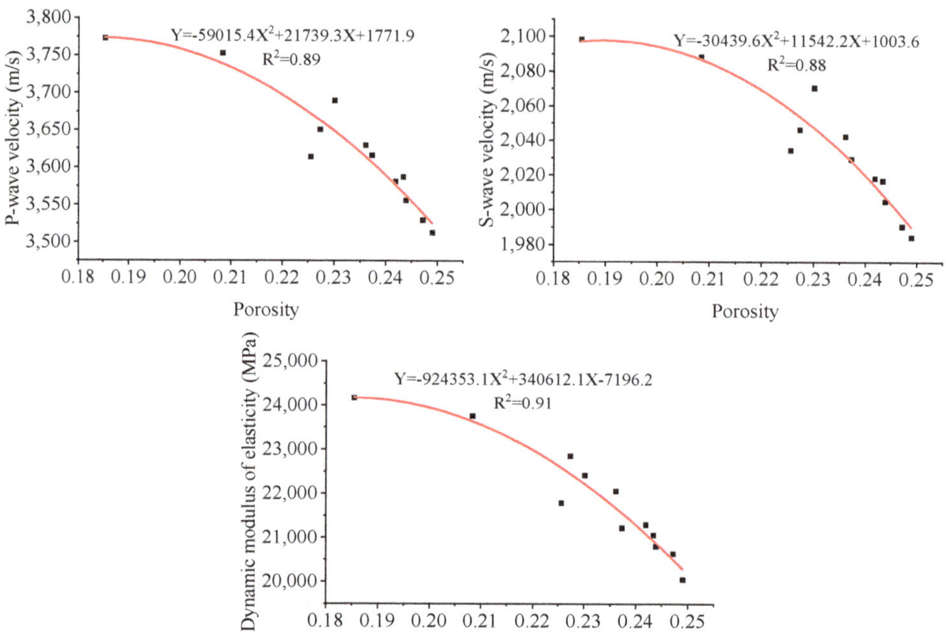

Figure 12. The relationship between the ultrasonic velocity (28 d) and porosity.

3.4. Mechanical Strength

3.4.1. Uniaxial Compressive Strength (UCS)

The UCS of specimens containing different fiber contents are presented in Figure 13. The increment in UCS is given in Table 6. It is observed that the maximum UCS value (30.08 MPa) occurs in the B0.25P0.05 group, which is 45.74% higher than in the control group. During the compression process, the fibers can bear a lot of loads and the friction between the fibers and the matrix prevents the formation and expansion of cracks [58,59]. For a single addition of BF or PANF, the UCS increases with increasing fiber content. Compared to the B0P0 group, the UCS of B0.25P0 and B0.50P0 increases by 13.78% and 17.38%, respectively. However, for mortar specimens with PANF, the increase in UCS is not significant, only 2.91%, 8.45% and 10.07%, respectively, because the bridging effect and cracking resistance effect are poor for a low PANF content. In addition, it is a consensus that the bridging effect of flexible fibers is more effective in tensile conditions rather than compressive conditions [60,61]. Obviously, the contribution of hybrid fibers to the increase in UCS is better than of a single fiber. According to previous research [62–64], flexible PANF (12 mm) in the mortar more easily forms a network structure and prevents the formation and propagation of microcracks, while rigid BF (6 mm) has a better stiffness to bear more shear loads and restrict the deformation of hardened specimens. Therefore, hybrid fibers can complement each other in terms of elastic modulus and they can form a better three-dimensional network structure. In addition, hybrid fibers with different lengths can restrict different crack widths under loading [65]. Therefore, the combination of hybrid fibers is better to enhance the UCS. However, the excessive hybrid fibers agglomerate together and absorb more water, reducing the fluidity and producing more pores inside the mortar [66]. Therefore, the UCS reduces if too many hybrid fibers are added; for example, B0.50P0.05 has a much lower UCS than B0.25P0.05. It can be concluded that the best hybrid fiber combination is B0.25P0.05 for this tailing mortar.

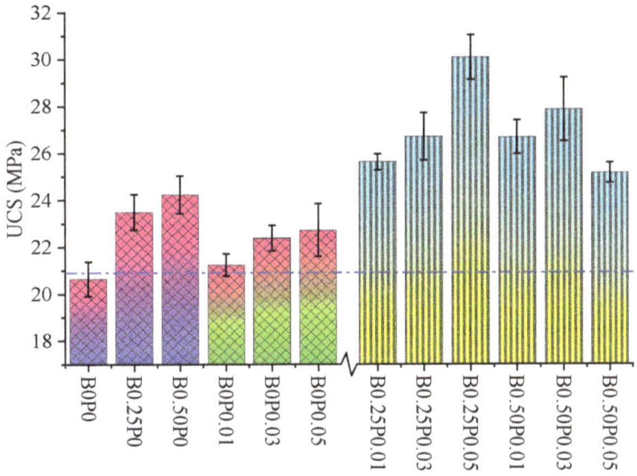

Figure 13. The UCS against incorporated fiber contents.

Table 6. The mechanical strength increments of mortar specimens with fibers.

Number	Increment		
	UCS	STS	FS
B0P0	0	0	0
B0.25P0	13.78%	9.04%	1.58%
B0.5P0	17.38%	10.99%	5.19%
B0P0.01	2.91%	3.85%	1.05%
B0P0.03	8.45%	10.29%	4.69%
B0P0.05	10.07%	14.49%	7.93%
B0.25P0.01	24.10%	28.55%	6.11%
B0.25P0.03	29.39%	33.41%	9.36%
B0.25P0.05	45.74%	52.33%	15.65%
B0.50P0.01	29.26%	28.84%	6.71%
B0.50P0.03	35.02%	36.25%	10.03%
B0.50P0.05	21.83%	22.67%	0.50%

The typical failure modes of the specimens after compressive tests are shown in Figure 14. The yellow line indicates the direction of crack propagation. The specimens with a low fiber content exhibit obvious shear failure characteristics and brittle failure mode. At its peak strength, the specimen ruptures suddenly and the main crack penetrates the specimen. With increasing fiber content, a three-dimensional network structure is formed which fills the internal pores and changes the propagation direction of microcracks. In addition, the main crack width decreases and many secondary cracks are observed, displaying better integrity and toughness [63]. However, excessive hybrid fiber has induced more defects, and a wider main crack occurs in the specimens during loading process. Therefore, the addition of hybrid fibers significantly improves the toughness of specimens. The failure mode also confirms that the best fiber combination is B0.25P0.05.

Figure 14. Typical failure mode of mortar specimens after compressive tests.

3.4.2. Split Tensile Strength (STS)

STS against fiber content is presented in Figure 15. It is evident that fibers significantly improve the STS of specimens. The increment ratio is from 3.85% to 52.33% (Table 6). The contribution of PANF to the STS is larger than that of the BF, which is opposite to the UCS. The longer flexible PANF more easily forms a network structure and possesses a better bridging effect; therefore, the bridging effect of PANF is more effective under tensile conditions. The detailed reinforced mechanism of PANF and BF can be explained as follows. PANF has a great dispersion of fiber bundles and distributes uniformly, which can delay the formation and propagation of microcracks during the hardening process. Under loading, the rigid BF can bear loads and limit the propagation of cracks. Figure 16 shows the microscopic distribution of PANF and BF in the mortar. This further confirms that PANF mainly inhibits the propagation of microcracks. PANF is prone to rupture in the crack area, and only when the PANF content increases to a certain amount are the cracking resistance and bridging effect obvious. Therefore, the hybrid fiber-reinforced specimens have the combination advantages of PANF and BF, and thus exhibit a better STS than the single fiber-reinforced specimens. For example, compared to the B0P0 group, the increment ratio of the B0.25P0.05 group is up to 52.33%. The greater improvement due to hybrid fibers was also found by Niu [62] and Sadrinejad [64]. The negative effect of excessive fibers on the STS are the same as on the UCS. The specimens in the B0.25P0.05 group have the maximum STS.

Figure 17 displays the typical failure modes of specimens after flexural tests. The cracks completely penetrate the B0P0 and B0P0.01 specimens after failure. With the increase in fiber content, the width and length of the crack decrease, and the propagation direction is significantly influenced by the fibers. Therefore, this implies that the fibers effectively prevent the development of cracks. However, the poor dispersion of too many hybrid fibers in B0.25P0.05 specimens cause the production of many pores, thus resulting in a wider and longer main crack.

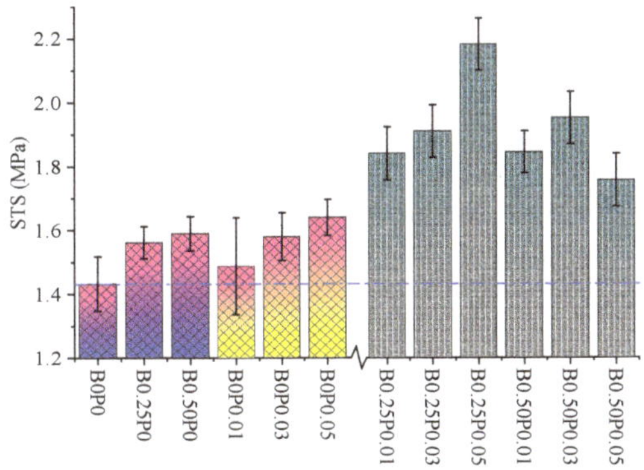

Figure 15. The STS against fiber content.

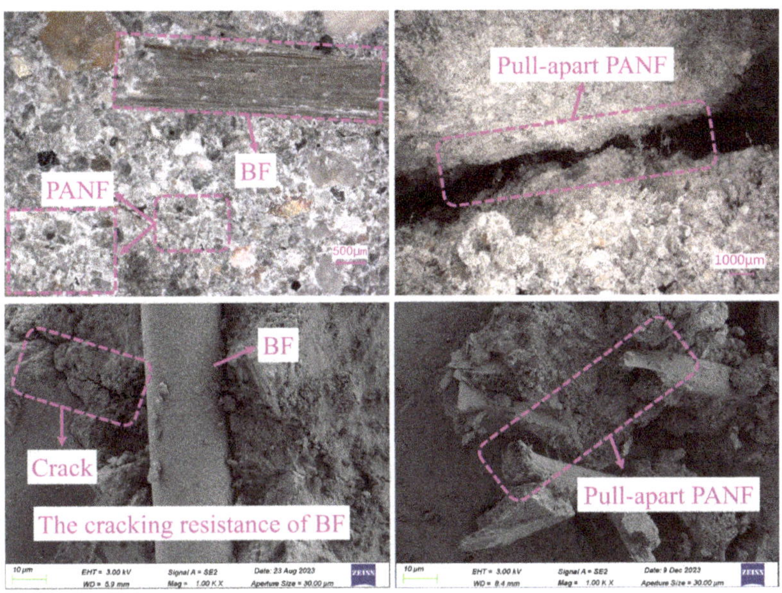

Figure 16. The distribution of PANF and BF in the mortar.

Table 7 shows the standards formulas of different countries for estimating the STS and FS based on the UCS. It is generally recognized that the relationship between the STS, FS and UCS of concrete can be expressed by using Equation (5). Based on the growth rate and previous standards in Table 7, $b = 0.75$ for STS and $b = 0.5$ for FS, respectively. The value of parameter a in Equation (5) is determined by fitting. Figure 18 displays the correlation between the STS and UCS of all mortar specimens. The correlation between the STS and UCS is significant, with $R^2 = 0.85$, which is also in good agreement with the results in Zheng [67] and Zheng [68].

$$f_{STS/FS} = a \times (f_{UCS})^b \quad (5)$$

3.4.3. Flexural Strength (FS)

The measured FS of single and hybrid fiber-reinforced specimens is illustrated in Figure 19. The specimens with fibers possess a larger FS than that of specimens in the control group without fibers. For single fiber-reinforced mortar, the FS displays an increasing trend with increasing fiber content. Overall, PANF causes greater improvements in FS than BF. In addition, the improvement in FS of hybrid fibers is greater than that of single fibers. The maximum FS is 6.33 MPa for B0.25P0.05, which has increased by 15.65% compared with B0P0. In addition, the agglomeration of hybrid fibers leads to an increase in porosity and then decreases the FS, which is similar to the change in the UCS and STS. It should be noted that the FS of B0P0.05 is higher than that of B.25P0.01 and B0.05P0.01; therefore, PANF is better than BF in enhancing FS.

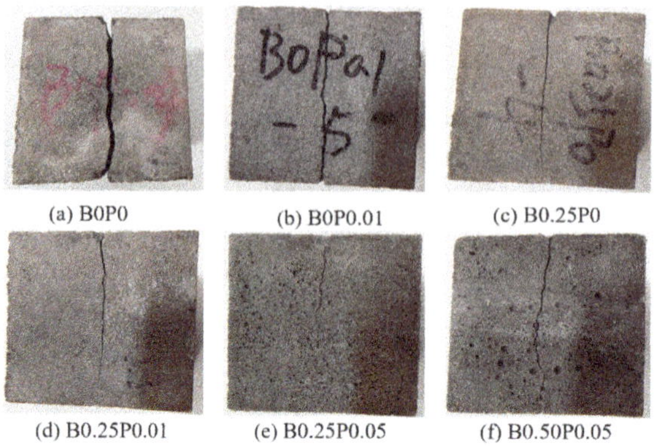

Figure 17. Typical failure modes of mortar specimens after split tensile tests.

Table 7. The standards formulas for estimating the STS and FS based on the UCS.

Standards	f_{STS}	Standards	f_{FS}
NZS 3101 [69]	$0.44 \times (f_{UCS})^{0.5}$	NZS 3101 [69]	$0.60 \times (f_{UCS})^{0.5}$
EC-04 [70]	$0.30 \times (f_{UCS})^{(2/3)}$	EC 04 [70]	$0.435 \times (f_{UCS})^{(2/3)}$
GB 50010 [71]	$0.19 \times (f_{UCS})^{0.75}$	DG/TJ [72]	$0.75 \times (f_{UCS})^{0.5}$
JSCE [73]	$0.23 \times (f_{UCS})^{(2/3)}$	IS 456 [74]	$0.70 \times (f_{UCS})^{0.5}$
ACI 318-11 [75]	$0.56 \times (f_{UCS})^{0.5}$	ACI 318-11 [75]	$0.62 \times (f_{UCS})^{0.5}$

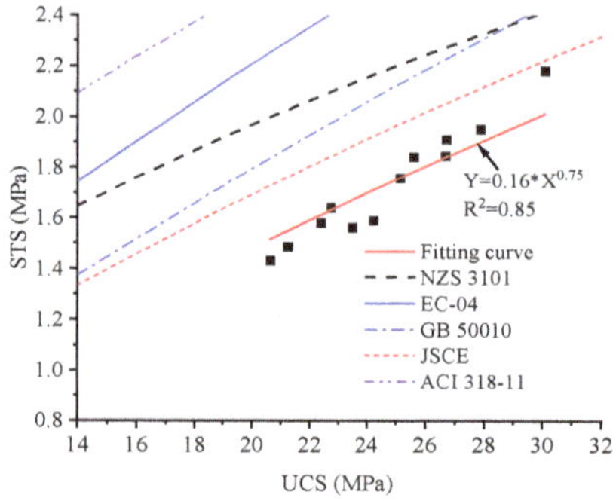

Figure 18. The relationship between the STS and UCS for the tailing mortar [69–71,73,75].

Figure 20 shows the failure modes of specimens after flexural tests. The B0P0 and B0P0.01 specimens exhibit brittle fracture characteristics. As the fiber content increases, the toughness of the specimen increases and the main cracks are narrow. For example, the main crack in B0.25P0.05 is almost invisible after failure and its FS is the largest.

Figure 21 displays the correlation between the FS and UCS for tailing specimens. Based on the slower growth rate of the experimental data, the best fit was chosen at $b = 0.5$. The correlation is significant, with $R^2 = 0.51$, which is consistent with previous findings in Zheng [67], Zheng [68] and Wang [26]. This implies that the change rule of all of the strength indexes is similar for the fiber types and contents, and the best fiber combination is B0.25P0.05.

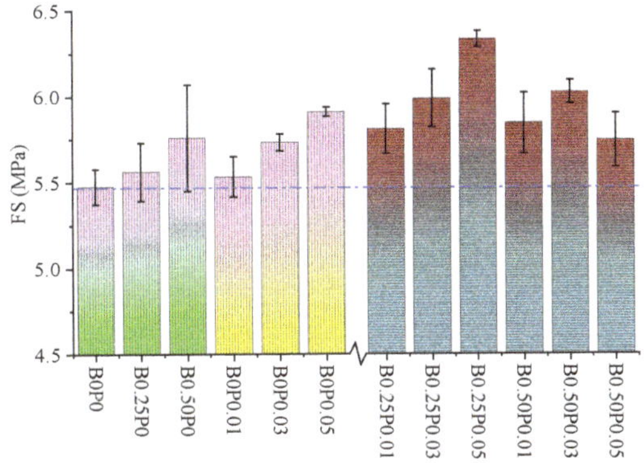

Figure 19. The FS against incorporated fiber content.

Figure 20. Typical failure mode of mortar specimens after flexural tests.

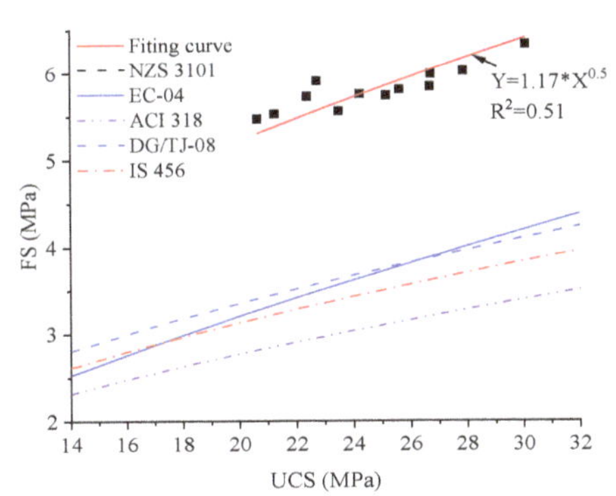

Figure 21. The relationship between the FS and UCS of mortar [69,70,72,74,75].

3.5. Quantification of the Synergy Effect of BF and PANF

Banthia [23] and Cui [76] proposed an index to evaluate the synergy effect of hybrid fibers in Equation (6). $M > 1$ represents a positive synergy effect, which means that the improvement due to hybrid fibers is larger than the sum of all fibers added alone. When $M \leq 1$, the synergy effect is insignificant or negative.

$$M = \frac{f_{hybrid,a+b} - f_0}{f_a + f_b - 2f_0} \quad (6)$$

where M is the synergy effect index; $f_{hybrid,a+b}$ is the mechanical strength of the hybrid fiber-reinforced mortar; f_a and f_b are the mechanical strength of the single fiber-reinforced mortar; f_0 is the mechanical strength of the mortar without fibers.

Table 8 shows the synergetic effect index of the HRTM on mechanical strengths. It is evident that the combinations of fibers have played a positive role in the growth of strengths except for the B0.50P0.05 group, which is inconsistent with the change in the UCS, STS and FS. The synergistic effect of the B0.25P0.05 group is evident on the UCS and STS, both of which are optimal. However, the best synergistic effect on the FS occurs in the B0.25P0.01 group, because the flexural strength of B0P0.01 is much smaller than that of B0P0.05, and thus B0P0.01 has a much larger improvement potential due to the addition BFs. In addition, the synergistic effect of B0.25P0.05 is still much better than the other hybrid combinations in terms of FS.

Table 8. The synergetic effect of hybrid fibers.

Number	M		
	UCS	STS	FS
B0.25P0.01	1.444	2.215	2.321
B0.25P0.03	1.322	1.728	1.492
B0.25P0.05	1.918	2.224	1.646
B0.5P0.01	1.442	1.943	1.075
B0.5P0.03	1.356	1.703	1.014
B0.5P0.05	0.795	0.890	0.371

3.6. Economic Analysis of HRTM Specimens

In this study, 50% SS was used as aggregate for the preparation of the HRTM. However, when HRTM specimens are utilized on a large scale in actual construction, 50% natural sand is used as aggregate at a much lower cost than SS (in Wuhan, SS: 10.18 CNY/kg and natural sand: 0.156 CNY/kg). Therefore, the price of natural sand was calculated in the economic analysis of HRTM specimens. The prices of cement (42.5), natural sand and SP in Wuhan from July 2022 to June 2023 are calculated in Table 9. In addition, the prices of dry-mixed mortar are as follows: M20 is 0.372 CNY/kg and M25 is 0.418 CNY/kg. The price of tailings is preliminarily determined based on transportation distance. Furthermore, according to the proportions in this experiment, the prices of fiber-reinforced specimens can be further obtained in Table 10. Based on the mixing proportion of previous tests [12], the price of the mortar with only SS (B0P0-100% SS) is calculated in Table 9. Therefore, compared to the dry mortar, the HRTM used in this experiment has a good economic benefit. The prices of the HRTM (B0P0.01, B0.25P0.03 and B0.25P0.05) are more cost-effective and they have greater performance compared to B0.50P0. Further, the prices of the HRTM (B0P0.01, B0.25P0.03 and B0.25P0.05) are much lower than the price of B0P0-100% SS. Through economic analysis, it is also recommended to use the tailing mortar with 0.25% BF and 0.05% PANF.

Table 9. The prices of original materials in Wuhan (2022~2023). (Unit: CNY/kg).

Materials	Cement	Natural Sand	IOTs	SP	BF	PANF
Price	0.448	0.156	0.030	5.200	12.500	9.000

Table 10. The prices of hybrid fiber-reinforced mortar. (Unit: CNY/kg).

Number	Price
B0P0-100% SS	0.2127
B0P0	0.1701
B0.25P0	0.2013
B0.50P0	0.2326
B0P0.01	0.1710
B0P0.03	0.1728
B0P0.05	0.1746
B0.25P0.01	0.2022
B0.25P0.03	0.2040
B0.25P0.05	0.2058
B0.50P0.01	0.2335
B0.50P0.03	0.2353
B0.50P0.05	0.2371

3.7. Microstructure Analysis

3.7.1. SEM

The microstructure of the interface between the fibers and the matrix was further investigated by conducting SEM tests (Figure 22). It can be observed that there are many pores in the tailing mortar. With the addition of fibers, some pores and cracks inside the mortar are filled and bridged, leading to a dense structure. In the B0P0.03 and B0.25P0 groups, the bridging effect is weak when not many fibers are contained, and obvious pores and cracks occur. In addition, if the content of fibers is small, it is to pull them out or break them. In the B0P0.05 and B0.50P0 groups, due to the increase in fiber content, the bridging effect and the three-dimensional network structure are significant. Furthermore, in the B0.25P0.05 group, the three-dimensional network structure created by the hybrid BF and PANF is more obvious. Therefore, a suitable content of fibers provides enough cracking resistance to enhance the mechanical strengths [77]. With increasing fiber content, the interfacial bonding and interfacial transition zone (ITZ) between the fibers and the matrix are improved. BF and PANF with good hydrophilicity can absorb some water; therefore, the water/cement ratio near their surface is higher than the cement matrix and thus promotes early cement hydration [78]. In the B0.50P0.05 group, too many fibers are contained, which results in poor dispersion of hybrid fibers, as shown in Figure 23. The bonding effect between the fibers and the matrix decreases and the final strength also reduces.

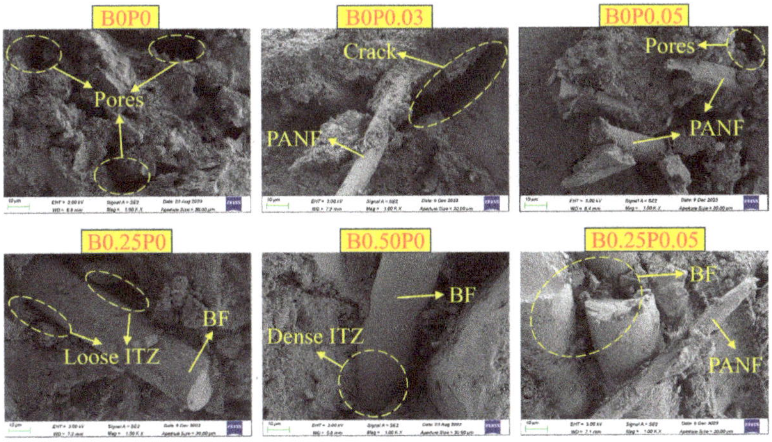

Figure 22. Typical SEM micromorphology images of mortar specimens.

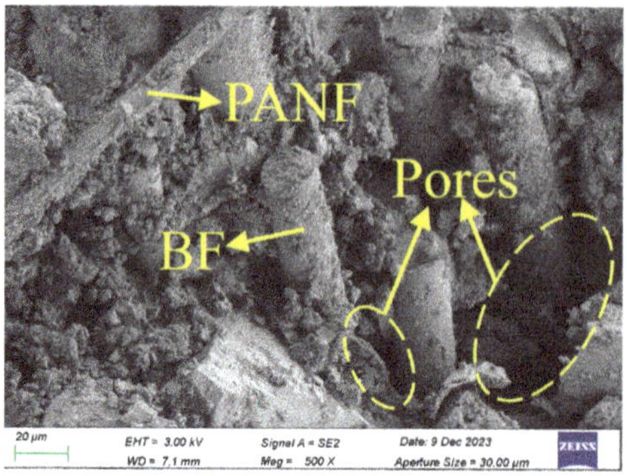

Figure 23. The agglomeration of excessive hybrid fibers.

3.7.2. XRD

Figure 24 displays the hydration product against the content of hybrid fibers. It can be observed that the main crystalline phases are SiO_2, C-S-H, $Ca(OH)_2$, $CaSO_4$, AFt and $CaCO_3$. The SiO_2 and $CaCO_3$ phases mainly exist in aggregates. C-S-H and $Ca(OH)_2$ are produced by the hydration reactions of cement. The production of $CaSO_4$ and AFt can be explained as follows. The high sulfur content in fine IOTs is prone to separate out and further be oxidized to sulfur ions, which can react with $Ca(OH)_2$ to form ettringite and gypsum. The addition of fibers does not change the type of hydration products, indicating that fibers are inert in the matrix. However, the amount of hydration products increases with the increasing content of hybrid fibers. Generally, the nucleation sites of the hydration products will increase after adding appropriate fibers, and thus promote the hydration reaction rate and produce more hydration products [30,79]. C-S-H and $Ca(OH)_2$ are beneficial for the increase in strength, and $CaSO_4$ and AFt can also promote strength in the early stages. Therefore, a higher content of the four hydration products plays a positive and significant role in strength. Figure 24 shows that the B0.25P0.05 group possessed the highest amount of hydration products. Therefore, its mechanical strengths are the highest. Conversely, in B0.50P0.05, the agglomeration of hybrid fibers decreases the hydration rate, leading to a reduction in hydration products.

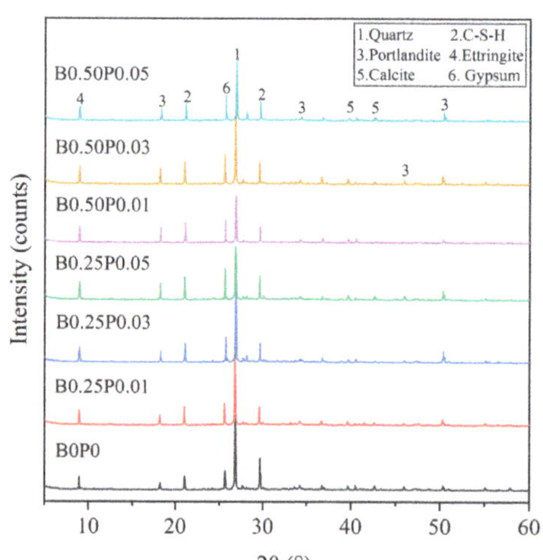

Figure 24. XRD patterns of mortar specimens.

4. Conclusions

This study investigated the physico-mechanical properties of tailing mortar reinforced with hybrid BF and PANF. In summary, the following conclusions can be drawn:

(1) The addition of BF and PANF reduces the fluidity of tailing mortar; however, fibers significantly enhance the overall performance of the HRTM. Furthermore, hybrid fibers cause greater improvements in the physico-mechanical properties than single fibers. For example, the improvement due to hybrid fibers in porosity and ultrasonic wave velocity is much better than that due to a single fiber. The optimum hybrid fiber combination for the CS, STS and FS is B0.25P0.05. Compared to the tailing mortar without fibers, the CS, STS and FS increase by 45.74%, 52.33% and 15.65%, respectively. The enhancement effect of the STS is the largest.

(2) The reinforced mechanism of the HRTM by adding fibers has been investigated. BF and PANF prevent the production and propagation of cracks and thus greatly improve the toughness of the mortar. BF (6 mm) has a better reinforcement effect that improves the UCS due to its high elastic modulus (63 GPa), while PANF (12 mm) is favorable to improve the STS and FS due to its better bridging effect and the creation of a denser network structure. However, when the fiber contents are excessive, the toughness decreases.

(3) The synergetic effect of hybrid fibers on the tailing mortar have been evaluated, showing that all of the hybrid fiber combinations have a positive effect, except for B0.50P0.05. The B0.25P0.05 group possesses the largest synergetic effect index (UCS: 1.918; STS: 2.224), and the synergistic effect of B0.25P0.05 is still much better (FS: 1.646) than the other hybrid combinations in terms of FS. Moreover, it is also a very economical choice (0.2058 CNY/kg).

(4) Observed through SEM tests, with increasing fiber content, the bond action between fibers and the matrix in the specimens is significant, which can improve the density of the tailing mortar. The XRD test shows that hybrid fibers cannot change the types of hydration products; however, they can provide more nucleation sites to promote the hydration reaction and improve the internal structure of the tailing mortar.

Therefore, it can be concluded that the most economical and high-performance combination of hybrid fibers for the HRTM is B0.25P0.05. In addition, it should also be noted that sulfur is harmful to the structure and long-term strength, and thus, the long-term performance of the HRTM should be investigated further.

Author Contributions: Conceptualization, J.X.; validation, Y.L., Z.C. and Y.S.; data curation, Y.L., Z.C. and Y.S.; writing—original draft preparation, Z.P. and S.H.; writing—review and editing, J.X., Z.P. and S.H.; visualization, Y.L., Z.C. and Y.S. All authors have read and agreed to the published version of the manuscript.

Funding: This work was supported by the National Natural Science Foundation of China (Grant No. 42377191 and 42072300) and the Project of the Natural Science Foundation of Hubei Province (Grand No. 2021CFA094). This work was supported by "The 14th Five Year Plan" Hubei Provincial advantaged characteristic disciplines (groups) project of Wuhan University of Science and Technology (Grant No. 2023A0303).

Data Availability Statement: Data will be made available upon request. The data are not publicly available due to privacy.

Conflicts of Interest: Author Jianbo Xu was employed by the company China First Metallurgical Group Co., Ltd. The remaining authors declare that the research was conducted in the absence of any commercial or financial relationships that could be construed as a potential conflict of interest.

References

1. Tiwari, A.; Singh, S.; Nagar, R. Feasibility Assessment for Partial Replacement of Fine Aggregate to Attain Cleaner Production Perspective in Concrete: A Review. *J. Clean. Prod.* **2016**, *135*, 490–507. [CrossRef]
2. Zhao, J.; Wang, Q.; Xu, G.; Shi, Y.; Su, Y. Influence of Macro-Synthetic Fiber on the Mechanical Properties of Iron Ore Tailing Concrete. *Constr. Build. Mater.* **2023**, *367*, 130293. [CrossRef]
3. Zhu, Q.; Yuan, Y.; Chen, J.; Fan, L.; Yang, H. Research on the High-Temperature Resistance of Recycled Aggregate Concrete with Irontailing Sand. *Constr. Build. Mater.* **2022**, *327*, 126889. [CrossRef]
4. Zhang, N.; Tang, B.; Liu, X. Cementitious Activity of Iron Ore Tailing and Its Utilization in Cementitious Materials, Bricks and Concrete. *Constr. Build. Mater.* **2021**, *288*, 123022. [CrossRef]
5. Kefeni, K.K.; Msagati, T.A.M.; Mamba, B.B. Acid Mine Drainage: Prevention, Treatment Options, and Resource Recovery: A Review. *J. Clean. Prod.* **2017**, *151*, 475–493. [CrossRef]
6. Han, X.; Wang, Y.; Zhang, N.; Meng, J.; Li, Y.; Liang, J. Facile Synthesis of Mesoporous Silica Derived from Iron Ore Tailings for Efficient Adsorption of Methylene Blue. *Colloids Surf. B* **2021**, *617*, 126391. [CrossRef]
7. Shettima, A.U.; Hussin, M.W.; Ahmad, Y.; Mirza, J. Evaluation of Iron Ore Tailings as Replacement for Fine Aggregate in Concrete. *Constr. Build. Mater.* **2016**, *120*, 72–79. [CrossRef]
8. Zhou, Y.; Yang, Z.; You, Z.; Wang, X.; Chen, K.; Guo, B.; Wu, K. Experimental Study on Fire Resistance of Concrete Beams Made with Iron Tailings Sand. *Buildings* **2022**, *12*, 1816. [CrossRef]
9. Andrews, A.; Nyarko, E.F.; Adjaottor, A.A.; Nsiah-Baafi, E.; Adom-Asamoah, M. Reuse and Stabilization of Sulphide Mine Tailings as Fine Aggregate for Construction Mortar. *J. Clean. Prod.* **2022**, *357*, 131971. [CrossRef]
10. Zhang, W.; Gu, X.; Qiu, J.; Liu, J.; Zhao, Y.; Li, X. Effects of Iron Ore Tailings on the Compressive Strength and Permeability of Ultra-High Performance Concrete. *Constr. Build. Mater.* **2020**, *260*, 119917. [CrossRef]
11. Zhao, S.; Fan, J.; Sun, W. Utilization of Iron Ore Tailings as Fine Aggregate in Ultra-High Performance Concrete. *Constr. Build. Mater.* **2014**, *50*, 540–548. [CrossRef]
12. Huang, S.; Pi, Z.; Cai, C.; Li, H. Utilization of High-Sulfur Iron Ore Tailings in Cement Mortar by Considering the Influence of Curing Temperature and Tailing Content. *J. Build. Eng.* **2023**, *74*, 106826. [CrossRef]
13. Li, T.; Wang, S.; Xu, F.; Meng, X.; Li, B.; Zhan, M. Study of the Basic Mechanical Properties and Degradation Mechanism of Recycled Concrete with Tailings before and after Carbonation. *J. Clean. Prod.* **2020**, *259*, 120923. [CrossRef]
14. Xu, F.; Wang, S.; Li, T.; Liu, B.; Zhao, N.; Liu, K. The Mechanical Properties and Resistance against the Coupled Deterioration of Sulfate Attack and Freeze-Thaw Cycles of Tailing Recycled Aggregate Concrete. *Constr. Build. Mater.* **2021**, *269*, 121273. [CrossRef]
15. Xu, F.; Wang, S.; Li, T.; Liu, B.; Li, B.; Zhou, Y. Mechanical Properties and Pore Structure of Recycled Aggregate Concrete Made with Iron Ore Tailings and Polypropylene fibers. *J. Build. Eng.* **2021**, *33*, 101572. [CrossRef]
16. Adesina, A. Performance of Cementitious Composites Reinforced with Chopped Basalt Fibres—An Overview. *Constr. Build. Mater.* **2021**, *266*, 120970. [CrossRef]
17. Quan, X.; Wang, S.; Liu, K.; Xu, J.; Zhang, K.; Zhao, N.; Li, B. Influence of Iron Ore Tailings By-Product on the Mechanical and Electrical Properties of Carbon Fiber Reinforced Cement-Based Composites. *J. Build. Eng.* **2022**, *45*, 103567. [CrossRef]
18. Zhao, K.; Lai, Y.; He, Z.; Liu, W.; Zhao, R.; Wang, Y.; Tian, X.; Nie, J. Study on Energy Dissipation and Acoustic Emission Characteristics of Fiber Tailings Cemented Backfill with Different Ash-Sand Ratios. *Process Saf. Environ. Prot.* **2023**, *174*, 983–996. [CrossRef]
19. High, C.; Seliem, H.M.; El-Safty, A.; Rizkalla, S.H. Use of Basalt Fibers for Concrete Structures. *Constr. Build. Mater.* **2015**, *96*, 37–46. [CrossRef]
20. Zhang, C.; Wang, Y.; Zhang, X.; Ding, Y.; Xu, P. Mechanical Properties and Microstructure of Basalt Fiber-Reinforced Recycled Concrete. *J. Clean. Prod.* **2021**, *278*, 123252. [CrossRef]
21. Chinchillas-Chinchillas, M.J.; Orozco-Carmona, V.M.; Gaxiola, A.; Alvarado-Beltrán, C.G.; Pellegrini-Cervantes, M.J.; Baldenebro-López, F.J.; Castro-Beltrán, A. Evaluation of the Mechanical Properties, Durability and Drying Shrinkage of the Mortar Reinforced with Polyacrylonitrile Microfibers. *Constr. Build. Mater.* **2019**, *210*, 32–39. [CrossRef]
22. Pakravan, H.R.; Latifi, M.; Jamshidi, M. Hybrid Short Fiber Reinforcement System in Concrete: A Review. *Constr. Build. Mater.* **2017**, *142*, 280–294. [CrossRef]
23. Banthia, N.; Majdzadeh, F.; Wu, J.; Bindiganavile, V. Fiber Synergy in Hybrid Fiber Reinforced Concrete (HyFRC) in Flexure and Direct Shear. *Cem. Concr. Compos.* **2014**, *48*, 91–97. [CrossRef]
24. Hay, R.; Ostertag, C.P. Influence of Transverse Cracks and Interfacial Damage on Corrosion of Steel in Concrete with and without Fiber Hybridization. *Corros. Sci.* **2019**, *153*, 213–224. [CrossRef]
25. Walton, P.L.; Majumdar, A.J. Cement-Based Composites with Mixtures of Different Types of Fibres. *Compos. Commun.* **1975**, *6*, 209–216. [CrossRef]
26. Wang, D.; Ju, Y.; Shen, H.; Xu, L. Mechanical Properties of High Performance Concrete Reinforced with Basalt Fiber and Polypropylene Fiber. *Constr. Build. Mater.* **2019**, *197*, 464–473. [CrossRef]
27. de Sá, F.R.G.; Silva, F.d.A.; Cardoso, D.C.T. Tensile and Flexural Performance of Concrete Members Reinforced with Polypropylene Fibers and GFRP Bars. *Compos. Struct.* **2020**, *253*, 112784. [CrossRef]
28. Branston, J.; Das, S.; Kenno, S.Y.; Taylor, C. Mechanical Behaviour of Basalt Fibre Reinforced Concrete. *Constr. Build. Mater.* **2016**, *124*, 878–886. [CrossRef]

29. Kizilkanat, A.B.; Kabay, N.; Akyüncü, V.; Chowdhury, S.; Akça, A.H. Mechanical Properties and Fracture Behavior of Basalt and Glass Fiber Reinforced Concrete: An Experimental Study. *Constr. Build. Mater.* **2015**, *100*, 218–224. [CrossRef]
30. Fiore, V.; Scalici, T.; Di Bella, G.; Valenza, A. A Review on Basalt Fibre and Its Composites. *Compos. B. Eng.* **2015**, *74*, 74–94. [CrossRef]
31. Wang, Y.; Kang, A.-H.; Wu, Z.-G.; Xiao, P.; Gong, Y.-F.; Sun, H.-F. Investigation of the Basalt Fiber Type and Content on Performances of Cement Mortar and Concrete. *Constr. Build. Mater.* **2023**, *408*, 133720. [CrossRef]
32. Zeng, Y.; Li, X.; Tang, A.; Zhao, N.; Li, L.; Jia, S. Axial Compressive Behavior of Basalt and Polyacrylonitrile Fibers Reinforced Lightweight Aggregate Concrete with Industrial Waste Ceramsite-Lytag after Freeze-Thaw Cycles. *J. Build. Eng.* **2023**, *76*, 107402. [CrossRef]
33. Li, Z.; Dong, J.; Chen, H.; Wu, Z.; Feng, K.; Zhang, G.; Cheng, S.; Jiang, T. Mechanical Behaviour and Acoustic Emission Characteristics of Basalt Fibre Mortar Rubble under Uniaxial Cyclic Compression. *Constr. Build. Mater.* **2023**, *393*, 132145. [CrossRef]
34. Guler, S.; Akbulut, Z.F. Workability, Physical & Mechanical Properties of the Cement Mortars Strengthened with Metakaolin and Steel/Basalt Fibers Exposed to Freezing-Thawing Periods. *Constr. Build. Mater.* **2023**, *394*, 132100.
35. Zeng, Y.; Tang, A. Comparison of Effects of Basalt and Polyacrylonitrile Fibers on Toughness Behaviors of Lightweight Aggregate Concrete. *Constr. Build. Mater.* **2021**, *282*, 122572. [CrossRef]
36. Liu, J.; Xiao, S.; Shen, Z.; Xu, L.; Zhang, L.; Peng, J. Study on the Oxidative Stabilization of Polyacrylonitrile Fibers by Microwave Heating. *Polym. Degrad. Stab.* **2018**, *150*, 86–91. [CrossRef]
37. Fan, S. Mechanical and Durability Performance of Polyacrylonitrile Fiber Reinforced Concrete. *Mater. Res.* **2015**, *18*, 1298–1303. [CrossRef]
38. Zheng, D.; Song, W.; Fu, J.; Xue, G.; Li, J.; Cao, S. Research on Mechanical Characteristics, Fractal Dimension and Internal Structure of Fiber Reinforced Concrete under Uniaxial Compression. *Constr. Build. Mater.* **2020**, *258*, 120351. [CrossRef]
39. Zeng, Z.; Li, C.; Chen, Z.; Ke, L. Study on Mechanical Properties and Optimum Fiber Content for Basalt/Polyacrylonitrile Hybrid Fiber Reinforced Concrete. *Adv. Mater. Sci. Eng.* **2022**, *2022*, 11. [CrossRef]
40. *GB 175-2007*; Standardization Administration of the People's Republic of China, Common Portland Cement. Standards Press of China: Beijing, China, 2007.
41. *GB/T 17671-2021*; Standardization Administration of the People's Republic of China, Test Method of Cement Mortar Strength (ISO Method). Standards Press of China: Beijing, China, 2021.
42. *GB/T 14684-2022*; Standardization Administration of the People's Republic of China, Sand for Construction. Standards Press of China: Beijing, China, 2022.
43. *JTG 3420-2020*; Ministry of Transport of the People's Republic of China, Testing Methods of Cement and Concrete for Highway Engineering. China Communication Press: Beijing, China, 2020.
44. *ASTM C597-09 (2009)*; American Society for Testing and Materials, Standard Test Method for Ultrasonic Pulse Velocity through Concrete. ASTM International: West Conshohocken, PA, USA, 2009.
45. Menta, P.K. Stability of Ettringite on Heating. *J. Am. Ceram. Soc.* **1972**, *55*, 55–57.
46. Shahjalal, M.; Islam, K.; Rahman, J.; Ahmed, K.S.; Karim, M.R.; Billah, A.M. Flexural Response of Fiber Reinforced Concrete Beams with Waste Tires Rubber and Recycled Aggregate. *J. Clean. Prod.* **2021**, *278*, 123842. [CrossRef]
47. Yang, L.; Jia, H.; Wu, A.; Jiao, H.; Chen, X.; Kou, Y.; Dong, M. Particle Aggregation and Breakage Kinetics in Cemented Paste Backfill. *Int. J. Miner. Metall. Mater.* 2023, *accepted*. [CrossRef]
48. Yang, L.; Jia, H.; Jiao, H.; Dong, M.; Yang, T. The Mechanism of Viscosity-Enhancing Admixture in Backfill Slurry and the Evolution of Its Rheological Properties. *Minerals* **2023**, *13*, 1045. [CrossRef]
49. Li, L.G.; Ouyang, Y.; Ng, P.-L.; Zeng, K.; Kwan, A.K.H. Influences of Fiber Length and Water Film Thickness on Fresh Properties of Basalt Fiber-Reinforced Mortar. *J. Zhejiang Univ.-Sci. A* **2021**, *22*, 344–356. [CrossRef]
50. Hossain, F.M.Z.; Pal, A.; Ahmed, K.S.; Bediwy, A.; Alam, M.S. Shear Behavior of Polypropylene Fiber-Reinforced Concrete Beams Containing Recycled Aggregate and Crumb Rubber. *J. Clean. Prod.* **2023**, *412*, 137370. [CrossRef]
51. Saloni, A.; Parveen, J.; Pham, T.M. Enhanced Properties of High-Silica Rice Husk Ash-Based Geopolymer Paste by Incorporating Basalt Fibers. *Constr. Build. Mater.* **2020**, *245*, 118422. [CrossRef]
52. Li, L.G.; Zhuo, H.X.; Zhu, J.; Kwan, A.K.H. Packing Density of Mortar Containing Polypropylene, Carbon or Basalt Fibres under Dry and Wet Conditions. *Powder Technol.* **2019**, *342*, 433–440. [CrossRef]
53. Li, L.; Zhao, Z.; Zhu, J.; Kwan, A.K.H.; Zeng, K. Combined Effects of Water Film Thickness and Polypropylene Fibre Length on Fresh Properties of Mortar. *Constr. Build. Mater.* **2018**, *174*, 586–593. [CrossRef]
54. Hwang, J.P.; Kim, M.; Ann, K.Y. Porosity Generation Arising from Steel Fibre in Concrete. *Constr. Build. Mater.* **2015**, *94*, 433–436. [CrossRef]
55. Zhang, Y.; Zheng, Y.; Du, C.; Hu, S.; Wang, Z. Hybrid Effects of Basalt and Polyvinyl Alcohol Fibers on the Mechanical Properties and Macro-Microscopic Analysis of Low-Heat Portland Cement Concrete. *J. Mater. Res. Technol.* **2023**, *25*, 608–632. [CrossRef]
56. Benaicha, M.; Jalbaud, O.; Hafidi Alaoui, A.; Burtschell, Y. Correlation between the Mechanical Behavior and the Ultrasonic Velocity of Fiber-Reinforced Concrete. *Constr. Build. Mater.* **2015**, *101*, 702–709. [CrossRef]
57. Ridengaoqier, E.; Hatanaka, S.; Palamy, P.; Kurita, S. Experimental Study on the Porosity Evaluation of Pervious Concrete by Using Ultrasonic Wave Testing on Surfaces. *Constr. Build. Mater.* **2021**, *300*, 123959.

58. Zhang, P.; Wang, K.; Wang, J.; Guo, J.; Ling, Y. Macroscopic and Microscopic Analyses on Mechanical Performance of Metakaolin/Fly Ash Based Geopolymer Mortar. *J. Clean. Prod.* **2021**, *294*, 126193. [CrossRef]
59. Miah, M.J.; Li, Y.; Paul, S.C.; Babafemi, A.J.; Jang, J.G. Mechanical Strength, Shrinkage, and Porosity of Mortar Reinforced with Areca Nut Husk Fibers. *Constr. Build. Mater.* **2023**, *363*, 129688. [CrossRef]
60. Li, K.F.; Yang, C.Q.; Huang, W.; Zhao, Y.B.; Wang, Y.; Pan, Y.; Xu, F. Effects of Hybrid Fibers on Workability, Mechanical, and Time-Dependent Properties of High Strength Fiber-Reinforced Self-Consolidating Concrete. *Constr. Build. Mater.* **2021**, *277*, 122325. [CrossRef]
61. Das, C.S.; Dey, T.; Dandapat, R.; Mukharjee, B.B.; Kumar, J. Performance Evaluation of Polypropylene Fibre Reinforced Recycled Aggregate Concrete. *Constr. Build. Mater.* **2018**, *189*, 649–659. [CrossRef]
62. Niu, D.; Su, L.; Luo, Y.; Huang, D.; Luo, D. Experimental Study on Mechanical Properties and Durability of Basalt Fiber Reinforced Coral Aggregate Concrete. *Constr. Build. Mater.* **2020**, *237*, 117628. [CrossRef]
63. Duan, M.; Qin, Y.; Li, Y.; Wei, Y.; Geng, K.; Zhou, H.; Liu, R. Mechanical Properties and Multi-Layer Perceptron Neural Networks of Polyacrylonitrile Fiber Reinforced Concrete Cured Outdoors. *Structures* **2023**, *56*, 104954. [CrossRef]
64. Sadrinejad, I.; Madandoust, R.; Ranjbar, M.M. The Mechanical and Durability Properties of Concrete Containing Hybrid Synthetic Fibers. *Constr. Build. Mater.* **2018**, *178*, 72–82. [CrossRef]
65. Teng, S.; Afroughsabet, V.; Ostertag, C.P. Flexural Behavior and Durability Properties of High Performance Hybrid-Fiber-Reinforced Concrete. *Constr. Build. Mater.* **2018**, *182*, 504–515. [CrossRef]
66. Nazir, K.; Canpolat, O.; Uysal, M.; Niş, A.; Kuranlı, Ö.F. Engineering Properties of Different Fiber-Reinforced Metakaolin-Red Mud Based Geopolymer Mortars. *Constr. Build. Mater.* **2023**, *385*, 131496. [CrossRef]
67. Zheng, W.; Wang, S.; Quan, X.; Qu, Y.; Mo, Z.; Lin, C. Carbonation Resistance and Pore Structure of Mixed-Fiber-Reinforced Concrete Containing Fine Aggregates of Iron Ore Tailings. *Materials* **2022**, *15*, 8992. [CrossRef]
68. Zheng, Y.; Zhuo, J.; Zhang, P.; Ma, M. Mechanical Properties and Meso-Microscopic Mechanism of Basalt Fiber-Reinforced Recycled Aggregate Concrete. *J. Clean. Prod.* **2022**, *370*, 133555. [CrossRef]
69. *NZS 3101*; Concrete Structures Standard: Part 1: The Design of Concrete Structures. Standards New Zealand: Wellington, New Zealand, 2006.
70. *EC-04. B.S.*; Eurocode 2: Design of Concrete Structures: Part 1-1: General Rules and Rules for Buildings. European Committee for Standardization: Brussels, Belgium, 2004.
71. *GB 50010-2010*; Chinese Standard: Code for Design of Concrete Structures. China Building Press: Beijing, China, 2010.
72. *DG/TJ-08*; Technical Code for Application of Recycled Aggregate Concrete. Shanghai Construction Standard Society (SCSS): Shanghai, China, 2017.
73. *JSCE, 2007*; Standard Specification for Concrete Structure. Japan Society of Civil Engineers: Tokyo, Japan, 2007.
74. *IS 456*; Plain and Reinforced Concrete Code of Practice. Bureau of Indian Standards Indian Standard: New Delhi, India, 2000.
75. *ACI 318-11*; Concrete Institute & International Organization for Standardization, Building Code Requirements for Structural Concrete and Commentary. American Concrete Institute: Farmington Hills, MI, USA, 2011.
76. Cui, K.; Xu, L.; Tao, T.; Huang, L.; Li, J.; Hong, J.; Li, H.; Chi, Y. Mechanical Behavior of Multiscale Hybrid Fiber Reinforced Recycled Aggregate Concrete Subject to Uniaxial Compression. *J. Build. Eng.* **2023**, *71*, 106504. [CrossRef]
77. Hong, L.; Chen, Y.D.; Li, T.D.; Gao, P.; Sun, L.Z. Microstructure and Bonding Behavior of Fiber-Mortar Interface in Fiber-Reinforced Concrete. *Constr. Build. Mater.* **2020**, *232*, 117235. [CrossRef]
78. Liang, N.; Geng, S.; Mao, J.; Liu, X.; Zhou, X. Investigation on Cracking Resistance Mechanism of Basalt-Polypropylene Fiber Reinforced Concrete Based on SEM Test. *Constr. Build. Mater.* **2024**, *411*, 134102. [CrossRef]
79. Chen, X.; Shi, X.; Zhou, J.; Chen, Q.; Li, E.; Du, X. Compressive Behavior and Microstructural Properties of Tailings Polypropylene Fibre-Reinforced Cemented Paste Backfill. *Constr. Build. Mater.* **2018**, *190*, 211–221. [CrossRef]

Disclaimer/Publisher's Note: The statements, opinions and data contained in all publications are solely those of the individual author(s) and contributor(s) and not of MDPI and/or the editor(s). MDPI and/or the editor(s) disclaim responsibility for any injury to people or property resulting from any ideas, methods, instructions or products referred to in the content.

buildings

MDPI

Article

Investigation of the Disintegration Characteristics of Neogene Mudstone at Different Burial Depths

Liang Peng [1,2], Wenxue Du [1,2], Ganggang Bai [1,2], Lahuancairang [1,2], Shixiang Yuan [1,2], Juntao Li [3] and Peng Feng [3,4,*]

[1] Qinghai Hydrogeology and Engineering Geology and Environgeology Survey Institute, Xining 810008, China
[2] Hydrogeological and Geothermal Geological Key Laboratory of Qinghai Province, Xining 810008, China
[3] School of Architecture and Civil Engineering, Chengdu University, Chengdu 610106, China
[4] Key Laboratory of Hydraulic and Waterway Engineering of the Ministry of Education, Chongqing Jiaotong University, Chongqing 400074, China
[*] Correspondence: fengpengfx@163.com

Abstract: The complex structure of Neogene mudstone plays an important role in geological disasters. A close relationship exists between the mechanisms of mudstone landslides and the disintegration characteristics of rocks. Therefore, understanding the disintegration characteristics of Neogene mudstone at different depths is crucial for enhancing engineering safety and assessing landslide stability. This study employed Neogene mudstone from different depths to perform disintegration and plastic limit experiments and revealed the sliding mechanisms of landslides involving Neogene mudstone, providing theoretical support for mitigating mudstone geological disasters. Our results demonstrate that Neogene mudstone from different depths experiences varied stress conditions and pore water pressure due to geological actions, significantly affecting the disintegration characteristics. By ignoring the factors of the slip surface, the slake durability index of mudstone decreases with increasing burial depth, while the plasticity limit index tends to rise. The influence of groundwater, geo-stress, and pore structure on Neogene mudstones at different depths results in overall weak stability and disintegration. Landslide occurrences are likely connected to the mechanical properties of mudstones at the slip surface, where a low slake durability index and higher plasticity index make the mudstones prone to fracturing, breaking, and disintegrating once in contact with water.

Keywords: Neogene mudstone; disintegration characteristics; mudstone landslides; slake durability index

1. Introduction

Neogene mudstone, frequently encountered in geotechnical engineering, has complex physical characteristics under the influence of multiple factors [1–3]. The plasticity and deformability of Neogene mudstones generally play a crucial role in geological disasters such as landslides, accompanying disintegration and weakening the strength of mudstone [4–7]. In addition, the burial depth also affects the disintegration of Neogene mudstone and the progression of landslides [8,9], with varying stress states, rock transformations, and geological effects at distinct levels [10,11]. Thus, a thorough investigation on the disintegration characteristics of Neogene mudstones at various burial depths is essential to better our comprehension of mudstone landslide mechanisms and the recognition of potential landslide hazards [12–15]. Researchers have investigated the disintegration characteristics of mudstone under varying conditions [16–19]. Luo et al. [20] conducted systematic research on the distribution and pore structure of mudstone, demonstrating a close relationship between the disintegration resistance of mudstone and its mineral composition. Huang et al. [21] performed disintegration tests on rock specimens, investigating their disintegration characteristics under dry–wet cycling conditions. Moreover, comparative studies on the disintegration characteristics of mudstone blocks in indoor environments, acidic conditions, and natural climates have been carried out [22,23]. In addition, Shen et al. [24] explored the morphological and fractal features of mudstone during the disintegration process.

Indeed, after disintegration, mudstones form powdery soil material, with the plasticity index acting as a crucial metric reflecting the viscosity and plastic behavior of soil material [25,26]. Shimobe et al. [27] examined the correlation between undrained shear strength, liquidity index, and moisture content for both remolded and undisturbed soils. Through experimental studies, no significant patterns were shown to exist between the temperature change and variations in liquid and plastic limits [28]. During liquid limit tests on mixed clays containing structured crystals and amorphous kaolinite, Stanchi [29] and Jefferson [30] showed a minimal temperature influence on liquid and plastic limits. Swaify et al. [31] focused their analysis on the effects of liquid limit, degree of dispersion, moisture retention, and saturated cations on the geomaterial properties. Davidson et al. [32] compared remolded and undisturbed soils, attributing differences in liquid and plastic limits to sample packaging, soil moisture, and remolding degree. Stanchi et al. [33] conducted a regression analysis on the impact of organic matter and carbon molecules on the liquid and plastic limits of soil, affirming that the cohesion of particles with diameters smaller than 0.425 mm influences the liquid and plastic limit.

At the microstructural scale, Tang et al. [34] investigated the microstructural damage of rock specimens under uniaxial compression based on scanning electron microscopy experiments, offering detailed analyses of the evolution and development characteristics of microcracks. Howarth et al. [35] examined the shapes of microscopic particles using rock slice imagery, while Campbell et al. [36] conducted a quantitative investigation on rock microstructure utilizing optical microscopy. In addition, based on the experiment on Neogene mudstone, the microstructural characteristics and softening mechanisms of soft rocks under various conditions have been explored [37–39]. The mineral composition and geological features of mudstone showed a significant influence on its resistance to disintegration, with weathering, diagenesis, hydrothermal activities, and geological transportation serving as the key factors affecting the engineering properties of mudstones [40]. Zhang et al. [41] analyzed the microstructural alterations in mudstone after rainfall and discovered the loosening mineral microstructure of mudstone and increasing porosity, singling out illite as a crucial agent of internal binding expansion and structural impairment.

So far, studies on the disintegration characteristics of Neogene mudstones at varying depths and inquiries into their liquid and plastic limits after disintegration have been relatively scant. Based on laboratory experiments and statistical analyses, this manuscript investigates the disintegration characteristics of Neogene mudstones from different depths and their consequent effects on mudstone landslide stability. After introduction, Section 2 introduces the specimen preparation and methodology utilized in this manuscript. Then, Sections 3 and 4 show the effects of burial depth on the disintegration characteristics and slake durability index of Neogene mudstones at macro and micro levels, respectively. The discussion of the formation of the Zhangjiawan landslide is established in Section 5. Section 6 summarizes the whole study.

2. Specimen Preparation and Methodology

2.1. Specimen Preparation

The Neogene mudstones utilized in this study were captured from the Zhangjiawan landslide with a burial depth ranging from 13 m to 17 m, as shown in Figure 1. Initially, we inserted drill rods to the designated depth and employed flush mediums such as mud or air to evacuate borehole cuttings, cool the drill bit, and sustain the integrity of the borehole walls. Subsequently, the rock cores were collected at predetermined depth intervals, and the retrieved cores were extracted from the barrels, followed by cataloging and tagging. We conserved the integrity of the rock cores by placing them into appropriate protective casings for further investigation. Cylindrical specimens with dimensions of $\Phi 50$ mm $\times$ 50 mm and a mass of 240 $\pm$ 20 g were prepared for the disintegration experiments. Each specimen can be labeled as S-D, where S stands for Neogene mudstone, and D represents the depth of mudstone. For instance, S-13 denotes Neogene mudstone at a depth of 13 m. Table 1 details the physical parameters of these samples.

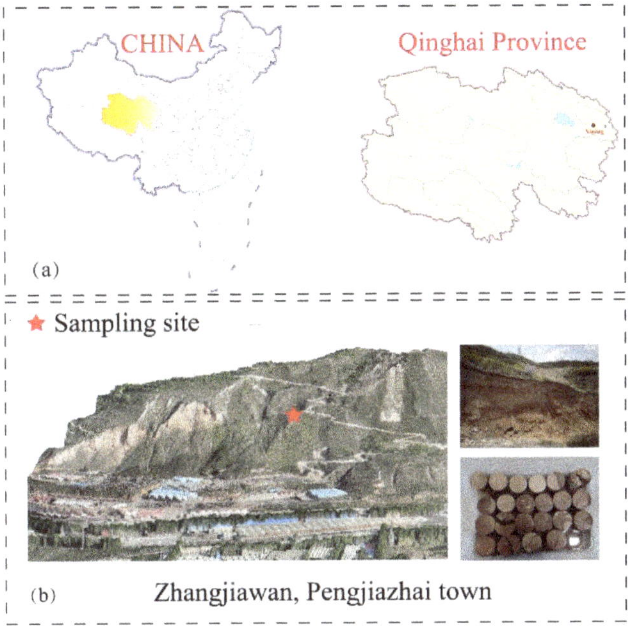

Figure 1. Preparation of Neogene mudstone specimens from Zhangjiawan landslide: (**a**) Qinghai province of China, (**b**) Sampling site.

Table 1. The geometrical parameters of Neogene mudstone at different depths.

Notation	Nature Weight (g)	Dry Weight (g)	High (cm)	Diameter (cm)	Density (g/cm^3)
S-13.0	248.68	221.70	5.420	5.037	2.30
S-13.2	240.54	216.62	5.163	5.056	2.32
S-13.4	226.29	202.95	4.895	5.044	2.31
S-13.7	230.44	204.60	5.129	5.011	2.28
S-14.0	229.79	203.87	5.039	5.015	2.31
S-14.3	230.52	205.59	5.076	5.041	2.28
S-14.6	231.09	205.88	5.106	5.043	2.27
S-16.7	227.25	192.09	5.123	5.010	2.25
S-17.0	251.95	222.88	5.532	5.045	2.28
S-17.3	242.89	215.29	5.298	5.046	2.29
S-17.5	229.69	205.28	4.985	5.047	2.30
S-18.0	223.75	199.27	5.114	5.000	2.23
S-19.0	231.71	206.62	5.136	4.916	2.38
S-19.3	244.08	219.46	5.318	5.043	2.30
S-19.6	237.85	213.14	5.140	5.045	2.31

An X-ray fluorescence spectrometer was employed to perform X-ray fluorescence spectroscopy on Neogene mudstone [42]. The principal chemical components of Neogene mudstone predominantly consist of SiO_2 and $CaCO_3$, accompanied by trace impurities, as illustrated in Figure 2.

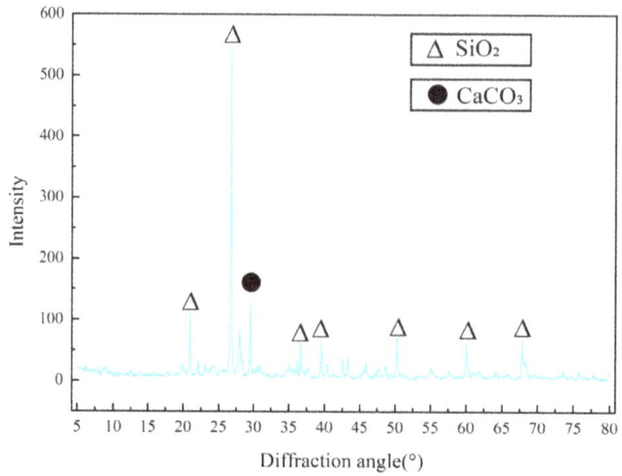

Figure 2. X-ray diffraction pattern of Neogene mudstone.

2.2. Methodology

2.2.1. Slake Durability Test

The slake durability index reflects the capability of a specimen to resist softening and disintegration, generally measured via slake durability tests. As shown in Figure 3d, the slake durability apparatus XYN-1 was adopted to determine the slake durability index [43,44]. The detailed experimental steps include: (1) Desiccating the samples at 100 °C for 24 h within an oven and then removing fragmented rocks and cooling the intact ones to ambient temperature. (2) Filling the trough with clear water, ensuring the water lies 20 mm beneath the rotary axis of the sieve drum. (3) After initiating the disintegration equipment, rotating at a velocity of 20 revolutions per minute for 10 min, and then drying the withdrawn sieve drum at 100 °C for another 24 h. (4) Weighing and recording the total mass of the rocks and the sieve drum. Following disintegration, sieving out and cataloging the residual fragments within the sieve drum, and thereafter reloading fragments with a particle size exceeding 2 mm back into the sieve drum. (5) Repeating the aforementioned steps until the sample exhibits consistent disintegration. The slake durability index can be calculated using Equation (1).

$$I_{dn} = \frac{m_n}{m_d} \times 100\% \quad (1)$$

where I_{dn} denotes the slake durability index after n standard cycles; m_n represents the initial dry mass of the sample; and m_d signifies the dry mass of the residual sample following the n standard cycle.

2.2.2. Plasticity Index Test

The plasticity index quantifies the difference between the liquid limit and the plastic limit of specimen under standard testing conditions, serving as a crucial parameter for characterizing the plastic deformation behavior of a specimen [45,46].

Determining the plasticity index of disintegrated mudstone can elucidate the extent and capability of its plastic deformation under stress. Plastic limit determination apparatus was employed to measure the liquid and plastic limits of the rock specimen, as shown in Figure 3e. The testing cone weighs 100 g and features an apex angle of 30 degrees. Initially, the dried specimens are passed through a 0.5 mm geotechnical sieve, the watered specimens are mixed thoroughly, and then they are sealed for over 18 h. Subsequently, the extracted samples are homogenized and placed into soil cups. The penetration depths of the samples are controlled below 5 mm, around 10 mm, and within 20 ± 0.2 mm to measure the plastic limit, intermediate value, and liquid limit, respectively. Once the penetration depth is achieved, the test point is reposited for a new measurement immediately, ensuring that the test points are at least 1 cm apart with an error margin of 0.5 mm. The average penetration depth is calculated at that point, suitable samples meeting the aforementioned penetration depth requirements are extracted, and then the moisture content is determined. Finally, based on the aforementioned value, the plasticity index of the specimen can be acquired in Equation (2).

$$P_I = L_L - P_L \tag{2}$$

where L_L and P_L represent the liquid limit and the plastic limit of the specimen; P_I represents the value of the plasticity index.

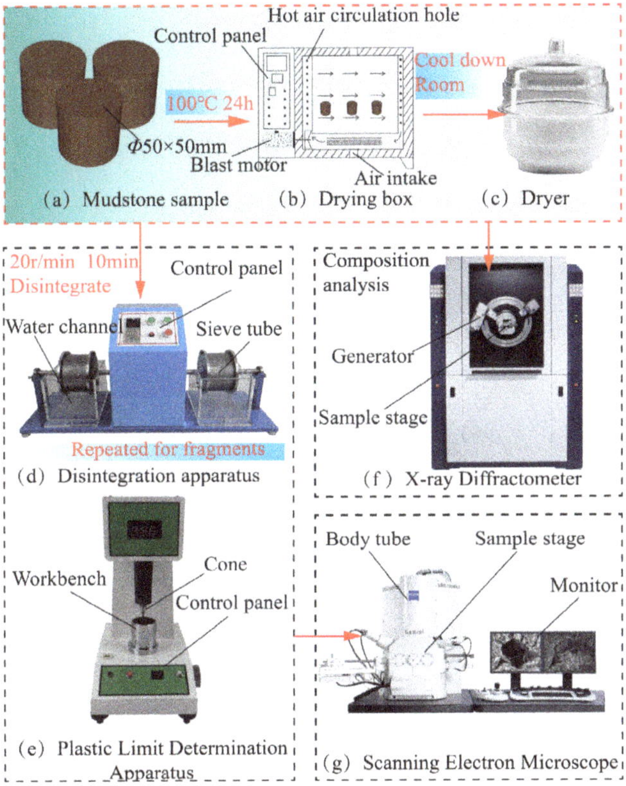

Figure 3. Experimental apparatus and experimental process for Neogene mudstone: (**a**) mudstone specimen, (**b**) dying box, (**c**) dryer, (**d**) disintegration apparatus, (**e**) plastic limit determination apparatus, (**f**) X-ray diffractometer, and (**g**) scanning electron microscope.

2.2.3. Scanning Electron Microscope System

The scanning electron microscope (SEM) system utilized in this study primarily consists of a vacuum system, an electron beam system, and an imaging system, as shown in Figure 3g. The electron beam system releases high-energy incident electrons that impact the surface of the material, directing a focused ultrafine beam onto the micro-observational surface and detecting the secondary electrons and backscattered electron signals induced by the electron beam system. Subsequently, signal processing enables the micro-observational surface of the specimen to be imaged [47,48]. This study employs SEM to perform the micro-scale scanning of Neogene mudstone, reaching a magnification of up to 10^6 times, offering an acceleration electric field for the projected electrons of up to 3×10^4 volts, and covering an effective scanning area of 10^{-2} mm for the specimens.

3. Results

3.1. Effects of Burial Depth on the Slake Durability Index of Neogene Mudstones

Figure 4 illustrates the slake durability index distribution of Neogene mudstone across various depths, with a noticeably higher value for depths ranging from 13 to 13.4 m; details are outlined in Table 2. Nevertheless, the slake durability index shows the lowest value at depths of 13.7 to 14 m. Overall, the slake durability index of Neogene mudstone trends downward with increasing depth. As the burial depth of mudstone escalates, groundwater leads to the absorption expansion and dissolution of Neogene mudstone. Simultaneously, Neogene mudstone experiences increasing permeation pressure and geo-stress, resulting in the alteration of its pore structure and promoting internal dissolution and fracturing, which in turn affects the slake durability index. The region between 13.7- and 14-m depth served as the sampling location for the landslide slip zone, which will undergo further discussion in Section 5.

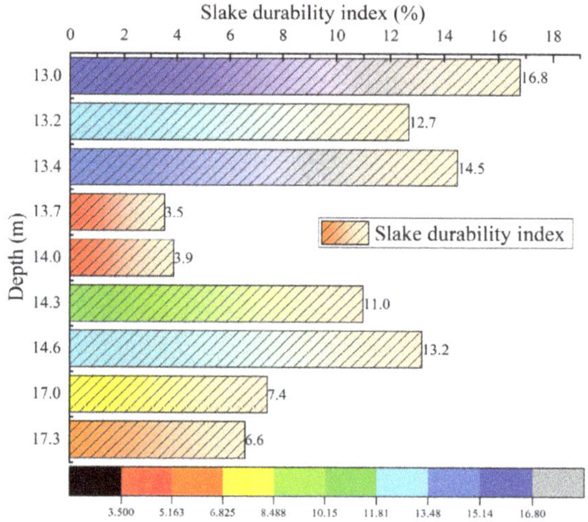

Figure 4. Slake durability index of Neogene mudstone under different depths; different color represents different value of the Slake durability index.

Table 2. The slake durability index of Neogene mudstones at different depths.

Notation	Sampling Site (m)	Slake Durability Index (%)
S-13.0	13.0–13.2	16.8
S-13.2	13.2–13.4	12.7
S-13.4	13.4–13.7	14.5
S-13.7	13.7–14.0	3.54
S-14.0	14.0–14.3	3.89
S-14.3	14.3–14.6	11.0
S-14.6	14.6–14.9	13.2
S-17.0	17.0–17.3	7.44
S-17.3	17.3–17.5	6.59

3.2. Effects of Burial Depth on the Plasticity Index of Neogene Mudstones

Figures 5 and 6 illustrate the liquid limits and plastic limits of Neogene mudstone at varied depths, showing notably lower values at depths ranging from 13 to 13.4 m, with the peaking value being between 13.7 and 14 m. In general, the higher liquid limits and plastic limits of Neogene mudstone can be characterized via the greater depth; details are shown in Table 3.

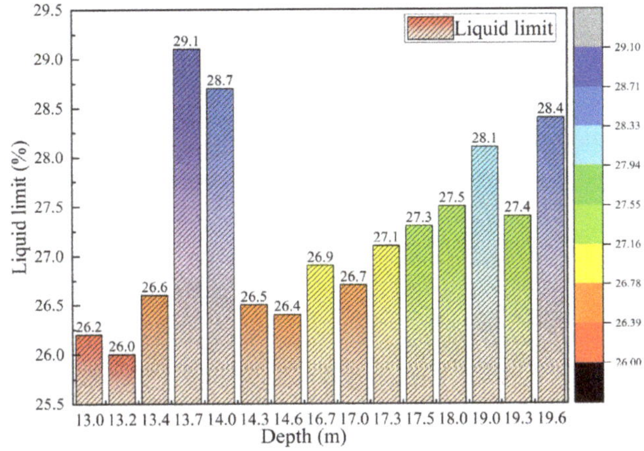

Figure 5. Effects of depth on the liquid limit of Neogene mudstone; different color represents different value of the liquid limit.

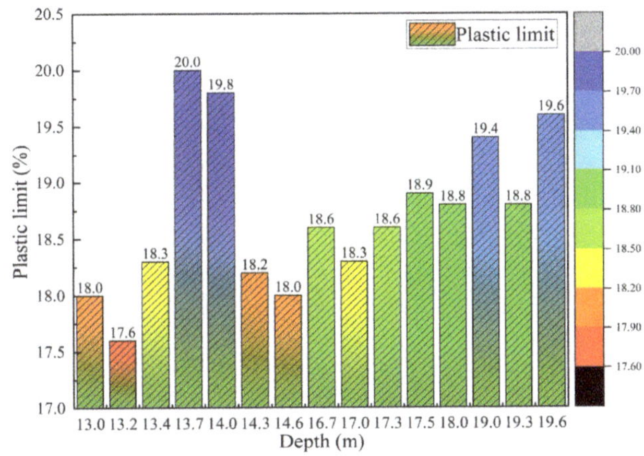

Figure 6. Effects of depth on the plastic limit of Neogene mudstone; different color represents different value of the plastic limit.

Table 3. The moisture content, liquid limit, plastic limit, and plasticity index of Neogene mudstone with different depths.

Notation	Moisture Content (%)	Liquid Limit (%)	Plastic Limit (%)	Plasticity Index (%)
S-13.0	3.4	29.1	20.0	9.1
S-13.2	0.9	26.1	17.8	8.3
S-13.4	3.2	26.7	19.2	7.5
S-13.7	2.8	26.3	17.4	8.9
S-14.0	1.2	27.9	19.8	8.1
S-14.3	3.6	25.4	18.1	7.3
S-14.6	2.1	24.9	17.9	7.0
S-16.7	1.1	26.5	18.7	7.8
S-17.0	1.6	26.9	19.3	7.6
S-17.3	2.5	25.8	17.6	8.2
S-17.5	3.9	25.7	17.8	7.9

Table 3. *Cont.*

Notation	Moisture Content (%)	Liquid Limit (%)	Plastic Limit (%)	Plasticity Index (%)
S-18.0	3.2	26.5	19.0	7.5
S-19.0	1.3	27.2	19.1	8.1
S-19.3	3.6	26.6	18.9	7.7
S-19.6	3.0	26.5	18.5	8.0

Figure 7 details the distribution of the plasticity index of Neogene mudstone at different depths, revealing a lower value from 13 to 13.4 m, and the highest index value was recorded between 13.7 and 14 m. Collectively, the plasticity index of Neogene mudstone shows an increasing trend with depth. Such results may be explicated by the diverse particle composition and structure at different mudstone depths, affecting the frictional interactions, bond strength, and deformation characteristics of Neogene mudstone, consequently impacting the plasticity index. As a matter of fact, elevated groundwater pressure and geo-stress at various depths can alter the pore structures of Neogene mudstone, leading to a higher water content in deeper strata, and thus also elevating the plasticity index.

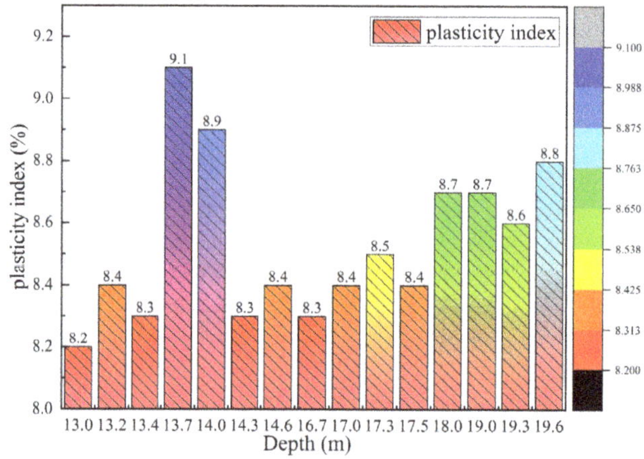

Figure 7. Effects of depth on the plasticity index of Neogene mudstone; different color represents different value of the plasticity index.

4. Microstructure Analysis

Figure 8 presents the SEM images of a typical Neogene mudstone sample after disintegration, revealing its microstructural characteristics. It can be observed that there is a rich content of hydrated clay minerals surrounding non-clay particles, creating a densely and unevenly agglomerated structure. The interior of these clay aggregates showcases a heterogeneous distribution of pores with significant size disparities. Large pores and a developed fractures segment fragment the cluster structure into several smaller sub-aggregates. Additionally, the clay minerals in the rock sample manifest as large particles composed of finer clay materials, intricately stacked, intertwined, and bonded with each other.

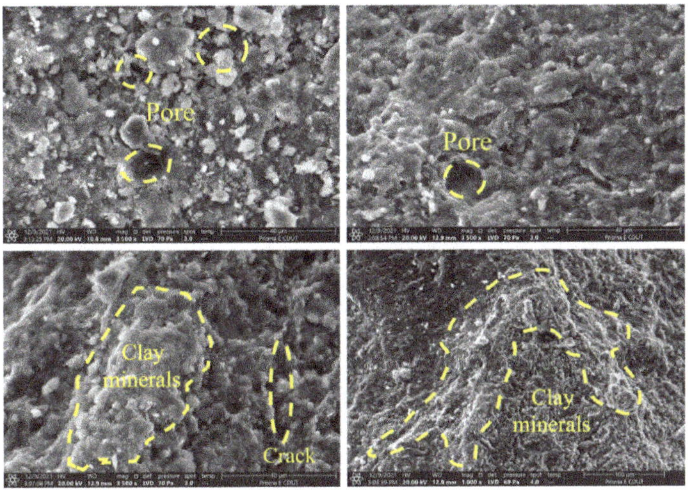

Figure 8. Typical SEM image of Neogene mudstone.

5. Discussion

The specimens in the experiments were derived from the Zhangjiawan landslide, with the surface layer being composed of loess and the lower part composed of fractured mudstone characterized by substantial porosity and susceptibility to softening and mudding. A weak sliding plane was formed once the water permeated the mudstone layer and led to rapid disintegration, softening, and mudding, as depicted in Figure 9. A slip zone can be found at approximately 14 m below the ground surface, where rock specimens S-13.7 and S-14.0 exhibit lower slake durability, signifying weaker abilities to resist softening and disintegration. Concurrently, S-13.7 and S-14.0, with a considerable plasticity index, are susceptible to plastic deformation.

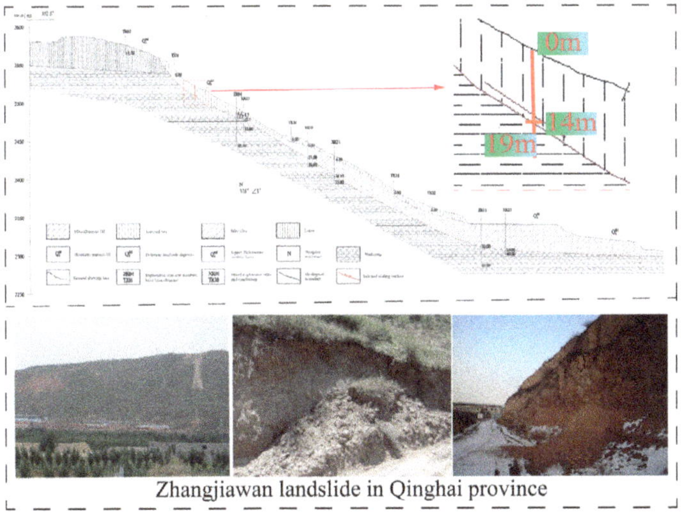

Figure 9. Longitudinal profile of the Zhangjiawan landslide and the sampling site.

In this case, mudstone in the slip zone exhibits instability, an aptitude for plastic deformation, and a tendency for disintegration. During the disintegration process, solid particles detach from the rock mass, leading to a loosened pore structure and diminishing cohesion, further escalating landslide risks. The disintegration of mudstone results in a loss of structural cohesiveness and reduction in shear strength, leading to an unstable slope condition. Subsequent to disintegration, mudstone often exhibits a rising porosity, generating larger pores and microcracks that enhance soil permeability. Mudstone in slip zones shows strong susceptibility to hygroscopic swelling, prone to expansion deformation upon moisture interaction, compromising the slope stability. Water can further enlarge and intensify existing cracks, with the potential to induce stress readjustment within the rock mass. In addition, particle displacement might also change the permeation properties of rock materials, engendering latent slip surfaces and multiplying the risk of landslides. At the micro level, the primary pore structure consists of the pores and fractures formed between coarse particles, predominantly along particle edges and interstices, which give rise to the integrity of the mudstone. With the diminishment of cementation, micro pores and fractures become finer and interconnected, leading to a loose and discontinuous microstructure in the mudstone. In this case, particle misalignment occurs frequently, the entire microstructure presents as fragile and loose, and a detrital configuration emerges. The disassociation of the coarse particle framework within this loosened microstructure signals the impending instability of the overall structure.

6. Conclusions

Neogene mudstone characterized by distinctive mechanical behavior, widely existing in a red-layer area, plays a significant role in the stability and safety of geotechnical engineering. Investigating the disintegration characteristics of Neogene mudstone at various burial depths can contribute to providing theoretical support for the deformation properties of mudstone with underground engineering. This study investigated the effects of burial depth on the disintegration characteristics of Neogene mudstone through laboratory experiments and revealed the intrinsic links between the disintegration characteristics of mudstone and landslide disasters at the sampling site. Ignoring the influence of slip surfaces, increasing burial depth correlates with a gradual decline in the slake durability index, while the plasticity index manifests an upward trend. Greater depths can amplify the effects of groundwater and geo-stress on the mechanical behaviors of mudstone, heightening the risks associated with wetting and softening. In addition, groundwater activity may lead to mudstone dissolution, along with a reduction in cohesion and shear strength, potentially resulting in plastic or compressive deformation and further diminishing the disintegration of mudstone. The porosity, morphology, and connectivity of Neogene mudstone can be influenced via increasing the burial depth, and alterations in pore structure can decrease the permeability and consequently affect the disintegration characteristics and deformation of Neogene mudstone. It should be noted that the Neogene mudstone samples utilized in this study originated from a restricted area. The laboratory conditions may not fully replicate the natural scenarios where mudstones have been subjected to over-extensive geological epochs. Thus, the mudstone specimens used in further investigations should be captured from a more extensive array of geological contexts. The complex chemical, physical, and biological processes in natural conditions can be emulated utilizing techniques such as solution immersion or cyclic temperature and humidity trials.

Author Contributions: Conceptualization and writing—original draft, L.P.; investigation and methodology, W.D.; data curation and validation, G.B. and L.; visualization and writing—reviewing and editing, S.Y. and J.L.; supervision and project administration, P.F. All authors have read and agreed to the published version of the manuscript.

Funding: The authors are grateful for the Qinghai Province Central Government Guided Fund for Local Science and Technology Development Project (2021ZY014), Qinghai Province "Kunlun Talents" Top-Tier Talent Project for High-End Innovation and Entrepreneurship (2022-KY-34), and the Open Research Subject of Key Laboratory of Hydraulic and Waterway Engineering of the Ministry of Education, Chongqing Jiaotong University (SLK2023A02).

Data Availability Statement: The original contributions presented in the study are included in the article, further inquiries can be directed to the corresponding author.

Conflicts of Interest: The authors declare no conflicts of interest.

References

1. Liu, J.; Xu, Q.; Wang, S.; Siva Subramanian, S.; Wang, L.; Qi, X. Formation and chemo-mechanical characteristics of weak clay interlayers between alternative mudstone and sandstone sequence of gently inclined landslides in Nanjiang, SW China. *Bull. Eng. Geol. Environ.* **2020**, *79*, 4701–4715. [CrossRef]
2. Sadisun, I.A.; Bandono; Shimada, H.; Ichinose, M.; Matsui, K. Physical disintegration characterization of mudrocks subjected to slaking exposure and immersion tests. *Indones. J. Geosci.* **2010**, *5*, 219–255. [CrossRef]
3. Wu, M.; Zhou, F.; Wang, J. Dem modeling of mini-triaxial test on soil-rock mixture considering particle shape effect. *Comput. Geotech.* **2023**, *153*, 105110. [CrossRef]
4. Mangano, G.; Alves, T.M.; Zecchin, M.; Civile, D.; Critelli, S. The Rossano-San Nicola Fault Zone evolution impacts the burial and maturation histories of the Crotone Basin, Calabrian Arc, Italy. *Pet. Geosci.* **2023**, *29*, 2022–2085. [CrossRef]
5. Mangano, G.; Ceramicola, S.; Alves, T.M.; Zecchin, M.; Civile, D.; Del Ben, A.; Critelli, S. A new large-scale gravitational complex discovered in the Squillace Embayment (central Mediterranean): Tectonic and geohazard implications. *Sci. Rep.* **2023**, *13*, 14695. [CrossRef] [PubMed]
6. Campilongo, G.; Campilongo, E.; Catanzariti, F.; Muto, F.; Ponte, M.; Critelli, S. Subsidence analysis by mean of DeGloT software: Application to the key-case of the Miocene-Quaternary Crotone Basin (Calabria, S. Italy). *Mar. Pet. Geol.* **2022**, *146*, 10596. [CrossRef]
7. Campilongo, G.; Ponte, M.; Muto, F.; Critelli, S.; Catanzariti, F.; Perri, F. Geotechnics, mineralogy and geochemistry of the Cutro Clay Formation, Calabria, Southern Italy. *J. Mediterr. Earth Sci.* **2022**, *14*, 11–26.
8. Li, H.; He, Y.; Xu, Q.; Deng, J.; Li, W.; Wei, Y.; Zhou, J. Sematic segmentation of loess landslides with STAPLE mask and fully connected conditional random field. *Landslides* **2023**, *20*, 367–380. [CrossRef]
9. Li, H.; He, Y.; Xu, Q.; Deng, J.; Li, W.; Wei, Y. Detection and segmentation of loess landslides via satellite images: A two-phase framework. *Landslides* **2022**, *19*, 673–686. [CrossRef]
10. Shi, J.S.; Wu, L.Z.; Qu, Y.X.; Li, B.; Wu, S.R. Neogene clay and its relation to landslides in the southwestern Loess Plateau, China. *Environ. Earth Sci.* **2018**, *77*, 1–11. [CrossRef]
11. Peng, J.; Leng, Y.; Zhu, X.; Wu, D.; Tong, X. Development of a loess-mudstone landslide in a fault fracture zone. *Environ. Earth Sci.* **2016**, *75*, 1–11. [CrossRef]
12. Kong, L.W.; Zeng, Z.X.; Bai, W.; Wang, M. Engineering geological properties of weathered swelling mudstones and their effects on the landslides occurrence in the Yanji section of the Jilin-Hunchun high-speed railway. *Bull. Eng. Geol. Environ.* **2018**, *77*, 1491–1503. [CrossRef]
13. Zhang, Z.; Liu, W.; Han, L.; Chen, X.; Cui, Q.; Yao, H.; Wang, Z. Disintegration behavior of strongly weathered purple mudstone in drawdown area of three gorges reservoir, China. *Geomorphology* **2018**, *315*, 68–79. [CrossRef]
14. Peng, J.; Tong, X.; Wang, S.; Ma, P. Three-dimensional geological structures and sliding factors and modes of loess landslides. *Environ. Earth Sci.* **2018**, *77*, 675. [CrossRef]
15. Conforti, M.; Muto, F.; Rago, V.; Critelli, S. Landslide inventory map of north-eastern Calabria (South Italy). *J. Maps* **2014**, *10*, 90–102. [CrossRef]
16. Zhang, Z.; Liu, W.; Cui, Q.; Han, L.; Yao, H. Disintegration characteristics of moderately weathered mudstone in drawdown area of Three Gorges Reservoir, China. *Arab. J. Geosci.* **2018**, *11*, 405–414. [CrossRef]
17. Su, X.; Tang, H.; Huang, L.; Shen, P.; Xia, D. The role of pH in red-stratum mudstone disintegration in the Three Gorges reservoir area, China, and the associated micromechanisms. *Eng. Geol.* **2020**, *279*, 105873. [CrossRef]
18. Feng, P.; Zhao, J.C.; Dai, F.; Wei, M.D.; Liu, B. Mechanical behaviors of conjugate-flawed rocks subjected to coupled static–dynamic compression. *Acta Geotech.* **2022**, *17*, 1765–1784. [CrossRef]
19. Feng, P.; Xu, Y.; Dai, F. Effects of dynamic strain rate on the energy dissipation and fragment characteristics of cross-fissured rocks. *Int. J. Rock Mech. Min. Sci.* **2021**, *138*, 104600. [CrossRef]
20. Luo, J.; Wu, Y.; Mi, D.; Ye, Q.; Huang, H.; Chang, Z.; Liu, X. Analysis of the distribution and microscopic characteristics and disintegration characteristics of carbonaceous rocks: A case study of the middle Devonian Luofu formation in western Guangxi of China. *Adv. Civ. Eng.* **2020**, *2020*, 8810648. [CrossRef]
21. Huang, K.; Kang, B.; Zha, F.; Li, Y.; Zhang, Q.; Chu, C. Disintegration characteristics and mechanism of red-bed argillaceous siltstone under drying–wetting cycle. *Environ. Earth Sci.* **2022**, *81*, 336. [CrossRef]
22. Gautam, T.P.; Shakoor, A. Comparing the slaking of clay-bearing rocks under laboratory conditions to slaking under natural climatic conditions. *Rock Mech. Rock Eng.* **2016**, *49*, 19–31. [CrossRef]
23. Fereidooni, D.; Khajevand, R. Utilization of the accelerated weathering test method for evaluating the durability of sedimentary rocks. *Bull. Eng. Geol. Environ.* **2019**, *78*, 2697–2716. [CrossRef]
24. Shen, P.; Tang, H.; Huang, L.; Wang, D. Experimental study of slaking properties of red-bed mudstones from the Three Gorges Reservoir area. *Mar. Georesources Geotechnol.* **2019**, *37*, 891–901. [CrossRef]

25. Sharma, B.; Bora, P.K. Plastic limit, liquid limit and undrained shear strength of soil—Reappraisal. *J. Geotech. Geoenviron. Eng.* **2003**, *129*, 774–777. [CrossRef]
26. Shakoor, A.; Gautam, T.P. Influence of geologic and index properties on disintegration behavior of clay-bearing rocks. *Environ. Eng. Geosci.* **2015**, *21*, 197–209. [CrossRef]
27. Shimobe, S.; Spagnoli, G. Relationships between undrained shear strength, liquidity index, and water content ratio of clays. *Bull. Eng. Geol. Environ.* **2020**, *79*, 4817–4828. [CrossRef]
28. Laguros, J.G. *Effect of Temperature on Some Engineering Properties of Clay Soils*; Highway Research Board: Washington, DC, USA, 1969; pp. 186–193.
29. Stanchi, S.D.; D'Amico, M.; Zanini, E.; Freppaz, M. Liquid and Plastic limits of mountain soils as a function of the soil and horizon type. *Catena* **2015**, *135*, 114–121. [CrossRef]
30. Jefferson, I.; Rogers, C.D.F. Liquid limit and the temperature sensitivity of clays. *Eng. Geol.* **1998**, *49*, 95–109. [CrossRef]
31. El Swaify, S.A.; Ahmed, S.; Swindale, L.D. Effects of adsorbed cations on physical properties of tropical red and tropical black earths: II. Liquid limit, degree of dispersion, and moisture retention. *J. Soil Sci.* **1970**, *21*, 188–198. [CrossRef]
32. Davidson, D.A. Problems in the determination of plastic and liquid limits of remoulded soils using a drop-cone penetrometer. *Earth Surf. Process. Landf.* **1983**, *8*, 171–175. [CrossRef]
33. Stanchi, S.; Catoni, M.; D'Amico, M.E.; Falsone, G.; Bonifacio, E. Liquid and plastic limits of clayey, organic C-rich mountain soils: Role of organic matter and mineralogy. *Catena* **2017**, *151*, 238–246. [CrossRef]
34. Tang, H.D.; Zhu, M.L. Timely spread characteristics of micro- and meso-scale fractures based on SEM experiment. *Environ. Earth Sci.* **2020**, *79*, 1–20. [CrossRef]
35. Howarth, D.F.; Rowlands, J.C. Quantitative assessment of rock texture and correlation with drillability and strength properties. *Rock Mech. Rock Eng.* **1987**, *20*, 57–85. [CrossRef]
36. Campbell, D.H.; Galehouse, J.S. Quantitative clinker microscopy with the light microscope. *Cem. Concr. Aggreg.* **1991**, *13*, 94–96.
37. Zhou, M.; Li, J.; Luo, Z.; Sun, J.; Xu, F.; Jiang, Q.; Deng, H. Impact of water–rock interaction on the pore structures of red-bed soft rock. *Sci. Rep.* **2021**, *11*, 7398. [CrossRef]
38. Xiaoshuai, X.; Huasong, C.; Xinhong, X.; Jing, W.; Jiawen, Z. Micro-structural characteristics and softening mechanism of red-bed soft rock under water-rock interaction condition. *J. Eng. Geol.* **2019**, *27*, 966–972.
39. Li, A.; Deng, H.; Wang, X.; Luo, J.; Zhang, J. Research on creep characteristics and constitutive model of red bed mudstone under saturated-dehydrated cycle. *J. Eng. Geol.* **2021**, *29*, 843–850.
40. Dhakal, G.; Yoneda, T.; Kato, M.; Kaneko, K. Slake durability and mineralogical properties of some pyroclastic and sedimentary rocks. *Eng. Geol.* **2002**, *65*, 31–45. [CrossRef]
41. Zhang, S.; Xu, Q.; Hu, Z. Effects of rainwater softening on red mudstone of deep-seated landslide, Southwest China. *Eng. Geol.* **2016**, *204*, 1–13. [CrossRef]
42. Wu, M.; Wang, J. Exploring particle breakage in sand under triaxial shearing using combined X-ray tomography and particle tracking method. *Géotechnique* **2023**, 1–40. [CrossRef]
43. Franklin, J.A.; Chandra, R. The slake-durability test. *Int. J. Rock Mech. Min. Sci. Geomech. Abstr.* **1972**, *9*, 325–328. [CrossRef]
44. Ulusay, R. *The ISRM Suggested Methods for Rock Characterization, Testing and Monitoring: 2007–2014*; Springer International Publishing: Berlin/Heidelberg, Germany, 2014; Volume 15, pp. 47–48.
45. Ahmad, A.; Lopulisa, C.; Baja, S.; Imran, A.M. The correlation of soil liquid limit and plasticity index for predicting soil susceptibility: A case study on landslides area in South Sulawesi. *IOP Conf. Ser. Earth Environ. Sci.* **2019**, *235*, 012007. [CrossRef]
46. Bhattarai, P.; Marui, H.; Tiwari, B.; Watanabe, N.; Tuladhar, G.R. Depth-wise variation of physical and mechanical properties of mudstone in relation to weathering. *J. Jpn. Landslide Soc.* **2007**, *44*, 79–89. [CrossRef]
47. Hanke, R.; Fuchs, T.; Salamon, M.; Zabler, S. X-ray microtomography for materials characterization. In *Materials Characterization Using Nondestructive Evaluation (NDE) Methods*; Woodhead Publishing: Cambridge, UK, 2016; pp. 45–79.
48. Zhou, W.; Apkarian, R.; Wang, Z.L.; Joy, D. Fundamentals of scanning electron microscopy (SEM). In *Scanning Microscopy for Nanotechnology*; Springer: New York, NY, USA, 2007; pp. 1–40. [CrossRef]

Disclaimer/Publisher's Note: The statements, opinions and data contained in all publications are solely those of the individual author(s) and contributor(s) and not of MDPI and/or the editor(s). MDPI and/or the editor(s) disclaim responsibility for any injury to people or property resulting from any ideas, methods, instructions or products referred to in the content.

buildings

MDPI

Article

The Mechanical Behavior and Constitutive Model Study of Coarse-Grained Soil under Cyclic Loading–Unloading in Large-Scale Plane Strain Conditions

Zhi Wang [1,*], Shuai Shao [2,*], Shengjun Shao [1,3] and Liguo Yang [4]

1 Institute of Geotechnical Engineering, Xi'an University of Technology, Xi'an 710048, China; sjshao@xaut.edu.cn
2 Department of Architecture and Urban Planning, Xi'an University of Technology, Xi'an 710048, China
3 Shanxi Key Laboratory of Loess Mechanics and Engineering, Xi'an University of Technology, Xi'an 710048, China
4 Department of Civil and Traffic Engineering, Yellow River Conservancy Technical Institute, Kaifeng 475003, China; 2010840291@yrcti.edu.cn
* Correspondence: jarowl@foxmail.com (Z.W.); shaoshuai@xaut.edu.cn (S.S.)

Abstract: To address loading and unloading issues in civil and hydraulic engineering projects that employ coarse-grained soil as fill material under plane strain conditions during construction and operation, cyclic loading–unloading large-scale plane strain tests were conducted on two types of coarse-grained soils. The effects of coarse-grained soil properties on shear behavior and various modulus relationships were analyzed. The research results showed that coarse-grained soils with better particle roundness exhibit significant shear dilation deformation; it was also found that low parent rock strength can lead to strain softening, and an increase in confining pressure suppresses shear dilation deformation. During the cyclic loading–unloading process, the initial unloading modulus (E_{iu}) > unloading–reloading modulus (E_{ur}) > initial reloading modulus (E_{ir}) > initial tangent modulus (E_i), with the unloading modulus considerably greater than the others. In finite element simulations and model calculations, it is essential to select appropriate modulus parameters based on the stress conditions of the soil to ensure calculation accuracy. In this work, an elastoplastic and nonlinear elastic theory was used to establish a cyclic loading–unloading constitutive model. By comparing the values obtained using this model with experimental measurements, it was found that the model can reasonably predict stress–strain variations during cyclic loading–unloading of coarse-grained soils under plane strain conditions.

Keywords: plane strain; coarse-grained soil; loading–unloading; constitutive model

1. Introduction

Coarse-grained soils possess excellent engineering properties such as high shear strength, large bearing capacity, good stability, and high permeability. Because of this, they are widely used in hydraulic and transportation engineering [1–3]. Civil and hydraulic projects, such as rockfill dams, slope protections, tunnels, and high embankments, involve the use of engineering structures, which have longitudinal lengths much greater than their transverse lengths. The longitudinal ends have constraints that prevent deformation in that direction, causing strain to develop more easily toward transverse sections where constraints are weaker. This strain boundary condition is known as the plane strain condition [4,5]. Therefore, it is more reasonable to study such engineering under plane strain conditions [6–9]. The plane strain conditions cause the soil mass to exhibit mechanical and strength characteristics, which are different from those exhibited under triaxial conditions [10,11]. Lu et al. [12] studied the principal stress in the direction of plane strain of soil and verified the rationality of the bilinear principal stress function. Pan et al. [13] carried out true triaxial tests and plane strain tests on sandstone rubble with different values for the intermediate principal stress ratio. Their results indicated that the contribution of intermediate principal stress to the enhancement of specimen strength was essentially reflected. Jiang et al. [14] conducted true triaxial tests and plane strain tests on rubble with four different lithologies; their results revealed strength parameters that were significantly greater than in conventional triaxial tests.

During the construction and operation stages of engineering projects, soil mass is usually subjected to single or multiple stress loading unloading effects. Examples include the rise and fall of water levels in reservoirs and rivers, the arrival and departure of heavy construction machinery, the piling up of excavated soil, and the asymmetric excavation of tunnels [15–18]. The repeated application and removal of loads can damage engineering structures and affect the stability of geotechnical bodies, impacting the efficient operation of engineering projects and potentially triggering engineering disasters. Ji et al. [19] analyzed the three-dimensional evolutionary mechanisms and deformation characteristics involved in reservoir bank collapses and suggested that the repeated rise and fall of the reservoir water level, along with the erosive action of water, are likely to lead to the collapse of the reservoir banks. Currently, research on the unloading and reloading of soil and rock is primarily focused on traditional conventional triaxial tests [20–27]. However, conventional triaxial tests cannot accurately reflect the complex stress path issues associated with changes in three-dimensional stress states. Therefore, studying the mechanical behavior of gravel under loading–unloading stress paths in a three-dimensional stress state can better reflect the stress state changes in coarse-grained soils. Wang et al. [28] conducted true triaxial tests on sandstone to study instability and failure characteristics in the surrounding rock under different disturbance stresses during the excavation unloading of deep underground tunnels. Wang et al. [29] conducted research on the deformation and failure characteristics of red sandstone via true triaxial loading and unloading tests, and the Mogi-Coulomb strength criterion can better reflect the strength failure characteristics of red sandstone during loading and unloading processes. Liu et al. [30] conducted undrained true triaxial tests on natural undisturbed clay and studied the permanent deformation of natural undisturbed clay under three-dimensional cyclic stress, and then established an empirical formula for predicting the permanent principal strain under three-dimensional cyclic stress. Wang et al. [31] conducted large-scale true triaxial loading and unloading tests on coarse-grained materials, analyzed the variation patterns of rebound shear modulus and proposed an estimation model for unloading rebound modulus.

In this study, a series of large-scale plane strain cyclic loading unloading tests were conducted on two types of coarse-grained soils with different particle circularities to address the safety and stability issues of the roadway and other plane strain types of engineering. The stress–strain response of the coarse-grained soil under a reciprocating stress path was tested. Based on the experimental results, the modulus of the coarse-grained soil under cyclic loading unloading stress path conditions was analyzed, and then a cyclic loading–unloading constitutive model with plane strain conditions was established. This model can predict the stress–strain behavior of reloading after unloading and accurately predict the deformation during high-stress unloading based on the unique mechanical characteristics of the soil mass during unloading.

2. Coarse-Grained Soil, Apparatus, and Scheme

2.1. Physical Properties of Coarse-Grained Soil

Two types of coarse-grained soil were collected from FengHe River beach in Shaanxi Province of China and from a certain transportation engineering unit; these were referred to as Soil A and Soil B, respectively. Soil A consisted of well-rounded gravel with a particle size range of 2 mm to 60 mm. Soil B contained angular gravel with a particle size range of 2 mm to 60 mm. The specific physical parameters are listed in Table 1. The test specimens were rectangular with dimensions of 300 mm length, 300 mm width, and 600 mm height. The coarse-grained soil and samples used for testing are shown in Figure 1.

Table 1. Physical parameters of coarse-grained soil.

Soil Sample	Sample Dry Density ρ_d (g·cm^{-3})	Specific Gravity *Gs*	Void Ratio *e*	Relative Density *Dr*
Soil A	1.96	2.64	0.344	0.89
Soil B	1.80	2.68	0.487	0.89

Figure 1. Sampling sites for the two types of coarse-grained soil, soil samples with different particle sizes, and test samples.

Both types of coarse-grained soil specimens were equally well graded. The grading curve is shown in Figure 2. The curvature coefficient (Cc) was 1.49, and the uniformity coefficient (Cu) was 6.26, categorizing the gravel as well-graded gravel (GW) according to engineering classification [32].

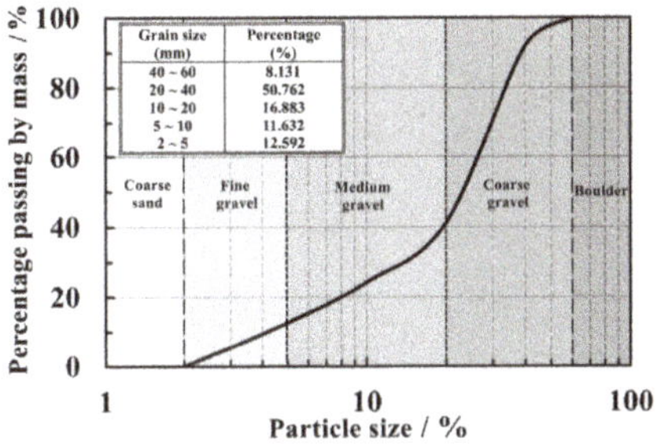

Figure 2. The grading curve of coarse-grained soil samples.

2.2. The Test Instruments and Plans

The planar strain testing apparatus was modified from a large-scale true triaxial device [33]. Using this apparatus, vertical loading is achieved via rigid bidirectional symmetric loading, and the horizontal confinement pressure is applied using flexible hydraulic latex capsules. The modification for planar strain in the pressure chamber involved the use of a pair of smooth steel plates instead of a pair of radial hydraulic bags. Behind each steel plate, there are six pressure sensors, and bolts mounted on the pressure sensors can control the distance between the steel plate and the inner wall of the pressure chamber. This enables the specimen to be maintained in a planar strain state while lateral confinement pressure is also applied. The modified planar strain pressure chamber is depicted in Figure 3. The test included confinement pressures of 100 kPa, 200 kPa, 300 kPa, and 400 kPa, with a vertical loading rate of 0.6 mm/min [32]. The vertical strain was loaded to levels of 0.5%, 1%, 3%, 6%, and 10%; after each of these levels was attained, the vertical stress began to unload at a rate of 0.06 mm/min [31]; this was followed by reloading until the confinement pressure was reached. The test was stopped when the vertical strain reached 15% [32].

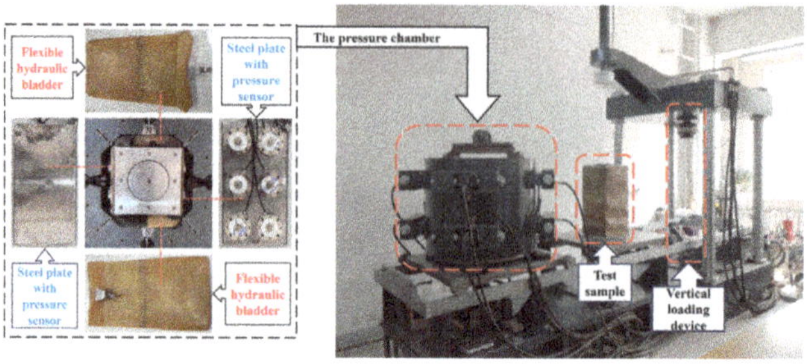

Figure 3. The large-scale true triaxial instrument and plane strain chamber.

3. Test Results and Discussion

3.1. The Volumetric Strain

Figure 4 illustrates the relationship between the ε_v (volumetric strain) and ε_1 (axial strain) on coarse-grained soil during the plane strain cyclic loading–unloading shear test. The volumetric strains on both types of coarse-grained soils exhibit characteristics of shear contraction followed by shear dilation. Soil A exhibits evident shear dilation at lower confining pressures, but this dilation diminishes as the confining pressure increases, although it tends to shift back to dilation with increased vertical strain. Soil B only exhibits noticeable shear dilation at 100 kPa, with shear contraction deformation at other confining pressures. It can also be observed that Soil A—with better roundness and a smaller void ratio—exhibits more pronounced shear dilation than Soil B. With a constant vertical loading rate, greater lateral dilation means that shear dilation is more likely to occur. Because the particles of Soil A are rounder and the sample has fewer voids under compression and shear, both intact and broken soil particles are more likely to move laterally in the direction of the lower confining stress, resulting in more significant lateral dilation. In contrast, Soil B, with its poorer particle roundness, exhibits more evident particle interlocking. This makes particle sliding more challenging, and the larger voids in the sample mean that both intact and broken particles are more likely to fill these internal voids, resulting in smaller lateral dilation.

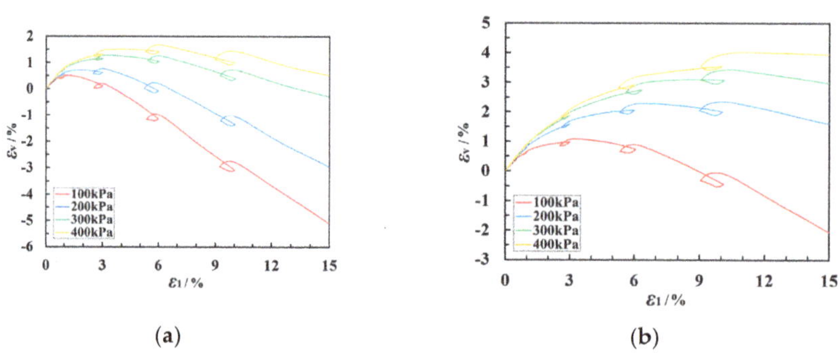

Figure 4. Cyclic loading–unloading ε_v-ε_1 curves for coarse-grained soil of the following types: (**a**) Soil A and (**b**) Soil B.

3.2. The Shear Stress–Strain Behavior

Figure 5 depicts the curves for the relationships between the q (generalized shear stress) and ε_s (generalized shear strain) during the cyclic loading–unloading tests for the two types of coarse-grained soils. The unloading and reloading segments of the q-ε_s curve form a hysteresis loop. The $\Delta\varepsilon_s$ (generalized shear strain unloading rebound deformation) at the beginning of unloading is minimal but rapidly increases as q continually decreases. A higher stress level leads to a larger rebound, resulting in a larger area enclosed by the hysteresis loop. The q-ε_1 curve for coarse-grained Soil A generally shows weak hardening, whereas the curve for Soil B consistently indicates strong hardening. The peak shear stress for Soil A is lower than that for Soil B. Soil A has a lower particle strength, making it more prone to brittle particle fracturing under vertical stress. Additionally, as described in the previous section, Soil B exhibits more significant shear contraction deformation, more pronounced compressive hardening, and a higher peak shear stress.

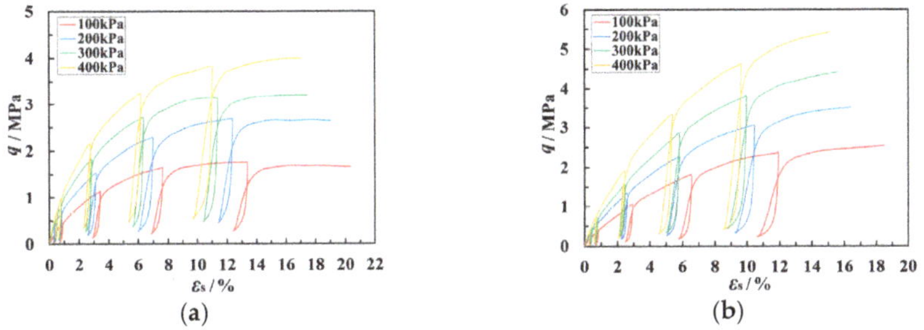

Figure 5. Cyclic loading–unloading q–ε_s curves for coarse-grained soil of the following types: (**a**) Soil A and (**b**) Soil B.

3.3. Modulus in Cyclic Loading–Unloading Stress Path

In finite element simulations and constitutive modeling calculations, it is necessary to select appropriate material parameters to ensure the accuracy of the calculations. For coarse-grained soils, during multiple unloading–reloading cycles, phenomena such as repeated changes in particle arrangement and particle fragmentation must be considered. These factors result in mechanical properties of coarse-grained soils that differ from those obtained using monotonic loading tests.

Figure 6 illustrates the moduli in cyclic loading–unloading tests. It can be seen that the E_i (initial tangent modulus) corresponds to the tangent modulus during the first loading segment (segment OA in Figure 6); the E_{iu} (initial unloading modulus) represents the tangent modulus during various unloading segments in the cyclic loading–unloading test (segments AC in Figure 6); E_{ir} denotes the initial reloading modulus during various reloading segments (segments CD in Figure 6); and the E_{ur} (unloading–reloading modulus) corresponds to the slope of the line connecting the two intersection points of the hysteresis loop (points B and C in Figure 6). The stress level S at each unloading point is calculated using Equation (1), and the variations in E_{iu}, E_{ir}, and E_{ur} for both coarse-grained soils under different confining pressures with S are shown in Figures 7 and 8, respectively.

$$S = \frac{q_{ul}}{q_f} \tag{1}$$

where q_{ul} is the generalized shear stress at the unloading point (kPa), and q_f is the peak generalized shear stress (kPa).

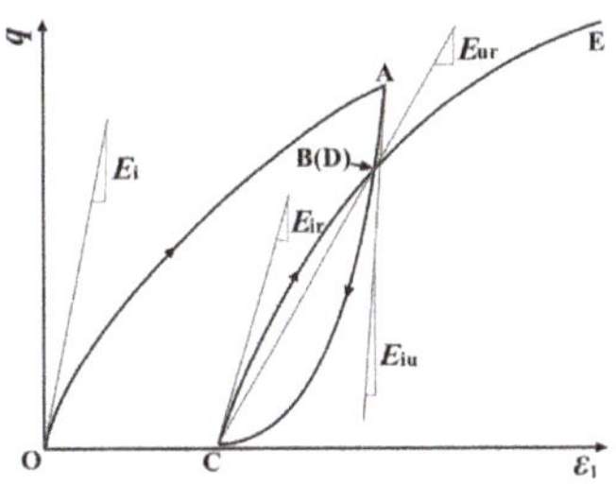

Figure 6. Schematic diagram of each modulus in the cyclic loading–unloading stress path.

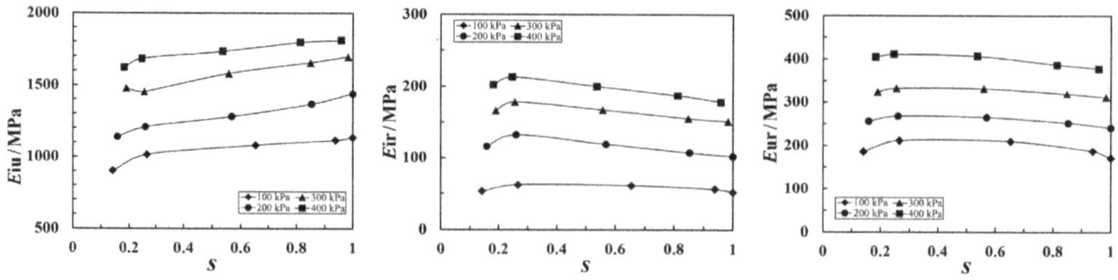

(a) (b) (c)

Figure 7. Relationship between modulus and *S* for Soil A with different moduli of (**a**) E_{iu}, (**b**) E_{ir}, and (**c**) E_{ur}.

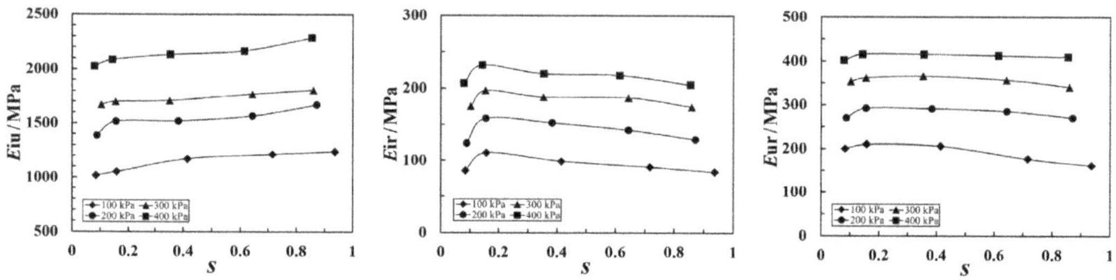

(a) (b) (c)

Figure 8. Relationship between modulus and *S* for Soil B with different moduli of (**a**) E_{iu}, (**b**) E_{ir}, and (**c**) E_{ur}.

As can be seen from Figures 7 and 8, the moduli of coarse-grained soil change with the stress level during cyclic loading–unloading tests. Both E_{ir} and E_{ur} initially increase and then decrease as *S* rises, and the peak value declines when the stress level is between 0.15 and 0.25. At lower stress levels, the specimen primarily undergoes compaction. The structural readjustment caused by unloading results in an increase in particle contacts, enhancing the specimen's stiffness and causing a slight rise in E_{ir} and E_{ur}. As the stress level rapidly increases, soil particles are more prone to breaking. The broken particles have lower strength than the intact ones and repeated reloading weakens particle strength, leading to a decrease in E_{ir} and E_{ur}. In contrast, E_{iu} increases with the rise in *S*. This is due to the cumulative increase in plastic strain from repeated unloading–reloading and particle fracturing, reducing the instantaneous vertical strain rebound upon unloading. The change in each modulus with stress level at a constant confining pressure is relatively minor; an average value can thus be taken as the modulus for that level of confining pressure. Values for the four moduli obtained from the plane strain cyclic loading–unloading tests for coarse-grained soil are summarized in Table 2.

Table 2. Modulus values for coarse-grained soil.

Soil Sample	**Soil A**				**Soil B**			
σ_3 (kPa)	100	200	300	400	100	200	300	400
E_i (MPa)	54	78	111	135	61	92	128	157
E_{ir} (MPa)	87	115	163	186	94	141	184	216
E_{ur} (MPa)	193	256	324	397	191	282	356	411
E_{iu} (MPa)	1049	1286	1573	1729	1139	1532	1731	2139

As can be seen in Table 2, values of all four moduli for the coarse-grained soil increase with the rise in confining pressure. Under different confining pressures, Soil B, which has higher strength, exhibits consistently higher modulus values compared with coarse-grained Soil A. At low confining pressures, E_{ir} is slightly less than E_i. As the confining pressure increases, E_{ir} gradually approaches E_i, with only a minor difference between them. Thus, in calculations, E_i can be used as an approximation for E_{ir}. The E_{ur} values for the two coarse-grained soil samples with consistent gradation show a small difference. However, E_{iu} is significantly higher than the other three moduli. Therefore, when calculating the unloading deformation, this modulus should be used separately to ensure accuracy.

The Janbu empirical equation, expressed as in Equation (2), describes the relationship between the modulus and the confining pressure.

$$E = Kp_a\left(\frac{\sigma_3}{p_a}\right)^n \tag{2}$$

where E is the modulus (kPa), K and n are modulus parameters, and P_a is the atmospheric pressure (kPa).

Taking the logarithm of both sides of Equation (2) yields Equation (3).

$$\lg\left(\frac{E}{p_a}\right) = n\lg\left(\frac{\sigma_3}{p_a}\right) + \lg K \tag{3}$$

In Equations (2) and (3), E can be taken as any of the four moduli, E_i, E_{iu}, E_{ir}, and E_{ur}, from the coarse-grained soil cyclic loading–unloading test. The fitting relationship between these four moduli and the confining pressure is illustrated by the curves in Figure 9.

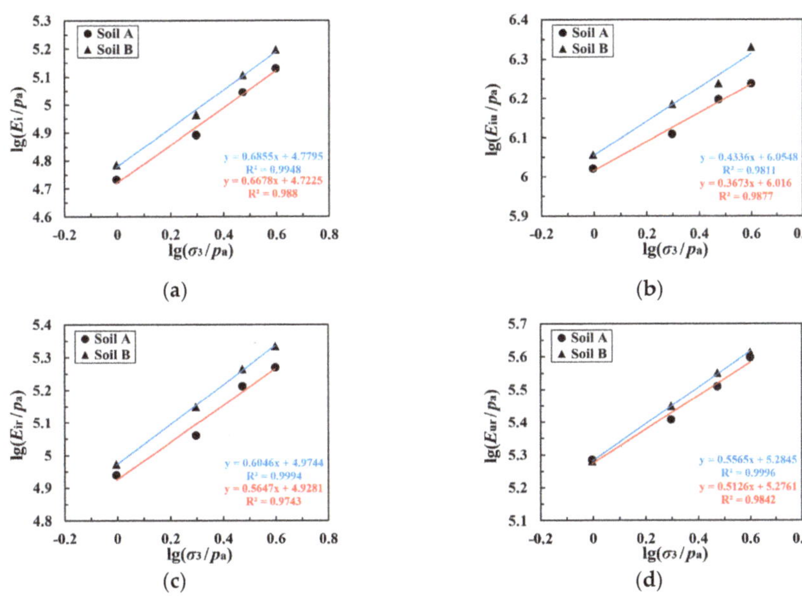

Figure 9. The fitting curves of the relationship between modulus and S for the two types of coarse-grained soil with different moduli of (**a**) E_i, (**b**) E_{iu}, (**c**) E_{ir}, and (**d**) E_{ur}.

As can be seen in Figure 9, $\lg(E_i/p_a)$, $\lg(E_{iu}/p_a)$, $\lg(E_{ir}/p_a)$, and $\lg(E_{ur}/p_a)$ each exhibit a linear relationship with $\lg(\sigma_3/p_a)$. We may say, therefore, that Equation (2) can reasonably predict the various moduli of coarse-grained soil under different confining pressure conditions in unload–reload tests. The parameters of the various moduli for the two types of coarse-grained soil are presented in Table 3.

Table 3. The modulus parameters of coarse-grained soil.

Modulus Parameters	Soil A	Soil B
K_i	52,784	60,187
K_{iu}	1,037,528	1,134,488
K_{ir}	84,742	94,275
K_{ur}	145,044	163,042
n_i	0.6678	0.6855
n_{iu}	0.3673	0.4336
n_{ir}	0.5647	0.6046
n_{ur}	0.6202	0.5908

4. A Cyclic Loading–Unloading Constitutive Model

4.1. The Idea of Establishing the Model

In Figure 10, point D on the reloading segment CF coincides with the unloading point B, with both points having the same stress q_α and strain ε_α. Upon reloading to point E, the stress q_β matches that at point A, but the strain ε_γ at point E is greater than the strain ε_β at point A. Conventional plasticity theory suggests that the position of the yield surface is related only to the historical maximum stress q_{max} and not to the stress path [34]. As depicted in Figure 11, during unloading, the yield surface contracts from A to C. The reduction in stress level causes soil particles to rebound and the particle fabric to redistribute. Upon reloading, the stress increment $d\sigma$ on the yield surface of point C, directed outward, results in new plastic strains. Consequently, points A and E in Figure 10 do not coincide, and so it is not entirely appropriate to consider segment CE as merely nonlinear elastic deformation. Both the initial loading and reloading deformations should be treated as elastoplastic deformations. For the present study, therefore, we employed a combined theory of elastoplasticity and nonlinear elasticity to establish a constitutive model for the unloading–reloading of coarse-grained soil under plane strain conditions.

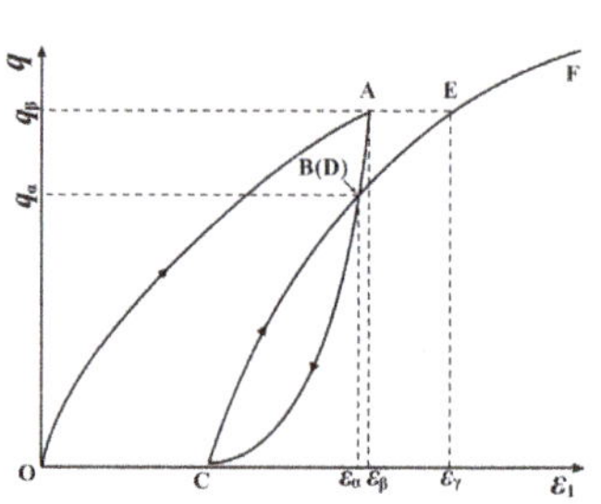

Figure 10. Schematic diagram of stress–strain relationship in cyclic loading–unloading test.

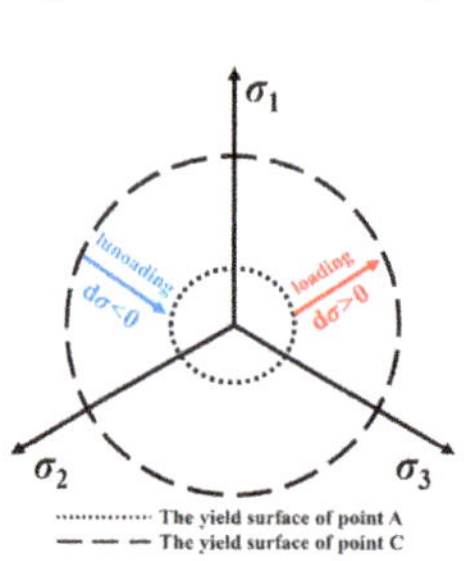

Figure 11. Schematic diagram of the yield surface in principal stress space.

4.2. The Loading (Reloading) Constitutive Model

As described in Section 4.1, the deformation in each loading phase of the cyclic loading–unloading test is elastoplastic deformation. Because coarse-grained soil is mainly composed of non-cohesive coarse particles, the Lade–Duncan model based on sandy soil may be considered more suitable to describe the stress–strain relationship during the loading phase. According to the concept of the elastoplastic constitutive model, the total strain increment can be obtained by summing the elastic strain increment and the plastic strain increment, as Equation (4).

$$d\varepsilon_{ij} = d\varepsilon_{ij}^{e} + d\varepsilon_{ij}^{p} \tag{4}$$

where $d\varepsilon_{ij}$ represents the total strain increment, $d\varepsilon_{ij}^{e}$ represents the elastic strain increment, and $d\varepsilon_{ij}^{p}$ represents the plastic strain increment.

The yield function, flow rule function, and hardening law function of the Lade–Duncan model are shown in Equations (5)–(7), respectively.

$$f = \frac{I_1^3}{I_3} - k = 0 \tag{5}$$

where I_1 is the first invariant of the stress tensor, I_3 is the third invariant of the stress tensor, and k is the hardening parameter.

$$g = I_1^3 - k_1 I_3 = 0 \tag{6}$$

where k_1 is the plastic potential parameter.

$$k = H(W^p) \tag{7}$$

where W^p is the plastic work.

4.2.1. Elastic Strain of the Loading Section

The elastic strain during loading is calculated using the generalized Hooke's law:

$$\begin{Bmatrix} \varepsilon_1^e \\ \varepsilon_2^e \\ \varepsilon_3^e \end{Bmatrix} = \frac{1}{E}\begin{bmatrix} 1 & -\nu & -\nu \\ -\nu & 1 & -\nu \\ -\nu & -\nu & 1 \end{bmatrix}\begin{Bmatrix} \sigma_1 \\ \sigma_2 \\ \sigma_3 \end{Bmatrix} \tag{8}$$

where ν is the Poisson's ratio, and E is the initial tangent modulus for the loading segment (MPa).

In Equation (8), the value of E for the initial loading segment is taken as E_i; for the reloading segment, it is taken as E_{ir}. As can be seen from Table 2, the difference between E_i and E_{ir} is small. To simplify calculations, E can be approximated as E_i. In the state of plane strain, $\varepsilon_2 = \varepsilon_2^p = \varepsilon_2^e = 0$, Equation (8) can now be simplified as Equation (9).

$$\begin{Bmatrix} \varepsilon_1^e \\ \varepsilon_3^e \end{Bmatrix} = \frac{1+\nu}{E_i}\begin{bmatrix} (1-\nu) & -\nu \\ -\nu & (1-\nu) \end{bmatrix}\begin{Bmatrix} \sigma_1 \\ \sigma_3 \end{Bmatrix} \tag{9}$$

4.2.2. The Hardening Parameter k and Plastic Work W^p

The plastic work is chosen as the hardening parameter in the hardening law; k and W^p have a hyperbolic relationship [34].

$$(k - k_t) = \frac{W^p}{a + bW^p} \tag{10}$$

where k_t is a value slightly greater than 27, a is the reciprocal of the initial slope of the hyperbola, and b is the asymptote of $(k - k_t)$ when W^p approaches positive infinity.

The value of a is related to the confining pressure and can be obtained by using Equation (11).

$$a = m\left(\frac{\sigma_3}{p_a}\right)^l \tag{11}$$

where m and l are dimensionless parameters.

Experimental results indicated that when the confining pressure is the same, the a values for each reloading segment are not significantly different and can be taken as an average value denoted as a_{rl}, with parameters m_{rl} and l_{rl}. However, the a value for the initial loading segment is different from that of the reloading segments and should be considered separately, being denoted as a_{il}, with parameters m_{il} and l_{il}.

The parameter b is related to the stress level S at the start point of unloading, and there is a power function relationship between them, as shown in Figure 12. This can be calculated using Equation (12).

$$b = S^c D \tag{12}$$

where D is a parameter related to the consolidation pressure; this can be calculated by Equation (13).

$$D = d\left(\frac{\sigma_3}{p_a}\right)^f \tag{13}$$

where c, d, and f are all dimensionless parameters.

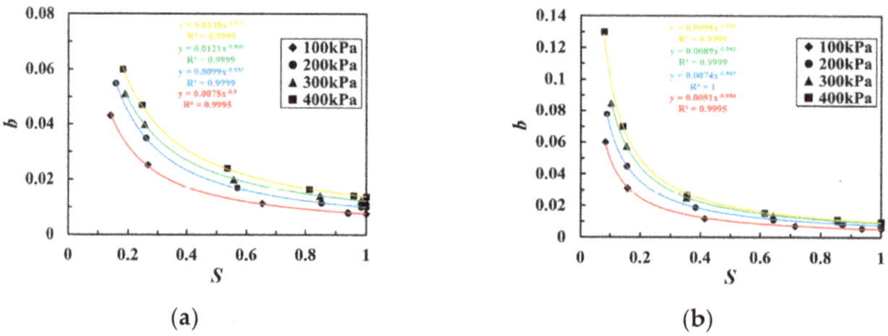

Figure 12. The fitting curve of the relationship between b and S for coarse-grained soil of the following types: (**a**) Soil A and (**b**) Soil B.

Taking the logarithm on both sides of Equation (12), we can obtain Equation (14). The fitting relationship between lg b and lg S is shown in Figure 13, where they have a relatively good linear relationship.

$$\lg b = c \lg S + \lg D \tag{14}$$

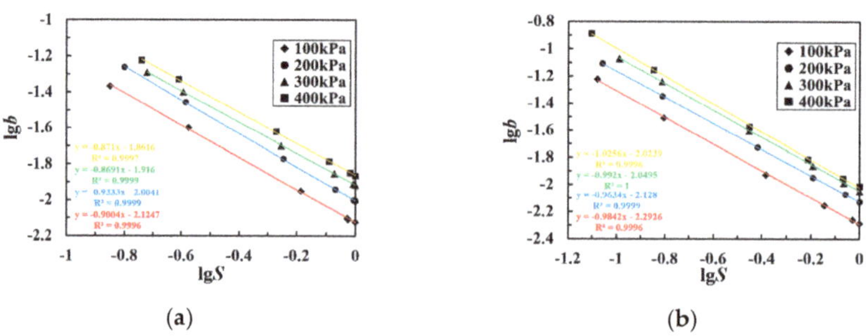

Figure 13. The fitting curve of the relationship between lg$b$ and lg$S$ for coarse-grained soil of the following types: (**a**) Soil A and (**b**) Soil B.

Similarly, taking the logarithm on both sides of Equation (13), we can obtain Equation (15).

$$\lg D = \lg d + f \lg(\sigma_3 / p_a) \tag{15}$$

The relationship between $\lg D$ and $\lg(\sigma_3/P_a)$ is shown in Figure 14; here, again, a relatively good linear relationship can be seen.

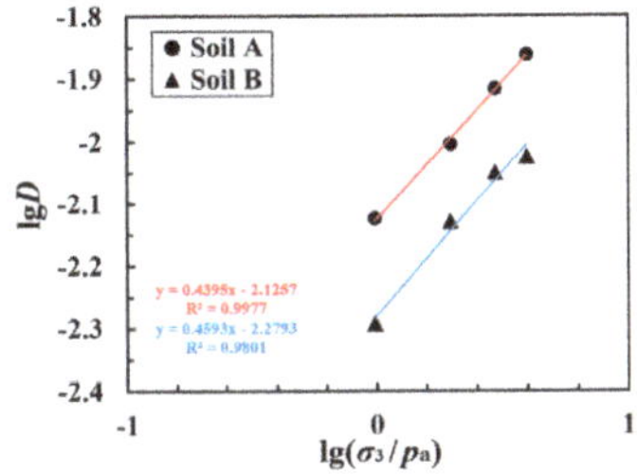

Figure 14. The fitting curve of the relationship between $\lg D$ and $\lg(\sigma_3/P_a)$.

By substituting Equation (15) into Equation (14), the parameter b can be calculated. The plastic work W^p can be calculated using Equation (16).

$$W^p = \int \sigma_{ij} d\varepsilon_{ij}^p \tag{16}$$

Differentiating Equation (16) yields Equation (17).

$$dW^p = \sigma_{ij} d\varepsilon_{ij}^p = p d\varepsilon_v^p + q d\varepsilon_s^p \tag{17}$$

where p is the average principal stress (kPa), ε_v^p is the plastic volumetric strain, and ε_s^p is the plastic generalized shear strain.

4.2.3. The Correction of Plastic Potential Parameter k_1

The plastic potential parameters can be indirectly obtained from the plastic dilatancy ratio v^p, which is defined in Equation (18).

$$\nu^p = -\frac{d\varepsilon_3^p}{d\varepsilon_1^p} \tag{18}$$

where $d\varepsilon_1^p$ and $d\varepsilon_3^p$ represent the increments of plastic strain in the vertical and confining pressure directions, respectively.

The Lade–Duncan model is based on conventional triaxial tests, where $\sigma_2 = \sigma_3$; however, under plane strain or true triaxial three-dimensional stress conditions, $\sigma_2 \neq \sigma_3$. Therefore, $d\varepsilon_1^p$ and $d\varepsilon_3^p$ need to be corrected and calculated via Equation (19).

$$\begin{cases} d\varepsilon_1^p = d\lambda \frac{\partial g}{\partial \sigma_1} = d\lambda \left(3I_1^2 - k_1\sigma_1\sigma_2\right) \\ d\varepsilon_3^p = d\lambda \frac{\partial g}{\partial \sigma_3} = d\lambda \left(3I_1^2 - k_1\sigma_2\sigma_3\right) \end{cases} \tag{19}$$

Substituting Equation (19) into Equation (18) yields Equation (20), from which k_1 can be obtained.

$$k_1 = \frac{3I_1^2(1+\nu^p)}{\sigma_2(\sigma_1 + \nu^p\sigma_3)} \tag{20}$$

4.2.4. The Plastic Multiplier $d\lambda$

From Equation (6), it is known that the flow rule g is a third-order homogeneous equation with respect to σ_{ij}. Therefore, Equation (17) can also be expressed as Equation (21) using Euler's theorem.

$$dW^{\mathrm{p}} = d\lambda \frac{\partial g}{\partial \sigma_{ij}} \sigma_{ij} = d\lambda 3g \tag{21}$$

Differentiating Equation (10) and substituting it into Equation (18) yields Equation (22), from which $d\lambda$ is obtained.

$$d\lambda = \frac{dW^{\mathrm{p}}}{3g} = \frac{adk}{3(I_1^3 - k_1 I_3)[1 - b(k - k_t)]^2} \tag{22}$$

Substituting Equation (22) into Equation (19) results in Equation (23), which allows for the calculation of the increments in plastic strain.

$$\left\{ \begin{array}{c} d\varepsilon_1^{\mathrm{p}} \\ d\varepsilon_3^{\mathrm{p}} \end{array} \right\} = \frac{adk}{3(I_1^3 - k_1 I_3)[1 - b(k - k_t)]^2} \left\{ \begin{array}{c} 3I_1^2 - k_1\sigma_1\sigma_2 \\ 3I_1^2 - k_1\sigma_2\sigma_3 \end{array} \right\} \tag{23}$$

In summary, the loading segment constitutive model parameters included are K_i, n_i, v, m_{rl}, l_{rl}, m_{il}, l_{il}, c, d, and f, i.e., a total of 10 parameters.

4.3. The Unloading Constitutive Model

In line with the development approach of the unloading–reloading constitutive model, a nonlinear elastic model was utilized to describe the nonlinear deformation of coarse-grained soil during substantial unloading. Taking the fifth unloading section of the cyclic loading–unloading test for Soil A at a confining pressure of 100 kPa as an example, as depicted in Figure 15, the q-ε_1 and ε_3-ε_1 curves in the unloading test resemble hyperbolas. However, the starting point of unloading does not lie at the origin of the stress–strain spatial coordinate system. Therefore, a coordinate system transformation is applied to the old coordinate systems of the q-ε_1 and ε_3-ε_1 curves. Via this transformation, a corrected E-v model is established to define the unloading constitutive relationship. The curves of $q' - \varepsilon'_1$ and $\varepsilon'_3 - \varepsilon'_1$ after coordinate transformation are shown in Figure 16.

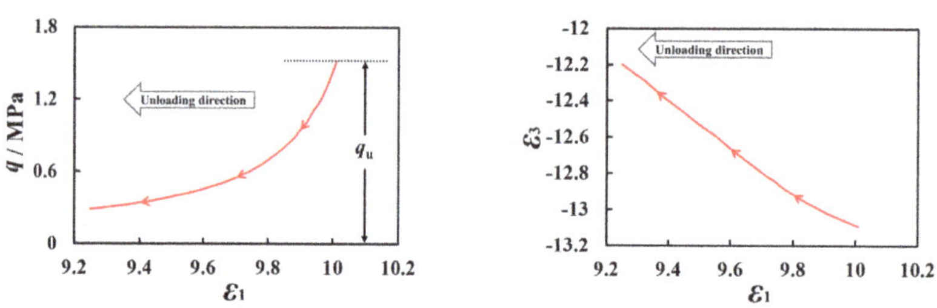

Figure 15. The unloading stress–strain curves in the original coordinate system with types of (**a**) q-ε_1 and (**b**) ε_3-ε_1. (The red arrow line represents the unloading curve and direction).

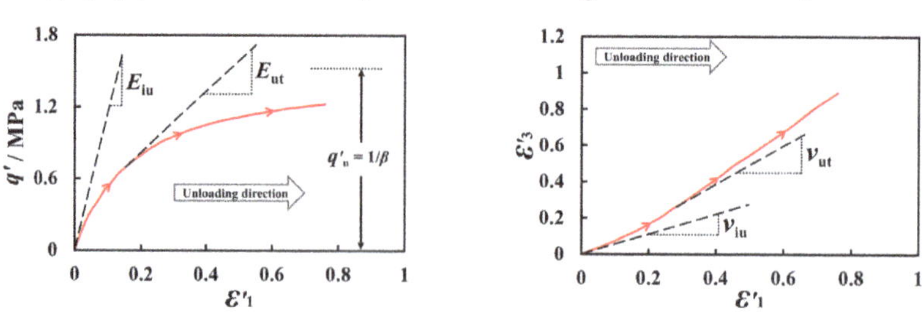

Figure 16. The unloading stress–strain curves in the new coordinate system with types of (**a**) $q' - \varepsilon'_1$ and (**b**) $\varepsilon'_3 - \varepsilon'_1$. (The red arrow line represents the unloading curve and direction).

4.3.1. The Unloading Tangent Modulus (E_{ut})

The E_{ut} may be defined as Equation (24).

$$E_{ut} = \frac{dq'}{d\varepsilon'_1} \tag{24}$$

The $q' - \varepsilon'_1$ curve in the new coordinate system can be described using Equation (25).

$$q' = \frac{\varepsilon'_1}{\alpha + \beta\varepsilon'_1} \tag{25}$$

where α and β are the model parameters, which can be obtained from Equation (26).

$$\begin{cases} \alpha = \frac{1}{E_{iu}} \\ \beta = \frac{1}{q_u} = \frac{1}{q'_u} \end{cases} \tag{26}$$

where E_{iu} is the initial unloading modulus, and q'_u is the ultimate unloading strength in the new coordinate system.

As can be seen from Figure 5, the unloading limit strength tends to approach zero in the original coordinate system. Therefore, in the new coordinate system, the q'_u value of the $q' - \varepsilon'_1$ curve can be taken as the q value of the initial unloading point. When calculating the unloading tangent modulus, the unloading limit strength can be directly obtained from the experiment, and thus only two parameters, K_{iu} and n_{iu}, are needed.

After differentiating Equation (25), Equation (27) can be obtained.

$$d\varepsilon'_1 = \frac{\alpha dq'}{(1 - \beta q')^2} \tag{27}$$

By substituting Equations (26) and (27) into Equation (24), E_{tu} can be calculated by Equation (28).

$$E_{tu} = \frac{1}{E_{iu}(1 - q'/q'_u)^2} \tag{28}$$

4.3.2. The Unloading Tangent Poisson's Ratio (ν_{tu})

As shown in Figure 16b, the unloading tangent Poisson's ratio (ν_{tu}) may be defined as Equation (29).

$$\nu_{ut} = -\frac{d\varepsilon'_3}{d\varepsilon'_1} \tag{29}$$

The $\varepsilon'_3 - \varepsilon'_1$ curve in the new coordinate system can be described by Equation (30).

$$\varepsilon'_1 = \frac{\varepsilon'_3}{f_u + D_u\varepsilon'_3} \tag{30}$$

where f_u and D_u are model parameters.

Different values of f_u and D_u can be obtained under different confining pressure conditions. f_u can be calculated using Equation (31), and D_u values—which are relatively close under different confining pressures—can be averaged.

$$f_u = G_u - F_u \lg\frac{\sigma_3}{p_a} \tag{31}$$

By differentiating Equation (30), Equation (32) can be obtained.

$$d\varepsilon'_3 = \frac{f_u d\varepsilon'_1}{(1 - D_u\varepsilon'_1)^2} \tag{32}$$

Substituting Equations (31) and (32) into Equation (29), ν_{ut} can be calculated by Equation (33).

$$\nu_{ut} = -\frac{d\varepsilon'_3}{d\varepsilon'_1} = \frac{G_u - F_u \lg(\sigma_3/p_a)}{(1 - D_u\varepsilon_1)^2} \quad (33)$$

The E_{tu}-ν_{tu} unloading constitutive model includes K_{iu}, n_{iu}, G_u, F_u, and D_u, i.e., a total of five model parameters. This is three fewer than the eight parameters in the E-v model, making the calculation more straightforward.

5. Validation of the Model

According to the cyclic loading–unloading constitutive model, the model parameters for Soil A and Soil B are listed in Table 4.

Table 4. The model parameters of coarse-grained soil.

Model Parameter	Soil A	Soil B
ν	0.2000	0.2100
l_{il}	0.5750	0.5183
m_{il}	114.5510	60.1866
l_{rl}	0.6785	0.6871
m_{rl}	708.2719	649.0826
c	−0.8935	−0.9913
d	0.0075	0.0053
f	0.4395	0.4593
G_u	0.8137	0.6076
F_u	−0.2016	−0.1822
D_u	0.3961	0.4692

K_i, n_i, K_{iu}, and n_{iu} are listed in Table 3.

The comparison between test measurement values (TMVs) and model calculations values (MCVs) of the q-ε_s and ε_v-ε_1 curves for the two types of coarse-grained soils is shown in Figures 17 and 18. This constitutive model can reasonably predict the stress–strain relationship of coarse-grained soils under plane strain cyclic loading–unloading conditions.

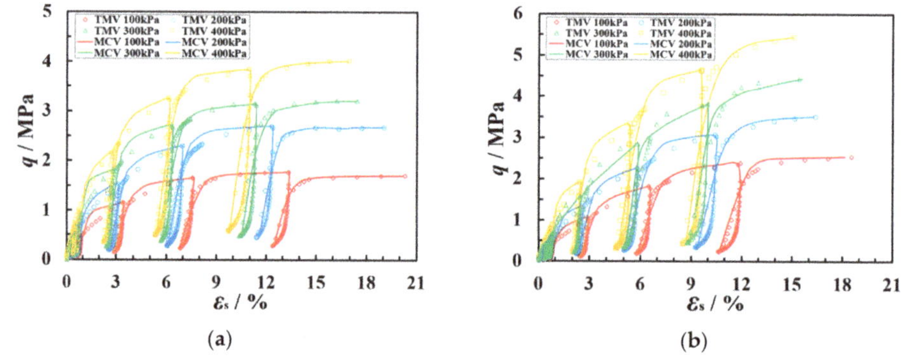

Figure 17. The q-ε_s curves of TMVs and MCVs for coarse-grained soil of the following types: (**a**) Soil A and (**b**) Soil B.

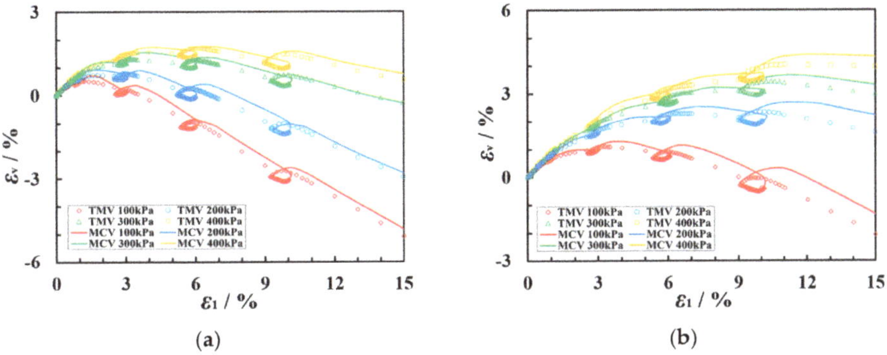

Figure 18. The ε_v-ε_1 curves of TMVs and MCVs for coarse-grained soil of the following types: (**a**) Soil A and (**b**) Soil B.

6. Conclusions

In the plane strain cyclic loading–unloading test, the more rounded the particles of the coarse-grained soil and the smaller the porosity, the more pronounced the dilative behavior. An increase in confining pressure restricts lateral expansion, reducing dilatancy. The strength of the parent rock is also a factor influencing the peak shear stress and the shape of the stress–strain curve. The greater the strength of the parent rock, the higher the peak shear stress, with the stress–strain curve exhibiting a more pronounced hardening behavior.

Under the same confining pressure conditions, both the E_{ir} and the E_{ur} initially increase and then decrease as the stress level and the number of unloading–reloading cycles increase. However, the initial reloading modulus E_{iu} continuously increases, and the moduli for Soil B are all greater than those for Soil A. All four moduli increase with an increase in confining pressure and exhibit a linear relationship with confining pressure on a logarithmic scale. At consistent confining pressures, E_{ir} is slightly greater than the initial tangent modulus E_i. E_{ur} is significantly greater than both E_i and E_{ir}, while E_{iu} is much larger than the other three moduli. When calculating significant unloading deformations, E_{iu} should be used to ensure accuracy.

Modifications were made to the Lade–Duncan model's plastic potential parameter k_1 for three-dimensional stress conditions, establishing a loading (reloading) model. After introducing the initial unloading modulus, coordinate system transformations were applied to correct the E-v model, establishing the E_{tu}-ν_{tu} unloading model. Consequently, a combined 15-parameter cyclic loading–unloading constitutive model under plane strain conditions was established. The physical meanings of the model parameters are clear, making it convenient for application in calculating complex stress path tests involving unloading–reloading. Comparisons between the model's calculated values and experimental values indicate that the model's predictions are relatively accurate.

Author Contributions: Investigation, Z.W.; methodology, S.S. (Shengjun Shao); writing—original draft, Z.W.; writing—review and editing, S.S. (Shuai Shao) and L.Y.; data curation, Z.W. and L.Y.; funding acquisition, S.S. (Shuai Shao) and S.S. (Shengjun Shao). All authors have read and agreed to the published version of the manuscript.

Funding: This research was supported by the National Natural Science Foundation of China (52108342), Basic Research Program of Natural Science in Shaanxi Province-Han-Wei Joint Found Project (2019JLP-21, 2021JLM-50), Shaanxi Water Science and Technology Program Project (2021slkj-12), Doctoral initial funding of Xi'an University of Technology (107-451122001).

Data Availability Statement: The experimental data supporting the conclusions are available from the corresponding author and senior author upon request.

Conflicts of Interest: The authors declare no conflicts of interest.

References

1. Zhai, Q.; Rahardjo, H.; Satyanaga, A.; Dai, G. Estimation of the soil-water characteristic curve from the grain size distribution of coarse-grained soils. *Eng. Geol.* **2020**, *267*, 105502. [CrossRef]
2. Wang, S.Y.; Zhan, Y.J.; Qu, T.M.; Qiu, T.; Wang, H.B. Effect of Gradation on Undrained Compressibility of Foam-Conditioned Coarse-Grained Soils. *Int. J. Geomech.* **2023**, *23*, 04023089. [CrossRef]
3. Guo, Q.G. *Engineering Characteristics and Application of Coarse-Grained Soil*; The Yellow River Water Conservancy Press: Zhengzhou, China, 1998.
4. Pan, X.D.; Hudson, J.A. Plane strain analysis in modelling three-dimensional tunnel excavations. *Int. J. Rock Mech. Min.* **1988**, *25*, 331–337. [CrossRef]
5. Guo, H.J.; Ji, M.; Zhao, W.S. Roadway support design based on in-situ stress and its asymmetrical distributions in a coal mine. *Arch. Min. Sci.* **2020**, *65*, 299–315.
6. Zhang, C.P.; Han, K.H. Collapsed shape of shallow unlined tunnels based on functional catastrophe theory. *Math. Probl. Eng.* **2015**, *2015*, 681257. [CrossRef]
7. Wilson, D.W.; Abbo, A.J.; Sloan, S.W.; Yamamoto, K. Undrained stability of rectangular tunnels where shear strength increases linearly with depth. *Can. Geotech. J.* **2017**, *54*, 469–480. [CrossRef]
8. Huang, M.S.; Wang, H.Y.; Yu, J.; Tang, Z. Undrained stability analysis of a plane strain circular tunnel using streamline velocity fields. *Int. J. Geomech.* **2019**, *19*, 1–7. [CrossRef]
9. Zlatanović, E.; Trajković-Milenković, M.; Lukić, D.; Brčić, S.; Šešov, V. A comparison of linear and nonlinear seismic tunnel-ground interaction analyses. *Acta Geotech. Slov.* **2016**, *13*, 27–42.
10. Shi, W.C.; Zhu, J.G.; Zhang, B.; Yu, T. Strength characteristics of coarse-grained soil under plane strain condition. *Chin. J. Geotech. Eng.* **2011**, *33*, 1974–1979.
11. Shi, W.C.; Zhu, J.G.; Zhang, K.Y.; Yu, T. Experimental study of deformation characteristics of coarse-grained soil under plane strain condition. *Rock Soil Mech.* **2013**, *34*, 101–108.
12. Lu, D.C.; Du, X.L.; Zhou, A.N.; Yao, Y.P. The principal stresses of soil mass in the direction of plane strain. *Adv. Mater. Res.* **2011**, *243–249*, 2657–2665. [CrossRef]
13. Pan, J.J.; Jiang, J.W.; Cheng, Z.L.; Xu, H.; Zuo, Y.Z. Large-Scale True Triaxial Test on Stress-Strain and Strength Properties of Rockfill. *Int. J. Geomech.* **2020**, *20*. [CrossRef]
14. Jiang, J.W.; Pan, J.J.; Cheng, Z.L.; Xu, H.; Tan, F. Experimental Study on the Applicability of Failure Criteria for Rockfill in Three-Dimensional Stress Conditions. *Int. J. Geomech.* **2021**, *21*. [CrossRef]
15. Liu, W.; Liang, J.W.; Xu, T. Tunnelling-induced ground deformation subjected to the behavior of tail grouting materials. *Tunn. Undergr. Space Technol.* **2023**, *140*, 105253. [CrossRef]
16. Dong, C.L.; Zhao, G.M.; Lu, X.Y.; Meng, X.R.; Li, Y.M.; Cheng, X. Similar simulation device for unloading effect of deep roadway excavation and its application. *J. Mt. Sci.-Engl.* **2018**, *15*, 1115–1128.
17. Sun, Y.G.; Gao, Y.F.; Liu, J.M.; Cao, J.T.; Huang, S.J. Experimental study on the softening of saturated clay of zhuhai tunnel under cyclic loading. *J. Yangtze River Sci. Res. Inst.* **2014**, *31*, 52–55.
18. Zhang, Y.W.; Xie, Y.L.; Weng, M.S. Centrifugal test on influence of asymmetric foundation excavation to an underlying subway tunnel. *Rock Soil Mech.* **2018**, *39*, 2555–2562.
19. Ji, F.; Shi, Y.; Li, R.; Zhou, H.; Wang, D.; Zhang, J. Progressive geomorphic evolution of reservoir bank in coarse-grained soil in East China–Insights from long-term observations and physical model test. *Eng. Geol.* **2021**, *281*, 105966. [CrossRef]
20. Zhu, J.G.; Wang, Y.L.; Jia, H.; Zhang, B. Experimental study on resilience behaviour of coarse-grained soils. *Chin. J. Geotech. Eng.* **2011**, *33*, 950–954.
21. Chu, F.Y.; Zhu, J.G.; Jia, H.; An, S.H. Experimental study of mechanical behaviour of coarse-grained soil in unloading and reloading. *Rock Soil Mech.* **2012**, *33*, 1061–1066.
22. Yu, H.; Shen, X.M.; Ye, Y.C.; Yang, J.; Zhu, C.H. Large-scale triaxial tests on dilatancy characteristics of lean cemented sand and gravel. *Front. Earth Sci.* **2021**, *9*, 799215. [CrossRef]
23. Zhang, P.F.; Liu, H.; Feng, Z.T.; Jia, C.F.; Zhou, R. A constitutive model of sandy gravel soil under large-sized loading/unloading triaxial tests. *Adv. Civ. Eng.* **2021**, *7*, 4998351. [CrossRef]
24. Ling, H.; Chen, S.S.; Zhai, Y.C.; Fu, H.; Shi, B.X. Experimental study on strength and deformation of filling materials of super-high dams with wide confining pressures under loading and unloading conditions. *Chin. J. Geotech. Eng.* **2021**, *43*, 27–33.
25. Wang, Y.M.; Wang, Y.Q.; Luo, S.; Liu, H.; Yi, G.S.; Peng, K. Influence of the Crack Angle on the Deformation and Failure Characteristics of Sandstone under Stepped Cyclic Uniaxial Compression with a Constant Lower Limit. *Mathematics* **2023**, *11*, 2187. [CrossRef]
26. Long, Y.; Sun, L.; Cai, Z.; Jiang, Z.; Wang, Z.; He, Q.; Bai, Z. Cyclic Loading and Unloading of Weakly Consolidated Sandstone with Various Water Contents. *Sustainability* **2023**, *15*, 13866. [CrossRef]
27. Tao, Z.F.; Wang, Z.Y.; Ling, J.M.; Tian, Y.; Cai, J.W.; Shi, R. Anisotropic resilient modulus model of granular materials based on particle characteristics. *Transp. Res. Rec.* **2021**, *2675*, 970–984. [CrossRef]
28. Wang, Y.; Du, W.; Zhang, D.; Yu, B. Effect of Loading and Unloading Rates on Sandstone Deformation and Dilatancy under True Triaxial Condition. *Sustainability* **2023**, *15*, 5105. [CrossRef]

29. Wang, S.; Wang, L.G.; Tian, J.S.; Fan, H.; Jiang, C.Y.; Ding, K. An experimental study on the effects of true triaxial loading and unloading stress paths on the mechanical properties of red sandstone. *Minerals* **2022**, *12*, 204. [CrossRef]
30. Liu, B.H.; Kong, L.W.; Xu, G.F.; Sun, Z.L. Effects of three-dimensional cyclic stresses on permanent deformation of natural undisturbed clay. *Int. J. Geomech.* **2022**, *12*, 04022220. [CrossRef]
31. Wang, Y.X.; Shao, S.J.; Wang, Z.; Liu, A.G. Experimental study on the unloading characteristics of coarse aggregate under true triaxial shear loading. *Chin. J. Rock Mech. Eng.* **2020**, *39*, 1503–1512.
32. Ministry of Water Resources of the PRC. *GB/T50123-2019*; Standard for Geotechnical Testing Method. China Planning Press: Beijing, China, 2019.
33. Shao, S.J.; Wang, Y.X.; Shao, S. A large-scale true triaxial apparatus with rigid-flexible-flexible boundary for granular materials. *Geotech. Test J.* **2021**, *44*, 1179–1196. [CrossRef]
34. Zheng, Y.R.; Kong, L. *Geotechnical Plastic Mechanics;* China Architecture & Building Press: Beijing, China, 2010.

Disclaimer/Publisher's Note: The statements, opinions and data contained in all publications are solely those of the individual author(s) and contributor(s) and not of MDPI and/or the editor(s). MDPI and/or the editor(s) disclaim responsibility for any injury to people or property resulting from any ideas, methods, instructions or products referred to in the content.

buildings

MDPI

Article

Field Tests and the Numerical Analysis of a Pile-Net Composite Foundation for an Intelligent Connected Motor-Racing Circuit

Xiaonan Wang [1,2] and Qitao Pei [2,*]

1 Wuhan Urban Construction Investment Development Group Co., Ltd., Wuhan 430023, China; xnwang1975@126.com
2 Wuhan Municipal Engineering Design & Research Institute Co., Ltd., Wuhan 430023, China
* Correspondence: pqt011@163.com

Abstract: In response to the problem of significant post-construction settlement that may occur in a motor racing circuit (MRC), two representative composite foundation testing areas, PHC pile (pre-tensioned spun high-strength concrete pile) and CFG pile (cement fly ash gravel pile), were selected for field tests to obtain the deformation law of pile–soil. Then, finite element numerical simulation was used to carry out back analysis on the geological mechanical parameters of the testing areas. The results showed that the error of soil settlement between the piles in the PHC pile and CFG pile testing areas were 8.2% and 9.6%, respectively, with good inversion precision. The obtained geological mechanical parameters can be used to predict the settlement of the rest of the MRC. On this basis, a finite element numerical model was constructed to analyze the bearing and deformation characteristics of the foundation of the MRC under five types of working conditions that may cause significant post-construction settlement. It showed that the settlement of the embankment was large in the middle and small on both sides after the consolidation of the embankment. The maximum settlement was about 27.0 mm, and the maximum longitudinal uneven settlement ratio of the embankment was 1.3/4000. The axial force of piles in the PHC pile and CFG pile composite foundations increased first and then decreased with depth. The maximum bending moment was located at the foot of slopes or at the boundary of strata, which was relatively small in the middle of the embankment. The deformation of the embankment and the bearing capacity of the piles could meet engineering requirements. This study has certain guiding significance for the design and construction of similar pile-net composite foundations.

Keywords: pile-net composite foundation; field test; soft soil; back analysis; numerical analysis

1. Introduction

A large number of deep soft clay layers are distributed in the southeastern coastal areas of China, the Yangtze River Basin, and other areas. Soft clay has the characteristics of low shear strength, strong compressibility, and high water content, and has poor engineering properties [1,2]. Building roads in these areas without the proper treatment of the roadbed can easily cause the problems of embankment instability, post-construction settlement, and differential settlement, seriously affecting the safety and comfort of the road's operation [3,4].

In recent decades, pile-net composite foundations have been widely used in the soft soil foundations of multiple projects (such as highways, railways, etc.) to improve the bearing capacity and stability of embankments, and reduce embankment settlement [5–8]. The composite foundation is composed of three parts: a reinforced cushion layer (geosynthetic materials and crushed stones), pile foundations, and soil between the piles. The key to the technology is to coordinate the pile soil load sharing ratio by laying a reinforced cushion layer in order to jointly bear the upper load and also create a good constraint effect on the lateral displacement near the slope foot [9]. However, the bearing mechanism of the upper embankment full of pile-net composite foundation is relatively complex, and it is necessary to comprehensively consider the soil arching effect [10,11], tensile membrane effect [12], and deformation coordination between the embankment, geogrid, and the composite foundation.

In order to better apply pile-net composite foundation technology to practical engineering, domestic and foreign scholars have made a lot of achievements in the last 20 years. In terms of laboratory model tests, Hewlett and Randolph [13] studied the spatial distribution characteristics of the soil arch in filling soil above the pile cap based on the assumption of the ultimate state of soil stress. By using the concrete piles to support embankments, Chen et al. [14] found that the stress concentration ratio between the piles and soil in composite foundations varied with the relative displacement of the piles and soil, ranging from 3.0 to 10.0. Castro and Sagaseta [15] concluded that the stress concentration ratio is equal to the elastic modulus ratio of the column to the soil without considering the lateral deformation of the soil. Liu and Zhang [16] analyzed the pile–soil interaction, the lateral deformation, and settlement characteristics of pile-net composite foundations in sloped soft soil via centrifugal tests. Some other scholars have conducted in-depth analyses on the load transfer in pile–soil interactions by using different physical model testing methods [17–20], believing that the load transfer mechanism can be determined by the soil arching effect in the embankment filling and the membrane effect imposed by the geogrid plate and crushed stones.

In terms of field testing, Liu et al. [21] studied the performance of PCP piles (prestressed concrete pipe piles) based on 10,333 tests on the aspects of the length and slenderness ratio, the bearing capacity and length as well as diameter, and controlled factors affecting pile top settlement. Young et al. [22] regarded the mechanism of the rigid pile-net composite foundation as an organic combination of the soil arching effect, grid tension, and pile–soil sharing effects, and proposed that the optimal pile spacing is three times the pile diameter. Zhang et al. [23] believed that the necessary condition for the occurrence of the soil arching effect is the differential settlement between the piles and soil. The modulus of the foundation soil decreases exponentially with the increase of settlement, and the variation range of the pile–soil load sharing ratio is 0.24–0.71. Yang and Liu [24] studied the settlement deformation of embankments, pore pressure dissipation, and pile–soil pressure distributions under a pile-net composite foundation.

In terms of numerical simulation, Zhang et al. [25] and Zhang et al. [26] analyzed the effect of sensitive factors such as the pile modulus, pile diameter, and cushion thickness on embankment settlement. Yoo [27] and Zhao et al. [28] studied the values of crushed stone and grid materials, and found that the modulus of crushed stone can be taken as 25 MPa in three-dimensional numerical calculations, and that the axial tensile stiffness of grid materials is 1000~4000 kN/m. Some scholars also consider the M-C model to be a more commonly used model with wider practicability [29], whereas the modified Cam-clay (MCC) model is usually used for soil simulation close to normal consolidated clay [30]. It is well known that the selection of constitutive model parameters has a great influence on the calculation results. Since the MCC model overestimates the peak friction angle of consolidated soil due to its elliptical yield surface [31], the prediction accuracy of its model is also affected by many factors (such as particle size, weathering degree, saturation, and the over-consolidation rate (OCR), etc.) [32,33], and the superposition of these factors may lead to the inaccuracy of the calculation results of the model. With the use of more advanced models, the number of parameters to be determined increases, which makes these models complicated to use, and a lack of knowledge of these parameters may result in inaccuracy in the obtained results. Therefore, enhancing the identification and rational application of constitutive model parameters is an effective way to improve the accuracy of numerical calculation.

In terms of theory, Kousik [34] and Zhuang and Wang [35] systematically studied the mechanism of the soil arching effect under a pile–soil composite foundation. Zhao et al. [36] derived the PSSR equation from the combined influence of the soil arching effect, tension membrane effect, and pile–soil interaction. Chen et al. [37] analyzed the settlement and negative friction of pile-net composite foundations based on one-dimensional nonlinear consolidation theory and a hyperbolic load transfer model. Yu et al. [38] proposed a general solution for the consolidation of composite foundations with concrete-cored sand gravel (CCSG) piles based on the assumption of equal strain. FILZ et al. [39] calculated the maximum displacement difference between piles and soil in typical unit bodies, and established deformation coordination equations in embankments, grids, and foundation soil. Luo and Lu [40] derived analytical formulas for the settlement deformation and pile–soil stress ratio of rigid pile composite foundations based on the differential equation of pile–soil load transfer.

The above scholars have mainly focused on the load transfer mode, pile–soil stress ratio, embankment load bearing, and deformation characteristics of pile-net composite foundations in homogeneous strata. However, there is little research on the bearing capacity and deformation characteristics of embankments under different transition zones of a pile-net composite foundation and multi-soil stratum with uneven distribution. Also, there is relatively sufficient research on pile-net composite foundations in fields such as highways, railways, airports, and housing construction, while there are few research results in deep soft soil areas such as motor racing circuits in ultra-high speed environments. In addition, the embankment settlement of motor racing circuits has a significant impact on traffic safety. On the one hand, the uneven settlement of embankments can lead to road deformation, uneven road surfaces, and unstable vehicle driving, thereby increasing the incidence of traffic accidents. On the other hand, embankment settlement leads to terrain deformation, affecting the road structure and posing a great threat to driving comfort and safety. Therefore, it is necessary to adopt stricter standards for controlling embankment settlement and deformation to ensure traffic safety. Considering that a motor racing circuit has the characteristics of high speed traffic (300 km/h or above), a long route (several kilometers long), strict settlement control requirements, and the risk of uneven settlement and foundation slip after filling and loading in deep soft soil areas, there is currently a lack of engineering experience to reference under similar complex conditions and with a high standard of control. It can be seen that the treatment of the soft soil foundation in this project is extremely difficult, which poses great challenges to the safe construction and subsequent normal operation of the project. So, this paper took the Wuhan intelligent connected motor racing circuit as its setting, and conducted a pile-net composite foundation treatment and quantitative analysis on the areas where there might be significant post-construction settlement and settlement difference in the engineering area, such as deep soft soil sections, sudden geological changes, and transition sections with different pile types. On this basis, the rationality of the soft soil foundation treatment scheme was comprehensively evaluated. The research results and engineering practice indicate that the comprehensive treatment design scheme for the pile-net composite foundation is reasonable and feasible. While meeting the requirements of economic and rapid construction, it provides a guarantee of the safe construction of the motor racing circuit and also provides reference for the treatment of similar deep soft soil foundations.

2. Project Overview

2.1. Project Introduction

The Wuhan intelligent connected motor racing circuit is located on the bank of the Dongjing River in Wuhan city, Hubei Province, China, with an area of approximately 875,000 square meters. The circuit is the only F2 class track in the world that is combined with a closed T5 class testing field. The position of the motor racing circuit is shown in Figure 1.

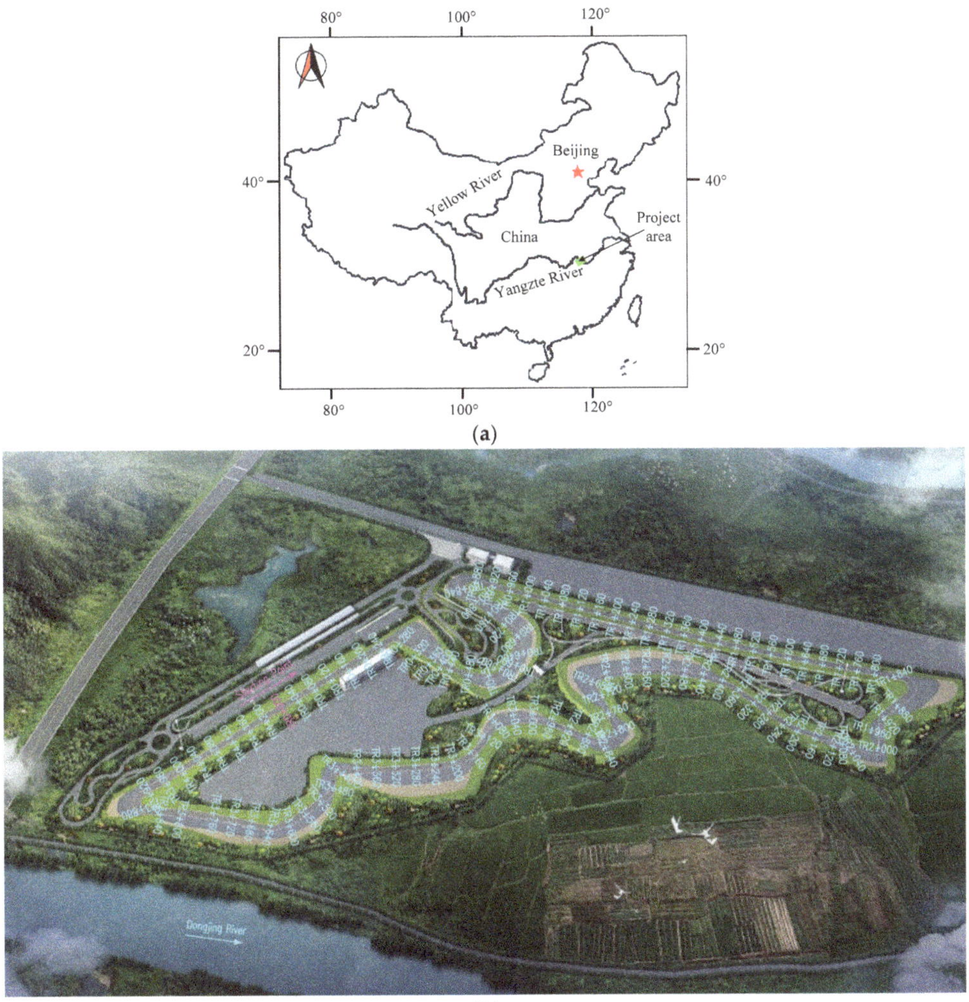

Figure 1. Project background, (**a**) regional map of the project in China. (**b**) Rendering of Wuhan International Circuit.

The main track of the motor racing circuit is a closed curve in a clockwise direction, with a mileage of K0 + 000~K4 + 237.21. The road has a total length of approximately 4.29 km, a width of approximately 50~190 m, and a speed limit of up to 300 km/h. The motor racing circuit is mainly used for high-speed control tests, high-speed network signal tests, and line-of-sight tests, etc., while also taking into account the performance competition of intelligent connected motor racing. Therefore, the tolerated settlement of the embankment in this project is relatively strict, with the post-construction settlement not exceeding 50 mm, and the longitudinal uneven settlement ratio of the embankment not exceeding 3/4000.

2.2. Geological Conditions

The racing track of this project mainly consists of ponds and ditches, with a small amount of wasteland distributed throughout. The shallow soil is pond mud, silt, cultivated soil, miscellaneous fill, etc., with an uneven distribution of soil quality. The site is located on the first terrace of the Yangtze River, belonging to the geomorphic unit of the river accumulation plain.

Within the depth range of this survey, the distribution and engineering geological characteristics of the main strata in the project are as follows: The 1-1 plain fill, mainly composed of plastic silty clay, is located on the shallow surface, with a layer thickness of 0.20–4.2 m. The 1-2 silt is in a flow plastic state, with a thickness of 0.7–4.5 m. The 2-1 clay is mainly in a plastic state, with a thickness of 0.4–4.8 m. The 2-2 silty clay is mainly in a soft-flow plastic state, with a thickness of 1.0–10.9 m. The 2a clay is mainly in a plastic state, with a thickness of 0.7–4.7 m. The 2b silt is in a moderately dense state, with a thickness of 0.6–5.4 m. The 2-3 clay is in a plastic to hard plastic state, with a thickness of 0.9–7.5 m. The 4 silty clay is mainly in a soft plastic state, with a thickness of 0.4–14.3 m. The 4b silt is in a moderately dense state, with a thickness of 0.5–9.2 m. The 5-1 silty clay is mainly in a plastic state, with local areas in a soft plastic state, with a thickness of 0.4–7.4 m. The 5-2 clay is in a plastic to hard plastic state, with a thickness of 0.6–7.3 m. The 5-3 clay is in a plastic to hard plastic state, with a thickness of 0.9–8.0 m. The 5-4 silicon clay is in a plastic to hard plastic state, with a thickness of 0.5–2.9 m. The 6-1 silt is in a moderately dense state, with a thickness of 0.2–15.0 m. The 6-2 silty sand is in a moderately dense state, with a thickness of 0.1–17.5 m.

The groundwater at the site mainly consists of upper stagnant water and pore-confined water. The upper stagnant water mainly exists in the pores of the first layer of topsoil. The water level is buried at a depth of 0~2.60 m and has an elevation of 19.27~22.18 m. Pore-confined water mainly exists in the three sandy soil layers, which are closely related to the hydraulic system of the Yangtze River. Due to the upper cohesive soil and muddy soil being relatively impermeable layers, there is basically no hydraulic connection between the confined water at the site and surface water, and its impact on the project will be negligible.

2.3. Foundation Treatment Method

To avoid the differential settlement caused by the filling of new and old embankments at the junction of ponds, ditches, and other areas, the area was excavated to a depth of 19.5 m before undergoing deep foundation treatment. The site was divided into a low filling area, medium filling area, and high filling area according to the height of the embankment filling. The low filling area, with a filling height less than 3.0 m, was treated with the CFG pile composite foundation. For medium or high filling areas greater than 3.0 m, the PHC pipe pile composite foundation treatment was adopted. The distribution of foundation treatment methods for the racing track is shown in Figure 2.

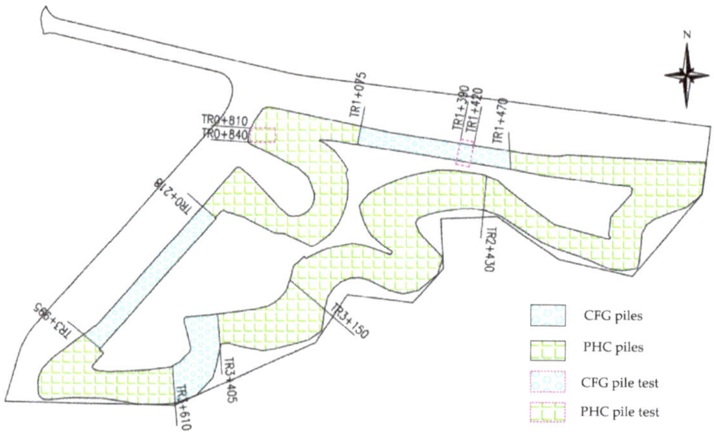

Figure 2. Schematic diagram of the embankment treatment methods for the racing track.

2.3.1. PHC Pile-Net Composite Foundation Treatment

The medium or high filling areas of the embankment were treated with a PHC pile-net composite foundation, as shown in Figure 3. According to geological conditions, the PHC pile has a wall thickness of 100 mm, arranged in a square with a pile spacing of 3.0 m. Also, it has a square pile cap of 1.5 m × 1.5 m × 0.35 m at the top, cast in place with C30 (compressive strength of 30 MPa) concrete. The top of the pile cap is laid with a 20 cm thick layer of crushed stone. After compaction, a number of steel and plastic geogrids (geogrids are laid at 20 cm intervals to the top of the embankment) are laid, and then filled with cement-stabilized soil and foamed light soil. The height of foamed light soil is determined according to the overall height of the embankment to ensure that the load at the top of pile is within the range of the target load. The length of a PHC pile is 15~25 m, and the designed characteristic value of a single pile's bearing capacity is 750~1100 kN.

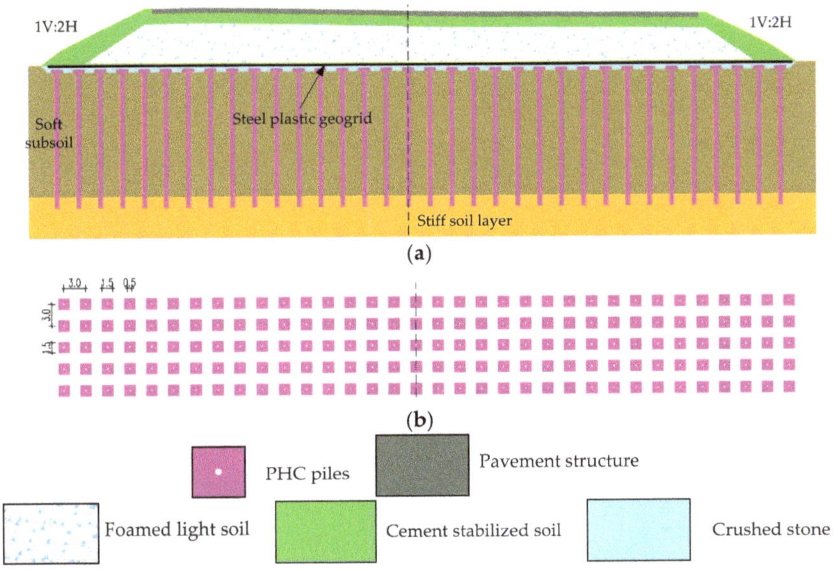

Figure 3. Treatment diagram of PHC pile-net composite foundation: (**a**) Cross section. (**b**) Pile location layout.

The PHC pile adopts the hammering method for pile sinking, and pile driving is carried out from the centerline of the track towards both sides. Before construction, the water in the pond should be drained. The silt at the bottom of the pond is removed down to 19.5 m, and a 40 cm layer of medium-coarse sand should be laid. After compaction, a 35 cm layer of graded crushed stone should be laid and compacted. Then, PHC pile construction can be carried out. Once the PHC pile is completed, the pile cap construction will be carried out, after excavation to 19.45 m.

In addition, before large-scale construction, PHC pile tests should be conducted on no less than three piles in each area. After 28 days of pile construction, the ultimate bearing capacity of a single pile can be determined by a static load test. The quality of the composite foundation should be mainly controlled by pile length, and the final penetration should not be less than 30 mm/10 blows.

2.3.2. CFG Pile-Net Composite Foundation Treatment

The low filling area of the embankment was treated with a CFG pile-net composite foundation, as shown in Figure 4. The CFG pile diameter is 0.5 m, arranged in a square with a pile spacing of 2.0 m. It has a square pile cap of 1.0 m × 1.0 m × 0.30 m at the top, cast in place with C30 concrete. The pile body is embedded 5 cm into the pile cap. A 20 cm thick layer of crushed stone is laid on the top surface of the pile cap. After compaction, a steel/plastic geogrid is laid (a geogrid is laid every 20 cm to the top of the embankment), and then filled with cement-stabilized soil. The height of the cement-stabilized soil is determined based on the overall height of the embankment, ensuring that the load on the pile top is within the range of the target load. The length of a CFG pile is 16 m~18 m, and the characteristic value of a single pile's bearing capacity is not less than 300 kN. The characteristic value of the composite foundation's bearing capacity is not less than 160 kPa, and the pile body's strength should reach C20 (compressive strength of 20 MPa), while the pile cap's strength should be C30.

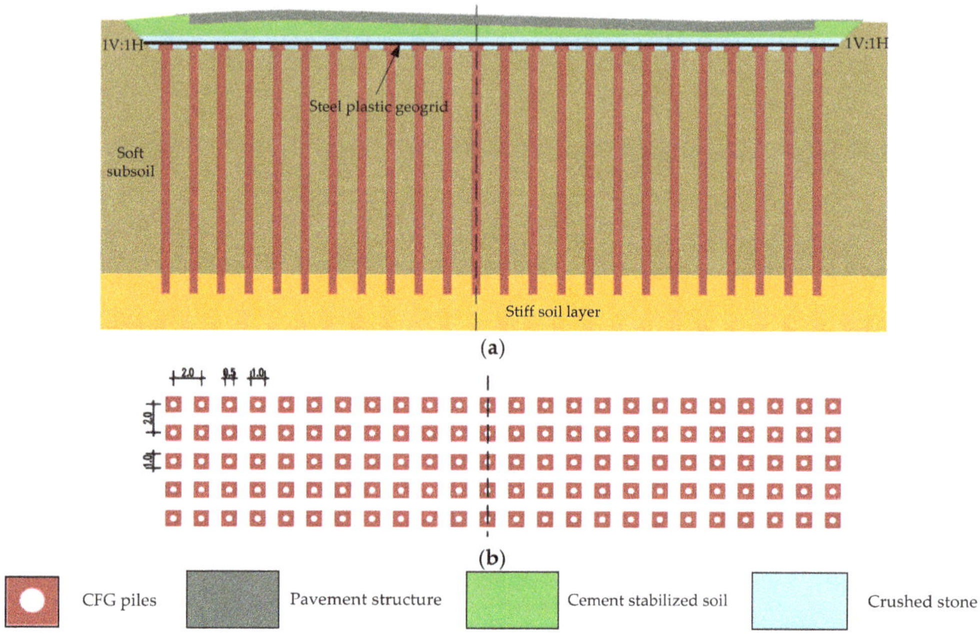

Figure 4. Treatment diagram of CFG pile-net composite foundation: (**a**) Cross section. (**b**) Pile location layout.

Before construction, unfavorable soil should be treated to a depth of 19.45 m, and then a 35 cm layer of graded crushed stone will be paved and compacted. Then, CFG pile construction can be carried out, and the pile's top elevation will be 19.5 m after excavation.

Similarly, before large-scale construction, CFG pile tests should be conducted on no less than three piles in each area. After 28 days of foundation pile construction, load tests should be used to determine the ultimate bearing capacity of a single pile and the ultimate bearing capacity of the composite foundation.

3. Field Tests

3.1. Test Sections

In order to obtain the pile–soil deformation characteristics of the pile-net composite foundation and provide a basis for the post-construction settlement prediction of the motor racing circuit, the Wuhan Institute of Rock and Soil Mechanics, Chinese Academy of Sciences, set up two representative test sections according to the thickness distribution difference of the soft soil and different pile types, as shown in Figure 2. In each test section, two monitoring cross sections were arranged, and the monitoring profiles of the different pile types are shown in Figures 5 and 6.

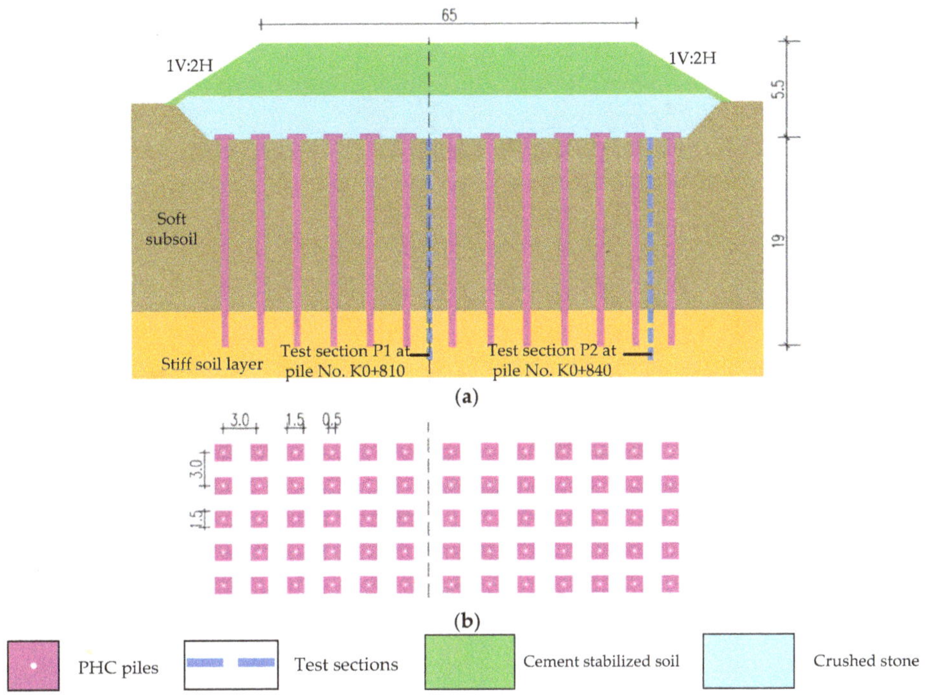

Figure 5. Loading schemes for PHC test: (**a**) Cross section. (**b**) Pile location layout.

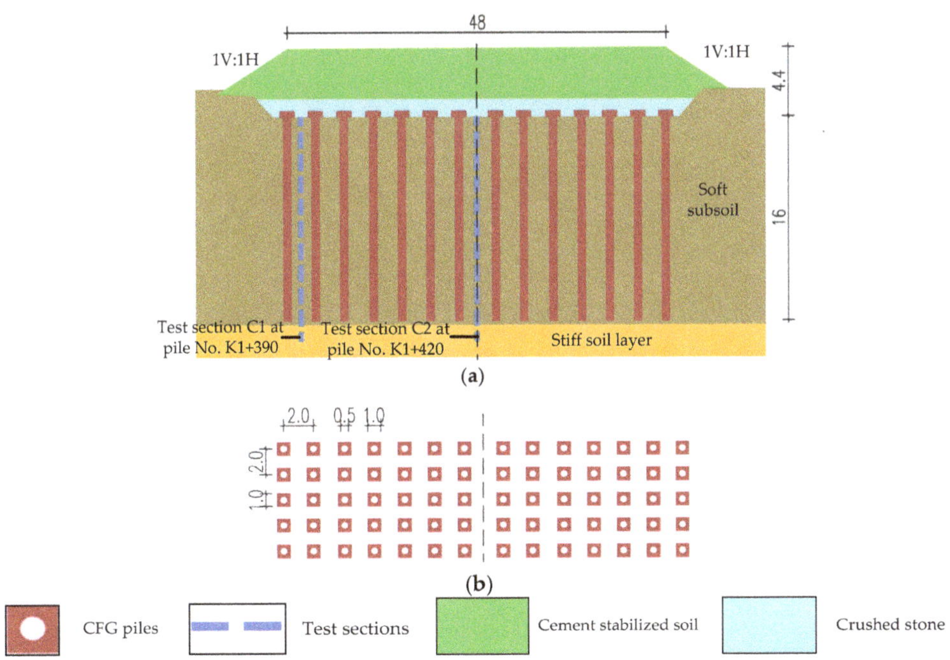

Figure 6. Loading schemes for CFG test: (**a**) Cross section. (**b**) Pile location layout.

(1) PHC test

Pile No. K0 + 810~K0 + 840 was treated with a PHC pile-net composite foundation with a length of 30 m and a width of 65 m. The P1 and P2 sections were located at the center and edge of the load, respectively. The distance between the P1 and P2 sections was 20 m.

(2) CFG test

Pile No. K1 + 390~K1 + 420 was subjected to a CFG pile composite foundation treatment with a length of 30 m and a width of 48 m. Similarly, the C1 and C2 sections were located at the edge and center of the load, respectively. The distance between the C1 and C2 sections was 20 m.

3.2. *Layout of Monitoring Points*

The settlement monitoring of this project is the key research content. The settlement of piles, settlement of soil between piles, peripheral settlement of the surcharge area, stratified settlement, surface horizontal displacement, deep horizontal displacement, pore water pressure, soil pressure monitoring, etc., should be tested. Since this section focuses on the monitoring of the settlement and deep horizontal displacement of pile tops and the soil between piles, the rest of the research is not discussed in detail, as recommended by study [41]. The monitoring sections are shown in Figures 5 and 6.

The settlement of the piles was observed by embedding settlement plates and test tubes at the pile head under the action of a pile load. Each test section selected five PHC piles or CFG piles, and there were a total of ten monitoring points, numbered ZT1~ZT10 (ZT stands for pile head). The monitoring points for the soil settlement between piles were arranged in the center of the pile group, designated as ZJ (ZJ stands for soil between piles), and there were ten monitoring points in total, numbered ZJ1~ZJ10. Deep horizontal displacement was observed via an inclinometer tube, and three deep horizontal displacement monitoring holes were arranged on the periphery of the monitoring section.

3.3. *Test Scheme and Duration*

After the test instrument has been checked, calibrated, and buried, the load test can be carried out. Based on the actual filling of the pile foundation, the actual load of the PHC and CFG pile-net composite foundations is 1.3 and 1.5 times that of the later working load of the pile foundation, respectively. Considering that the geogrid is easily damaged during construction, which affects the full performance of the reinforcement effect of the embankment, the geogrid material is taken as a safety reserve in the test and the subsequent analysis of the project, and its influence on the embankment is ignored. The loading profile of each test section is shown in Figures 5 and 6.

The project began in early December 2020 and finished in mid-December. Monitoring started after the full load was completed, and the monitoring was finished on 16 March 2021, lasting about 100 days. During the loading period and the 100 days after loading, there was no obvious soil collapse except for local small-scale soil collapses caused by continuous rainfall.

3.4. *Test Results and Analysis*

3.4.1. Settlement of the Pile Top and Soil between Piles

According to the measured data, the variation curves of typical section settlement over time under different pile tests are drawn, as shown in Figure 7. As can be seen from the figure, for the PHC pile-net composite foundation, the maximum settlement of pile head ZT3 was 13.6 mm, and the maximum settlement of soil ZJ4 between piles was 50.41 mm. The settlement of soil between the piles was 3.5~3.7 times that of adjacent pile heads. Similarly, for the CFG pile-net composite foundation, the maximum settlement of pile head ZT9 was 13.34 mm, and the maximum settlement of soil ZJ8 between piles was 39.65 mm. The settlement of soil between the piles was 2.7~3.0 times that of adjacent pile heads.

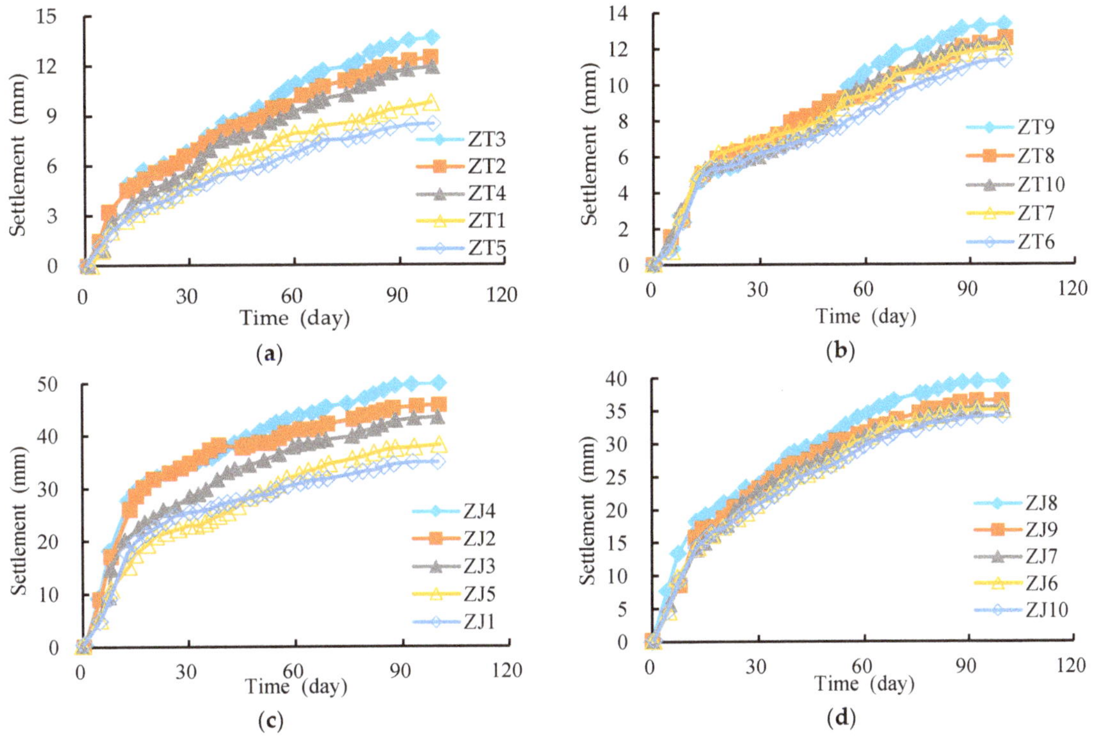

Figure 7. Relationship between soil cumulative settlement at the pile top or between piles and time under different pile tests: (**a**) Pile top of PHC test. (**b**) Pile top of CFG test. (**c**) Soil between piles of PHC test. (**d**) Soil between piles of CFG test.

Further analysis showed that in both the PHC pile and CFG pile composite foundation treatment areas, the settlement of the pile head and soil between the piles increased rapidly during loading, and decreased immediately after loading, but the cumulative settlement value still increased in a hyperbolic manner. By early March of 2021, the slope of the settlement–time curve began to decrease, that is, the settlement rate began to slow down.

3.4.2. Deep Horizontal Displacement

The deep horizontal displacement in the load area was monitored. Taking the test data at an interval of 20 days as the object, the variation curve of the horizontal displacement monitored by the inclinometer tube over time is shown in Figure 8. It can be seen from the figure that during the monitoring period, the lateral displacement of the soil in the load area did not exceed 1.5 mm, and the direction of lateral displacement was spread outward from the load center. On the whole, no obvious lateral compression deformation occurred in the soil of the PHC pile or CFG pile areas, and the lateral displacement gradually decreased with the increase of depth.

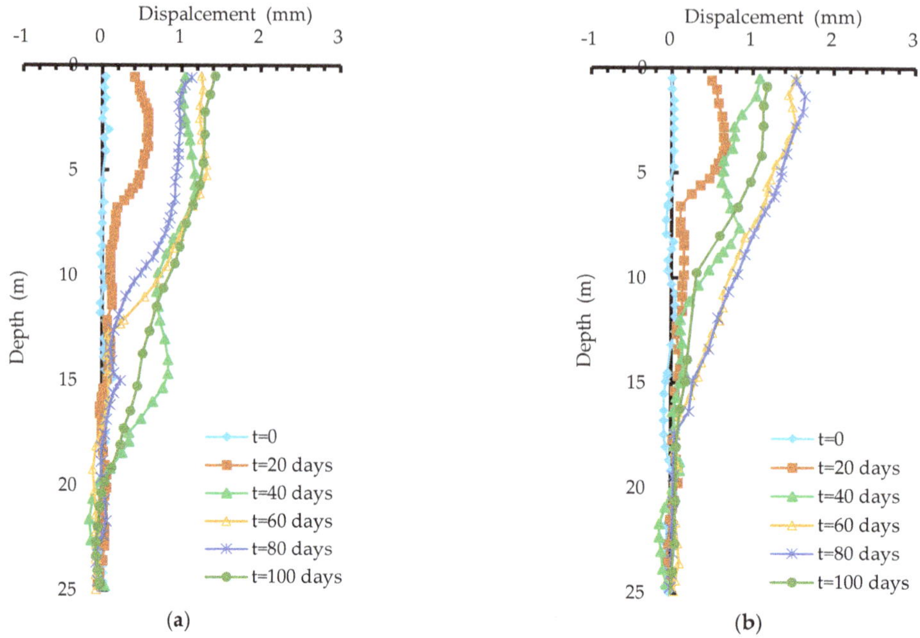

Figure 8. Curves of horizontal displacement of an inclinometer pipe with time under different pile tests: (**a**) PHC test. (**b**) CFG test.

4. Back Analysis of Stratum Parameters and the Reinforcement Treatment of Embankments

According to reference [41], the pile-net composite foundation in this test area generated a maximum excess pore water pressure of 80 kPa due to overloading, and the pore water pressure gradually dissipated with time. By early March 2021, there was still excess pore water pressure close to 20 kPa that had not yet dissipated. In theory, settlement may still have occurred during the test section, so it is necessary to predict the subsequent settlement.

4.1. The Numerical Model and Mesh Grids

The large commercial software 'MIDAS GTS NX 2022 R1 [42]' was used for the numerical simulation of this project. The calculation zones are located in sections of the PHC test and CFG test, and the numerical models are 100 m × 32 m (X × Y) and 100 m × 30 m (X × Y), respectively. The X-axis is parallel to the road axis and is positive along the road. The Y-axis extends from elevation −6.5 m to the top of the load. As shown in Figure 9, the grid model is set up based on topography, geological conditions, and loading materials, etc. Figure 9a shows the finite element model of the PHC pile-net composite foundation, which is divided into 13,005 grid elements and 13,831 nodes, respectively. Similarly, Figure 9b shows the calculation model of the CFG pile-net composite foundation, with a total of 11,628 grid elements and 13,108 nodes.

Considering the calculation range of the model and the arrangement of the pile body, the solid model in this paper is divided using quadrilateral grid elements. Through continuously adjusting the grid size (such as 1 m, 2 m, 3 m, 4 m, 5 m) and comparing the displacement calculation values with the measured ones, it was found that when the grid size of the soil strata, the piles, and pile cap are 2 m and 1 m, respectively, the calculation results are in good agreement with the measured values, and the calculation efficiency is optimal.

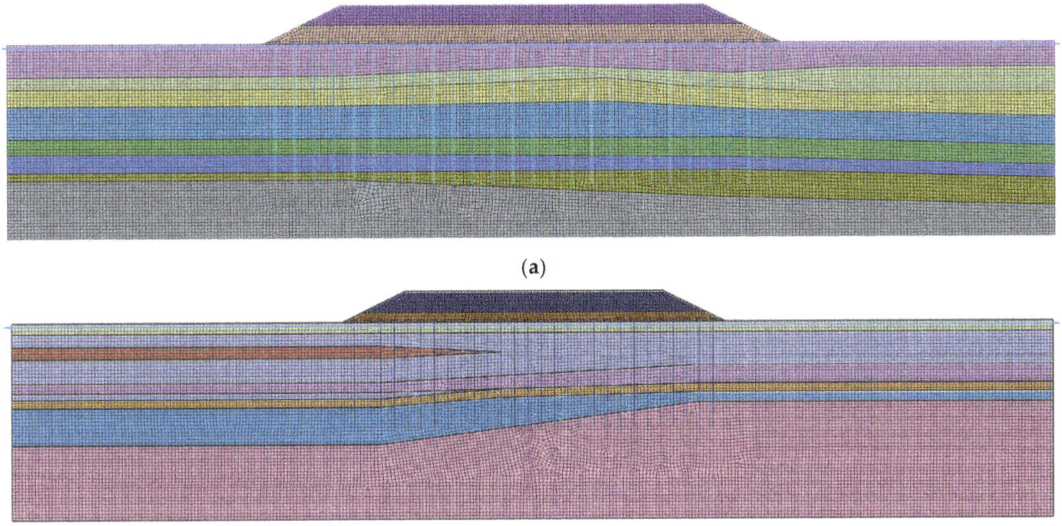

(a)

(b)

Figure 9. Finite element mesh model: (a) PHC pile-net composite foundation. (b) CFG pile-net composite foundation.

4.2. Constitutive Model and Computational Conditions

In the numerical calculation, the solid element is used to describe the behavior of the soil, crushed stone cushion, and embankment filling. The Mohr–Coulomb yield criterion is used for the constitutive model and soil consolidation is also considered. The elastic constitutive model is adopted for the pile cap, pile body, and pavement structure. The physical and mechanical parameters of the soil, loading materials, and pavement structure are shown in Tables 1 and 2, respectively. The structural parameters of the piles and their caps are shown in Table 3.

Table 1. Physical and mechanical parameters of the soil.

Category	Unit Weight /(kN/m^3)	Permeability Coefficient /(10^{-7}m/s)	Compression Modulus /MPa	Poisson's Ratio	Cohesion /kPa	Friction Angle /(°)
1-1 plain fill	19.9	1.0	5.3	0.3	12	10
1-2 silt	16.8	0.1	2	0.42	10	4
2-1 clay	19.5	0.2	5.2	0.32	18	10
2-2 silty clay	17.8	0.1	3	0.39	12	6
2-3 clay	19.9	0.2	7	0.4	22.8	12.4
2a clay	19.2	10	6.6	0.41	23	12
2b silt	19.5	0.6	5.7	0.35	15	21
4 silty clay	18.9	0.1	4.5	0.39	13.5	9.3
4b silt	19.6	50	8.5	0.35	14	23
5-1 silty clay	19.1	0.1	5.5	0.36	18.8	11
5-2 clay	19.2	0.2	7.5	0.38	26.5	13.2
5-3 clay	19.5	0.2	9.2	0.27	39.5	14.7
5-4 Silty clay	19.6	0.1	6.2	0.35	21	11
6-1 Silt	19.7	50	10	0.3	10	20
6-2 Silty sand	20	500	17	0.3	0	34

Table 2. Physical and mechanical parameters of loading materials and pavement structure.

Category	Unit Weight /(kN/m^3)	Elastic Modulus /MPa	Poisson's Ratio	Cohesion /kPa	Friction Angle /(°)
Crushed stone	22	50	0.25	0	30
Embankment filling	19	40	0.25	23	11
Pavement structure	25	3000	0.2	-	-

Table 3. Structural parameters of piles and their caps.

Category	Unit Weight /(kN/m^3)	Elastic Modulus /GPa	Poisson's Ratio	Normal Stiffness /(10^6 kN/m^3)	Shear Stiffness /(10^6 kN/m^3)
CFG pile	22	10~20	0.2	1~4.5	0.09~0.4
PHC pile	25	15~30	0.2	1.3~12	0.1~1.1
Pile caps	25	30	0.2	/	/

The normal displacement at the bottom of the model, as well as the horizontal displacement at the periphery of the model, are all constrained. The action of tectonic stress is ignored, only self-weight is considered.

4.3. Back Analysis of Stratum Parameters

The soil mechanical parameters mentioned above are obtained from laboratory tests, and the influence of pile construction and embankment filling disturbance has not been considered. In addition, the structure parameters of the piles are all empirical values. Therefore, it is necessary to invert the soil mechanical parameters and pile modulus in the project by combining the settlement monitoring data of the piles and soil between piles, so as to obtain more practical calculation parameters, considering construction disturbance.

The flow chart of the back analysis of soil and pile parameters is shown in Figure 10, mainly involving several aspects, which are considered using the following approaches:

(1) The selection of monitoring information on displacement. Prior to back analysis computation, proper monitoring information on displacement needs to be selected so as to obtain reasonable equivalent soil mechanical parameters and a pile modulus. In general, the selection should follow the following principles: (a) the monitoring information needs to be accurate and reliable without abnormal changes; (b) the monitoring information should be complete, information recorded by pre-buried monitoring equipment should be selected as a priority to carry out the back analysis of complete displacement; (c) the monitoring information needs to be adequate and representative; and (d) the monitoring information should change slightly and no invalid values should be used. Based on the above selection principles, the monitored data of soil settlement between piles at the two test sections (see Section 3.1) are selected. In this way, the monitoring points involved in the back analysis of the mechanical parameters of soil can be easily determined.

(2) Determination of inversion parameters. Soil strata at the test sections mainly include 2-2 silty clay, 2a clay, 4 silty clay, and 6-1 silt, and the parameters of the pile modulus have a significant impact on the embankment settlement. Namely, six parameters highly sensitive to embankment settlement are subjected to back analysis, including the deformation moduli of 2-2 silty clay, 2a clay, 4 silty clay, and 6-1 silt, and the elastic modulus of PHC piles and CFG piles. Therefore, in total six parameters need to undergo back analysis.

(3) The back analysis method. By optimizing the back propagation (BP) network structure through a genetic algorithm (GA), the GA-BP method can be formed [43]. This method has great advantages in its nonlinear mapping and global search ability, and has been widely applied in the back analysis of parameters. In order to obtain reasonable soil and pile parameters, the uniform design method and the GA-BP method are comprehensively used for parameters' inversion. First, the scope of the parameters

for back analysis were determined based on the results of field tests, laboratory experiments, and engineering experience. Then, the uniform design theory was used to design experiments through combinations of multiple parameters and levels, thus obtaining multiple groups of samples for orthogonal analysis. Afterwards, the above samples were imported into a neural network optimized by the genetic algorithm to perform network training and attain the nonlinear mapping relation between the parameters for back analysis and the monitored displacement. Finally, the measured values were input into the trained network to obtain the mechanical parameters for back analysis. According to calculation results from back analysis, the deformation moduli of 2-2 silty clay, 2a clay, 4 silty clay, and 6-1 silt within the disturbance range were 6.2 MPa, 13.2 MPa, 9.0 MPa, and 21.0 MPa. The elastic moduli of the PHC piles and CFG piles are 19.5 GPa and 12.0 GPa, respectively.

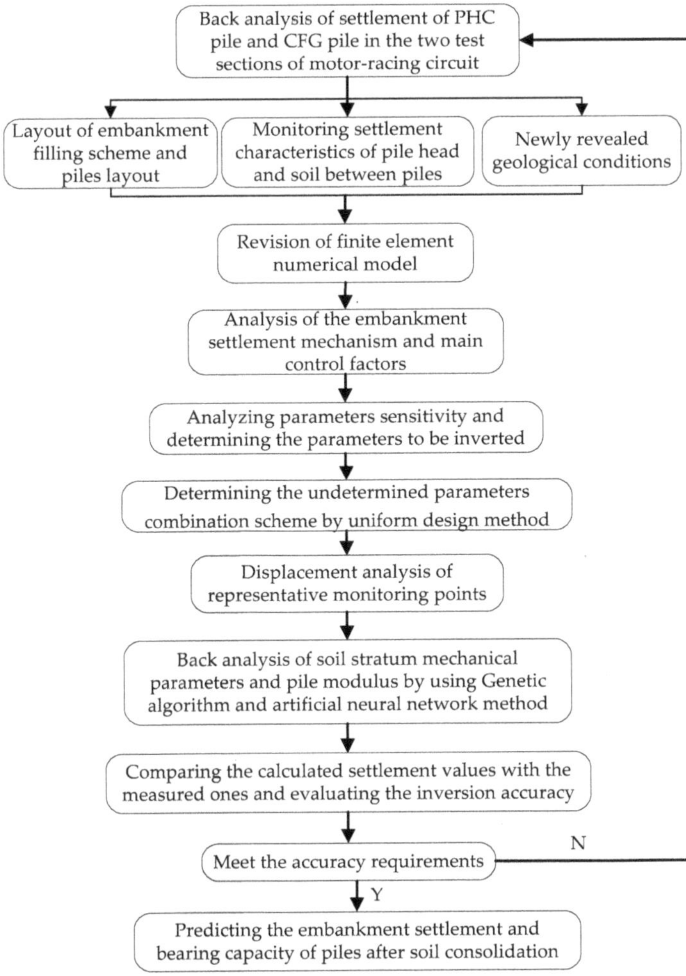

Figure 10. Schematic diagram of the back analysis of soil mechanical parameters and pile moduli.

4.4. Evaluation of the Results Using the Back Analysis Method

Figure 11 plots the comparison between the calculated and measured values of soil settlement between piles under different pile-net composite foundation treatments. Figure 12 shows the contour maps of settlement after soil consolidation under different pile-net composite foundation treatments. It can be seen that the variation trend of the measured data and the calculated value are basically consistent. The results can be specifically listed as follows: (1) when the PHC pile test was finished, the measured maximum settlement of the soil between piles was 50.4 mm, and the calculation value via back analysis was 54.5 mm, with an error of 8.2%. After the soil consolidation was completed, the maximum settlement was 70 mm. (2) Similarly, once the CFG pile test was finished, the measured maximum settlement of the soil between the piles was 39.65 mm, and the calculation value was 43.4 mm, with an error of 9.6%. After soil consolidation was basically completed, the maximum settlement reached 59 mm. (3) Whether the PHC pile or CFG pile test, the calculation values from back analysis were slightly greater than the measured ones, and the maximum error did not exceed 10%, indicating that the inversion accuracy was good and could meet the engineering requirements.

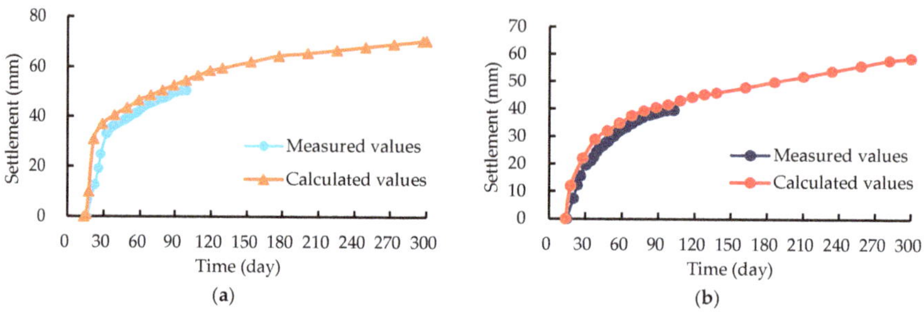

Figure 11. Comparison between the calculated and measured values of soil settlement between piles under different pile-net composite foundations: (**a**) PHC pile-net composite foundation. (**b**) CFG pile-net composite foundation.

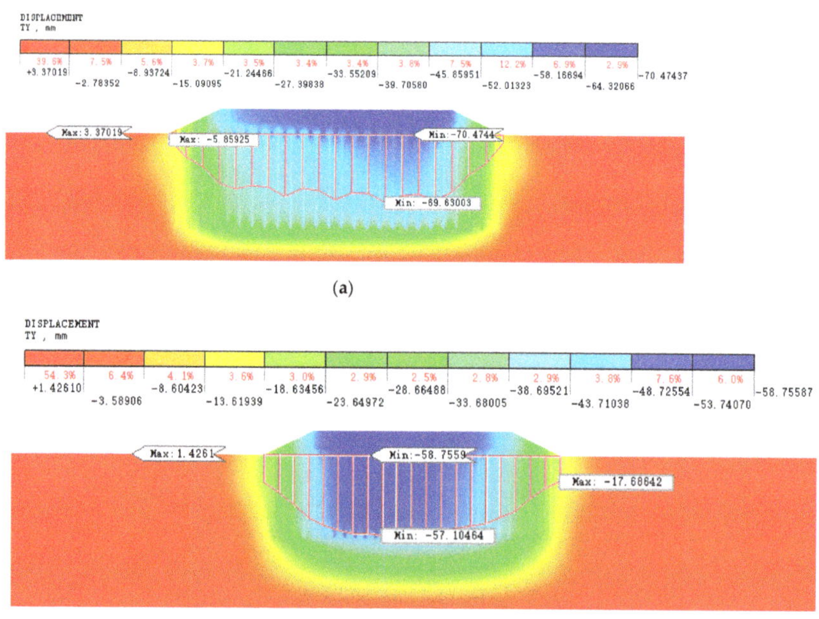

Figure 12. Contour maps of settlement after soil consolidation under different pile-net composite foundations: (**a**) PHC pile-net composite foundation. (**b**) CFG pile-net composite foundation.

However, due to the fact that after the field test there was still an excess pore water pressure of nearly 20 kPa that had not dissipated, there might still be settlement in the test sections, which was consistent with the trend of the calculation results. Furthermore, the maximum settlement of the soil between piles through the PHC and CFG pile-net composite foundation treatments could reach 70 mm and 59 mm, respectively, which were greater than the specified 50 mm settlement standard. Therefore, it is necessary to reinforce embankments in these areas.

4.5. Embankment Reinforcement Treatment and Its Effectiveness

4.5.1. Embankment Reinforcement Treatment

In the PHC pile test area, the bottom of the PHC pile has not entered the stable stratum, and the bearing capacity of the pile-net composite foundation is insufficient due to poor geological conditions. The cross-section of the PHC pile reinforcement is shown in Figure 13a. The layout of the pile positions in the reinforcement section is shown in Figure 13b. The reinforcement treatment methods are described as follows: (1) The driving of the PHC piles was repeated and penetration control adopted, that is, the final penetration should not be less than 30 mm/10 blows. (2) The soil between piles was reinforced by cement–soil mixing piles, which passed through 2-2 silty clay and 2a clay for about one meter.

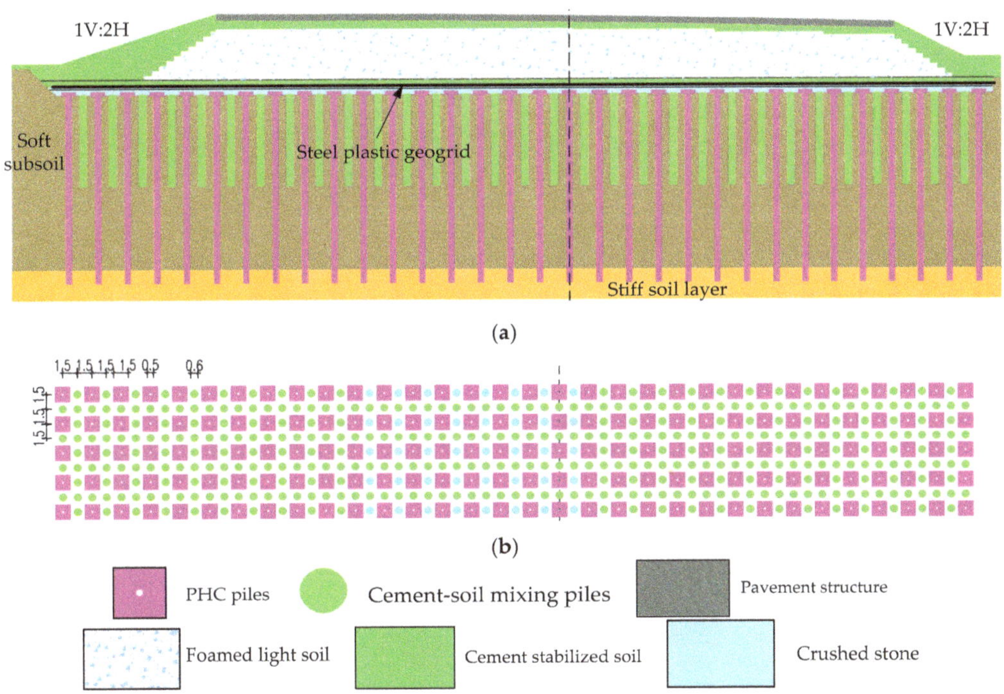

Figure 13. Reinforcement diagram of pile-net composite foundation embankment in the PHC test section: (**a**) Cross section. (**b**) Pile location layout.

Similarly, for the CFG pile test area, due to the poor geological conditions, the bearing capacity of the CFG pile-net composite foundation was insufficient. The cross-section of the CFG pile reinforcement is shown in Figure 14a. The layout of the pile positions in the reinforcement section is shown in Figure 14b. In order to improve the bearing capacity of the composite foundation, additional CFG piles need to be installed in these areas, and the pile bottom should penetrate the sand to about one meter.

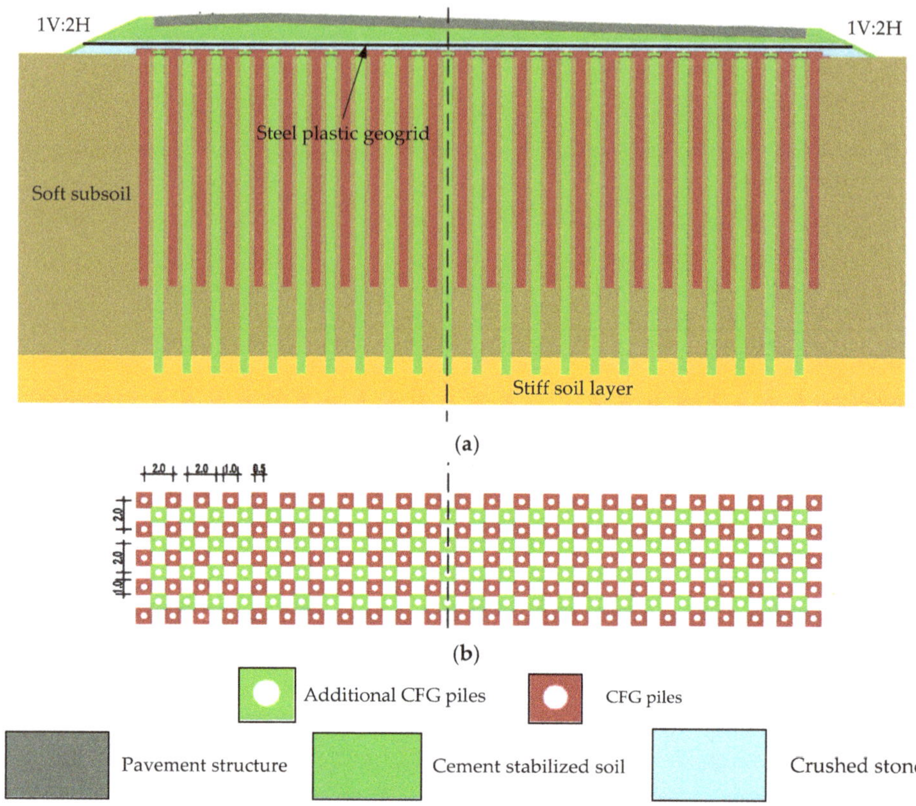

Figure 14. Reinforcement diagram of pile-net composite foundation embankment in CFG test section: (**a**) Cross section. (**b**) Pile location layout.

4.5.2. On Site Implementation Effect

In order to obtain the settlement after embankment reinforcement, four observation points of the pile top settlement and four observation points of the soil settlement between piles were set up in the PHC pile test area. In addition, in the CFG pile test area, two settlement observation points in the middle of the embankment and two monitoring points in the shoulder of embankment were also set up. By November 2021, the filling of the embankment had been completed, the pore water pressure had started to dissipate and the settlement was basically stable.

The cumulative settlement curve of each point after reinforcement in the PHC pile test area was shown in Figure 15a. As shown in the figure, after the completion of embankment construction, the cumulative minimum settlement point was ZD3, with a settlement value of 7.11 mm. The cumulative maximum settlement point was ZJ4, with a settlement value of 21.01 mm. Similarly, the cumulative settlement curve of each point in the CFG pile test area is shown in Figure 15b. It can be seen that after the completion of embankment construction, the cumulative minimum settlement point was LJ1, with a settlement value of 13.77 mm. The cumulative maximum settlement point was LZ2, with a settlement value of 15.88 mm. The cumulative settlement at each measuring point in the two test areas meets the design requirements.

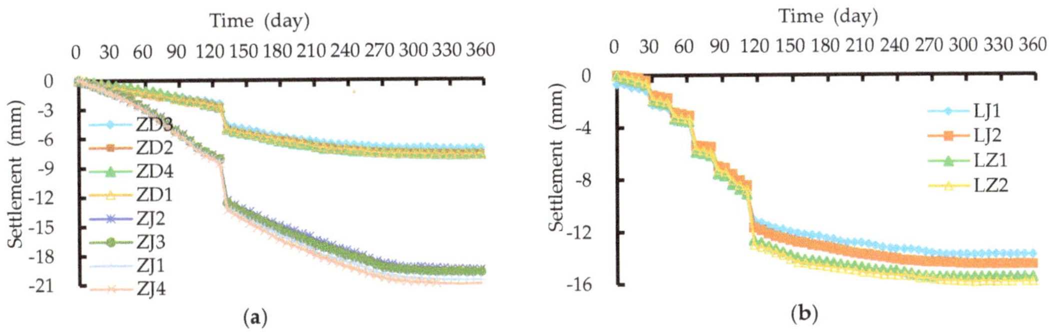

Figure 15. Accumulated settlement curves of the pile top and soil between piles after reinforcement in different pile test areas: (**a**) PHC pile test. (**b**) CFG pile test.

5. Numerical Simulation Analysis of the Typical Working Conditions of the Intelligent Connected Motor Racing Circuit

The motor racing circuit mainly adopted CFG pile and PHC pile-net composite foundations. In order to evaluate the differential settlement caused by different embankment treatment methods in the transitional area and the area where the settlement is likely to be large (such as sudden changes in the strata), five typical working conditions were selected for the finite element numerical simulation analysis based on geological conditions, pile-net size and layout, and embankment settlement standards. The specific working conditions and pile numbers were (1) Condition 1: the PHC pile area in the cross-section at pile number TR3 + 260; (2) Condition 2: the CFG pile area in the cross-section at pile number TR1 + 440; (3) Condition 3: the PHC pile area along the longitudinal section between pile number TR1 + 700 and TR1 + 800; (4) Condition 4: the CFG pile area along the longitudinal section between pile number TR3 + 460 and TR3 + 560; and (5) Condition 5: the PHC pile and CFG pile transition area between pile number TR0 + 160 and TR0 + 260. The typical working condition positions are shown in Figure 16.

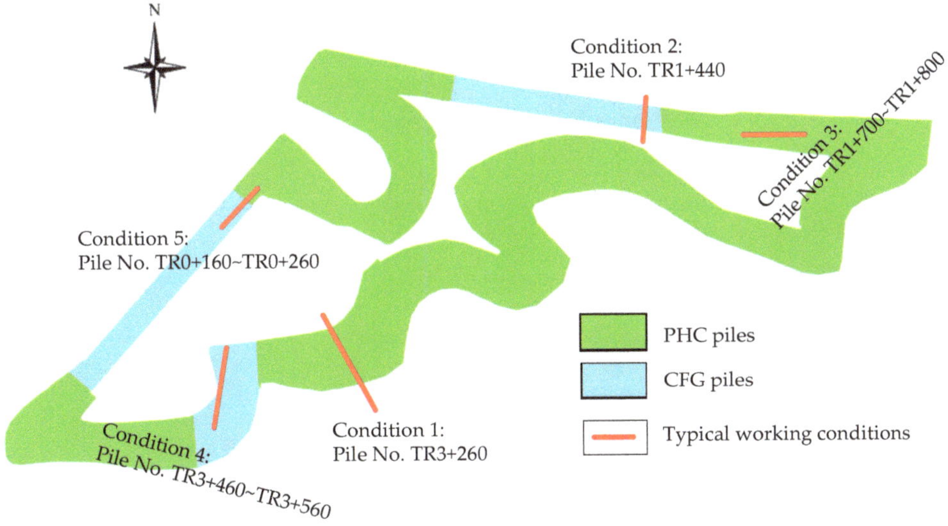

Figure 16. Typical working condition positions of the motor racing circuit.

5.1. Model Construction and Mesh Grids

Typical two-dimensional numerical models are established based on soil strata, embankment filling materials, pile caps, pile bodies, the pavement structure, etc. In order to reduce the influence of the boundary effect of the model on the calculation results, the width of the model should be 2~3 times the width of the embankment, and the depth of the model is should be as 1.5~2 times that of the pile length [44]. The Y-axis is positive along the elevation direction, and the X-axis is positive along the road pile number direction. The size of the numerical model is 100 m × 40 m (X × Y).

The finite element numerical models under typical working conditions are shown in Figure 17. Figure 17a,b plot the cross-sectional calculation models of the PHC pile-net composite foundation under Condition 1 and the CFG pile-net composite foundation under Condition 2, respectively. Figure 17c,d plot the longitudinal section calculation models of the PHC pile-net composite foundation under Condition 3 and the CFG pile-net composite foundation under Condition 4, respectively. Figure 17e plots the transition zone model of the PHC pile and CFG pile-net composite foundations under Condition 5.

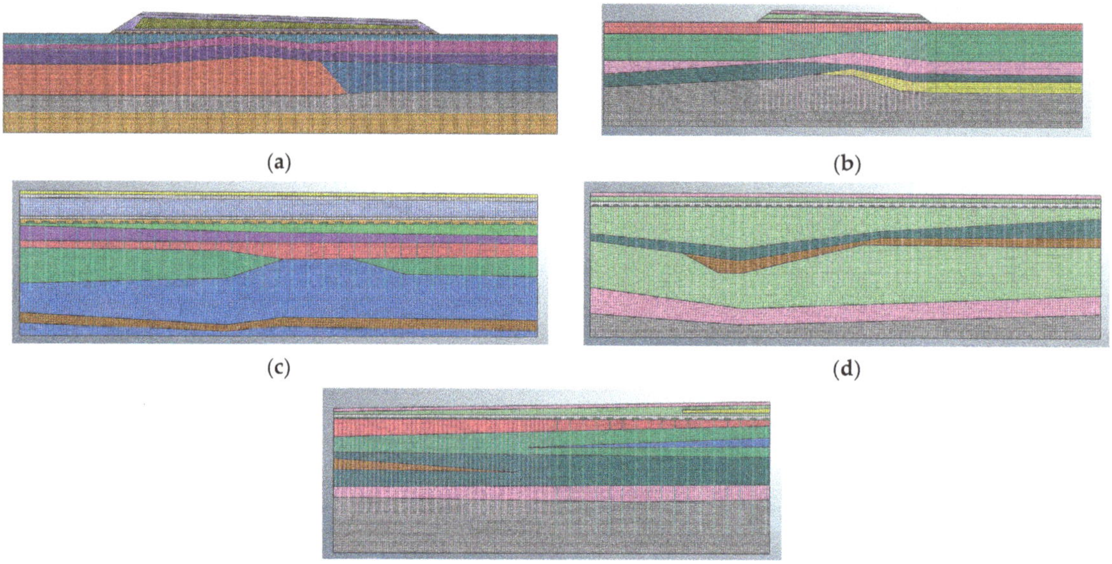

Figure 17. Calculation models under typical working conditions: (**a**) Condition 1. (**b**) Condition 2. (**c**) Condition 3. (**d**) Condition 4. (**e**) Condition 5.

5.2. The Constitutive Model and Calculation Conditions

Solid elements were used to simulate each soil stratum and embankment filling in the model, and the Mohr–Coulomb elasto-plastic criterion was adopted in the constitutive model. The elastic constitutive model was adopted for the pile cap, pile body, and pavement structure. Some soil stratum parameters, such as 2-2 silt clay, 2a clay, 4 silty clay, and 6-1 silt, as well as the elastic modulus of the PHC pile and CFG pile bodies, were obtained via back analysis (see Section 4.3). The other parameters were consistent with Section 4.2.

The bottom displacement of the model and the horizontal displacement on both sides of the model were constrained. The groundwater level in the model was set to one meter below the surface. Since the embankment was built in a soft soil stratum with a shallow groundwater level, the excess pore water pressure would suddenly increase and effective stress would decrease during the filling process, so the generation and dissipation of the excess pore water pressure should be considered. Consolidation analysis was adopted here. The hydraulic boundary conditions were set as follows: both sides of the model were undrained boundaries, and the surface and bottom of the model were drained boundaries.

5.3. Analysis of Calculation Results

5.3.1. The Effect of the Pile-Net Composite Foundation on Settlement

Figure 18a,b show the settlement displacement contour maps of the pile-net composite foundation along the cross-section of the embankment under Condition 1 (PHC pile) and Condition 2 (CFG pile) at the end of soil consolidation, respectively. The figures also plot the surface settlement curves under different working conditions. It can be seen that within the range of the foot of the embankment, the settlement curve of the embankment approximately presents an inverted trapezoidal feature of being large in the middle and small on both sides. The maximum settlement was approximately located in the middle of the embankment's loading, and the value was positively correlated with the height of the loading. At the foot of the embankment, as it moved away from the foot of the slope, the vertical displacement of the embankment gradually changed from settlement to upward uplift. This is mainly due to the sliding trend of the embankment slope towards the outside caused by loading, and the foot of the embankment slope was subjected to lateral compression. Further analysis showed that in both Condition 1 and Condition 2, there was a significant soil arching effect between the piles, which transferred the overlying load to the pile at the arch foot through the soil arching, thereby improving the bearing capacity of the pile-net composite foundation. In addition, there was relative deformation between the pile and the soil, which was mainly caused by the difference in stiffness. After soil consolidation was completed, the maximum settlement deformation of the embankment in Condition 1 and Condition 2 was 32.1 mm and 18.1 mm, respectively.

Similarly, Figure 18c,d plot the contour maps of the settlement displacement of the pile-net composite foundation along the longitudinal section of the embankment under Condition 3 (PHC pile) and Condition 4 (CFG pile) at the end of soil consolidation, respectively. The figures show the surface settlement curves under different working conditions. From Figure 18c, it can be seen that the settlement curve of the longitudinal section of the embankment was approximately rectangular, indicating that there was little difference in the settlement deformation of the foundation under the same pile-net composite foundation when the loading and the pile length were constant. The settlement mainly depended on geological conditions, and the maximum settlement was located in areas with poor geological conditions. However, the settlement curve of the embankment's longitudinal section in Figure 18d can be divided into two rectangles, with differential settlement occurring at the boundary. This is mainly due to the significant variation in the strata of this section (as shown in Figure 17d), and the length of the CFG pile in the longitudinal section of the embankment had changed, resulting in differences in settlement between the two. In addition, the soil arching effect between piles was more obvious, further indicating the soil reinforcement effect of the pile-net composite foundation. After the completion of soil consolidation, the maximum settlement deformation of the foundation under Condition 3 and Condition 4 was 33.8 mm and 19.8 mm, respectively.

Figure 18e shows the contour map of the settlement displacement of the pile-net composite foundation along the longitudinal section of the embankment under Condition 5 (the transition zone between PHC and CFG piles) at the end of soil consolidation. Similarly, the figure also shows the surface settlement curve under Condition 5.

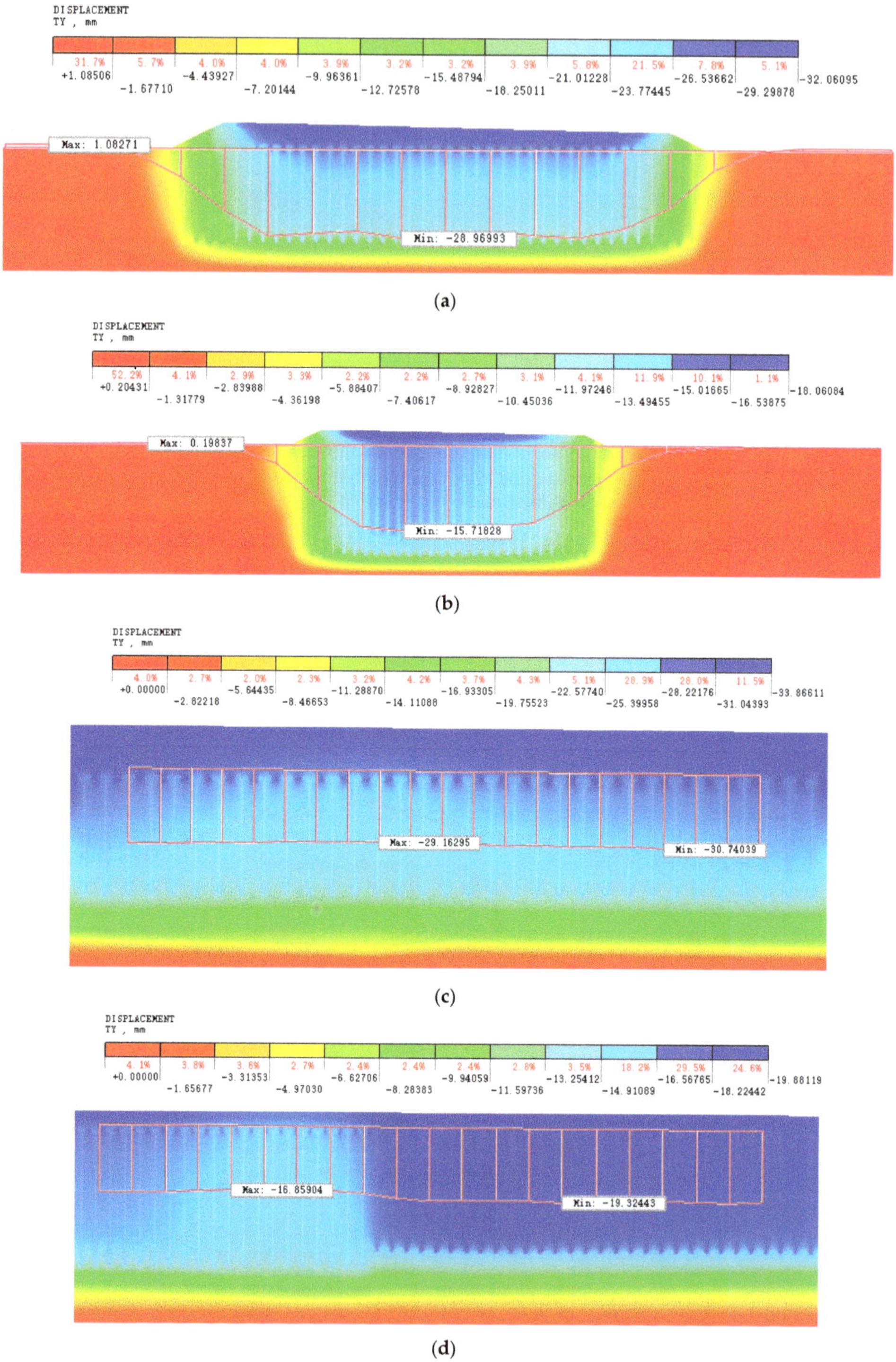

Figure 18. *Cont.*

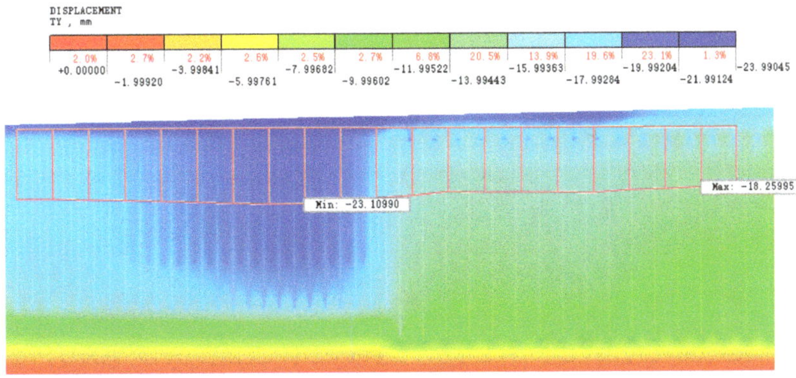

(e)

Figure 18. Settlement contour maps and curves of the embankment bottom under different working conditions at the end of soil consolidation: (**a**) Condition 1. (**b**) Condition 2. (**c**) Condition 3. (**d**) Condition 4. (**e**) Condition 5.

From the figure, it can be seen that the settlement curve of the longitudinal section of the embankment can be divided into two rectangles, and there was a significant difference in settlement at the boundary of the two rectangles. This is mainly due to the fact that this section belonged to a transitional zone and two foundation treatment methods, PHC pile and CFG pile, were adopted. The different stiffness of the two types of pile led to the difference in settlement. Further analysis showed that the deformation of soil with the PHC pile was significantly smaller than that with the CFG pile, indicating that the pile-net composite foundation with a higher stiffness had better settlement control. In addition, whether it was the PHC pile or CFG pile, there was a significant soil arching effect between piles, indicating that the pile spacing in this project was relatively reasonable, fully exerting the comprehensive reinforcement effect of a pile-net composite foundation. After soil consolidation was completed, the maximum settlement deformation of the embankment under Condition 5 was located at the boundary between the two types of pile, with a value of 30 mm.

5.3.2. The Effect of the Pile-Net Composite Foundation on Lateral Deformation

In order to analyze the lateral displacement characteristics of the soil at the bottom of the embankment after soil consolidation, the PHC pile-net composite foundation under Condition 1 was taken as an example.

Figure 19 shows the contour map of the lateral displacement of soil along the cross-section of the embankment of the pile-net composite foundation under typical working conditions. The figure also shows the variation curve of the lateral displacement of typical soil with depth. As shown in the figure, the lateral displacement of the deep soil stratum in the middle of the embankment was the smallest, about 2.2 mm, and the lateral displacement gradually decreased with increasing depth. However, the maximum lateral deformation occurred near the foot of the embankment slope, about 5~10 mm.

In addition, the lateral deformation at the foot of embankment decreased first, then increased and then decreased with the depth. The lateral deformation at the foot of the embankment on both sides showed an approximately symmetrical trend with depth. This is mainly because the soil stratum in the loading area was not significantly deformed due to lateral constraints, while there was a free surface near the foot of the embankment. Under the loading effect, stress concentration occurred near the foot of the slope, resulting in a sliding trend of the embankment slope. The lateral deformation of the soil stratum was significantly increased due to lateral compression. When it exceeded a certain depth, the effect of local stress concentration was weakened, resulting in the lateral displacement gradually decreasing with the increase of depth.

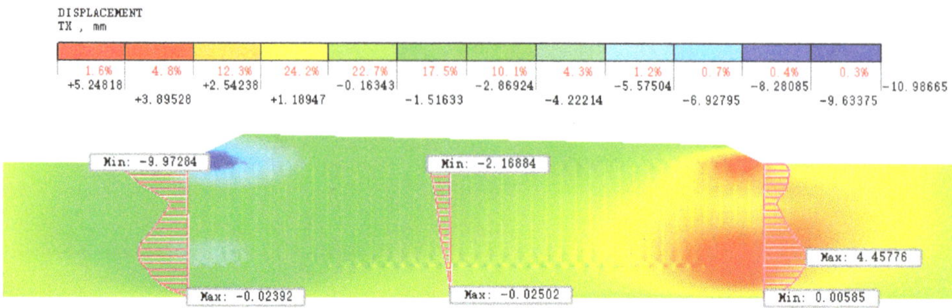

Figure 19. Contour map of the lateral displacement of soil along the cross-section of the embankment under Condition 1 after soil consolidation, and the variation curve of the lateral displacement of soil at typical locations with depth.

5.3.3. The Effect of the Pile-Net Composite Foundation on the Bearing Capacity of Piles

In order to analyze the influence of the pile-net composite foundation on the axial force at the end of soil consolidation, the typical working conditions 1 (PHC pile) and 5 (transition zone between PHC piles and CFG piles) were used as examples for research.

Figure 20a plots the contour map of the axial force of the pile along the cross-section of the embankment under Condition 1 after soil consolidation. As shown in the figure, the axial force near the foot of the embankment on both sides was relatively small, with a value of about 66 kN, while the axial force of the pile located within the range of the top of the slope on both sides was relatively large, with a maximum axial force of 382.3 kN. This is mainly due to the existence of a free surface near the foot of the embankment slope. Under the loading effect, the embankment slope exhibited a sliding trend, causing a certain deflection of the direction of the pile's force in this area. Then, the horizontal force component appeared, which made the piles prone to being damaged by shear and bending moments. However, the vertical axial force value was not significant. In addition, as the depth increased, the axial force of the pile first increased and then decreased. This is mainly due to the fact that soil settlement near the surface was significantly greater than that of the pile in the loading area, and there was a differential settlement between the two. The pile was not only subjected to gravity and the load transmitted by the heaped soil, but also bore the downward pull force (i.e., negative frictional resistance) of the soil. The force on the pile gradually increased with depth. When exceeding a certain depth (the critical depth is called the neutral plane), the settlement of pile was greater than that of surrounding soil, and the pile was subjected to upward frictional resistance. The force on the pile gradually decreased with depth. This was basically consistent with the results of Abushar et al. [45].

Figure 20b shows the axial force diagram of the pile along the longitudinal section of the embankment under Condition 5 after soil consolidation. As shown in the figure, whether it was PHC pile or CFG pile, the axial force of the pile showed a distribution law of first increasing and then decreasing with the increase of depth. The maximum axial force was located in the middle and lower parts of the pile, which was consistent with the above conclusion. In addition, when the loading was constant, the axial force of the PHC pile was relatively small, with a maximum of about 35 kN, while the axial force of the CFG pile could reach a maximum of 560 kN. Further analysis showed that, due to the settlement difference between the two types of piles at the boundary, the friction force was more complex after superposition, resulting in an increase in the axial force of the longer PHC pile and a decrease in the axial force of the shorter CFG pile. When exceeding a certain horizontal range, the influence of the superimposed friction at the boundary on the pile gradually weakened.

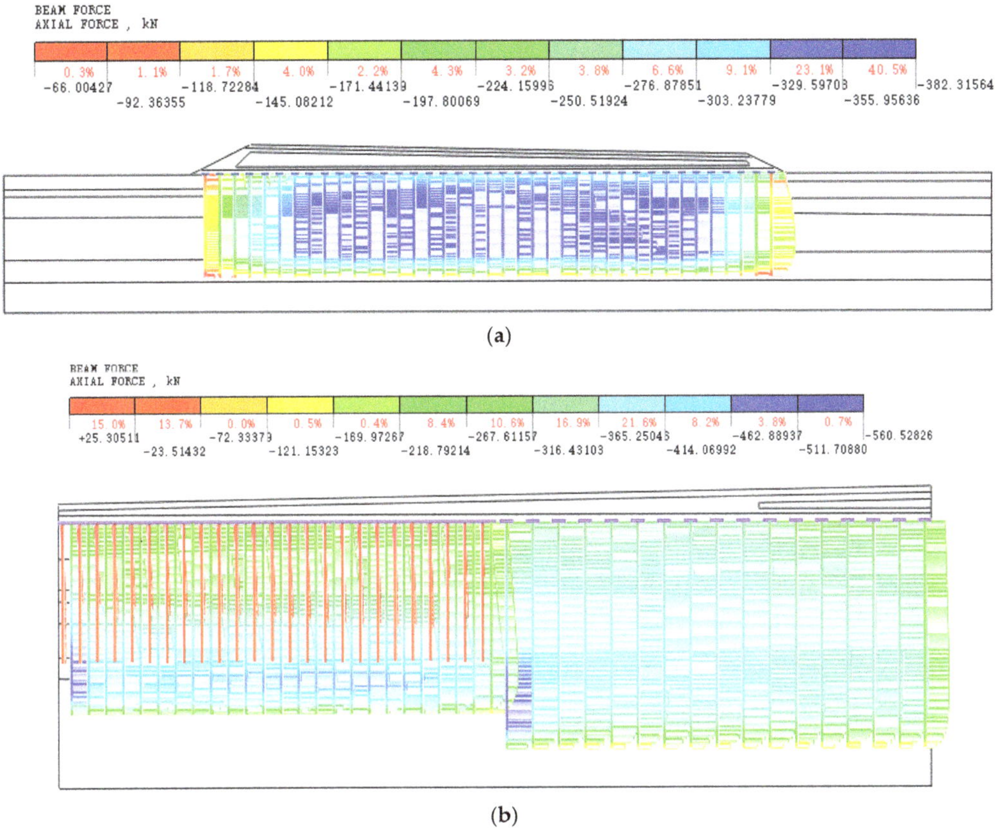

Figure 20. Pile axis diagram under different working conditions after consolidation: (**a**) Condition 1. (**b**) Condition 5.

Similarly, in order to analyze the influence of the pile-net composite foundation on the bending moment of piles at the end of soil consolidation, Figure 21 shows the variation of the bending moment of piles with depth under different working conditions.

Figure 21a,b show the bending moment diagrams of the pile along the cross-section of the embankment under Condition 1 (PHC pile) and Condition 2 (CFG pile) after soil consolidation. As shown in the figure, the piles at the foot of the embankment slopes on both sides were subjected to significant lateral horizontal forces due to loading. Then, there was a sliding trend in the embankment slope, resulting in a large bending moment of the piles in this area. At the same time, the bending moment turns at the soft soil interface. The pile near the middle of the embankment was mainly subjected to a vertical load, and its bending moment was relatively small, generally not exceeding 3 kN·m. After soil consolidation was completed, the maximum bending moment values for Condition 1 and 2 were 21.8 kN·m and 29.5 kN·m, respectively.

Figure 21c,d show the bending moment diagrams along the longitudinal section of the embankment under Condition 3 (PHC pile) and Condition 4 (CFG pile) at the end of soil consolidation. As shown in the figure, the bending moment values are relatively large in areas with significant geological differences and at the foot of the slope, while the bending moment values of the pile are relatively small beyond a certain horizontal distance. After soil consolidation was completed, the maximum bending moment values for Condition 3 and 4 were 28.1 kN·m and 16.6 kN·m, respectively.

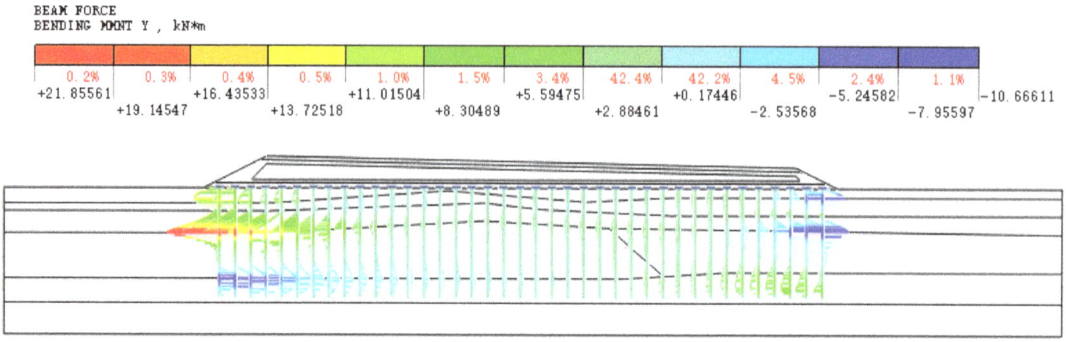

(**a**)

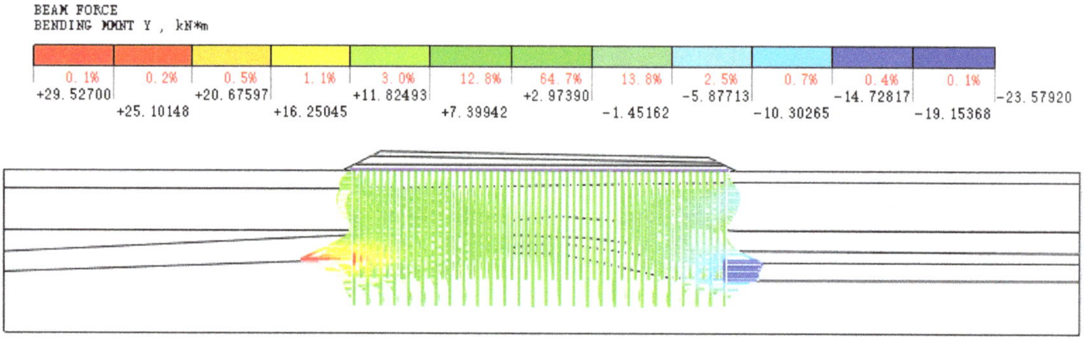

(**b**)

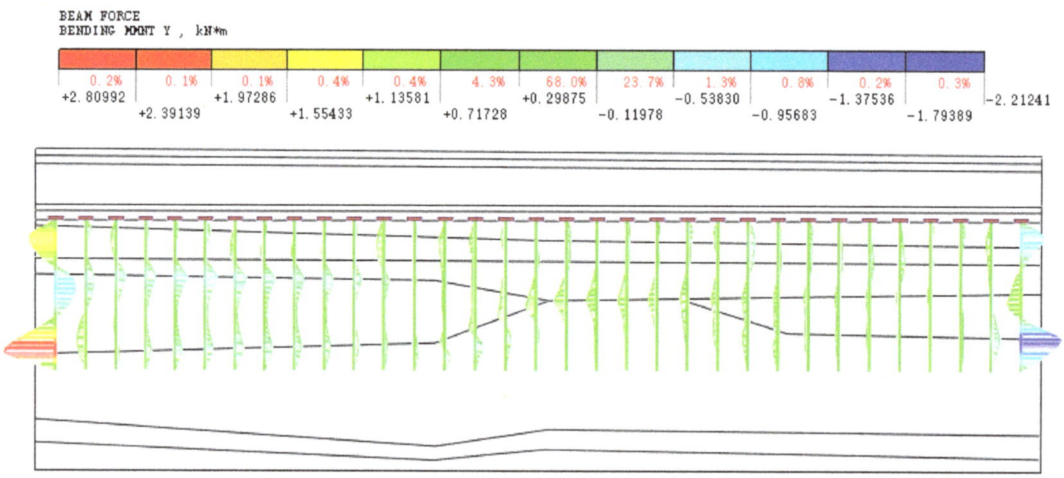

(**c**)

Figure 21. *Cont.*

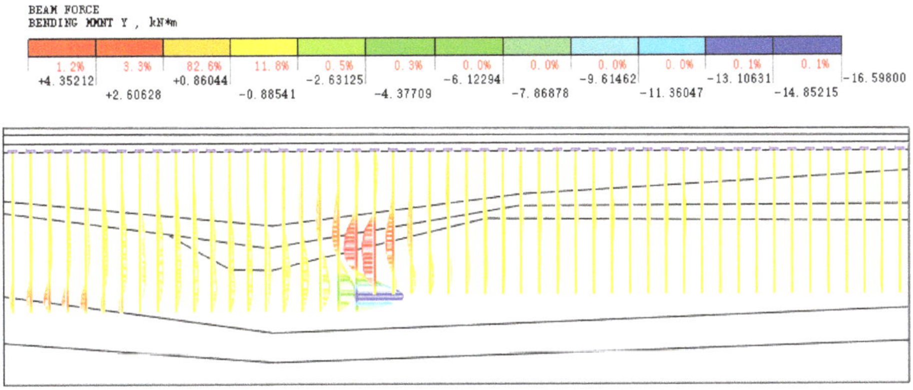

(**d**)

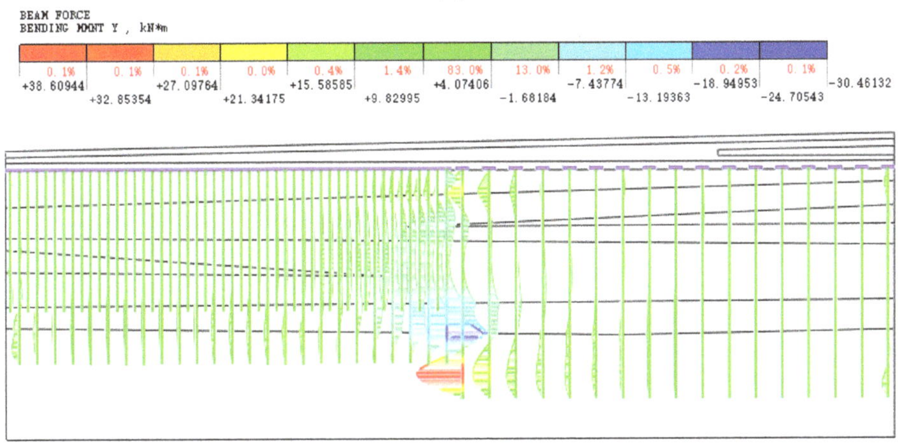

(**e**)

Figure 21. Bending moment diagram of piles under different working conditions at the end of consolidation: (**a**) Condition 1. (**b**) Condition 2. (**c**) Condition 3. (**d**) Condition 4. (**e**) Condition 5.

Figure 21e plots the bending moment diagram of the pile along the longitudinal section of the embankment under Condition 5 after soil consolidation. As shown in the figure, the bending moment at the boundary between the PHC pile and CFG pile was relatively large, reaching up to 38.6 kN·m. However, after exceeding a certain horizontal distance, the bending moment of both types of piles was relatively small, generally not exceeding 2.0 kN·m. Therefore, the maximum bending moment of the pile was mainly located at the foot of the slope or at the boundary of different types of piles, and turned on the interface of the soil strata. The bending moment of the piles in other parts bore less force.

5.4. Comprehensive Evaluation of Settlement Control

Based on the above results, we can obtain the post-construction settlement and the longitudinal uneven settlement ratio (LUSR) of embankments under typical working conditions after soil consolidation. Then, the treatment scheme of the pile-net composite foundation of the motor racing circuit can be evaluated.

Considering that the embankment is filled layer by layer, the total settlement of the embankment can be calculated, but the post-construction settlement is difficult to distinguish. According to reference [46], the post-construction settlement of the embankment is 0.8 percent of the total settlement. In addition, the maximum LUSR of the embankment can be obtained by calculating the section with significant curve changes according to the settlement curve of the embankment (see Figure 18c,d). Based on the calculation results in Section 5.3.1, the embankment settlement and LUSR under typical working conditions after soil consolidation are listed in Table 4. It can be seen from the table that after soil consolidation was completed, the maximum post-construction settlement of the embankment under the most unfavorable conditions of the project was 27 mm, located in the range of pile no. TR1 + 700~TR1 + 800, where the PHC pile-net composite foundation treatment was adopted. The largest LUSR of the embankment was 1.3/4000, located in the range of pile number TR0 + 160~TR0 + 260, which was the transition zone between the PHC pile and CFG pile-net composite foundations. In addition, in terms of pile bearing capacity, combined with the calculation results in Section 5.3.3, it can be seen that when soil consolidation was completed, the maximum axial force values of the PHC pile and CFG pile were 382.3 kN and 560 kN, respectively, which were mainly located in the middle and lower parts of the piles at the top of the embankment slope. The maximum bending moment values were 21.8 kN·m and 38.6 kN·m, respectively, which were mainly located at the foot of the embankment and the boundary between different types of piles.

Table 4. Settlement and LUSR of the embankment under typical working conditions after soil consolidation.

Working Conditions	Pile-Net Composite Foundation	Pile No.	Total Settlement/mm	Post-Construction Settlement /mm	The Longitudinal Uneven Settlement Ratio (LUSR)
Condition 1	PHC pile	TR3 + 260	32.1	25.7	/
Condition 2	CFG pile	TR1 + 440	18.0	14.4	/
Condition 3	PHC pile	TR1 + 700~TR1 + 800	33.8	27.0	0.2/4000
Condition 4	CFG pile	TR3 + 460~TR3 + 560	19.8	15.8	0.55/4000
Condition 5	PHC pile and CFG pile	TR0 + 160~TR0 + 260	24.0	19.2	1.3/4000

It can be concluded that after soil consolidation, the maximum post-construction settlement of the embankment was 27.0 mm < 50.0 mm, and the LUSR of the embankment was 1.3/400 < 3/4000, both of which met the settlement control standards of the embankment. Moreover, the axial force and bending moment of the pile were within the range of their compressive and bending strength, indicating that the pile-net composite foundation treatment scheme adopted in this project was feasible. It provides a guarantee of the safe construction and operation of the project.

6. Discussion

Through field tests, the back analysis of stratum parameters and pile moduli, and the numerical simulation of mechanical characteristics of pile-net composite foundations under five typical working conditions considering the construction process, the embankment settlement and differential settlement after soil consolidation in the motor racing circuit were deeply studied, and the following understandings could be obtained:

In terms of settlement and lateral deformation, for soft soil strata, whether the PHC pile or CFG pile-net composite foundation treatment was adopted, the settlement of the soil between piles was significantly greater than the settlement of the pile head. Namely, in this project, the settlement of soil between piles was 2.7 to 3.7 times that of adjacent pile head, and the worse the strata conditions, the greater the difference between the settlement of soil between piles and the pile head.

When it comes to the pile-net composite foundation treatment of weak soil embankments, a rigid pile, such as PHC pile or CFG pile, is commonly used. For the areas that do not meet the requirements, the method of high pressure grouting reinforcement near the pile body is often adopted. Under the process of high pressure grouting, although a reinforcement effect can be obtained, some failure phenomena such as deformation, tilt, and even damage to the pile body often occur. In order to solve this problem effectively, this paper puts forward a safe and effective remedy, which is adding PHC pile and CFG pile, and using a cement–soil mixing pile between the rigid piles. This method makes full use of the squeezing effect, which can significantly reduce post-construction settlement and avoid its influence on the adjacent pile. Compared with the traditional remedial method, the proposed method in this paper has the advantages of faster construction speed and a better effect, and has been successfully applied in practical engineering applications.

In terms of pile–soil interaction parameters, due to the deep soft soil in the project area, in order to fully utilize the bearing capacity of the pile-net composite foundation, the piles were arranged reasonably to ensure that the overall stress on the pile and soil was uniform. A pile-net composite foundation mainly bears vertical loading. So, the parameters that have a significant impact on the settlement and deformation control of the embankment are the elastic modulus of the pile and the deformation modulus of the soil strata. Considering that soil strata in the test sections mainly include 2-2 silty clay, 2a clay, 4 silty clay, and 6-1 silt, it is necessary to invert the deformation modulus of the above soil strata, as well as elastic modulus of the PHC piles and CFG piles. Based on the laboratory test results, the relationship between the compressive modulus and deformation modulus of the strata, construction disturbance, the distribution range of the deformation modulus of the strata, and the elastic modulus of the CFG and PHC piles can be preliminarily determined. Then, based on the measured data, the above parameters can be inverted and analyzed to obtain more reasonable pile–soil interaction parameters. Research has shown that the soil deformation modulus and pile modulus have a significant impact on embankment settlement. The larger the values of both the soil deformation modulus and pile modulus, the smaller the embankment settlement, which is basically consistent with the actual situation.

In terms of pile bearing capacity, regardless of whether the PHC pile- or CFG pile-net composite foundation was used for the deep embankment treatment, the axial force of the piles increased first and then decreased with depth, and the maximum axial force was located in the middle and lower part of the piles. This is mainly because the soil settlement near the surface was obviously greater than the settlement of piles in the loading area, and there was differential settlement between the two. The downward pull (i.e., negative friction resistance) transmitted by gravity and heaped soil made the force of pile gradually increase with depth. When the depth exceeded a certain level, the settlement of the pile was greater than that of the surrounding soil, and the pile was subjected to upward friction resistance, and the force of the pile gradually decreased with depth. In addition, the maximum bending moment was located at the foot of the slope and the boundary between different types of piles, which was smaller in the middle of the embankment, and the bending moment turned at the interface of soft soil. This is mainly due to the sliding trend of the embankment at the foot of the slope, resulting in a larger bending moment of the pile, while the pile near the middle of the embankment was mainly subjected to a vertical load, and its bending moment was relatively small. Moreover, when the pile-net composite foundation was adopted for embankment treatment, the axial force and bending moment of the pile were generally not large. The maximum axial force of the project was 560 kN, and the maximum bending moment was 38.6 kN·m. The axial force and bending moment were both within the range of the compressive and bending strength of the pile.

Further analysis showed that no matter whether PHC pile or CFG pile, there was an obvious soil arching effect between piles, and the soil arching effect was conducive to enhancing the reinforcement effect of composite foundation. In practical engineering, the pile spacing should be rationally arranged. In this project, when the pile spacing was 3–4 times the pile diameter, the soil arching effect of the pile-net composite foundation was the best. It is worth mentioning that the pore water pressure dissipated slowly due to the short field test time, the interference of the construction load, and the heaped soil. After the completion of the field test, there was still nearly 20 kPa of pore water pressure, which failed to fully reflect the settlement characteristics. Therefore, there might be some differences between the calculated results and the actual ones, which could be further discussed in the future.

7. Conclusions

Taking the Wuhan intelligent connected motor racing circuit as our setting, a method of combining field tests with numerical simulation was proposed to identify, treat, and evaluate embankment settlement and differential settlement under typical working conditions, such as deep soft soil sections, sudden geological changes, and transition sections, with different pile types. The main conclusions are as follows:

(1) Field tests showed that for medium or high filling areas, the maximum settlement of the pile head and soil between piles was 13.6 mm and 50.41 mm, respectively, and the settlement of soil between piles was 3.5~3.7 times that of adjacent pile heads. For low filling areas, the maximum settlement of the pile head and soil between piles was 13.34 mm and 39.65 mm, respectively. The cumulative settlement value of the pile head and soil between piles increased in a hyperbolic form. The lateral displacement direction of the soil strata in the loading area was spread outward from the loading center, and the lateral displacement did not exceed 1.5 mm.

(2) Based on field test data, the GA-BP method, combining the uniform design method, was used to carry out back analysis on the mechanical parameters of the soil strata and pile modulus. The results showed that the settlement calculation values were basically consistent with the measured ones, with an error of no more than 10%, indicating good inversion accuracy. In addition, after soil consolidation, the predicted maximum settlement values of the PHC pile and CFG pile in the two pile test sections were about 70 mm and 59 mm, respectively. By adding PHC piles, CFG piles, and reinforcing cement–soil mixing piles between the piles, the maximum settlements were 21.01 mm and 15.88 mm, respectively, meeting the settlement control requirements.

(3) By selecting five typical working conditions, a numerical analysis was conducted on the settlement and pile bearing capacity of the embankment in the engineering area. The calculation showed that after the soil consolidation was completed, the maximum settlement of the embankment was 27.0 mm, and the maximum LUSR of the embankment was 1.3/4000. The maximum lateral displacement in the middle of the embankment is about 2.2 mm. The maximum axial forces of the PHC piles and CFG piles were 382.3 kN and 560 kN, respectively, and the maximum bending moments were 21.8 kN·m and 38.6 kN·m. The axial forces and bending moments of the piles were within their compressive and flexural strength ranges. It can be concluded that the pile-net composite foundation treatment methods adopted in this project were feasible, and the embankment settlement and differential settlement were controllable, which can meet safety and operational requirements.

However, the current research on pile-net composite foundations in soft soil areas mainly focuses on the bearing characteristics and deformation control effects of a certain soil stratum. The theory, laboratory equipment, and field tests for predicting embankment settlement in the different transition zones of pile-net composite foundations and multiple spatial variation zones of a soft soil stratum are not yet mature, and further study is needed.

Author Contributions: X.W.: methodology, investigation, and supervision; Q.P.: methodology, data curation, validation, writing, and funding acquisition. All authors have read and agreed to the published version of the manuscript.

Funding: This research was funded by the National Science Foundation of China (No. 51609018).

Data Availability Statement: The data used to support the findings of this study are included in the article.

Conflicts of Interest: Author Xiaonan Wang was employed by the company Wuhan Urban Construction Investment Development Group Co., Ltd. and Author Qitao Pei was employed by the company Wuhan Municipal Engineering Design & Research Institute Co., Ltd. The companies were not involved in the study design, collection, analysis, interpretation of data, the writing of this article or the decision to submit it for publication.

References

1. Yin, J.H.; Feng, W.Q. A new simplified method and its verification for calculation of consolidation settlement of a clayey soil with creep. *Can. Geotech. J.* **2017**, *54*, 333–347. [CrossRef]
2. Zhu, H.H.; Zhang, C.C.; Mei, G.X.; Shi, B.; Gao, L. Prediction of one-dimensional compression behavior of Nansha clay using fractional derivatives. *Mar. Georesources Geotechnol.* **2017**, *35*, 688–697. [CrossRef]
3. Zhang, S.M.; Jian, G.L.; Liao, Y.L. Effect of the strengthening area and the slope rate on bearing and deforming behaviors of CFG pile-geogrid composite foundations. *Chin. J. Rock Mech. Eng.* **2019**, *38*, 192–202.
4. Zheng, G.; Zhao, J.P.; Zhou, H.Z. State-of-the-art review for techniques on ground improvement of highway and railway. *Chin. J. Ground Improv.* **2021**, *3*, 91–99. (In Chinese)
5. Wu, P.C.; Yin, J.H.; Feng, W.Q.; Chen, W.B. Experimental study on geosynthetic-reinforced sand fill over marine clay with or without deep cement mixed soil columns under different loadings. *Undergr. Space* **2019**, *4*, 340–347. [CrossRef]
6. Reshma, B.; Rajagopal, K.; Viswanadham, B.V.S. Centrifuge model studies on the settlement response of geosynthetic piled embankments. *Geosynth. Int.* **2020**, *27*, 170–181. [CrossRef]
7. Wu, P.C.; Feng, W.Q.; Yin, J.H. Numerical study of creep effects on settlements and load transfer mechanisms of soft soil improved by deep cement mixed soil columns under embankment load. *Geotext. Geomembr.* **2020**, *48*, 331–348. [CrossRef]
8. Esen, A.F.; Woodward, P.K.; Laghrouche, O.; Čebašek, T.M.; Brennan, A.J.; Robinson, S.; Connolly, D.P. Full-scale laboratory testing of a geosynthetically reinforced soil railway structure. *Transp. Geotech.* **2021**, *28*, 100526. [CrossRef]
9. Arulrajah, A.; Abdullah, A.; Bo, M.W.; Bouazza, A. Ground improvement techniques for railway embankments. *Proc. Inst. Civ. Eng. Ground Improv.* **2009**, *162*, 3–14. [CrossRef]
10. Terzaghi, K. *Theoretical Soil Mechanics*; Wiley: New York, NY, USA, 1943.
11. Low, B.K.; Tang, S.K.; Choa, V. Arching in piled embankments. *J. Geotech. Eng.* **1994**, *120*, 1917–1938. [CrossRef]
12. Giroud, J.P.; Bonaparte, R.; Beech, J.F.; Gross, B.A. Design of soil layer-geosynthetic systems overlying voids. *Geotext. Geomembr.* **1990**, *9*, 11–50. [CrossRef]
13. Hewlett, W.J.; Randolph, M.F. Analysis of piled embankments. *Ground Eng.* **1988**, *21*, 12–18.
14. Chen, Y.M.; Cao, W.P.; Chen, R.P. An experimental investigation of soil arching within basal reinforced and unreinforced piled embankments. *Geotext. Geomembr.* **2008**, *26*, 164–174.
15. Castro, J.; Sagaseta, C. Consolidation and deformation around stone columns: Numerical evaluation of analytical solutions. *Comput. Geotech.* **2011**, *38*, 354–362. [CrossRef]
16. Liu, F.; Zhang, J.J. Centrifuge test on deformation characteristics of pile-geogrid composite foundation in soft soil under slope. *Chin. J. Rock Mech. Eng.* **2018**, *37*, 209–219.
17. Blanc, M.; Rault, G.; Thorel, L.; Almeida, M. Centrifuge investigation of load transfer mechanisms in a granular mattress above a rigid inclusions network. *Geotext. Geomembr.* **2013**, *36*, 92–105. [CrossRef]
18. Fagundes, D.F.; Almeida, M.S.; Thorel, L.; Blanc, M. Load transfer mechanism and deformation of reinforced piled embankments. *Geotext. Geomembr.* **2017**, *45*, 1–10. [CrossRef]
19. Girout, R.; Blanc, M.; Thorel, L.; Dias, D. Geosynthetic reinforcement of pile-supported embankments. *Geosynth. Int.* **2018**, *25*, 37–49. [CrossRef]
20. Das, A.K.; Deb, K. Experimental and 3D numerical study on time-dependent behavior of stone column–supported embankments. *Int. J. Geomech.* **2018**, *18*, 04018011. [CrossRef]
21. Liu, J.W.L.; Zhang, Z.M.; Zhao, Y.B.; Lin, C.G. Statistical research on performance of prestressed concrete pipe piles. In Proceedings of the 2011 International Conference on Electric Technology and Civil Engineering (ICETCE), Lushan, China, 22–24 April 2011; pp. 711–714.
22. Oh, Y.I.; Shin, E. C Reinforcement and arching effect of geogrid-reinforced and pile-supported embankment on marine soft ground. *Mar. Georesources Geotechnol.* **2007**, *25*, 97–118. [CrossRef]
23. Zhang, C.; Jiang, G.; Liu, X.; Buzzi, O. Arching in geogrid-reinforced pile-supported embankments over silty clay of medium compressibility: Field data and analytical solution. *Comput. Geotech.* **2016**, *77*, 11–25. [CrossRef]
24. Yang, M.; Liu, S. Field tests and finite element modeling of a prestressed concrete pipe pile-composite foundation. *KSCE J. Civ. Eng.* **2015**, *19*, 2067–2074. [CrossRef]
25. Zhang, S.; Tang, C. Subgrade Settlement of High Speed Railway in Seasonal Frozen Soil Area. *J. Northeast. Univ. (Nat. Sci.)* **2013**, *34*, 1202–1205.
26. Zhang, F.; Liu, Y.; Xu, Z.Y.; Li, Z.Y. Factors influencing subgrade post-construction settlement of CFG pile composite foundation in Wuhan-Guangzhou high-speed railway. *J. Southwest Jiaotong Univ.* **2015**, *50*, 783–788.
27. Yoo, C. Settlement behavior of embankment on geosynthetic-encased stone column installed soft ground—A numerical investigation. *Geotext. Geomembr.* **2015**, *43*, 484–492. [CrossRef]
28. Zhao, L.S.; Zhou, W.H.; Fatahi, B.; Li, X.B.; Yuen, K.V. A dual beam model for geosynthetic-reinforced granular fill on an elastic foundation. *Appl. Math. Model.* **2016**, *40*, 9254–9268. [CrossRef]

29. Jamsawang, P.; Yoobanpot, N.; Thanasisathit, N.; Voottipruex, P.; Jongpradist, P. Three-dimensional numerical analysis of a DCM column-supported highway embankment. *Comput. Geotech.* **2016**, *72*, 42–56. [CrossRef]
30. Nunez, M.A.; Briançon, L.; Dias, D. Analyses of a pile-supported embankment over soft clay: Full-scale experiment, analytical and numerical approaches. *Eng. Geol.* **2013**, *153*, 53–67. [CrossRef]
31. Hájek, V.; Mašín, D.; Boháč, J. Capability of constitutive models to simulate soils with different OCR using a single set of parameters. *Comput. Geotech.* **2009**, *36*, 655–664. [CrossRef]
32. Yan, W.M.; Li, X.S. Mechanical response of a medium-fine-grained decomposed granite in Hong Kong. *Eng. Geol.* **2012**, *129*, 1–8. [CrossRef]
33. Abad, S.A.N.K.; Tugrul, A.; Gokceoglu, C.; Armaghani, D.J. Characteristics of weathering zones of granitic rocks in Malaysia for geotechnical engineering design. *Eng. Geol.* **2016**, *200*, 94–103. [CrossRef]
34. Deb, K. A mathematical model to study the soil arching effect in stone column-supported embankment resting on soft foundation soil. *Appl. Math. Model.* **2010**, *34*, 3871–3883. [CrossRef]
35. Zhuang, Y.; Wang, K. Finite element analysis on the dynamic behavior of soil arching effect in piled embankment. *Transp. Geotech.* **2018**, *14*, 8–21. [CrossRef]
36. Zhao, L.S.; Zhou, W.H.; Geng, X.; Yuen, K.V.; Fatahi, B. A closed-form solution for column-supported embankments with geosynthetic reinforcement. *Geotext. Geomembr.* **2019**, *47*, 389–401. [CrossRef]
37. Chen, R.P.; Zhou, W.H.; Chen, Y.M. Influences of soil consolidation and pile load on the development of negative skin friction of a pile. *Comput. Geotech.* **2009**, *36*, 1265–1271. [CrossRef]
38. Yu, J.; Cai, Y.; Qi, Z.; Guan, Y.; Liu, S.; Tu, B. Analytical analysis and field test investigation of consolidation for CCSG pile composite foundation in soft clay. *J. Appl. Math.* **2013**, *2013*, 795962. [CrossRef]
39. Filz, G.M.; Sloan, J.A.; McGuire, M.P.; Smith, M.; Collin, J. Settlement and vertical load transfer in column-supported embankments. *J. Geotech. Geoenvironmental Eng.* **2019**, *145*, 04019083. [CrossRef]
40. Luo, Q.; Lu, Q.Y. Settlement calculation of rigid pile composite foundation considering pile-soil relative slip under embankment load. *China J. Highw. Transp.* **2018**, *31*, 20–30. (In Chinese)
41. Institute of Rock and Soil Mechanics, Chinese Academy of Sciences. *Monitoring Project for Closed Test Field of Wuhan New Energy and Intelligent Connected Motor-Racing Circuit*; Institute of Rock and Soil Mechanics, Chinese Academy of Sciences: Wuhan, China, 2021. (In Chinese)
42. Wang, H.T.; Tu, B.X.; Su, P.; Gu, X.K. *Numerical Simulation Technology and Engineering Applications of MIDAS GTS NX*; Dalian University of Technology Press: Dalian, China, 2020. (In Chinese)
43. Pei, Q.T.; Li, H.B.; Liu, Y.Q.; Song, Q.J.; Li, N.N. Two-stage back analysis of initial geostress field of dam areas under complex geological conditions. *Chin. J. Rock Mech. Eng.* **2014**, *33*, 2779–2785. (In Chinese)
44. Chen, Z.Y. *Theory Methods and Programs of Soil Slope Stability Analysis*; China Water & Power Press: Beijing, China, 2003. (In Chinese)
45. Abusharar, S.W.; Zheng, J.J.; Chen, B.G. Finite element modeling of the consolidation behavior of multi-column supported road embankment. *Comput. Geotech.* **2009**, *36*, 676–685. [CrossRef]
46. *JTG/T D31-02-2013*; Technical Guidelines for Design and Construction of Highway Embankment on Soft Ground. Chinese Standard: Beijing, China, 2013. (In Chinese)

Disclaimer/Publisher's Note: The statements, opinions and data contained in all publications are solely those of the individual author(s) and contributor(s) and not of MDPI and/or the editor(s). MDPI and/or the editor(s) disclaim responsibility for any injury to people or property resulting from any ideas, methods, instructions or products referred to in the content.

buildings

MDPI

Article

Experimental Study on Dynamic Characteristics of Saturated Soft Clay with Sand Interlayer under Unidirectional and Bidirectional Vibration

Sui Wang [1,2,3,*], Yuanqiang Cai [1], Liyong Zhang [2], Yongjian Pan [2], Bin Chen [3], Peng Zhao [3] and Yuanming Fang [3]

1 College of Civil Engineering, Zhejiang University of Technology, Hangzhou 310014, China; caiyq@zju.edu.cn
2 Zhejiang Engineering Survey and Design Institute Group Co., Ltd., Ningbo 315012, China; lyzhang111@126.com (L.Z.); pyjwj1963@163.com (Y.P.)
3 School of Civil and Transportation Engineering, Ningbo University of Technology, Ningbo 315211, China; chenbin@nbut.edu.cn (B.C.); zhaopeng02022@163.com (P.Z.); fym_1999@163.com (Y.F.)
* Correspondence: wangsui10610@163.com

Abstract: The marine and alluvial plains along the southeastern coast of China are widely distributed in sandy formations, including smaller sand lenses and interlayers. The interlayers of sand have a significant impact on the mechanical properties of soft clay. In this paper, a large number of undrained unidirectional and bidirectional cyclic loading tests for soft clay with sand interlayers were carried out by a dynamic triaxial test system. Test results show that, under unidirectional and bidirectional cyclic vibration, the area of the hysteresis loop decreases and the slope of the connecting line at both ends of the hysteresis loop increases with the increasing of frequency. For the same vibration frequency, the area of the bidirectional vibration hysteresis loop and the slope of the connecting line at both ends are smaller than that of the unidirectional cyclic vibration. Under the same dynamic stress ratio, cumulative axial deformation caused by unidirectional and bidirectional vibration increases with the increasing frequency. Under unidirectional vibration, dynamic elastic modulus decreases at first, and then increases with the increasing frequency. For the same frequency, dynamic elastic modulus of the sample increases with the increase in cycles. Due to the effect of radial cyclic stress, the curves of dynamic elastic modulus and damping ratio with frequency under bidirectional vibration are opposite to those under unidirectional vibration.

Keywords: soft clay; sand interlayer; unidirectional and bidirectional vibration; cyclic stress ratio; vibration frequency; cumulative deformation

1. Introduction

Soft clay is a special soil mass formed by marine and lacustrine sediments in coastal, lake, swamp, and river bay areas in modern times. It is generally characterized by high water content, high compressibility, low strength, poor permeability, and high sensitivity and rheology [1–7]. According to the survey results of Ningbo metro, a large number of loose sand interlayer with a thickness of about 10cm are in the soft clay of metro tunnel engineering. According to the design code, this kind of interlayer has almost no influence on the evaluation of foundation liquefaction, but on the other hand, its foundation liquefaction can cause dramatic changes in stratum deformation, which will cause the subway tunnel to bear an additional huge horizontal thrust and cause damage to the subway tunnel. In the previous seismic response analysis, it is considered that the seismic action is mainly horizontal shear. The stress condition is mainly simulated by the stress on the 45° plane of the soil element under the condition of uniform consolidation under the cyclic load of unidirectional excitation. Yasuhara [8] and Matsui [9] have conducted in-depth research on the strength and stiffness attenuation of soil after cyclic loading, and pointed out the influence of cyclic cycles. Moreover, in terms of material constitutive model representation, scientific publications have proposed yield functions that represent very reliably the cyclic behavior of clays [10,11]. To sum up, only the effect of horizontal shear is considered in the seismic design, which is not safe for the actual project, and the effect of vertical seismic load should also be considered.

A large number of tests have been carried out on the dynamic characteristics of soft clay under cyclic loading at home and abroad. In the 1960s, Seed and Lee [12] proposed that the action of seismic wave can be simplified as the action of applying equivalent reciprocating horizontal shear stress to the soil under the condition that the vertical stress remains unchanged. Wichtmann [13] considered the dynamic characteristics of soil under the coupling action of vertical cyclic normal stress and horizontal cyclic normal stress under traffic load, which confirmed that the existence of horizontal cyclic normal stress greatly affected the dynamic characteristics of soil; under unidirectional vibration, the pore pressure and deformation of the specimen accumulated more rapidly, and thus the failure was reached earlier. Kammerer [14] studied the pore pressure characteristics of saturated sand under bidirectional vibration, and the results showed that the influence of bidirectional vibration on the dynamic pore pressure was significantly greater than that of unidirectional vibration, and the dynamic pore pressure changed more greatly under bidirectional vibration. Hyde [15] conducted a series of cyclic undrained tests on silty clay samples. The test results show that cyclic loading on normal and weakly overconsolidated soils will cause a significant increase in overconsolidation. The research results of Silvestri [16] and Narasimha [17] show that the effective consolidation pressure, deformation, and undrained strength of clay under cyclic loading are affected by the loading frequency. Liang [18] proposed a constitutive model of saturated soft clay under dynamic load based on the concept of boundary surface plasticity and considering the anisotropy of materials by using the node invariants of second-order stress tensor and clay structure tensor. However, there is no relevant research on the dynamic characteristics of saturated soft clay with interlayer sand under the condition of bi-directional excitation cyclic load.

Therefore, in this paper, a bidirectional dynamic triaxial test system was used to carry out bidirectional excited cyclic vibrations under different radial cyclic stress ratios, different frequencies, and different waveforms on the undisturbed soft clay with a sand interlayer in the Ningbo metro excavation; the influence of interlayer sand was analyzed on the dynamic characteristics of saturated soft clay.

2. Research Contents and Methods

2.1. Sample Equipment and Materials

A British GDS (Global Digital System) bidirectional dynamic triaxial test system was adopted, as shown in Figure 1. During the test, GDS-LAB software v2.6.7 in the computer could collect the stress, strain, pore pressure, and other data during the vibration of the sample.

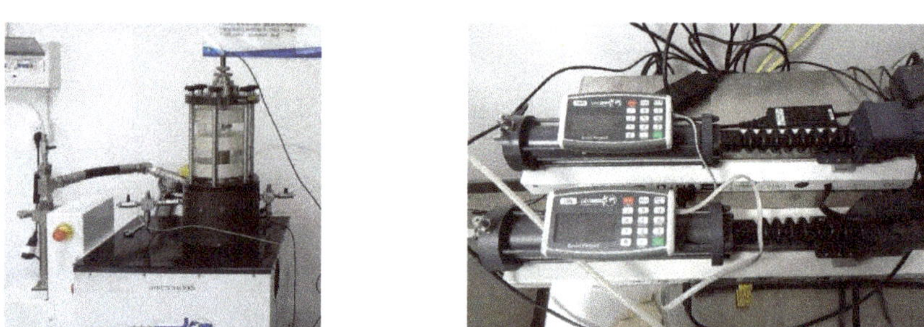

(**a**) Dynamic confining pressure and servo motor (**b**) Pressure control system diagram machine

Figure 1. GDS bidirectional dynamic triaxial test system.

Preparation of undisturbed soil sample: The undisturbed soil is cut into $\Phi \times H = 38\,\text{mm} \times 76\,\text{mm}$ soil samples according to the specifications, as shown in Figure 2. Since the soil sample is composed of soft clay and a large number of sand interlayer with small thickness inside, it is necessary to avoid delamination of the soil sample and failure of cutting. The basic physical parameters of soil samples are shown in Table 1.

Figure 2. Soft clay sample with intercalated sand. (Taken from Binjiang Waterfront Project, No. zk167-02, 515-518).

Table 1. Basic physical parameters of soil sample.

Property	Value
Density $\rho/(\text{g}\cdot\text{cm}^{-3})$	1.82
Plastic limit I_p	15.6
Liquid limit $w_L/\%$	40.1
Dry density $\rho_d/(\text{g}\cdot\text{cm}^{-3})$	1.36
Moisture content $\omega(\%)$	34.22
Cohesion c (kPa)	19.5
Internal friction angle $\varphi(°)$	29.74

The sample soil is taken from the undisturbed soil in the Ningbo metro excavation, Zhejiang Province, and the depth of the soil sample is 15–20 m. Interlayer sand soft clay is grayish yellow, soil is soft, and has a rotten smell. Among them, the interlayer sand is fine and smooth. The basic physical parameters of soil samples are shown in Table 1.

2.2. Comparison of Single and Bidirectional Test Principles

A uniaxial excitation cyclic triaxial test keeps the lateral confining pressure unchanged, and apply $\pm\sigma_d/2$; therefore, the cyclic shear stress applied to the 45° inclined plane of the specimen is still $\pm\sigma_d/2$; however, the normal stress becomes $\sigma_0 \pm \sigma_d/2$, as shown in Figure 3.

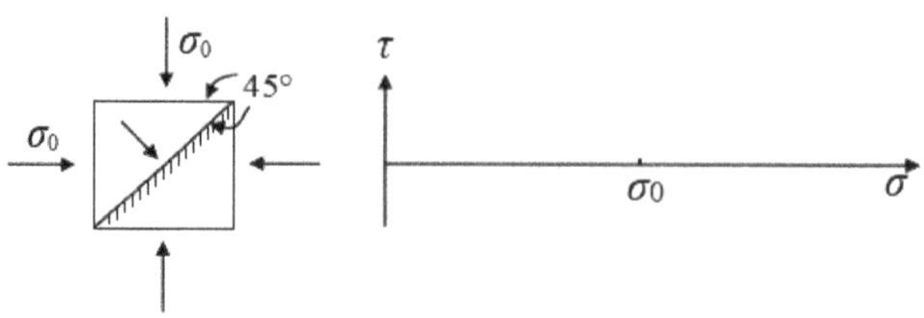

(**a**) Stress state of soil during consolidation

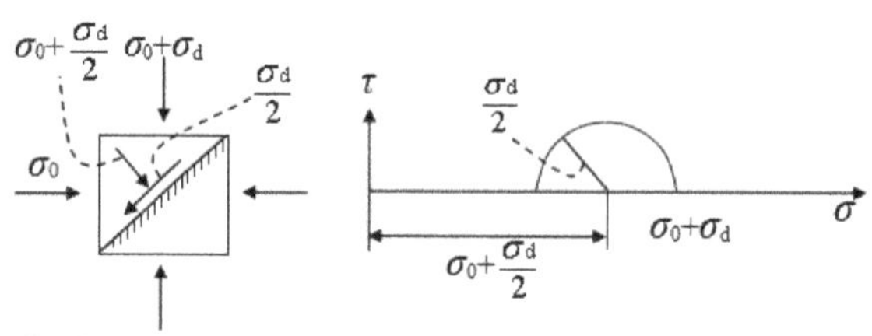

(**b**) Stress state of soil during half cycle of compression

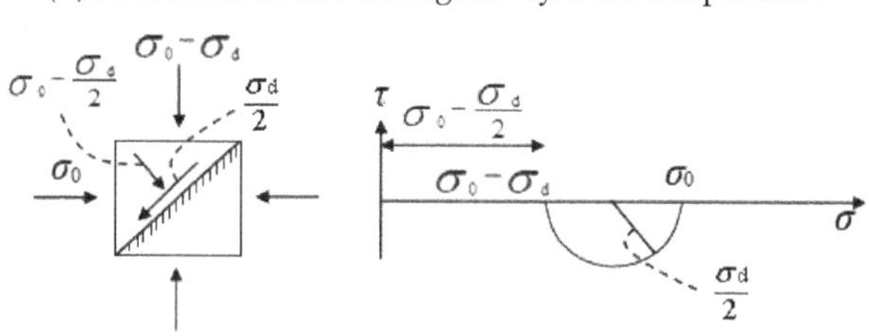

(**c**) Stress state of soil during half cycle of tension

Figure 3. The principle of unidirectional test.

The biaxial excitation cyclic triaxial test is to apply $\pm\sigma_d/2$ cyclic stress of, and $\mp\sigma_d/2$ cyclic stress; therefore, the normal stress on the 45° plane of the specimen remains unchanged to simulate the stress state of soil level during earthquake. The test principle is shown in Figure 4.

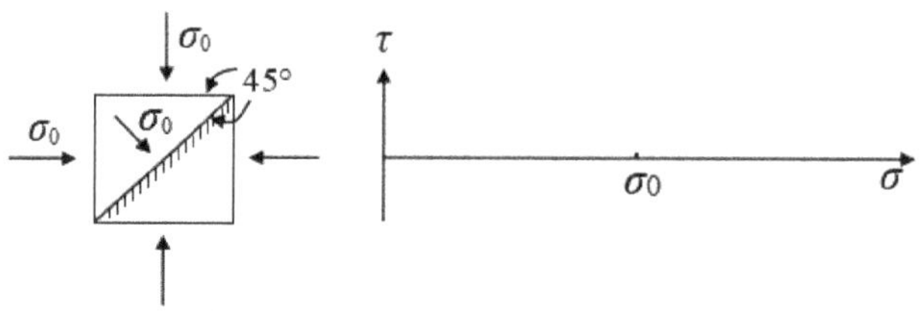

(**a**) Stress state of soil during consolidation

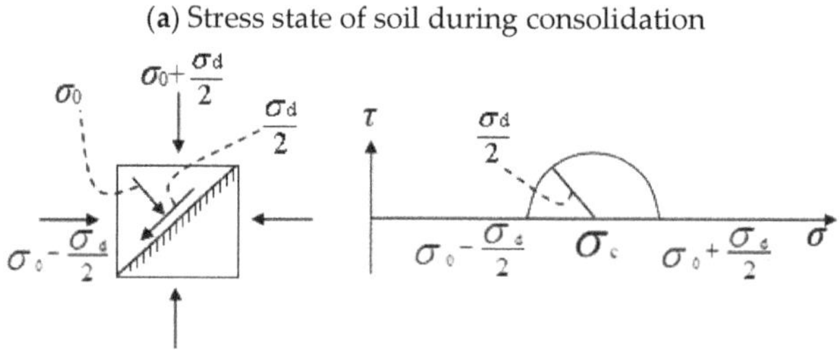

(**b**) Stress state of soil during half cycle of compression

Figure 4. *Cont.*

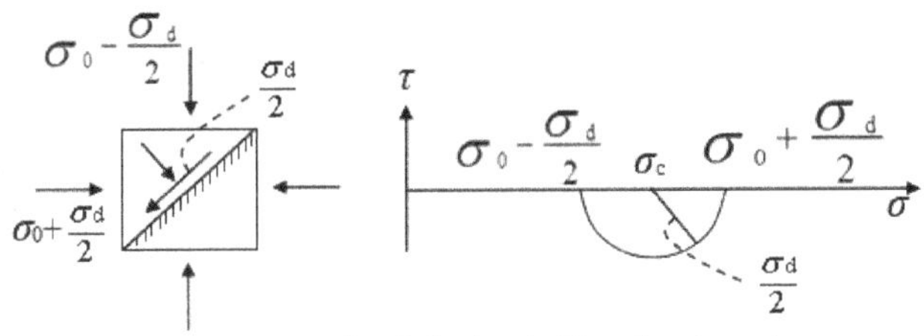

(c) Stress state of soil during half cycle of tension

Figure 4. The principle of bidirectional test.

2.3. Test Method

(1) Sample preparation: Undisturbed samples with a diameter of 38 mm and a height of 76 mm are used in the test. The samples are prepared according to the soil test specification.

(2) Saturation stage: The samples are put into the vacuum saturator to vacuum. When the vacuum pressure is exposed to negative pressure, maintain the vacuum pressure in the cylinder for more than 1 h and stand for more than 10 h. In order to ensure that the saturation reaches more than 95%, the samples are saturated by back pressure before the experiment until the saturation meets the requirements [19].

(3) Consolidation stage: isobaric consolidation is adopted in the test, and the consolidation confining pressure is 100 kPa.

(4) Vibration stage: The loading mode of stress control is adopted. The loading waveform is generated by the servo system, and different waveforms can be imported through the test system. The detailed test scheme is shown in Table 2.

Table 2. Scheme of dynamic triaxial test.

ID	Frequency f/Hz	Confining Pressure σ'_3/kPa	r_c	Rc	N
D-1	0.2	100	0.1	0	10,000
D-2	0.5	100	0.1	0	10,000
D-3	1.0	100	0.1	0	10,000
D-4	2	100	0.1	0	10,000
D-5	5	100	0.1	0	10,000
S-1	0.2	100	0.1	0.05	10,000
S-2	0.5	100	0.1	0.05	10,000
S-3	1.0	100	0.1	0.05	10,000

In this test, the GDS bidirectional dynamic triaxial test system is adopted to carry out experimental research on dynamic deformation characteristics of saturated intercalated sand soft clay. In order to measure the stress level of cyclic axial eccentric stress and cyclic confining pressure, the axial cyclic stress ratio r_c and radial cyclic stress ratio Rc are introduced [20,21]:

$$rc = \sigma1^{\mathrm{ampl}}/2\sigma'3 \tag{1}$$

$$Rc = \sigma3^{\mathrm{ampl}}/2\sigma'3 \tag{2}$$

where $\sigma_1{}^{\mathrm{ampl}}$, $\sigma_3{}^{\mathrm{ampl}}$ represents the amplitude of cyclic axial stress and lateral stress, respectively, σ'_3 represents the effective lateral stress after consolidation. Since the maximum application frequency of bidirectional coupling of the test instrument is 5 Hz, and the vibration frequency of cyclic axial stress and cyclic confining pressure cannot be controlled independently during the test, a higher vibration frequency will cause the waveform of cyclic confining pressure to be unstable, which will have a certain impact on the test results. Therefore, based on the above considerations, the frequency of bidirectional cyclic vibration is set to be the same, and it is proposed to apply three frequencies during the test: 0.2 Hz, 0.5 Hz, 1.0 Hz.

3. Analysis of Test Results

3.1. Comparison of Single and Bidirectional Cyclic Vibration for Different Frequencies

The hysteresis loop mainly reflects the deformation characteristics of the sample, stiffness degradation, and energy dissipation of the specimen under repeated cyclic loading, which is the basis for determining the restoring force model and nonlinear seismic response analysis [22–24]. Under the condition of unidirectional and bidirectional vibration, three groups of hysteretic curves with different vibration cycles at different frequencies are selected for comparison, as shown in Figures 5 and 6.

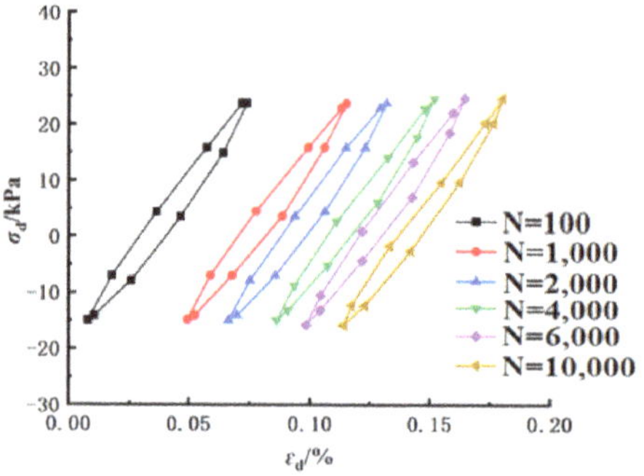

(**a**) f = 0.2 Hz

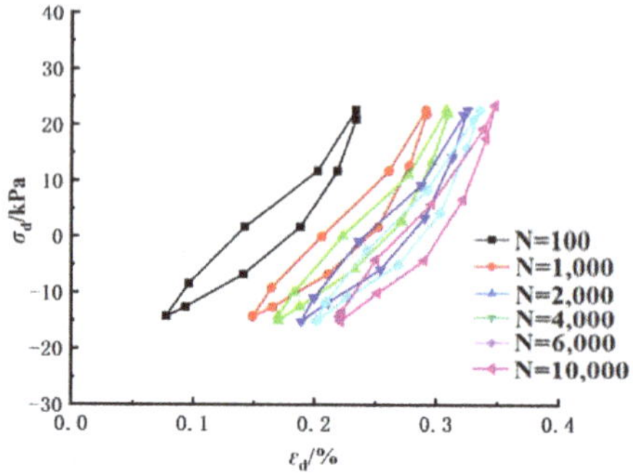

(**b**) f = 0.5 Hz

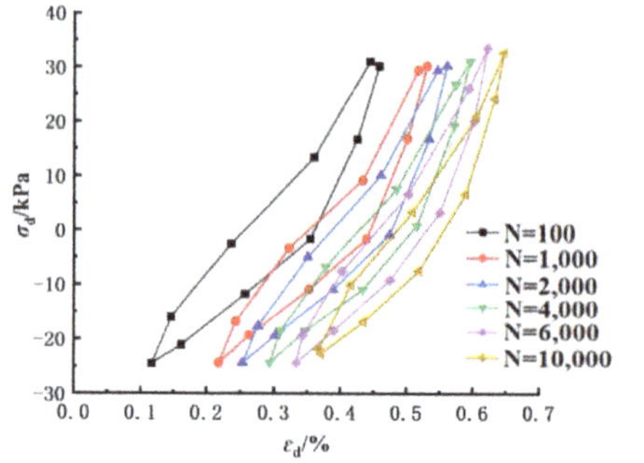

(**c**) f = 1.0 Hz

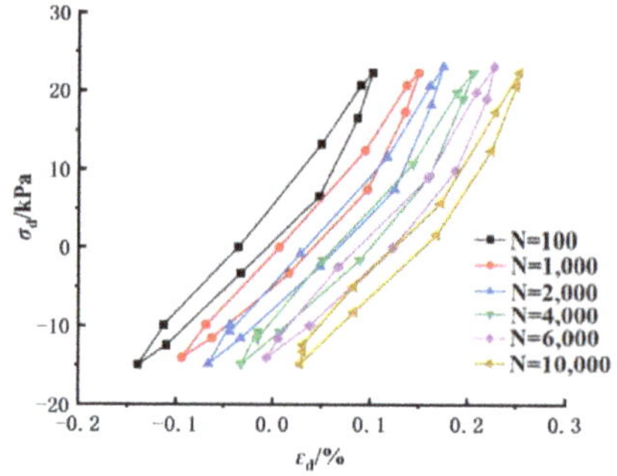

(**d**) f = 2 Hz

Figure 5. Hysteresis loops for different frequencies under unidirectional cyclic loading.

Through the comparison of hysteretic cycles with different frequencies under the condition of unidirectional and bidirectional vibration, it can be found that under the condition of unidirectional and bidirectional cyclic vibration, with the increase in frequency, the area of the hysteretic cycle tends to decrease; however, the slope of the connecting line at both ends of the hysteretic cycle tends to increase, indicating that with the increase in frequency, the soft clay with interlayer sand does not have enough time to deform and stretch. Therefore, the sample is constantly compacted, and the deformation rate decreases with the increase in frequency. For the same frequency, the area of bidirectional vibration hysteresis loop is smaller than that of unidirectional vibration hysteresis loop. The slope of the connecting line between the two ends of the bidirectional vibration hysteresis loop curve is less than that of the unidirectional vibration hysteresis loop curve.

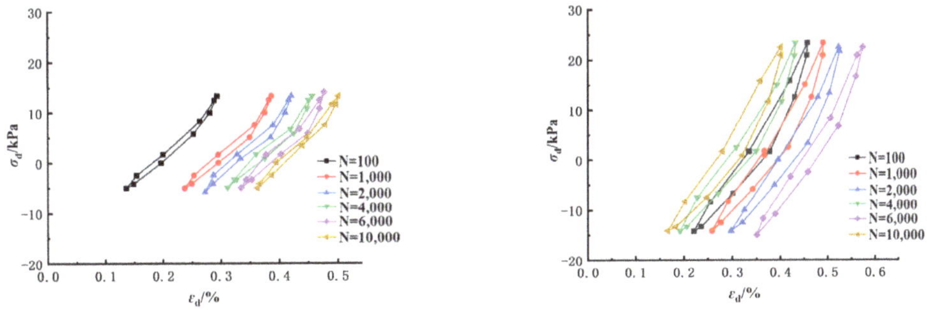

(a) $f = 0.5$ Hz

(b) $f = 1.0$ Hz

Figure 6. Hysteresis loops for different frequencies under bidirectional cyclic loading.

Figure 7 shows the effective stress path curve of the sandwich sand soft clay under different vibration frequencies. It can be seen from the figure that when the vibration frequency is 0.5–1.0 Hz, with the increase in the vibration frequency, the amplitude of the stress path of the cyclic vibration of the sample gradually increases from right to left, which is caused by the increase in the internal pore pressure and the acceleration of the average effective stress attenuation of the soil under the low-frequency load. When the vibration frequency is 2.0–5.0 Hz, the average effective stress attenuation slows down. In addition, under the action of a low-frequency load, the deviator stress of the sample does not attenuate; under a high-frequency cyclic load, the deviator stress reaches the peak at the initial moment and the deviator stress attenuates with the increase in the number of vibration cycles. At the same time, it can be seen from the figure that the stress path under different vibration frequencies is under the effective stress path K'_f of static load, and with the increase in vibration frequency, the stress path under cyclic load gradually approaches the K'_f line.

Figure 8 shows that when the cyclic stress ratio $r_c = 0.225$, the vibration frequency change in the load affects the influence of ε_d-N relationship curve. It can be seen from the figure that the vibration frequency also has a significant impact on the dynamic strain of soft clay. When the number of cycles is the same, the dynamic strain of the soil sample decreases gradually with the increase in vibration frequency. This is mainly because with the increase in frequency, the pore pressure of saturated interbedded sand and soft clay cannot rise in time, resulting in the failure of soil samples under a large number of cycles. In addition, the figure also shows that the magnitude of vibration frequency has little influence on the dynamic strain of saturated soft clay with sand interlayer. From the variation trend of the driven strain, it can be concluded that the critical cyclic stress ratio of saturated intercalated sand and soft clay will increase when the cyclic load frequency is large, while the critical cyclic stress ratio of saturated intercalated sand and soft clay will decrease when the cyclic load frequency is low.

Figure 9 shows the relationship between the dynamic pore pressure ratio u/p'_0 and the number of cycles under different loading frequencies when the cyclic stress ratio $r_c = 0.225$. It can be seen from the figure that under the same load action times, the smaller the frequency of the soil sample, the greater the pore pressure, and the smaller the number of cycles, the failure will occur. The main reason for this phenomenon is that the pore water pressure has enough time to rise under the low-frequency cyclic load. However, under the cyclic load with high frequency, due to the existence of sand interlayer in the sample, the sand interlayer has the effect of water resistance and permeability reduction on the pore water in the soil sample, resulting in the pore water pressure rising in time. When the load frequency is 5 Hz, after 5000 cycles, the pore pressure develops slowly, and the pore pressure change mode is significantly different from the load frequency of 0.5 Hz and 1.0 Hz.

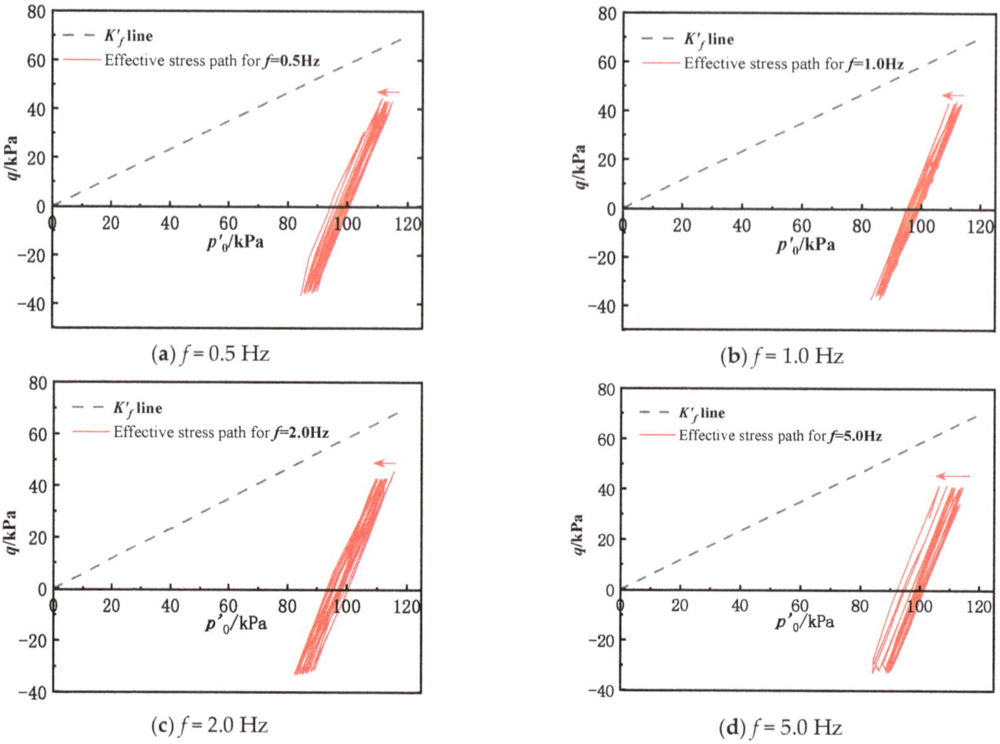

(a) $f = 0.5$ Hz (b) $f = 1.0$ Hz (c) $f = 2.0$ Hz (d) $f = 5.0$ Hz

Figure 7. Effective stress path curve of sandwich sand soft clay under different frequencies.

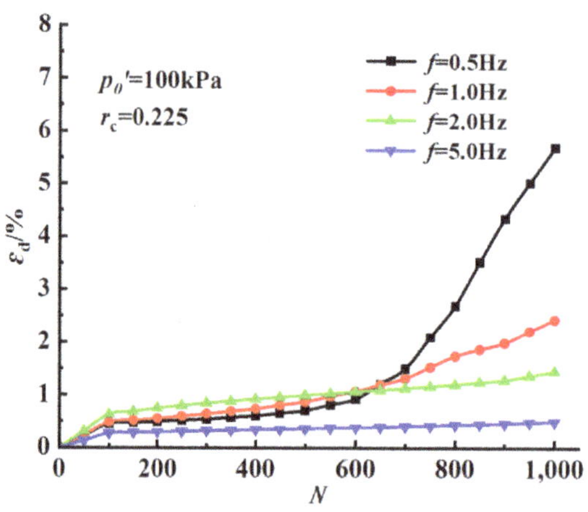

Figure 8. Double-amplitude dynamic strain of soil samples with cycles for different frequencies.

Figure 10 shows the change curve of shear modulus and damping ratio of soil samples with frequency. It can be seen from Figure 10a that under different cycles, the dynamic shear modulus of the sample gradually attenuates with the increase in load frequency. At the same time, under the same load frequency, the dynamic shear modulus of the sample tends to increase with the increase in the number of cyclic vibration, and the increase amplitude of the dynamic shear modulus increases with the increase in the number of cycles at f = 0.5 Hz. It can be seen from Figure 10b that the damping ratio of sandwich sand and soft clay shows the same change rule, and increases with the increase in load frequency, and under the same load frequency, the energy consumed by the soil sample to resist deformation increases with the increase in the number of cycles.

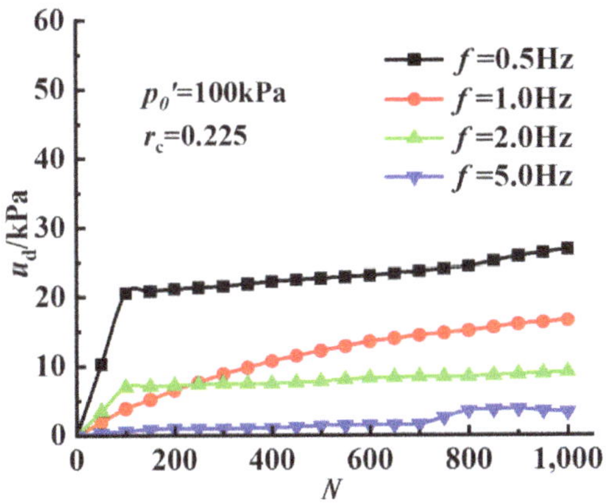

Figure 9. Dynamic pore pressure ratio of soil samples with cycles for different frequencies.

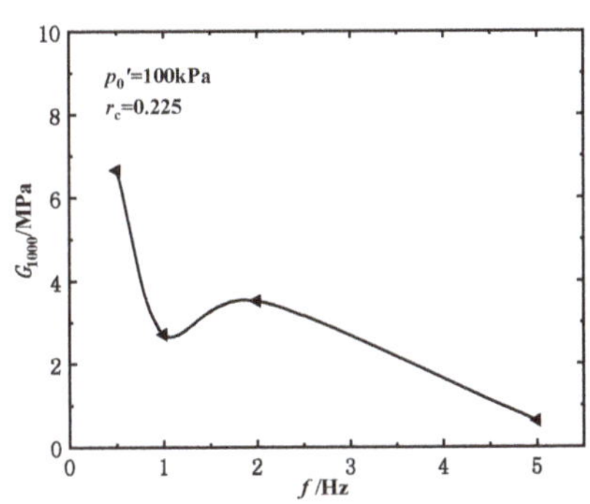

(**a**) Curve of shear modulus versus frequency

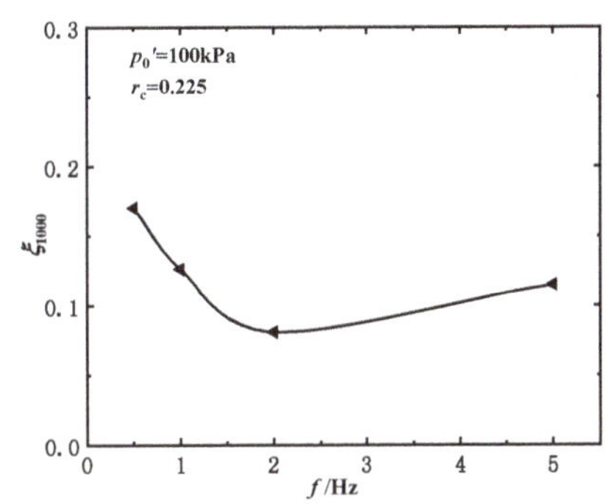

(**b**) Curve of damping ratio versus frequency

Figure 10. Variation of shear modulus and damping ratio with frequency.

3.2. Dynamic Characteristics of Soft Clay under Unidirectional and Bidirectional Cyclic Vibration

According to the test scheme (Table 2), the dynamic triaxial tests under different cyclic vibration conditions are carried out, and the relationship curves of the cumulative axial deformation, damping ratio, dynamic elastic modulus, and dynamic pore water pressure of saturated sandwich sand soft clay samples with the development of vibration times are obtained, as shown in Figure 11.

As shown in Figure 11a, under the same confining pressure, the specimen is subjected to static deviating stress during unidirectional vibration, which makes the cumulative axial deformation of the specimen develop rapidly at the initial stage of vibration. With the increase in vibration cycles, the growth of cumulative axial deformation slows down, and the change range of cumulative axial deformation caused by unidirectional vibration with the increase in frequency is greater than that of bidirectional vibration. Under the same dynamic stress ratio, the cumulative axial deformation caused by unidirectional and bidirectional vibration increases with the increase in frequency. This is because the normalized pore pressure increases with the increase in frequency under unidirectional and bidirectional cycles, which reduces the normalized effective stress accordingly so that the cumulative axial deformation increases.

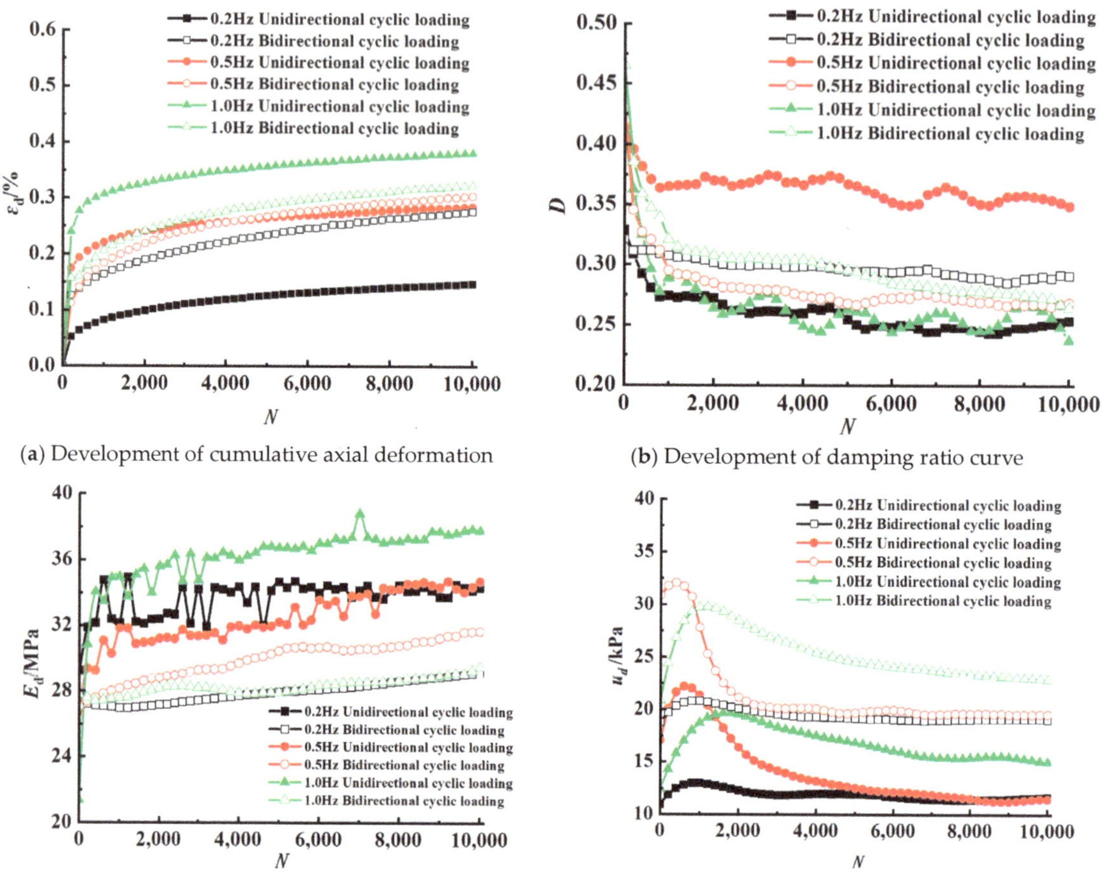

(**a**) Development of cumulative axial deformation (**b**) Development of damping ratio curve

(**c**) Development of dynamic elastic modulus curve (**d**) Development of dynamic pore pressure

Figure 11. Development curve of dynamic parameters with vibration times.

As shown in Figure 11b, under unidirectional vibration, the damping ratio for the same axial dynamic stress ratio and different frequencies decreases with the increase in vibration times, the damping ratio of the first 2000 cycles decreases sharply, and the rate of decline of damping ratio decreases with the increase in vibration cycles. In addition, the greater the frequency, the faster the rate of decline of the damping ratio. In the case of bidirectional vibration, the rate of decline of the damping ratio is basically the same at different frequencies, and the amplitude of decline of the damping ratio is roughly the same.

As shown in Figure 11c, under unidirectional vibration, the dynamic elastic modulus of the sample with different frequencies develops rapidly at the initial stage of vibration, and the growth rate gradually slows down with the increase in the number of vibrations. This is because the soil body bears repeated cyclic loads under the amplitude, and the internal structure of the soil body has not changed significantly. Under bidirectional vibration, the dynamic elastic modulus changes in a small range and reaches the maximum value when f = 0.2 Hz.

As shown in Figure 11d, under the condition of unidirectional vibration, the dynamic pore pressure increases with the increase in vibration frequency, and the dynamic pore pressure reaches the peak at the vibration number N = 1000 and gradually tends to be flat with the increase in vibration number. Under the condition of bidirectional vibration, when the vibration times are less than 1000, the pore pressure of f = 0.5 Hz develops faster than that of other frequencies.

Therefore, when the sample is subjected to dynamic load, different vibration modes have complex effects on the dynamic characteristics of the sample. The initial static deviator stress of unidirectional vibration will promote the development of cumulative plastic strain, and the tensile stress produced by bidirectional vibration will hinder the development of dynamic elastic modulus.

3.3. Dynamic Elastic Modulus and Damping Ratio under Different Vibration Times

The relationship curve between dynamic elastic modulus and frequency under different vibration cycle conditions is shown in Figure 12. Under the same axial dynamic stress ratio, the dynamic elastic modulus of unidirectional vibration decreases first and then increases with the increase in frequency. When f < 0.5 Hz, the strength of the soft clay is significantly reduced by sand interlayer, and the energy consumption inside the soil increases with the increase in frequency, and reaches the peak value when f = 0.5 Hz. For the same frequency, the dynamic elastic modulus of the sample increases with the increase in the number of vibrations. Under the condition of the same axial cyclic stress ratio and radial cyclic stress ratio, the dynamic elastic modulus of bidirectional vibration shows a change trend of first increasing and then decreasing with the increase in frequency, which is just opposite to the unidirectional vibration, indicating that the radial cyclic stress during bidirectional vibration enhances the ability of soil to resist deformation. Therefore, it can be concluded that when f < 0.5 Hz, the sand interlayer can strengthen the strength of the soft clay, the dynamic elastic modulus increases with the increase in vibration times.

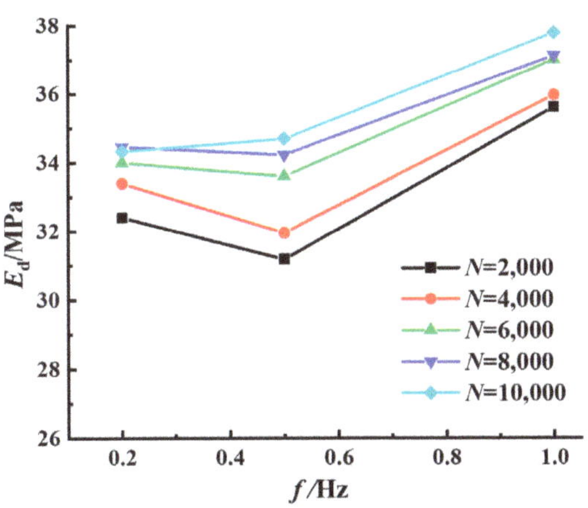

(**a**) Unidirectional vibration

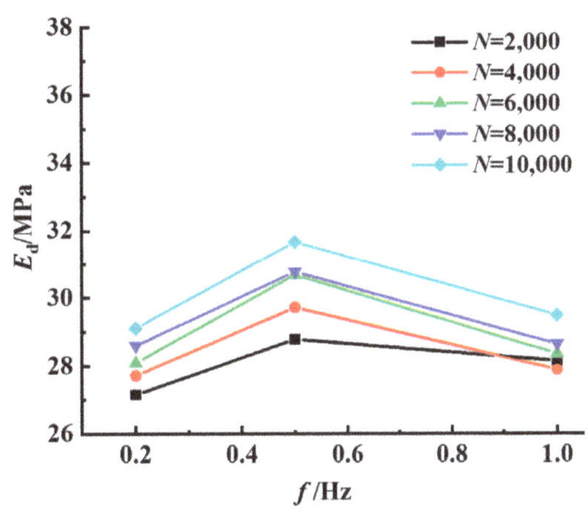

(**b**) Bidirectional vibration

Figure 12. Variation of dynamic elastic modulus with frequency.

The relationship curve between damping ratio and frequency under different vibration cycle conditions is shown in Figure 13. It can be seen that under the same axial dynamic stress ratio, the damping ratio of unidirectional vibration increases first and then decreases with the increase in frequency, and reaches the maximum value when f = 0.5 Hz. Under the same axial cyclic stress ratio and radial cyclic stress ratio, due to the action of lateral cyclic dynamic stress, the damping ratio of bidirectional vibration shows a trend of first decreasing and then slowly increasing with the increase in frequency. Therefore, it can be concluded that under different cyclic loads, the change in frequency has a great impact on the saturated soft clay containing sand interlayer.

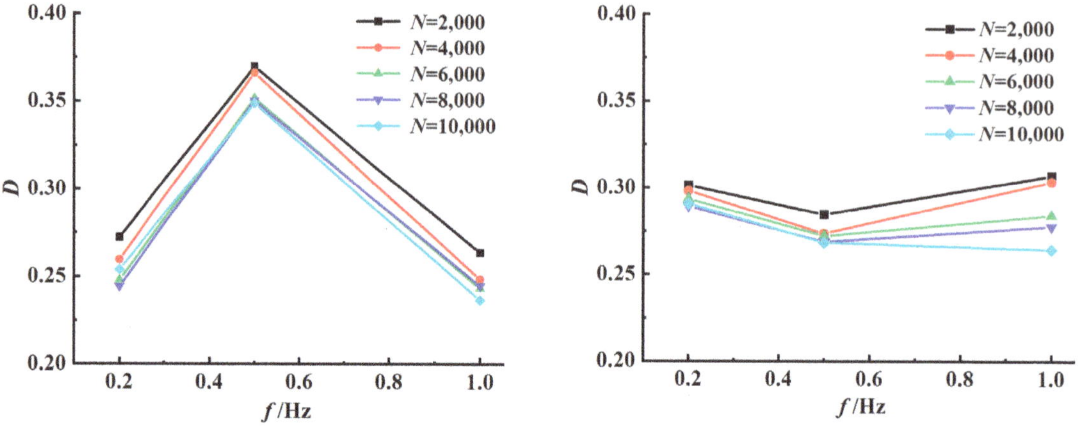

(a) Unidirectional vibration

(b) Bidirectional vibration

Figure 13. Variation of damping ratio with frequency.

4. Conclusions

A large number of dynamic triaxial tests were conducted and the effects of frequency, stress ratio on shear modulus, and damping ratio were studied. The following conclusions can be drawn:

(1) Under the condition of unidirectional and bidirectional cyclic vibration, the area of the hysteresis loop tends to decrease with the increase in frequency, and the slope of the connecting line at both ends of the hysteresis loop tends to increase. For the same frequency, the area of the bidirectional vibration hysteresis loop and the slope of the connecting line at both ends are smaller than that of the unidirectional cyclic vibration.

(2) Under the same dynamic stress ratio, the cumulative axial deformation caused by unidirectional and bidirectional vibration increases with the increasing of frequency. Under the same cycle, the dynamic pore pressure increases with the increase in vibration frequency, and the dynamic pore pressure reaches the peak at the cycle $N = 1000$, and gradually tends to be flat with the increase in cycle.

(3) For the same frequency, the dynamic elastic modulus of the sample increases with the increase in cycles, and the damping ratio decreases, and reaches the maximum value when $f = 0.5$ Hz. Due to the effect of radial cyclic stress, the curves of dynamic elastic modulus and damping ratio with frequency under bidirectional vibration are just opposite to those under unidirectional vibration, indicating that radial cyclic stress can enhance the ability of soil samples to resist deformation.

Author Contributions: Conceptualization, S.W., Y.C., L.Z., Y.P. and B.C.; methodology, S.W., Y.C., L.Z., Y.P., B.C. and P.Z.; formal analysis, Y.C.; data curation, S.W. and P.Z.; writing—original draft preparation, S.W. and Y.F.; writing—review and editing, S.W. and Y.C. All authors have read and agreed to the published version of the manuscript.

Funding: The work presented in this paper was sponsored by the Supported by Systematic Project of Key Laboratory of New Technology for Construction of Cities in Mountain Area (No. LNTCCMA-20200104), the Systematic Project of Guangxi Key Laboratory of Disaster Prevention and Structural Safety (No. 2019ZDK005), the Initial Scientific Research Fund of Young Teachers in Ningbo University of technology (2019A610394), the Ningbo Public Welfare Science and Technology Planning Project (No. 2019C50012). These financial supports are gratefully acknowledged.

Data Availability Statement: The data used to support the findings of this study are included in the article.

Conflicts of Interest: The authors declare no conflict of interest.

References

1. Qian, J.-G.; Du, Z.-B.; Yin, Z.-Y. Cyclic degradation and non-coaxiality of soft clay subjected to pure rotation of principal stress directions. *Acta Geotech.* **2017**, *13*, 943–959. [CrossRef]
2. Guo, L.; Chen, J.; Wang, J.; Cai, Y.; Deng, P. Influences of stress magnitude and loading frequency on cyclic behavior of K0-consolidated marine clay involving principal stress rotation. *Soil Dyn. Earthq. Eng.* **2016**, *84*, 94–107. [CrossRef]
3. Qian, J.-G.; Wang, Y.-G.; Yin, Z.-Y.; Huang, M.-S. Experimental identification of plastic shakedown behavior of saturated clay subjected to traffic loading with principal stress rotation. *Eng. Geol.* **2016**, *214*, 29–42. [CrossRef]
4. Guo, L.; Wang, J.; Cai, Y.; Liu, H.; Gao, Y.; Sun, H. Undrained deformation behavior of saturated soft clay under long-term cyclic loading. *Soil Dyn. Earthq. Eng.* **2013**, *50*, 28–37. [CrossRef]
5. France, J.W.; Sangrey, D.A. Effects of Drainage in Repeated Loading of Clays. *J. Geotech. Eng. Div.* **1977**, *103*, 769–785. [CrossRef]
6. Talesnick, M.; Frydman, S. Irrecoverable and Overall Strains in Cyclic Shear of Soft Clay. *Soils Found.* **1992**, *32*, 47–60. [CrossRef] [PubMed]
7. Andersen, K.H.; Brown, S.F.; Foss, I.; Poul, J.H.; Rosenbrand, W.F. Effect of cyclic loading on clay behavior. In Proceedings of the Design and Construction of Offshore Structures, London, UK, 7 July 2015; pp. 75–79.
8. Yasuhara, K.; Murakami, S.; Song, B.-W.; Yokokawa, S.; Hyde, A.F.L. Postcyclic Degradation of Strength and Stiffness for Low Plasticity Silt. *J. Geotech. Geoenvironmental Eng.* **2003**, *129*, 756–769. [CrossRef]
9. Matsui, T.; Bahr, M.; Abe, N. Estimation of shear characteristic degradation and stress-stain relationship of satuarated clay after cyclic loading. *Soil Found.* **1992**, *32*, 161–172.
10. Kavvadas, M.; Amorosi, A. A constitutive model for structured soils. *Géotechnique* **2000**, *50*, 263–273.
11. Savvides, A.A.; Papadrakakis, M. A computational study on the uncertainty quantification of failure of clays with a modified Cam-Clay yield criterion. *SN Appl. Sci.* **2021**, *3*, 659.
12. Seed, H.B.; Lee, K.L. Liquefaction of Saturated Sand during Cyclic Loading. *J. Soil Mech. Found. Div. ASCE* **1966**, *92*, 105–134. [CrossRef]
13. Wichtmann, T.; Niemunis, A.; Triantafyllidis, T. On the influence of the polarization and the shape of the strain loop on strain accumulation in sand under high-cyclic loading. *Soil Dyn. Earthq. Engine* **2007**, *27*, 14–28. [CrossRef]
14. Kammerer, A.M.; Seed, B.; Wu, J. Pore water development in liquefiable soils under bi-directional loading conditions. In Proceedings of the 13th World Conference on Earthquake Engineering, Vancouver, BC, Canada, 1–6 August 2004; pp. 697–704.
15. Hyde, A.F.L.; Ward, S.J. Effect of cyclic loading on the undrained shear strength of a silty clay. *Mar. Geotechnol.* **1986**, *6*, 299–314. [CrossRef]
16. Silvestri, V.; Bouteldja, M. On the consolidation response of a sensitive clay under cyclic loading. In Proceedings of the Canadian Geotechnical Conference, Saskatoon, SK, Canada, 27–29 September 1993; pp. 455–464.
17. Narasimha, R.S. Behaviour of marine clay under wave type of cyclic loading. In Proceedings of the International Conference on Off-shore Mechanics and Arctic Engineering-OMAE, Cancun, Mexico, 8–13 June 2003.
18. Liang, R.Y.; Ma, F. Anisotropic Plasticity Model for Undrained Cyclic Behavior of Clays. I: Theory. *J. Geotech. Eng.* **1992**, *118*, 229–245. [CrossRef]
19. Rehman, Z.U.; Khalid, U.; Ijaz, N.; Mujtaba, H.; Haider, A.; Farooq, K.; Ijaz, Z. Machine learning-based intelligent modeling of hydraulic conductivity of sandy soils considering a wide range of grain sizes. *Eng. Geol.* **2022**, *311*, 106899. [CrossRef]
20. Cai, Y.; Gu, C.; Wang, J.; Juang, C.H.; Xu, C.; Hu, X. One-Way Cyclic Triaxial Behavior of Saturated Clay: Comparison between Constant and Variable Confining Pressure. *J. Geotech. Geoenvironmental Eng.* **2013**, *139*, 797–809. [CrossRef]
21. Idriss, I.M.; Dobry, R.; Singh, R.D. Nonlinear Behavior of Soft Clays during Cyclic Loading. *J. Geotech. Eng. Div.* **1978**, *104*, 1427-1447. [CrossRef]
22. Rehman, Z.U.; Luo, F.; Wang, T.; Zhang, G. Large-Scale Test Study on the Three-Dimensional Behavior of the Gravel–Concrete Interface of a CFR Dam. *Int. J. Géoméch.* **2020**, *20*, 04020046. [CrossRef]
23. Rehman, Z.U.; Zhang, G. Cyclic behavior of gravel–steel interface under varying rotational shear paths. *Can. Geotech. J.* **2021**, *58*, 305–316. [CrossRef]
24. Rehman, Z.U.; Zhang, G. Shear coupling effect of monotonic and cyclic behavior of the interface between steel and gravel. *Can. Geotech. J.* **2018**, *56*, 876–884. [CrossRef]

Disclaimer/Publisher's Note: The statements, opinions and data contained in all publications are solely those of the individual author(s) and contributor(s) and not of MDPI and/or the editor(s). MDPI and/or the editor(s) disclaim responsibility for any injury to people or property resulting from any ideas, methods, instructions or products referred to in the content.

buildings

MDPI

Article

Experimental Study on Mechanical Properties of Structured Clay under Different Unloading Rates and Unloading Stress Paths

Lu Li, Meng Zang *, Rongtang Zhang and Haijun Lu

School of Civil Engineering and Architecture, Wuhan Polytechnic University, Wuhan 430023, China
* Correspondence: zangmeng@whpu.edu.cn

Abstract: Consolidated undrained triaxial shear tests were performed on undisturbed saturated structured clay at three unloading rates (0.1, 0.25, and 2.5 kPa/min) using a GDS triaxial system to determine the effects of different unloading rates and unloading stress paths on the stress–strain relationship, pore pressure variation, and failure strength characteristics of Zhanjiang structured clay. Microstructural changes in the clay were observed during shear tests at different unloading rates. Furthermore, the obtained stress–strain relationship indicates strain-softening under different unloading stress paths. Under the same axial strain, a larger unloading rate caused a larger deviatoric stress. Under the same conditions, the higher the confining pressure, the greater the peak pore pressure, the smaller the unloading rate, the greater the pore pressure development, and the greater the variation in the pore pressure. Moreover, the undrained shear strength increased with an increase in the unloading rate from 0.1 to 2.5 kPa/min. The change in the unloading rate had a greater effect on the undrained strength under the passive tensile path than that under the passive compression path. The microstructure of the Zhanjiang structured clay changed after shear tests at different unloading rates, exhibiting various degrees of adjustment in the particle arrangement, contact relations, pore sizes, and shapes.

Keywords: structured clay; unloading stress path; unloading rate; unloading failure strength

1. Introduction

The mechanical properties of soft clay, such as deformation and strength, are related to stress paths [1,2] and corresponding rate effects [3–5] owing to the characteristics of high clay grain content, high moisture content, and large pore ratio, among others. The excavation of a foundation pit is a typical unloading project in geotechnical engineering. The deformation characteristics, mechanical parameters, and failure mechanisms of soils significantly differ when they undergo different unloading stress paths and unloading rates to reach the same stress state [6].

Numerous studies on the mechanical properties of soils under unloading stress paths have been conducted, and remarkable achievements have been made. Liu et al. [7] demonstrated that the stress–strain relationship for soft soils is affected by unloading stress paths. Yan et al. [8] found that excess pore pressure in saturated soft clay under unloading conditions effects the excavation of foundation pit and heave of pit bottom through a comparative study between triaxial unloading experiments and theoretical research. Lim et al. [9] demonstrated that the unloading mechanical parameters of soft soils can be used to accurately calculate the excavation deformations of foundation pits under undrained conditions. Huang et al. [1] conducted K_0 consolidation undrained shear tests on Shenzhen soft soil under the influence of unloading stress paths; the results showed variation in the unloading strength and initial shear modulus of soils under unloading stress paths, and the deformation properties of soils were greatly influenced by the unloading stress paths. Jia et al. [10] performed axial loading and lateral unloading shear tests on Wenzhou soft clay using a stress-path triaxial apparatus and proposed a four-parameter stress–strain curve model based on the deformation mechanism of structured soils and incremental linear elasticity theory. Huang et al. [11] reported that the unloading strength parameters and unloading modulus of marine sedimentary soft soils were significantly influenced by the unloading stress paths, and an increase in excess pore water pressure weakened the unloading strengths, particularly when the cohesion was stronger. The above studies indicate that mechanical properties including the stress–strain relationship, pore pressure characteristics, deformation and strength characteristics at failure, and the initial unloading modulus of soils are significantly related to the unloading stress paths.

In the 1930s, Buisman [12] reported that the stress, strain, and strength of soils exhibited a non-negligible rate-dependent relationship via indoor experiments. Some researchers have conducted rate-dependent tests on different clays [13–15] and concluded that the undrained shear strength increases with increasing strain rates, thus considering the shear strength rate parameters for quantitative description. For example, Whitman et al. [14] obtained the strength rate parameter of 3.5% for Mississippi alluvial soils, but the magnitude of the effects of the increasing strain rate on the undrained shear strength varied greatly. Zhu et al. [15] performed shear tests on saturated marine clay under compression and tension path conditions at three different strain rates and found that the effect of the shear rate on the undrained shear strength of the tension path was greater than that of the compression path. Wu et al. [16] conducted K_0-consolidation undrained triaxial tests on marine clay considering the effects of stress states and strain rates and concluded that the higher the strain rate, the higher the undrained shear strength, and the lower the pore pressure of soils with different over-consolidation ratios. Cai's [17] research results demonstrated that strong structured clay presents a unique shear rate mechanical effect, and its consolidation undrained shear strength does not change monotonically with shear rate but is characterized by first decreasing and then increasing with increasing shear strain rate, and there is a critical rate phenomenon. These previous research outcomes indicate that the strength characteristics of soils are strongly related to the shear rate. However, these tests generally examined the strain rate effects under a specific stress history, which is not entirely applicable to actual excavation projects, such as those involving slopes and foundation pits, where stress rate problems exist. Meanwhile, Li et al. [18] analyzed the mechanical properties of undisturbed expansive clay under the influence of different unloading rates and unloading stress paths using a GDS stress-path triaxial apparatus. The results showed that the pore pressure of soils under reduced compression and reduced extension stress paths remained negative, the undrained shear strength increased with an increase in unloading rate, and the failure pattern of expansive clay was related to the shear rate and microcracks. Yang et al. [19] performed undrained unloading tests of isotropic consolidation on a soft dredger fill and developed formulas for the initial tangent modulus and unloading failure strength considering the influence of the unloading rates and unloading stress paths. Song et al. [20] conducted triaxial tests on remolded deep clay under the effect of stress paths and stress rates. As a result, the strength of soils exhibited a strong rate dependence under different stress paths, and the damage caused by unloading conditions on soil was more abrupt than that caused by loading conditions.

The aforementioned studies comprehensively considered the influence of unloading stress paths and unloading rates on the mechanical properties of soils; however, most focused on soft soils with weak structural properties. Zhanjiang clay is known for its strong structural properties, with an open flocculation structure and strong cementation of free oxides at the microscopic level, as well as an abnormal combination of poor physical properties and good mechanical property indices at the macroscopic level. To date, a certain understanding of the static and dynamic properties as well as microscopic mechanisms of Zhanjiang clay has been achieved [21–24], whereas the mechanical properties of strong structured clay under the influence of different unloading stress paths and unloading rates are less researched. Therefore, in this study, a series of undrained unloading tests was conducted under isotropic consolidation conditions regarding various unloading stress paths and unloading rates to simulate different unloading methods and speeds in foundation pit excavation projects as well as to investigate the influence law of different unloading stress paths and unloading rates on the strain, pore pressure, and strength characteristics of strong structured clay, with the aim of providing a reference and theoretical basis for the selection of geotechnical engineering parameters in structured clay areas.

2. Materials and Methods

2.1. Test Material

Test soil samples were obtained from the foundation pit in Zhanjiang City (Figure 1), at the depth of 8–10 m. The soil sample was grey-brown and homogeneous, and the horizontal (normal) direction had weak laminar surface development. The physical and mechanical indices of the test soil samples are presented in Table 1; the moisture content, particle density, liquid and plastic limits, and permeability test were determined as per ASTM D2216-19, ASTM D854-92, ASTM D4318-17e1, and ASTM D1586-11, respectively [25–28], indicating that the Zhanjiang clay possesses inferior physical properties, such as high natural moisture content, large void ratio, high liquid–plastic limit, and small vertical permeability coefficient, but superior mechanical properties, including high unconfined compressive strength. The results of the unconfined compressive strength tests on undisturbed and remolded Zhanjiang clay are displayed in Figure 2, the results were determined as per ASTM D2166-16 [29], and the unconfined compressive strengths of undisturbed soil and remolded soil were 114 and 16.4 kPa, respectively, with the sensitivity of 6.95. According to the classification suggested by Touiti et al., Zhanjiang clay should be considered a medium-sensitive clay [30]. Undisturbed Zhanjiang soil samples exhibit significant brittleness characteristics, as deformation was observed under compression and the peak stress was realized at small axial strain; with the apparent failure surface, the soil sample resistance decays rapidly. The strength of the remolded soil gradually increases with the development of deformation and eventually results in ductile bulging failure [24].

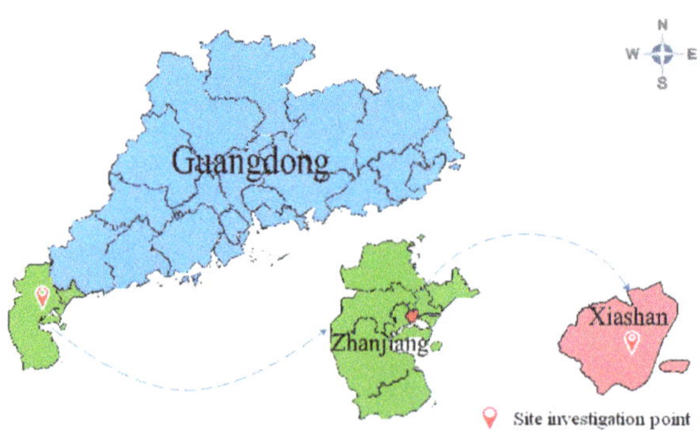

Figure 1. Sampling location in Zhanjiang City, Guangdong Province, China.

Table 1. Physical and mechanical indices for Zhanjiang structured clay.

Particle Density/(g·cm^{-3})	Moisture Content w/%	Liquid Limit w_L/%	Plastic Limit w_p/%	Plasticity Index I_p	Void Ratio e_0	Vertical Permeability Coefficient K/(10^{-7} cm/s)	Unconfined Compressive Strength q_u/kPa	Structural Yield Stress σ_k/kPa
2.69	43.85	54.2	20.8	33.4	1.2	2.6	114	300–400

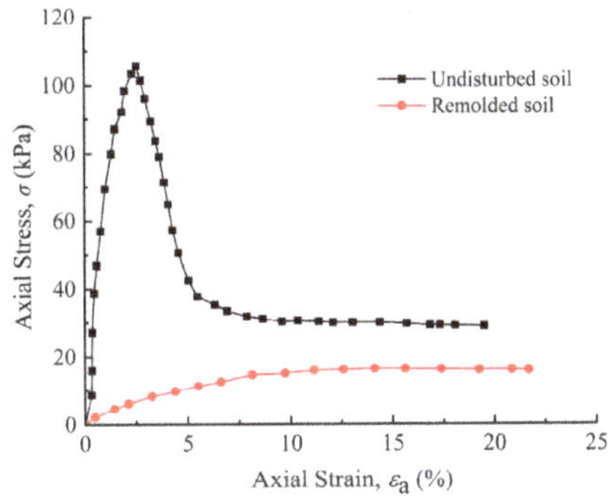

Figure 2. Unconfined compressive stress–strain curve for Zhanjiang structured clay.

The X-ray diffraction (XRD) results of the Zhanjiang structured clay are presented in Figure 3 and Table 2. Clay minerals (<0.005 mm) accounted for more than 60% of the particle composition; therefore, the Zhanjiang clay was classified as soft clay.

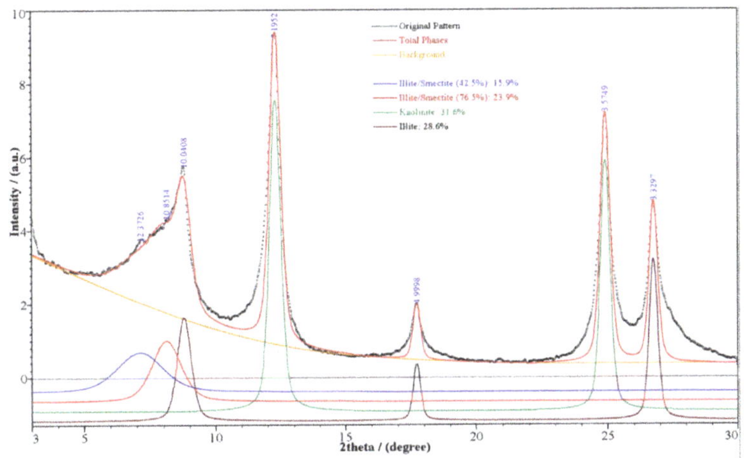

Figure 3. XRD result for Zhanjiang structured clay.

Table 2. Mineral composition of Zhanjiang structured clay.

Relative Mineral Content/%						
Quartz	K-Feldspar	Na-Feldspar	Illite	Kaolinite	Illite-Smectite Mixed LayerI	Illite-Smectite Mixed LayerII
62	3	3	8.96	10.24	5.12	7.68

2.2. Determination of Unloading Stress Path

The stress paths experienced by the soil in various parts during the excavation of a foundation pit are complex and vary significantly with excavation depth, excavation method, and support form of the foundation pit. In this study, four stress paths were considered, experienced by the soil in typical parts (A, B, and C) of the active and passive zones of the pit, and employed as the test paths, as depicted in Figure 4.

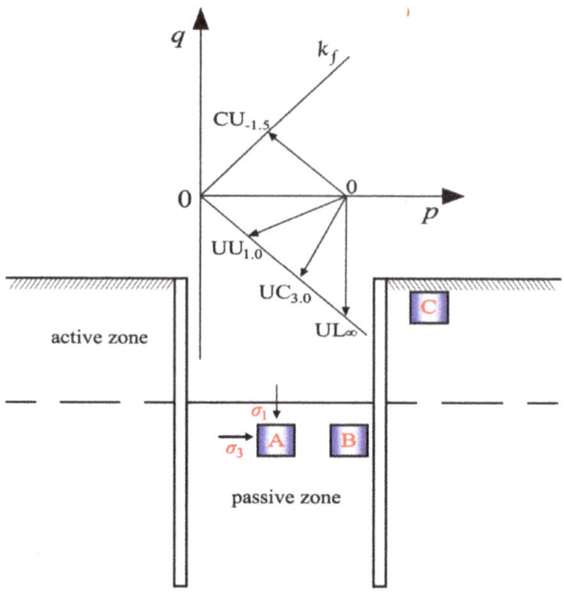

Figure 4. Schematic of soil stress states and unloading stress paths in foundation pit projects.

In the passive earth pressure zone, for soil unit A near the excavation surface of the foundation pit, the vertical unloading is the most obviously affected by the excavation of the overlying soils. This indicates that the axial stress unloading and the horizontal stress remain unchanged (σ_1 decreases, σ_3 remains unchanged), so the unloading stress path expressed as $UC_{3.0}$ and the horizontal stress may decrease with uplifting of the pit bottom. In addition, there will be both the axial stress and horizontal stress unloading (σ_1 decreases, σ_3 decreases), so the unloading stress path is expressed as $UU_{1.0}$. Furthermore, offset of the foundation pit support into the pit can lead to the axial stress unloading and the horizontal stress loading (σ_1 decreases, σ_3 increases) of soil unit B, and the unloading stress path is expressed as UL_{∞}. In the active earth pressure zone, with the excavation of the foundation pit, the horizontal stress decreases and the axial stress remains unchanged (σ_3 decreases, σ_1 remains unchanged) for the soil on the outside of the foundation support, i.e., soil unit C; the unloading stress path is expressed as $CU_{-1.5}$. Accordingly, four representative unloading paths were selected: $UC_{3.0}$, $UU_{1.0}$, UL_{∞}, and $CU_{-1.5}$. The stress paths were named as L for loading, C for constant load, U for unloading, and the former letter for the vertical stress state, the latter letter for the horizontal stress state, and the footnote for the stress path slope $\eta = \Delta q/\Delta p$ [23].

2.3. Selection of Unloading Rate

During the excavation of a foundation pit, the mechanical properties of soils are influenced by not only the stress path, but also the speed of excavation (i.e., different unloading rates), particularly the strength and deformation properties of the soil. In this study, we initially set several groups of different unloading rates based on various unloading stress paths through indoor tests. The specific unloading rate selection scheme is shown in Table 3.

Table 3. Preliminary selection scheme for unloading rates.

Group	Unloading Rate, v/kPa·min^{-1}	Confining Pressure/kPa	Unloading Stress Path
I	0.01, 0.025, 0.05	100	$UC_{3.0}$ (σ_1 decreases, σ_3 remains unchanged)
II	0.25, 0.5, 1	100	$UC_{3.0}$ (σ_1 decreases, σ_3 remains unchanged)
III	0.1, 0.25, 2.5	100	$UC_{3.0}$ (σ_1 decreases, σ_3 remains unchanged)

The stress–strain relationship curves obtained from the test scheme in Table 3 are shown in Figure 5.

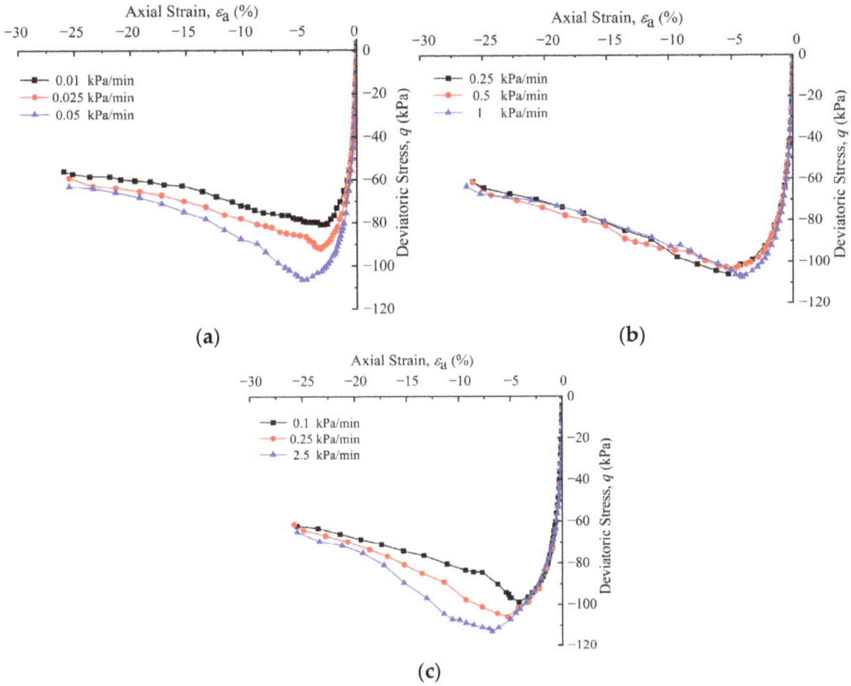

Figure 5. Stress–strain curves obtained under different unloading rates. (**a**) I: 0.01, 0.025, 0.05 kPa·min^{-1}; (**b**) II: 0.25, 0.5, 1 kPa·min^{-1}; (**c**) III: 0.1, 0.25, 2.5 kPa·min^{-1}.

From Figure 5, it can be seen that the stress–strain relationship curves for the Zhanjiang structured clay under different unloading rates exhibit strain softening. In the cases of unloading rates of 0.01, 0.025, 0.05 kPa/min and 0.25, 0.5, 1 kPa/min, the stress–strain curves corresponding to 0.25, 0.5, and 1 kPa/min are not sensitive to this unloading rate. Therefore, the unloading rate is insignificant in engineering practice and was not examined further in this study. Since a foundation pit project generally requires rapid excavation and support, and the soil experiences the unloading stress path predominantly under undrained conditions, the final unloading rates were taken as 0.1, 0.25, and 2.5 kPa/min considering the test period and the actual excavation rate of the foundation pit project. The experimental scheme is summarized in Table 4.

Table 4. Unloading rate test programs.

Zone	Consolidation Confining Pressure/kPa	Stress Path Slope/η	Unloading Stress Path	Unloading Rate v/kPa·min^{-1}	Drainage Condition
Active earth pressure zone	50/100/200	−1.5	$CU_{-1.5}$ (σ_1 remains unchanged, σ_3 decreases)	0.1, 0.25, 2.5	Undrained
Passive earth pressure zone	50/100/200	1.0	$UU_{1.0}$ (σ_1 decreases, σ_3 decreases)	0.1, 0.25, 2.5	
		3.0	$UC_{3.0}$ (σ_1 decreases, σ_3 remains unchanged)		
		∞	UL_{∞} (σ_1 decreases, σ_3 increases)		

2.4. Test Method

The test apparatus was a GDS stress-path triaxial apparatus, and the specimen was trimmed using a wire saw, which was a cylinder with a height of 80.0 mm and diameter of 39.1 mm. The specimen encased in the rubber membrane was sealed at the cap (the receiving element) and base (the transmitting element) and positioned in a triaxial compression chamber, after which pressure was applied. Vacuum suction saturation and back pressure saturation were prepared before testing, and an undrained triaxial shear test was performed according to the set stress path after isotropic consolidation. The test was completed when the axial strain reached 20%, and the test results were determined according to ASTM D4767-11 [31].

3. Results and Discussion

There are two types of specimen failures under unloading conditions: compression and elongation failures. The analysis shows that among the four stress paths in Table 4, the failure mode of the specimen under $CU_{-1.5}$ path is axial compression failure, while that of the specimen under the other three stress paths is axial elongation failure, or "tensile failure". The failure modes of specimens are provided in Figure 6.

(a) (b) (c)

Figure 6. Typical failure modes of specimens. (**a**) Undisturbed sample before test; (**b**) compression failure; (**c**) tensile failure.

3.1. Effect of Unloading Rate on Stress-Strain Relationship

The deviatoric stress–axial strain curves for the Zhanjiang structured clay under different unloading stress paths are shown in Figures 7–10. Deviatoric stress $q = \sigma_1 - \sigma_3$: in the passive compression path, q is positive ($CU_{-1.5}$); in the passive tensile paths, q is negative ($UC_{3.0}$, UL_{∞}, $UU_{1.0}$), while the axial strain is positive in the compression path and negative in the tensile path.

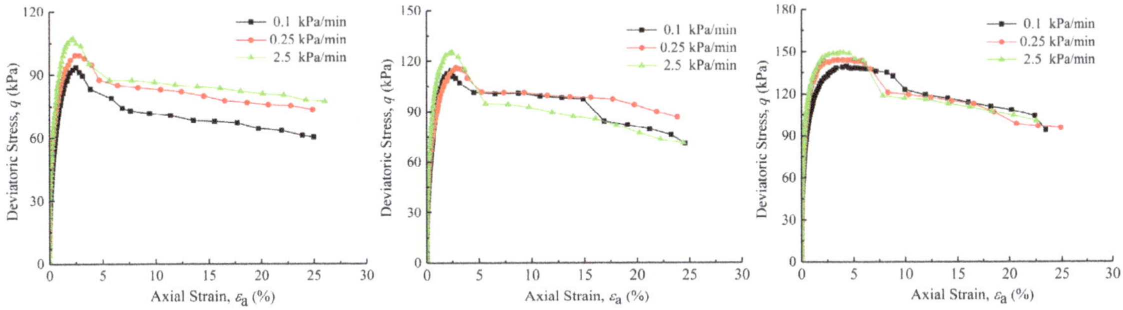

(**a**) (**b**) (**c**)

Figure 7. Strain–stress curves for $CU_{-1.5}$ path under different confining pressure. (**a**) 50 kPa; (**b**) 100 kPa; (**c**) 200 kPa.

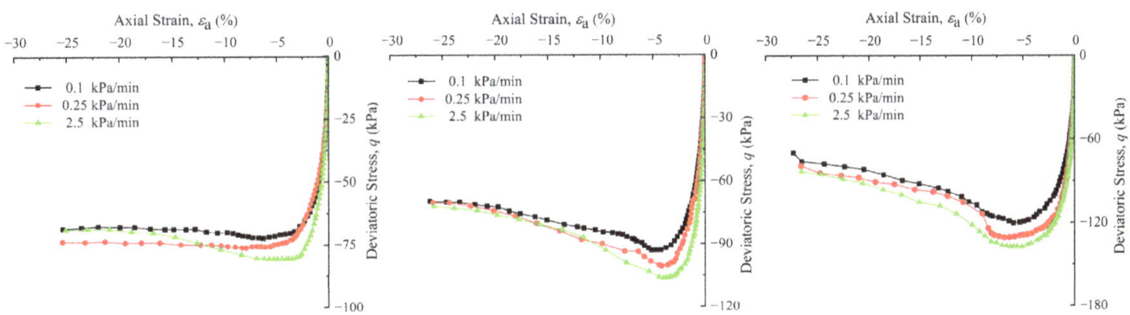

(**a**) (**b**) (**c**)

Figure 8. Strain–stress curves for $UU_{1.0}$ path under different confining pressure. (**a**) 50 kPa; (**b**) 100 kPa; (**c**) 200 kPa.

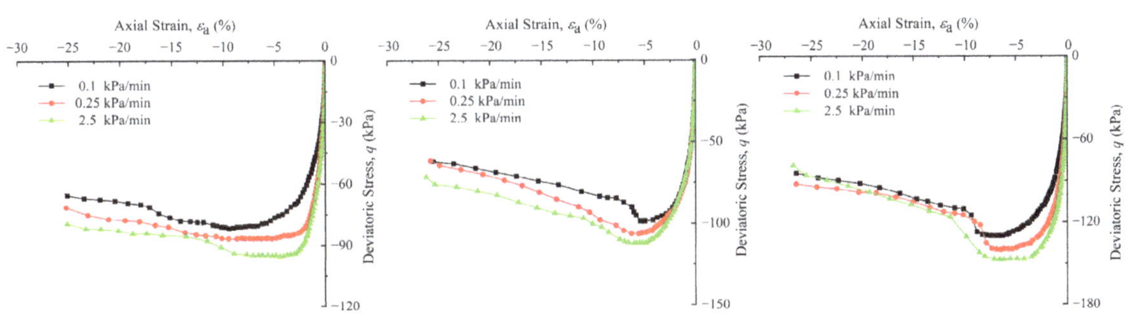

(**a**) (**b**) (**c**)

Figure 9. Strain–stress curves for $UC_{3.0}$ path under different confining pressure. (**a**) 50 kPa; (**b**) 100 kPa; (**c**) 200 kPa.

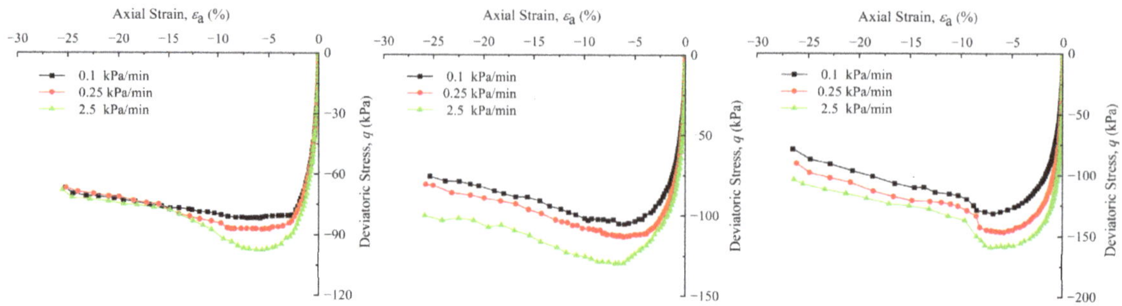

(a) (b) (c)

Figure 10. Strain–stress curves for UL_{∞} path under different confining pressure. (**a**) 50 kPa; (**b**) 100 kPa; (**c**) 200 kPa.

From Figures 7–10, it can be seen that the stress–strain curves for Zhanjiang structured clay exhibit strain softening at different unloading rates under both passive compression and passive tensile paths; more specifically, with an increase in axial strain, the deviatoric stress first increases rapidly and then gradually decreases after reaching the peak stress. Under the passive compression path, the smaller the confining pressure, the more significant the strain softening; whereas, under the passive tensile path, the larger the confining pressure, the more evident the strain softening. For the same axial strain, the larger the unloading rate, the greater the deviatoric stress of the soil under the same unloading stress path; further, with increased unloading rate, the soil obtained a higher secant modulus, which decreased significantly with increase in unloading time. Therefore, quick excavation with instantly available support is advisable during corresponding construction.

Under the same stress path and confining pressure, the stress–strain curves for the specimen at different unloading rates are more concentrated in the initial stage of unloading, at low deviatoric stress and soil deformation, demonstrating that the stress–strain curves were not sensitive to the unloading rate at this stage. Moreover, with an increase in the unloading amount, when the deviatoric stress reaches the failure stress of the soil and the strain increases rapidly, the soil undergoes large deformation failure, and the strength of the soil decays rapidly. Furthermore, for the same consolidation pressure, the peak deviatoric stress of the stress–strain curves is related to the unloading rate; the larger the unloading rate, the larger the failure stress of the soil. This is because when the unloading rate is high, the soil particles do not have enough time to move during shearing, and it is difficult to overcome the slip, rotation, rolling, and transposition resistance between particles in a relatively short period. Only when the energy gathers to the effective stress then the resistance of the particles is overcome, forming a dislocation and producing deformation, which resulted in large macroscopic failure stress. When the unloading rate is low, the soil particles have more time for directional adjustment or displacement in the longer unloading period to achieve rolling and directional arrangement of the particles; thus, the failure stress was smaller.

The stress–strain curves under the $CU_{-1.5}$ unloading path clearly differ from those for the other three groups of stress paths. The deformation of the specimen under the $CU_{-1.5}$ path exhibited axial compression, and the axial strain when the specimens reached failure was between 2% and 4%. In foundation pit excavation projects, the soil outside the foundation pit support in the active earth pressure zone is damaged when a small strain is generated. This indicates that the soil outside the pit can reach the active limit equilibrium state under a small lateral deformation. Therefore, the horizontal displacement of the retaining wall should be effectively controlled in actual engineering to avoid sliding damage to the soil behind the retaining wall. Furthermore, the deformation of the specimens under the $UC_{3.0}$, UL_{∞}, and $UU_{1.0}$ paths showed axial elongation, and the axial strains when the specimens reached failure under the passive tensile paths were mostly concentrated around

4~9%. In actual engineering, soil uplift deformation at the bottom of the pit should be monitored to avoid soil uplift damage in the foundation pit, i.e., the strain of the soil inside the foundation pit in the passive earth pressure area is larger when damage is produced.

3.2. Effect of Unloading Rate on Pore Water Pressure

The pore pressure–strain relationship curves of the specimens under different unloading paths are displayed in Figures 11–14. It can be seen from the figures that the confining pressure influences the development of pore pressure: under the same conditions, the larger the confining pressure, the larger the peak pore pressure or the final pore pressure value. The change law for the pore pressure with strain is the result of the combined effect of the pore pressures generated by elastic expansion and plastic deformation during shearing.

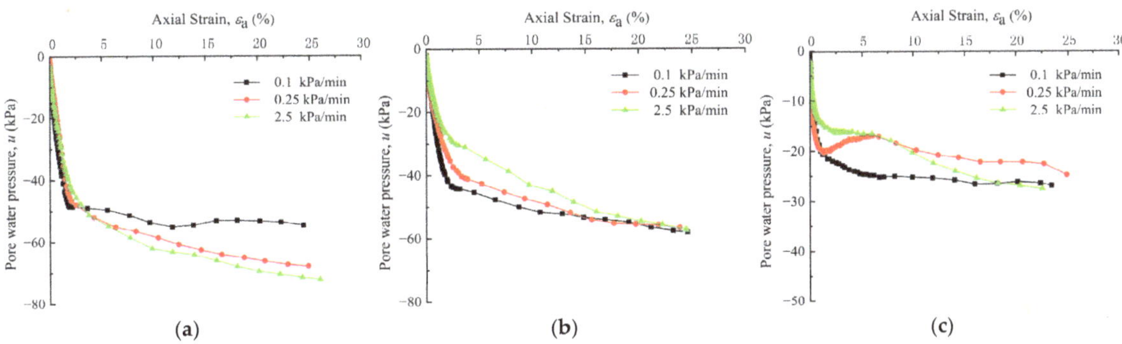

Figure 11. Pore pressure–strain curves for $CU_{-1.5}$ path under different confining pressure. (**a**) 50 kPa; (**b**) 100 kPa; (**c**) 200 kPa.

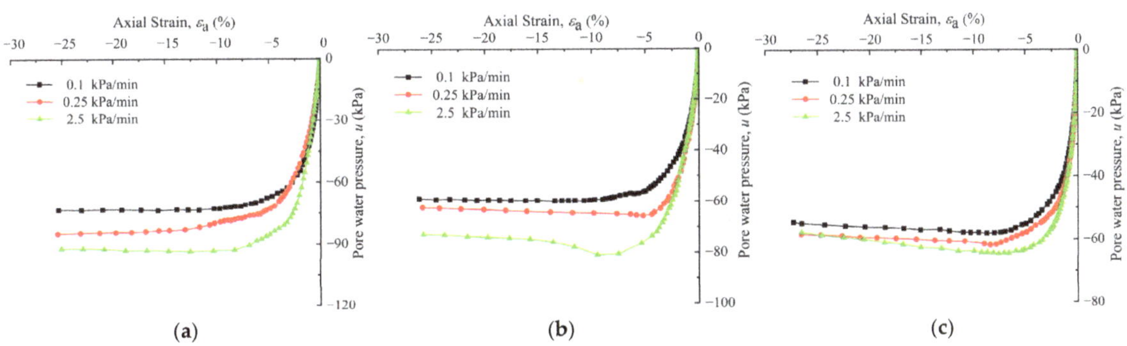

Figure 12. Pore pressure–strain curves for $UU_{1.0}$ path under different confining pressure. (**a**) 50 kPa; (**b**) 100 kPa; (**c**) 200 kPa.

The pore water pressure change laws are quite different under various stress paths. Under the $CU_{-1.5}$ passive compression path, the pore water pressure remained negative with the development of strain. At the beginning of unloading, the pore water pressure accumulated rapidly with an increase in tensile strain, and after the strain exceeded a certain value, the growth of the negative pore water pressure gradually slowed. This is because the negative pore water pressure generated by the elastic expansion at the beginning of unloading was greater than the positive pore water pressure generated by the unloading compression. Thus, the pore water pressure decreased rapidly, and with the increase in the deviatoric stress, the growth rate of the pore water pressure generated by the elastic expansion and plastic deformation of the soil slowed, and the pore water pressure tended to a stable state. Moreover, under the same conditions, the larger the unloading rate, the slower the pore water pressure development at the beginning of unloading. This is because, in the initial stage of unloading, the faster the unloading speed, the less time the pore water pressure generated by elastic expansion has to grow. As the unloading continued, the failure stress of the soil increased, and the final pore water pressure increased.

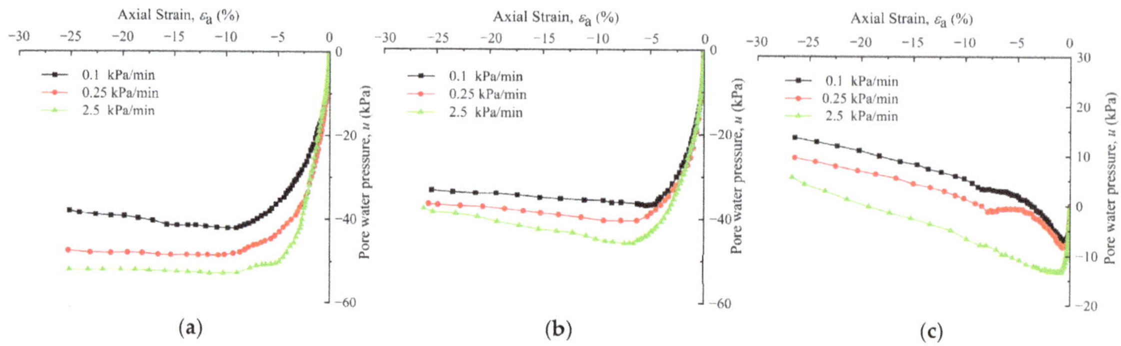

Figure 13. Pore pressure–strain curves for $UC_{3.0}$ path under different confining pressure. (**a**) 50 kPa; (**b**) 100 kPa; (**c**) 200 kPa.

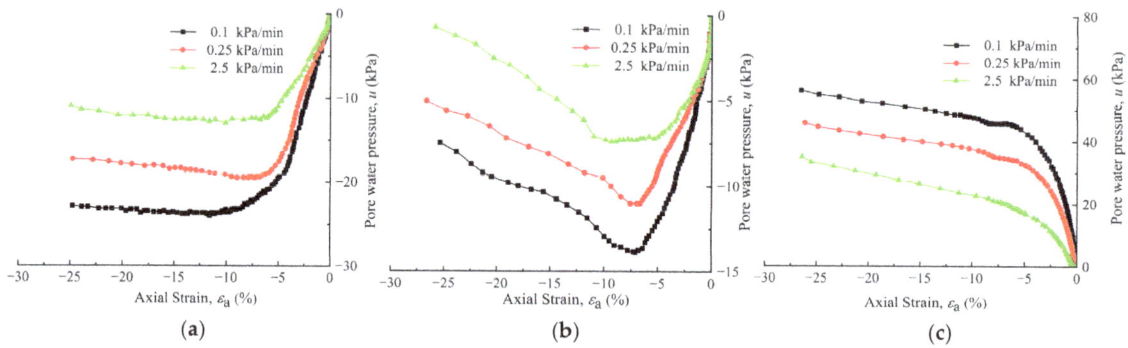

Figure 14. Pore pressure–strain curves for UL_∞ path under different confining pressure. (**a**) 50 kPa; (**b**) 100 kPa; (**c**) 200 kPa.

Under the $UU_{1.0}$ passive tensile path, the development law for the pore water pressure with axial strain during shearing under different confining pressures is essentially the same. Moreover, the pore water pressure generated during the entire unloading process remained negative, and the pore water pressure decreased rapidly with the increase in axial strain, then slowed and gradually stabilized. This is due to the unloading of the lateral stress. The soil sample will have a certain degree of elastic expansion and the pore water pressure will decrease, which will be influenced by the axial stress. Specifically, when the unloading stress path is $UU_{1.0}$, both the axial and lateral stresses decreased throughout the shearing process; thus, the pore water pressure decreased significantly and remained negative.

Under the $UC_{3.0}$ and UL_∞ passive tensile paths, when the confining pressure was small, i.e., 50 and 100 kPa, the negative pore water pressure first increased and then decreased with increasing axial strain during the shearing process. Before reaching the yield point of the soil, the negative pore water pressure generated by the elastic expansion was larger than the positive pore water pressure generated by the plastic deformation; therefore, the pore water pressure first decreased rapidly and then gradually increased, owing to the deviatoric stress. After reaching the yield point, the pore water pressure generated by the plastic deformation caused by the shearing failure gradually became larger than that generated by the elastic expansion. Thus, the pore water pressure increased rapidly, and the pore water pressure–strain curves exhibit a corresponding turning point. Moreover, when the confining pressure was larger (e.g., 200 kPa), the pore water pressure of the specimens changed from negative to positive. The negative pore water pressure generated by the elastic expansion was smaller than the positive pore water pressure generated by the plastic deformation during shearing, and the pore water pressure tended to increase monotonically with the axial strain. Furthermore, the smaller the unloading rate under the same conditions, the more fully the pore pressure develops, and the greater the pore pressure variation.

In summary, during the unloading process, the negative pore pressure gradually increased at the start of unloading, and the axial strain varies when the pore pressure reached its maximum value under different unloading paths. The unloading rate affected the growth path and dissipation rate of negative pore pressure. Owing to the high moisture content and low permeability coefficient of the soft clay, the unloading of the pit excavation will generate a negative excess pore pressure in the pit bottom and surrounding soil. With the dissipation of the negative pore pressure, the effective stress in the soil at the bottom of the pit gradually decreases, which then causes the expansion and uplift of the soil at the bottom of the pit. Therefore, during the unloading construction, attention should be paid to engineering disasters caused by negative pore water pressure dissipation.

3.3. Effect of Unloading Rate on Stress Path

The total and effective stress paths of the Zhanjiang structured clay under different unloading paths are depicted in Figures 15–18. It can be seen from the figures that when the unloading path is the same, the change trends of the total stress paths under different consolidation pressures are basically the same, and all of them develop according to the stress path in the set *p*-*q* plane. The test consistency was good, indicating that the results of this experiment were reliable. Additionally, the development laws for the effective stress paths of the Zhangjiang structured clay under different unloading rates are essentially the same. In the initial shearing, the effective stress path was linear, and the difference in the stress rate was manifested in the stress path by the different ratios of dq/dp', which was at the stage of a large increase in the deviatoric stress in the smaller strain range of the stress–strain curves. Due to the low permeability of clay, there is hysteresis in the growth of the pore water pressure, resulting in small differences in the pore water pressure at different unloading rates; therefore, the effective stress paths overlap in the initial stage of shearing. As shearing continued, the pore water pressure in the specimen began to increase rapidly; therefore, the effective stress path and the total stress path at different stress paths began to separate. Moreover, with decreases in the unloading rate, the deflection range of the effective stress path became continually larger, which coincides with the largest variation range in the pore pressure at the unloading rate of 0.1 kPa/min in the pore water pressure and strain relationship graph.

The effective stress path of the specimen under the $CU_{-1.5}$ and $UU_{1.0}$ unloading stress paths remains on the right side of the total stress path, which is because the pore water pressure of the specimens under these paths is always negative. Additionally, under the $UC_{3.0}$ and UL_{∞} unloading stress paths, the pore water pressure of the specimens is negative at small confining pressure, and the effective stress path is on the right side of the total stress path; however, when the confining pressure is large, the pore water pressure changes from negative to positive with development of the axial strain, and the effective stress path gradually inclines from the right side to the left side.

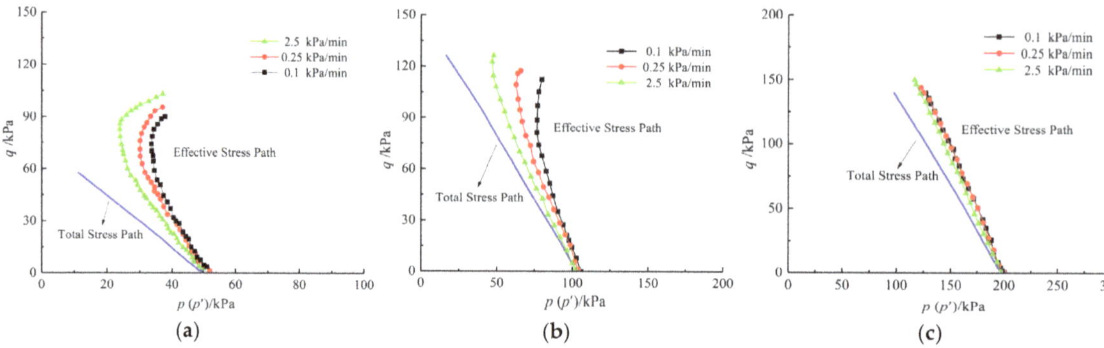

Figure 15. Effective stress path and total stress path under $CU_{-1.5}$ path for different confining. (**a**) 50 kPa; (**b**) 100 kPa; (**c**) 200 kPa.

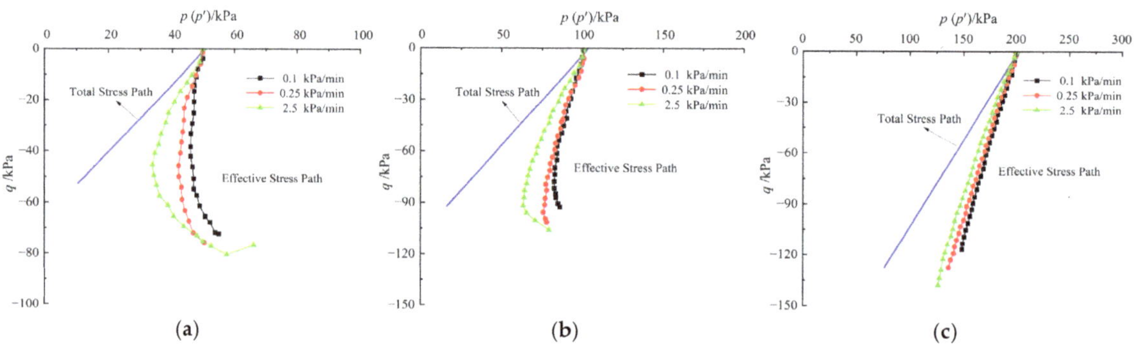

Figure 16. Effective stress path and total stress path under $UU_{1.0}$ path for different confining. (**a**) 50 kPa; (**b**) 100 kPa; (**c**) 200 kPa.

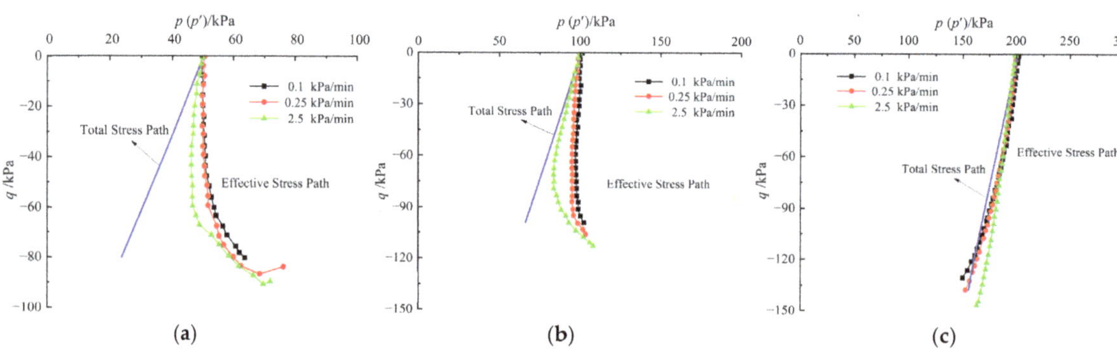

Figure 17. Effective stress path and total stress path under $UC_{3.0}$ path for different confining. (**a**) 50 kPa; (**b**) 100 kPa; (**c**) 200 kPa.

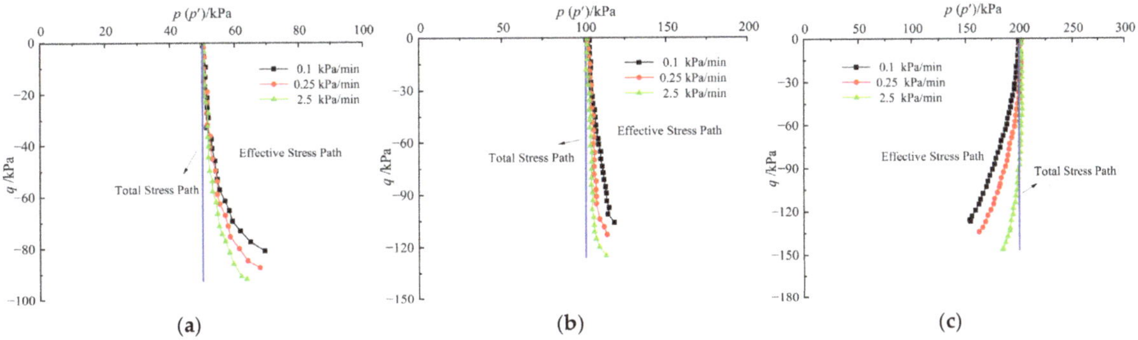

Figure 18. Effective stress path and total stress path under UL_{∞} path for different confining pressure. (**a**) 50 kPa; (**b**) 100 kPa; (**c**) 200 kPa.

3.4. Expressions for Unloading Rate Parameters

The relationship curves for the failure strength and unloading rate of specimens under different unloading paths are presented in Figure 19. It can be observed from the figures that the unloading strength of the specimens varies in magnitude at different consolidation pressures, and the larger the confining pressure, the greater the influence of the unloading rate on the unloading strength of the soil. The unloading strength of the soil under the $UU_{1.0}$ unloading path was the smallest, followed by that under $UC_{3.0}$, and the largest was under the UL_{∞} path. Additionally, under the same unloading path, the failure strength of the soil increased with increasing unloading rate, and the failure strength of the soil increased more when the unloading rate increased from 0.1 to 0.25 kPa/min, while the failure strength of the soil increased relatively less when the unloading rate increased from 0.25 to 2.5 kPa/min.

A linear relationship was determined between the undrained shear strength and the unloading rate. Under the same unloading path, the soil unloading strength increased with the unloading rate, and the cohesion varied significantly with an increase in the unloading rate; however, the internal friction angle did not vary significantly. The cohesion and internal friction of clay do not work simultaneously. Cohesion is characterized by brittle properties and can reach its maximum value under a small strain, whereas internal friction can only be fully exerted after considerable deformation has occurred. Moreover, the cohesion and internal friction exerted different degrees under different strain conditions, which were controlled by the shear rate. In addition, there may be differences in the strain between the cohesion and internal friction that fully functioned under different shear rate conditions. When the unloading rate is small (e.g., 0.1 kPa/min), the cohesion and internal friction of the soil act synchronously in a smaller proportion, and the failure strength is lower. Moreover, when the unloading rate increases to a certain level (e.g., 0.25 kPa/min), the proportion of the cohesion and internal friction of the soil acting synchronously increases, and the soil in the shear process, must overcome the rotation, rolling, and displacement resistance of the particles in addition to overcoming the slippage between the particles, which leads to a greater increase in the failure strength. Moreover, with a further increase in the unloading rate (e.g., 2.5 kPa/min), although it is also necessary to overcome the particle rotation, rolling, and other resistances, the proportion of cohesion and internal friction acting synchronously decreases, and the degree of growth of the failure strength decreases.

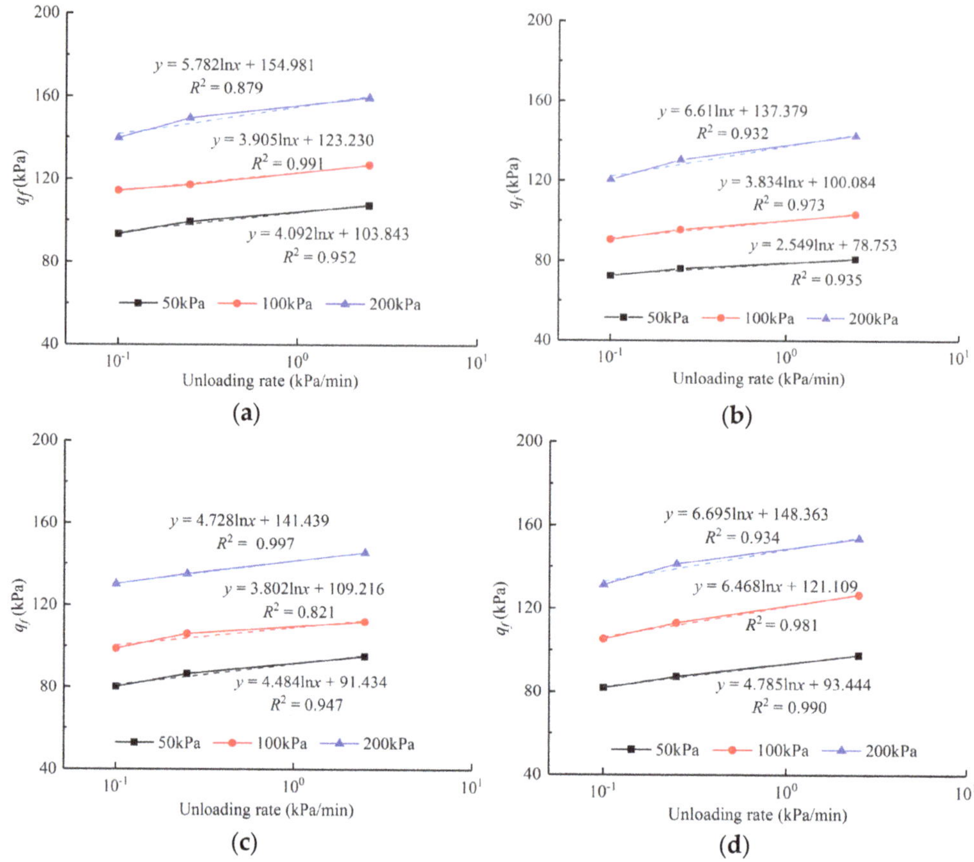

Figure 19. Relationship curves for the failure strength and unloading rate of specimens under different unloading paths. (**a**) $CU_{-1.5}$; (**b**) $UU_{1.0}$; (**c**) $UC_{3.0}$; (**d**) UL_{∞}.

To quantitatively describe the effect of the unloading rate on the undrained strength, we refer to the equation of the undrained shear strength unloading rate proposed by Li et al. [18], expressed as:

$$\rho_{\dot{q}_{a0}} = \frac{\frac{\Delta S_u}{S_{u0}}}{\Delta(\lg \dot{q}_a)} \times 100\% \tag{1}$$

where $\dot{q}_{a0}$ is the reference unloading rate (here, $\dot{q}_{a0} = 0.1$ kPa/min); $\rho_{\dot{q}_{a0}}$ is the unloading rate parameter under the reference unloading rate $\dot{q}_{a0}$; S_{u0} is the undrained shear strength under the reference unloading rate $\dot{q}_{a0}$; ΔS_{u0} represents the increase in undrained strength corresponding to the increase in the unloading rate $\Delta(\lg \dot{q}_a)$ in logarithmic coordinates.

The calculated results for unloading rate parameter $\rho_{0.1}$ under different unloading paths are shown in Table 5. For the consolidation pressures of 50, 100, and 200 kPa, $\rho_{0.1}$ under the $CU_{-1.5}$ passive compression path is 6.09%, 4.22%, and 2.92%, respectively, and the average value is 4.41%. Meanwhile, the average values of $\rho_{0.1}$ under the $UU_{1.0}$, $UC_{3.0}$, and UL_{∞} passive tensile paths and different consolidation pressures are 5.68%, 6.06%, and 9.06%, respectively. This indicates that the change in the unloading rate had a greater effect on the undrained strength under the passive tensile path than under the passive compression path.

Table 5. Shear test results for specimens under different unloading paths.

Unloading Stress Path	σ_0/kPa	Unloading Rate v/kPa·min^{-1}	Failure Strength qf/kPa	Unloading Rate Parameter $\rho_{0.1}$/%	S_u/σ_0
$CU_{-1.5}$	50	0.1	93.58		0.94
	50	0.25	99.36	6.09	0.99
	50	2.5	107.25		1.07
	100	0.1	114.59		0.57
	100	0.25	117.33	4.22	0.59
	100	2.5	126.19		0.63
	200	0.1	139.71		0.35
	200	0.25	143.95	2.92	0.36
	200	2.5	149.50		0.37
$UU_{1.0}$	50	0.1	72.26		0.72
	50	0.25	76.09	4.95	0.76
	50	2.5	80.84		0.81
	100	0.1	93.17		0.47
	100	0.25	100.48	5.90	0.50
	100	2.5	106.36		0.53
	200	0.1	120.01		0.30
	200	0.25	130.52	6.20	0.33
	200	2.5	137.87		0.34
$UC_{3.0}$	50	0.1	80.12		0.80
	50	0.25	86.60	7.82	0.87
	50	2.5	95.15		0.95
	100	0.1	98.88		0.49
	100	0.25	105.94	5.13	0.53
	100	2.5	111.05		0.56
	200	0.1	130.32		0.33
	200	0.25	140.09	5.23	0.35
	200	2.5	146.68		0.37
UL_{∞}	50	0.1	81.87		0.82
	50	0.25	87.43	8.03	0.87
	50	2.5	97.65		0.98
	100	0.1	105.37		0.53
	100	0.25	113.32	9.62	0.57
	100	2.5	129.70		0.65
	200	0.1	129.30		0.32
	200	0.25	146.39	9.52	0.37
	200	2.5	158.84		0.40

The relationship between the normalized undrained shear strength and the unloading rate was obtained by taking unloading paths $CU_{-1.5}$ and $UU_{1.0}$ as examples, as shown in Figure 20. As shown in the figure, the normalized shear strength under the unloading conditions is influenced by the stress path and unloading rate; thus, the mechanical parameters should be selected appropriately in the engineering design.

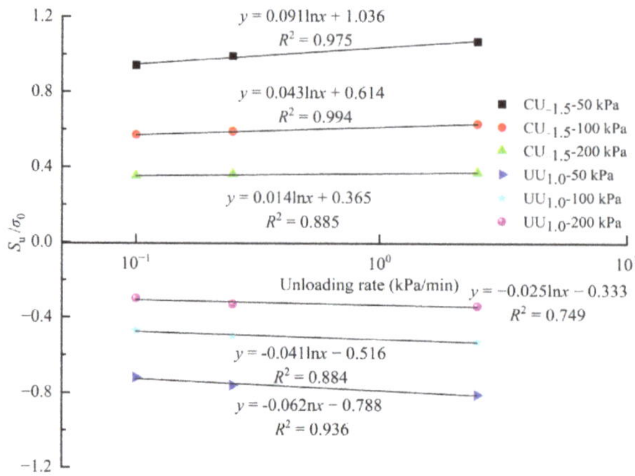

Figure 20. Relationship between normalized undrained shear strength and unloading rate.

4. Microstructure Analysis

An FEI QUANTA 200 environmental scanning electron microscope was used to conduct scanning electron microscopy (SEM) tests near the shear bands of the undisturbed samples and samples after shear tests at different unloading rates. The specimen for the SEM tests was a cube of 10 mm × 10 mm × 10 mm. Before SEM was performed, a fresh section of the sample was sprayed with gold to increase the electrical conductivity. An image with a representative field-of-view was selected for each sample, as shown in Figure 21.

The SEM image of the undisturbed soil in Figure 21a shows curly lamellar mineral particles stacked in sheets. The structure of the Zhanjiang structured clay actually comprises a lamellar-stacked particle unit made of many single lamellae stacked together; specifically, a clay domain consisting of granular clastic minerals and a few monolithic clay mineral particles. The flattened lamellar pile and the single lamellar are dominated by edge–face contacts and edge–edge contacts, while a small amount of face–face contacts form a directional disorderly open flocculation structure. Additionally, the pores in the Zhanjiang structured clay existed mostly in the form of intergranular pores and a small number of isolated pores. The isolated pores were large and discontinuously distributed, mostly in the shape of circles or ellipticals, whereas the intergranular pores appeared as sub-stable polygons.

From Figure 21b, it is clear that when specimens were shear damaged at the rate of 0.1 kPa/min, the arrangement between the basic units became closer due to the slow unloading rate. It is also apparent that the arrangement of soil particles maintained some lamellar single-sheet stacks, and some of the clay mineral aggregates broke into fine soil particles. Between the particles, there were mainly face–face contacts, a small number of edge–edge contacts, and the arrangement was not directional. Moreover, intragranular pores remained, and some of the pores gradually broke into smaller ones.

From Figure 21c, it can be seen that when the specimens were shear damaged at the rate of 0.25 kPa/min, with an increase in the unloading rate, the breakage of soil particles became more significant than that at 0.1 kPa/min, and the soil particles comprised a small number of lamellar forms and broken fine particles.

As shown in Figure 21d, when the specimens were shear-damaged at the rate of 2.5 kPa/min, owing to the fast unloading rate, the soil skeleton of the specimen was rapidly destroyed by shearing. The flocculation structure in the natural state disappeared, replaced by the face-to-face and point-to-face contacts of the large-volume unit body, and the soil particles of the specimen became finer.

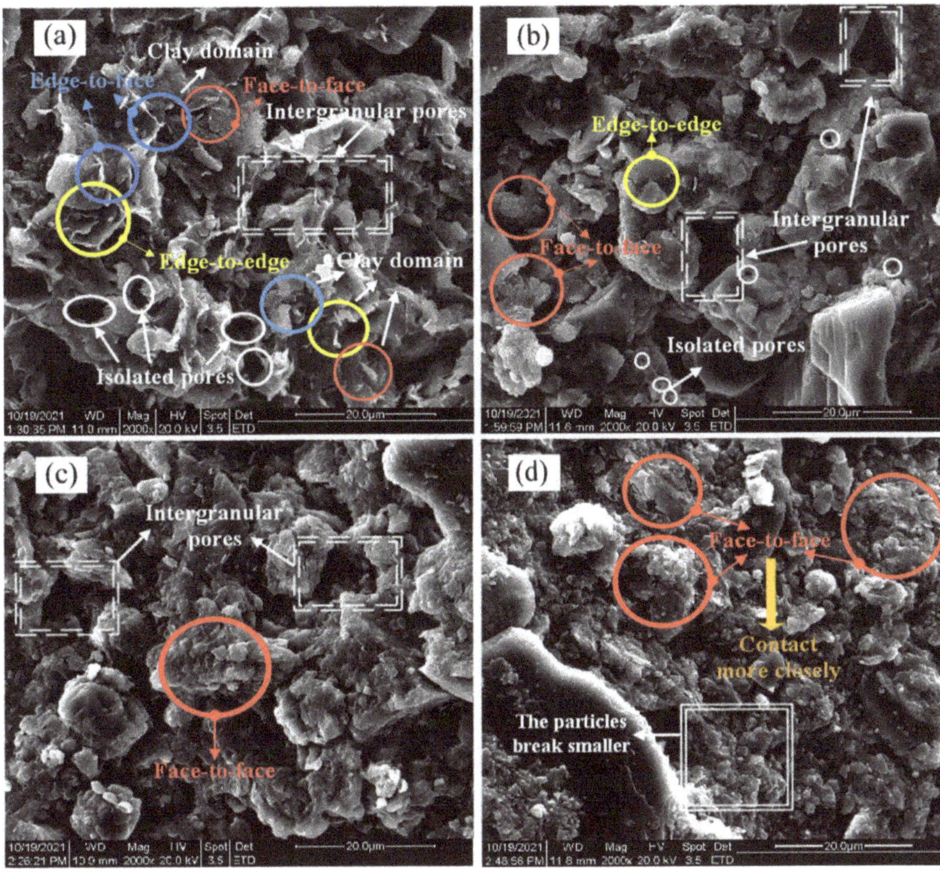

Figure 21. SEM images of specimens at different unloading rates under $CU_{-1.5}$, σ_0 = 50 kPa: (**a**) Undisturbed sample; (**b**) 0.1 kPa/min; (**c**) 0.25 kPa/min; (**d**) 2.5 kPa/min.

The microstructures of the Zhanjiang structured clay changed after shearing tests under different unloading rates, and the clay had no obvious sheet-pile-structured units owing to the destruction of connections, and the amount of intragranular pores decreased.

5. Conclusions

- The stress–strain relationship curves of the Zhanjiang structured clay exhibited strain softening under unloading states. Under the same axial strain, the greater the unloading rate, the greater the deviatoric stress. Under the same unloading path, the larger the confining pressure, the more significant the influence of the unloading rate on the soil. Moreover, the degree of deformation of the specimen under different unloading paths varied: the failure mode of the specimen under the $CU_{-1.5}$ passive compression path was axial compression failure, while those of the specimens under the $UU_{1.0}$, $UC_{3.0}$, and UL_{∞} passive tensile paths were axial tensile failure.
- The pore water pressure characteristics of Zhanjiang structured clay are closely related to the unloading rate and unloading path. Under the $CU_{-1.5}$ and $UU_{1.0}$ unloading paths, the development of the pore water pressure with strain remained negative, and the effective stress path remained on the right side of the total path. While under the $UC_{3.0}$ and UL_{∞} unloading paths, the development of the pore water pressure of specimens significantly differed due to the influence of the confining pressure.

Furthermore, under the same conditions, the greater the confining pressure, the greater the peak pore water pressure, the smaller the unloading rate, the more fully the pore water pressure developed, and the greater the range of pore water pressure change. Therefore, during unloading construction, attention should be paid to engineering disasters caused by negative pore water pressure dissipation.

- The undrained shear strength of the Zhanjiang structured clay exhibited a semi-logarithmic linear increasing relationship with increasing unloading rate. Introducing unloading rate parameter $\rho_{0.1}$, the undrained strength under the $CU_{-1.5}$ passive compression path increased by 4.41% as the unloading rate increased from 0.1 to 2.5 kPa/min. The increases in undrained strength under the $UU_{1.0}$, $UC_{3.0}$, and UL_{∞} passive tensile paths were 5.68%, 6.06%, and 9.06%, respectively. Thus, the influence of the unloading rate on the passive tensile paths is greater than that on the passive compression paths.
- The microstructure of the strong structured clay sample from Zhanjiang changed after shear tests at different unloading rates. The particle arrangement, contact relations, pore size, and shape varied to different degrees. During shearing, the Zhanjiang structured clay continuously self-adjusts its internal structure.

Author Contributions: L.L.: conceptualization, methodology, data curation, and writing—original draft. M.Z.: conceptualization, methodology, writing—review and editing, and investigation. R.Z.: methodology and investigation. H.L.: funding acquisition. All authors have read and agreed to the published version of the manuscript.

Funding: This work was funded by the National Natural Science Foundation of China (Grant No. 11802215) and the Natural Science Foundation of Hubei Province of China (Grant No. 2022CFA011), and the authors are grateful for this financial support.

Data Availability Statement: If readers require the data from this article, please contact the corresponding author.

Conflicts of Interest: The authors declare no conflict of interest.

References

1. Huang, W.; Li, J.J.; Lu, Y.M.; Li, D.S.; Mou, Y.Q.; Wu, X.H.; Jiang, Z.J.; Li, Z.Y. Mechanical properties of soft soil considering the influence of unloading stress paths. *Adv. Civ. Eng.* **2021**, *2021*, 8813882. [CrossRef]
2. Huang, W.; Wen, K.J.; Li, J.J.; Fu, X.M.; Liu, S.H.; Li, Y.; Li, L.; Amini, F. Mechanical properties of soft soils experiencing lateral unloading under initial excess pore water pressure. *Arab. J. Geosci.* **2019**, *13*, 718. [CrossRef]
3. Han, J.; Yin, Z.Y.; Dano, C.; Hicher, P.Y. Effect of strain rate on the adhesive bond shearing resistance of stiff clay. *Transp. Geotech.* **2021**, *27*, 100479. [CrossRef]
4. Torisu, S.S.; Pereira, J.M.; De Gennaro, V.; Delage, P.; Puech, A. Strain-rate effects in deep marine clays from the Gulf of Guinea. *Géotechnique* **2012**, *62*, 767–775. [CrossRef]
5. Le, T.; Airey, D.; Standing, J.R. Influence of shearing strain rate on the mechanical behaviour of three structured clays. *Géotechnique* **2022**, 1–56. [CrossRef]
6. Lambe, T.W. Stress path method. *J. Soil Mech. Found.* **1967**, *93*, 268–277. [CrossRef]
7. Liu, G.B.; Hou, X.Y. Unloading stress-strain characteristics of soft soils. *Tunnel Rail Transit.* **1997**, *2*, 16–23. (In Chinese)
8. Yan, S.W.; Zhang, J.J.; Tian, Y.H.; Sun, L.Q. Pore pressure characteristics in isotropic consolidated saturated clay under unloading conditions. *J. Mar. Sci. Tech.* **2016**, *24*, 3. [CrossRef]
9. Lim, A.; Ou, C.Y. Stress paths in deep excavations under undrained conditions and its influence on deformation analysis. *Tunn. Undergr. Sp. Tech.* **2017**, *63*, 118–132. [CrossRef]
10. Jia, M.C.; Luo, W.S.; Zhou, Y.H.; Zhao, S.; Zhang, Z. A model of undrained stress-strain curves considering stress path and strain softening. *Int. J. Geomech.* **2021**, *21*, 04021213. [CrossRef]
11. Huang, W.; Wen, K.J.; Liu, D.Y.; Dong, Q.; Li, J.J.; Li, Y.; Li, L. Experimental study on influence of excess pore water pressure and unloading ratio on unloading mechanical properties of marine sedimentary soft soils. *Ocean Eng.* **2020**, *195*, 106680. [CrossRef]
12. Buisman, A.S. Results of long duration settlement tests. In Proceedings of the 1st International Conference on Soil Mechanics and Foundation Engineering, Cambridge, UK, 22–26 June 1936; Volume 1, pp. 103–107.
13. Sheahan, T.C.; Ladd, C.C.; Germaine, J.T. Rate-dependent undrained shear behavior of saturated clay. *J. Geotech. Eng.* **1996**, *122*, 99–108. [CrossRef]
14. Whitman, R.V.; Richardson, A.M. Effect of strain-rate upon undrained shear resistance of a saturated remoulded fat clay. *Géotechnique* **1963**, *13*, 310–324. [CrossRef]

15. Zhu, J.G.; Yin, J.H. Strain-rate-dependent stress-strain behavior of overconsolidated Hong Kong marine clay. *Can. Geotech. J.* **2000**, *37*, 1272–1282. [CrossRef]
16. Wu, S.S.; Lok, T.; Xu, Y.W.; Wang, D. Rate-dependent behaviour of a saturated clay under different stress histories and K_0 conditions. *Acta Geotech.* **2020**, *15*, 1501–1512. [CrossRef]
17. Cai, Y.; Kong, L.W.; Guo, A.G.; Tuo, Y.F. Effects of shear strain rate on mechanical behavior of Zhanjiang strong structured clay. *Rock Soil Mech.* **2006**, *27*, 1235–1240. (In Chinese)
18. Li, X.M.; Kong, L.W.; Guo, A.G. Experimental study of the influence of unloading rate on the shear mechanical properties of undisturbed expansive clay. *Rock Soil Mech.* **2019**, *40*, 3758–3766. (In Chinese)
19. Yang, A.W.; Yang, S.K.; Zhang, Z.D. Experimental study of mechanical properties of dredger fill under different unloading rates and stress paths. *Rock Soil Mech.* **2020**, *41*, 2891–2912. (In Chinese)
20. Song, F.N.; Huang, X.; Luo, T.T.; Zou, J.Q.; Fu, R. Strain energy evolution and damage characteristics of deep clay under different stress rates. *J. Cent. South Univ.* **2022**, *29*, 2005–2018. [CrossRef]
21. Zhang, X.W.; Wang, G.; Liu, X.Y.; Xu, Y.Q.; Kong, L.W. Microstructural analysis of pore characteristics of natural structured clay. *Bull. Eng. Geol. Environ.* **2022**, *81*, 473. [CrossRef]
22. An, R.; Kong, L.W.; Shi, W.Z.; Zhang, X.W. Stiffness decay characteristics and disturbance effect evaluation of structured clay based on in-situ tests. *Soils Found.* **2022**, *62*, 101184. [CrossRef]
23. Li, L.; Zang, M.; Zhang, R.T.; Lu, H.J. Deformation and strength characteristics of structured clay under different stress paths. *Math. Probl. Eng.* **2022**, *2022*, 9266206. [CrossRef]
24. Zang, M.; Kong, L.W.; Zhang, R.T.; Wang, W.W. Study on characteristics of shear modulus of Zhanjiang structured clay during triaxial shearing. *Geotech. Geol. Eng.* **2021**, *39*, 603–616. [CrossRef]
25. *ASTM D2216*; Standard Test Methods for Laboratory Determination of Water (Moisture) Content of Soil and Rock by Mass. American Society for Testing and Materials: West Conshohocken, PA, USA, 2019.
26. *ASTM D854*; Standard Method of Test for Specific Gravity of Soils. American Society for Testing and Materials: West Conshohocken, PA, USA, 2019.
27. *ASTM D4318*; Standard Test Methods for Liquid Limit, Plastic Limit, and Plasticity Index of Soils. American Society for Testing and Materials: West Conshohocken, PA, USA, 2017.
28. *ASTM D1586*; Standard Method of Test for Penetration Test and Split-Barrel Sampling of Soils. American Society for Testing and Materials: West Conshohocken, PA, USA, 2018.
29. *ASTM D2166*; Standard Method of Test for Unconfined Compressive Strength of Cohesive Soil. American Society for Testing and Materials: West Conshohocken, PA, USA, 2016.
30. Touiti, L.; Bouassida, M.; Van, I.W. Discussion on the Tunis soft clay sensitivity. *Geotech. Geol. Eng.* **2009**, *27*, 631–643. [CrossRef]
31. *ASTM D4767*; Standard Test Method for Consolidated Undrained Triaxial Compression Test for Cohesive Soils. American Society for Testing and Materials: West Conshohocken, PA, USA, 2020.

Disclaimer/Publisher's Note: The statements, opinions and data contained in all publications are solely those of the individual author(s) and contributor(s) and not of MDPI and/or the editor(s). MDPI and/or the editor(s) disclaim responsibility for any injury to people or property resulting from any ideas, methods, instructions or products referred to in the content.

buildings

MDPI

Article

Research on Dynamic Pile-Driving Formula Parameters and Driving Feasibility of Extra-Long PHC Pipe Piles

Xiaomin Liu, Yonggang Xiao *, Junlong Zhou, Longbo Ge and Ziwen Song

Technical Center, China Construction Sixth Engineering Bureau Corp., Ltd., Tianjin 300012, China
* Correspondence: xiaoyonggang2010@163.com

Abstract: Prestressed high-strength concrete (PHC) pipe pile has the advantages of high single pile bearing capacity, a wide range of applications, good driving resistance, fast construction speed, etc. It has been widely used in high-rise buildings, bridges, ports, and other industries. The application of extra-long PHC pipe piles with a length of more than 50 m is increasing. However, there are few studies on the drivability and hammering criteria of extra-long PHC piles. To analyze the drivability of extra-long piles and predict their bearing capacity, in this paper, high-strain dynamic tests were carried out on 14 test sections with the pile foundation of Temburong Bridge in Brunei as the research background. The hammer stop control criteria calculated according to the Hiley formula would lead to excessive hammering. Three types of damage occurred during construction: pile shaft breakage, weld tearing, and pile head breakage. The weight and drop height of the piling hammer selected for this project were appropriate, and the extra-long test piles can be hammered to the design depth. The values of C_p (Compression of the pile) and n (the efficiency of the blow) were fitted based on the dynamic test data, which provided a more accurate reference for the selection of subsequent piling parameters of the project. It provides a more accurate calculation method for predicting the bearing capacity of extra-long PHC piles and provides control criteria for pile stopping and a scientific basis for their design and construction.

Keywords: PHC pipe piles; driving feasibility; dynamic pile-driving formula; high-strain dynamic test

1. Introduction

A pile load test can control the piling quality and verify whether the pile design is reasonable, and promote the continuous development of foundation design and construction practice [1–8]. The traditional prestressed high-strength concrete (PHC) pipe pile has the advantages of the high bearing capacity of a single pile, wide application range, good driving resistance, and fast construction speed. At present, it has been widely used in high-rise buildings, bridges, ports, and other industries [9–15].

High-strain dynamic testing has been widely used in engineering piles, has achieved good results, and has also been generally recognized by the engineering community. During the whole process of PHC pile driving, the pile-driving wave equation can be used to simulate and calculate important information such as the bearing capacity of a single pile, the tensile and compressive stress of the pile body, and the number of blows, so as to help the designer verify and determine the pile type and bearing capacity required by the project [16–20]. After the high-strain pile-driving test, the feasibility analysis can be carried out to obtain the estimated values of the tensile and compressive stress change of the pile body, pile-driving resistance, and bearing capacity [21]. The feasibility analysis can simulate the pile stress during the pile-driving process and provide a basis for the selection of pile-driving equipment. In addition to detecting the integrity of the pile body, the high-strain method can also detect the vertical compressive bearing capacity of a single pile. Zhang et al. carried out high strain detection and static load tests on the pile foundation of an offshore wind power project [22]. The results showed that the number of simulated hammer blows by the high-strain method was similar to the actual hammer blows, which proved the reliability of the high-strain piling procedure to analyze the vertical compressive load-bearing capacity of pile foundations. Liu et al. collected the high-strain method and static load method test results of 76 PHC pipe piles in the field for comparison [23]. The results showed that 84.2% of the tested piles had the bearing capacity error of the high-strain method controlled within ±10%. Some of the bearing capacity measured by the high-strain method was greater than the bearing capacity of the static load method, and some were less than the bearing capacity of the static load method. The probability of these two situations occurring was roughly the same. Therefore, the high-strain method can accurately test the piling effect of engineering piles, and it can be used to optimize the pile design and piling process in the pile test stage and engineering piling stage.

Using the relevant measured data of the high-strain method, the relevant parameters of the Hiley formula can be calculated. A typical expression of the Hiley formula is as follows:

$$P_u = WHn/(S + C/2) \tag{1}$$

Zhang et al. compared the predicted geotechnical pile-bearing capacity using the Hiley formula and the high-strain method [24], and the results showed that the pile-bearing capacity values derived from the Hiley formula were consistently lower than those derived from the high-strain method and static load tests. Triantafyllidis proposed a simple method to calculate the mass of moving piles under impact action and compared the results of this method with an example of long pile construction in Hong Kong. Four formulas including Hiley's formula were used to estimate the pile-bearing capacity by Arsyad et al. The results showed that the calculated values of Hiley's formula were underestimated. Four formulas including Hiley's formula were used to estimate the pile-bearing capacity by Arsyad et al. [25]. The results showed that the calculated values of Hiley's formula were underestimated. Tokhi proposed a method to improve the widely used Hiley kinetic equation, which can evaluate the bearing capacity of piles more accurately [26].

In summary, it can be seen that the Hiley formula does not accurately predict the bearing capacity of piles, and for extra-long piles, the predicted values are even more inaccurate. There are few studies on the drivability and hammering criteria of extra-long PHC piles. It is essential to explore the pile-driving formula applicable to using hydraulic hammers to drive extra-long piles. In order to analyze the drivability of super-long piles and predict their bearing capacity, in this paper, high-strain dynamic tests were carried out on 14 test sections with the pile foundation of Temburong Bridge in Brunei as the research background.

The Temburong Sea-crossing Bridge in Brunei crosses Brunei Bay, which connects the Brunei-Muara area and the Temburong area, as shown in Figure 1. This line is 30 km in length, of which the CC4 bid section is 12 km in length, including 940 spans of fully prefabricated viaducts, 25 spans of partially prefabricated viaducts, and 3 spans of cast-in-place continuous beam bridges. The CC4 bid section crosses the virgin forest protection area. There are many swamps, and the weak layer is thicker. Most of the bridge bases are precast tubular piles, and their pile numbers are P269 to P1209. The rocks near P1209 are relatively shallow. The outer diameter of the prefabricated pipe piles is 900 mm, and the inner diameter is 620 mm, with a total of 7536. Its total length is about 5.5×10^5 m, the average length is about 72 m, and the longest is 90 m. All piles were driven by hydraulic hammering.

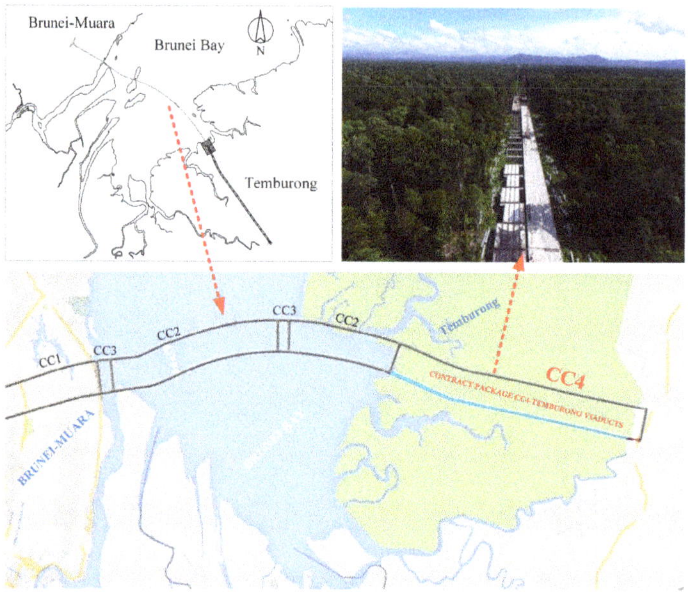

Figure 1. The geographical location of the Brunei Temburong bridge.

2. Materials and Methods

A test section of piling was carried out prior to the piling of the Brunei Temburong Bridge. High-strain dynamic load tests were carried out for extra-long PHC pipe piles to test and record the pile penetration, elastic deformation, bearing capacity, hammering efficiency, and other parameters. The pile number of the prefabricated pipes pile was P269 to P1237, and 14 test sections were set at different locations. There were 14 piers in the test section, marked as PPT1–PPT14. All piles in the test section were tested using the high-strain dynamic method during the hammering and sinking of the piles. The location and work direction of 14 test piles is shown in Figure 2. The whole line was divided into 4 working faces for construction. They were P269–P516, P754–P517, P755–P989, and P1237–P990. The P269, P754, and P1237 were three starting working faces.

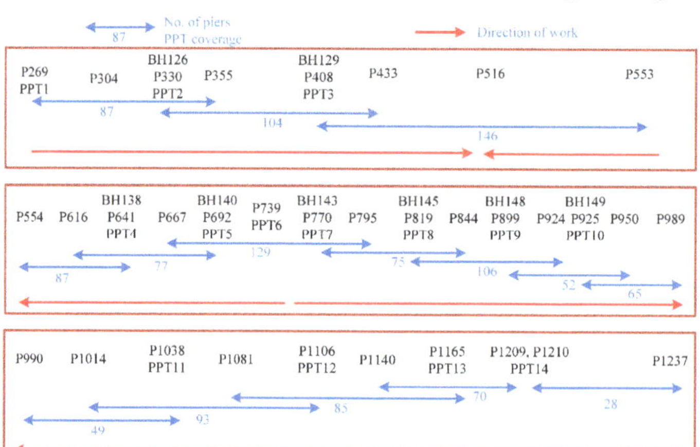

Figure 2. The location and work direction of 14 test piles.

2.1. Piling Dynamic Load Tests [27]

Dynamic load tests are used to provide on-site estimates of the bearing capacity of piles. The Pile Driving Analyzer (PDA) system is specially used for dynamic load testing and drive monitoring.

Dynamic monitoring is achieved by installing two sets of sensors at the pile head. Strain and acceleration sensors (two of each type) are installed below the pile head. These sensors are directly installed on the opposite sides of the pile above the ground. After connecting the strain and acceleration sensors to the pile head and PDA system, these sensors record the relationship between the pile top force and velocity with time when the pile head is vertically hammered. This information is stored for future redisplay and further analysis.

The PDA program records parameters including estimated driving resistance, estimated static resistance, energy transmitted through sensors, and many other parameters. The details of the test pile are shown in Table 1. From the table, it can be seen that the hammers that provide impact force for dynamic load tests are hydraulic hammers of 200 kN and 160 kN, with a drop distance of 0.6 m to 1.2 m and a pile length of 39 m to 80 m.

Table 1. The details of the test piles.

Pile Number	Pile Length (m)	AS-Build Toe Level (m)	Holding Layer	Hammer Weight (kN)	Drop Height (m)	DVL (kN)
PPT1	78	−70.0	Stiff clay	200	0.9	3000
PPT2	80	−76.0	Hard clay	200	0.9	3000
PPT3	80	−75.5	Dense sand	200	0.9	3000
PPT4	70	−65.3	Stiff clay	200	0.9	2600
PPT5	68	−60.7	Dense sand	200	0.9	2600
PPT6	78	−69.3	Dense sand	200	1.2	2600
PPT7	59	−52.3	Dense sand	200	0.9	2600
PPT8	72	−64.7	Stiff clay	200	0.9	2600
PPT9	70	−59.2	Medium dense sand	200	0.9	2600
PPT10	69	−60.2	Stiff clay	200	0.9	2600
PPT11	59	−50.4	Hard clay	160	0.9	2400
PPT12	55	−45.0	Hard clay	160	0.9	2400
PPT13	58	−44.2	Stiff clay	160	0.9	2400
PPT14	39	−34.4	Mudstone	160	0.6	2400

Where DVL is Design Verification Load.

The instruments and equipment for this high-strain test included hammering equipment, sensors, signal acquisition, and analyzers. The connection diagram of testing instruments and equipment is shown in Figure 3. The hammering equipment used was the hammer used for driving PHC pipe piles. The installation of two acceleration sensors and two strain sensors met the specification requirements.

The CAPWAP analysis program uses wave equations to analyze pile-driving parameters, which are used to calculate pile movement and stress during the pile-driving process. Using the CAPWAP program, the force and speed of hammering recorded in the selected field were used as the input of signal matching. The CAPWAP program established a pile-soil model to analyze and recorded pile-top force/time and pile-top velocity/time parameters. The model of the pile consisted of a block mass, springs, dampers, and surrounding soil. The CAPWAP program calculated parameters such as force and velocity of the model pile, and then compared them with actual measured parameters such as force and velocity. If the difference between the two was large, the pile model and the soil resistance model were continuously adjusted until a good match was obtained. Once the matching was achieved, a good dynamic model of the pile can be obtained. Then, this dynamic model will be used to determine the distribution of pile parameters along the length direction of the pile, and the CAPWAP program can be used to predict the load and settlement of the pile.

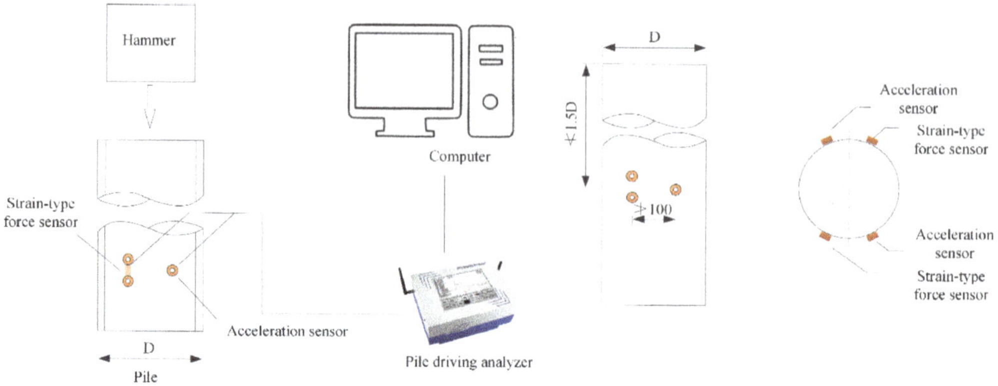

Figure 3. The connection diagram of testing instruments and equipment.

In this test, the base piles of 14 piers were performed as a high-strain dynamic test. The test parameters of the test piles are shown in Table 1.

2.2. The Calculation of the Final Set (S)

Among many pile-driving formulas provided in the literature, the Hiley formula is the most commonly used pile-driving formula at present. When it is impossible to determine the stopping standard by field pile test, the reference value of final penetration can be determined according to the calculation results of the Hiley formula and the driving experience value of similar pile foundation conditions. The Hiley formula is expressed as follows:

$$P_u = \frac{k \times W \times H}{S + (C_c + C_p + C_q)/2} \times \frac{W + e^2 \times W_p}{W + W_p} \quad (2)$$

The formula needs to measure the temporary compression (C_p) of the pile and the temporary compression value (C_q) of the soil mass at the end of the last 10 blows at the end of pile driving. The maximum bearing capacity of the pile can be calculated according to the actual pile length and the measured elastic compression value of the pile and soil (C_p + C_q). If the design bearing capacity of the pile is not greater than the calculated value, the required bearing capacity of the pile is considered to be met.

In this project, when the final set reaches the preset value, in order to conduct high-strain dynamic tests, two sensors were installed symmetrically at the top of the piles. Through the sensor, the force and acceleration of piles can be obtained under dynamic loads. According to PDA analysis of related data, it is easy to obtain factors such as pile stress, hammer energy, and ultimate bearing capacity of piles. According to the data obtained by the dynamic measurement, the parameter η value in the Hiley formula (the effective transfer coefficient of the hammer energy) and the C (the total elastic deformation of the pile-soil system) value can be calculated.

The final set (S) of this project was determined by the Hiley formula, and its expression was as follows:

$$S = \frac{kWH(W + e^2 W_p)}{P_u(W + W_p)} - \frac{C_c + C_p + C_q}{2} \quad (3)$$

In this table, k is the mechanical efficiency of the hammer, which represents the reduction coefficient of falling hammer efficiency [28–30]. When the hammer is an ideal free-fall hammer, k = 1; when the hammer is not freely falling, k < 1, usually k = 0.8~0.9. e is the coefficient of restitution, defined as the ratio of the initial velocity and the final velocity of the material after impact. e is 0.25 for reinforced-concrete piles with helmets, and e is 0.5 for steel piles. However, for PHC pipe piles, the specifications do not specify how to select e. Based on engineering experience, e is taken between 0.35 and 0.5.

Take the pipe pile with a pile length of 76 m and a bearing capacity of 4620 kN as an example to calculate the S value under different falling distances. During the S value calculation, multiple parameters need to be assumed. For this project, it is assumed that k is 0.8, e is 0.35, and H is 1.0–1.5 m. The calculation results are shown in Table 2. The E' is effective hammer energy per blow, and its value is kWH. The C is $C_c + C_p + C_q$. It can be seen from the table that the calculation value of S is 36.7 mm to 45.7 mm.

Table 2. Calculation of the final set of piles at a different drop height.

Length (m)	H (m)	E' (kN.m)	C (mm)	S (mm)	Notes
76	1.0	128.0	7	44.8	P_u = 4620 kN W = 160 kN W_p = 689 kN k = 0.8 e = 0.35
76	1.1	140.8	10	37.7	
76	1.2	153.6	10	45.7	
76	1.3	166.4	12	43.7	
76	1.4	179.2	15	36.7	
76	1.5	192.0	16	39.6	

In order to clearly clarify the relationship between the final set and the height of the falling hammer, Formula (2) is represented by different curves, as shown in Figure 4. It can be seen from the figure that for piles with the same parameters such as L, k, e, and other parameters, the final set increases with the height of the falling hammer. The calculation value of the final set decreases with the elastic deformation of the pile-soil system. However, for on-site piling, it is very inconvenient to use Table 2 or Figure 4 to determine the final set. After the parameters of P_u, k, and e are given assumptions, it is very convenient to make a form to query the final set, as shown in Table 3. This form can query the final set under different pile lengths and different elastic compression.

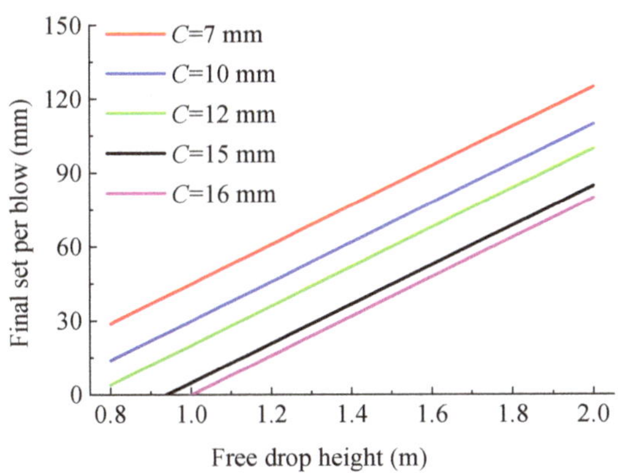

Figure 4. Relationship between final set per blow and free drop height.

Table 3 is applicable to W = 200 kN, H = 1.0 m, and P_u = 4620 kN. The set values are obtained based on assumed $(C_p + C_q)$ = (9–18) mm. The assumed $(C_p + C_q)$ value will be verified when the actual $(C_p + C_q)$ value is obtained on-site later. The C_c value takes 3 mm and k takes 0.8. When a driving cap with cushion is used, e is 0.35. In the table, the value of n is $(W + W_p \times e^2)/(W + W_p)$. Taking L = 76 m as an example, the calculated final set is 35 mm if $C_p + C_q$ = 12 mm is assumed.

Table 3. Proposed set values table based on pile lengths and selected ($C_p + C_q$).

Pile Length (m)	n	Temporary Compression ($C_p + C_q$ mm)									
		9	10	11	12	13	14	15	16	17	18
50	0.387						49	44	39	34	29
51	0.384						47	42	37	32	27
52	0.381						46	41	36	31	26
53	0.377					50	45	40	35	30	25
54	0.374					49	44	39	34	29	
55	0.371					48	43	38	33	28	
56	0.368					47	42	37	32	27	
57	0.365					46	41	36	31	26	
58	0.362				50	45	40	35	30	25	
59	0.359				49	44	39	34	29		
60	0.357				48	43	38	33	28		
61	0.354				47	42	37	32	27		
62	0.351				46	41	36	31	26		
63	0.349			50	45	40	35	30	25		
64	0.346			49	44	39	34	29			
65	0.344			49	44	39	34	29			
66	0.341			48	43	38	33	28			
67	0.339			47	42	37	32	27			
68	0.337			46	41	36	31	26			
69	0.334		50	45	40	35	30	25			
70	0.332		50	45	40	35	30	25			
71	0.330		49	44	39	34	29				
72	0.328		48	43	38	33	28				
73	0.326		47	42	37	32	27				
74	0.324		47	42	37	32	27				
75	0.322		46	41	36	31	26				
76	0.320	50	45	40	35	30	25				
77	0.318	50	45	40	35	30	25				
78	0.316	49	44	39	34	29					
79	0.314	48	43	38	33	28					
80	0.312	48	43	38	33	28					

In the Hiley formula, n is the efficiency of the blow, which represents the ratio of energy after impact to the striking energy of the ram [31,32]. The expression is as follows:

$$n = \frac{W + e^2 W_p}{W + W_p} = \frac{1 + e^2 W_p / W}{1 + W_p / W} \tag{4}$$

When $W > W_p e$ or

$$n = \frac{W + e^2 W_p}{W + W_p} - \left(\frac{W - e W_p}{W + W_p} \right)^2 \tag{5}$$

When $W < W_pe$. For extra-long PHC pipe piles, the selection of parameter n is not applicable. Therefore, it is necessary to study a simple and applicable method for obtaining n values.

3. Results and Discussion

3.1. Results and Analysis of Dynamic Load Pile-Driving Tests

When the final set meets the hammer requirements, the pile analyzer (referred to as PDA) was used for dynamic testing to test whether the bearing capacity of the pile reaches the design standard. Otherwise, it is necessary to continue driving the pile until the bearing capacity meets the design requirements.

The dynamic load test was used to provide field estimates of the mobilized static load-carrying capacity of the piles. In addition, it can be used to check pile structural integrity and to obtain field data for later computer signal matching to determine capacity and soil resistance. The S and $(C_p + C_q)$ values were determined by conventional manual measurement methods. The pile penetration record was a curve drawn by a pencil where the workers were near the target pile. The pencil tracked the motion of the pile on a piece of paper under the driver's hammer, as shown in Figure 5. The on-site staff explained the figure to check whether the piling results were in accordance with the final penetration derived from the pile formula. Otherwise, the piling must be repeated and tested until the results are satisfactory. If the actual penetration is not greater than the calculated value of S, it is considered that the penetration of the pile meets the requirements. Taking the PPT1 pile as an example, the calculated S value according to the formula is 57 mm, while as shown in Figure 5, the actual recorded S value during pile driving is 48 mm. Obviously, the penetration of this pile meets the calculation requirements.

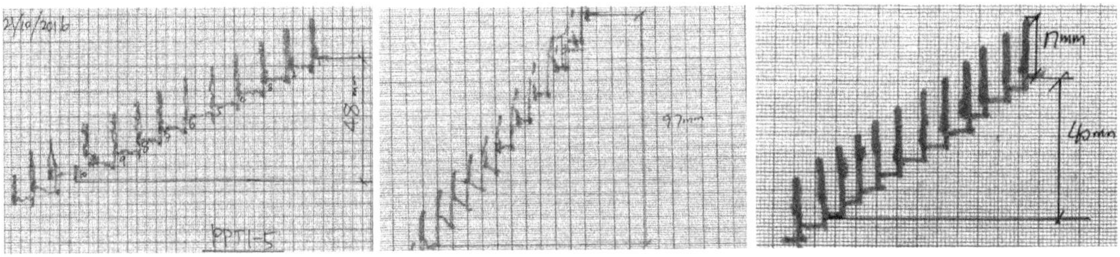

Figure 5. Piling penetration record of partial test piles.

The dynamic test results of the test piles are shown in Table 4.

Table 4. The test results of the test piles.

Pile Number	S (mm)	$C_p + C_q$ (mm)	CSX (MPa)	TSX (MPa)	EMX (kN·m)	P_u (kN)	ETR (%)
PPT1	48	15	15.6	3.8	45	4726	31.8
PPT2	177	18	28.2	4.8	96	3677	53.3
PPT3	115	12	24.0	3.0	59	3966	32.8
PPT4	57	10	31.1	1.4	120	5414	66.7
PPT5	78	12	28.0	1.1	103	7337	57.2
PPT6	81	16	41.8	2.8	184	6560	76.4
PPT7	97	13	35.3	3.6	152	5368	84.4
PPT8	40	17	36.7	5.1	150	5327	83.3
PPT9	54	14	25.7	1.7	104	4658	57.8
PPT10	39	15	30.0	3.1	114	6349	63.3
PPT11	33	11	28.8	7.4	68	4764	47.2
PPT12	42	9	35.9	4.3	137	5030	95.1
PPT13	102	10	37.8	4.0	122	5445	84.7
PPT14	35	10	14.4	1.3	86	4761	89.6

Where: EMX is Maximum Energy Transferred to the pile; CSX is Maximum Compression Stress; TSX is Maximum Tension Stress; ETR is Hammer Efficiency.

It can be seen from Table 4 that the maximum value of the final entry S value is 177 mm, and the minimum value is 33 mm. According to the design requirements, the ultimate compressive bearing capacity of the piles should be greater than 2 times the DVL. It can be seen from the table that the P_u value of most piles is greater than 2 times DVL. After 28 days of rest, the ultimate bearing capacity of all piles can meet the design requirements. It can be seen from Table 4 that the maximum value of the final set (S) value is 177 mm, and the minimum value is 33 mm. The changing range of the S value is large, and the distribution is more discrete. The variation of the $C_p + C_q$ value is from 9 to 18 mm, and its distribution is relatively concentrated. In addition, under the same driving conditions, the hammering efficiency varies greatly. In addition, there was a large difference in hammering efficiency under the same driving conditions.

3.2. *Analysis of Driving Feasibility*

The maximum compressive stress of the pile shaft refers to the maximum compressive stress generated by the hammering force at the top of the pile. There are many influencing factors, including pile length, pile section, pile materials, soil, pile cushion, pile hammer weight, and drop height. Reasonable control of the maximum stress in the pile can prevent the pile from being cracked. The maximum value of CSX is 41.8 MPa, and the maximum value of TSX is 7.4 MPa. The concrete grade used for the PHC pipe piles in this project was C50. None of the maximum compressive stresses exceeded the ultimate compressive strength of concrete, while the maximum tensile stresses in some of the piles exceeded the ultimate tensile strength of concrete. Under repeated hammering action, some of the pile heads or pile bodies were damaged. A total of three types of damage were found to have occurred during construction by means of PDA tests and visual observation: pile shaft breakage, weld tearing, and pile head breakage, as shown in Figure 6. The occurrence of these failures during the pile testing process was caused by various reasons, including hammer deflection, insufficient penetration, and unsuitable pile pads. The pile hammer was hydraulically driven, with the oil pipe hanging on one side, causing the hammer body to tilt due to unilateral pulling. The axis of the hammer deviates significantly from the axis of the PHC pipe pile, causing severe shaking of the pile body during construction. In addition, some parameters in the Hiley formula have assumed values that differ significantly from the actual values, resulting in the calculated penetration S value being too small. Furthermore, some pile cushion materials used paper cushions, which contained 17 layers of paper. Through practical testing, the effectiveness of the paper cushion is not good, and it is easy to burn under high impact temperatures, which can cause uneven thickness of the pile cushion, resulting in an uneven hammering force on the pile head.

(a) (b) (c)

Figure 6. Failure types of the PHC pipe pile driving. (**a**) Pile shaft breakage; (**b**) Weld tearing; (**c**) Pile head breakage.

Before piling, every 0.25 m on the pile body was marked, and the number of blows per 0.25 m would be automatically recorded by the instrument. The number of blows in each group of test piles had a good correlation, and the number of blows in the same group of test piles was very close. The relationship between the stratigraphic parameters; the number of pile-driving hammer blows; and the depth for PPT1, PPT6, and PPT14 piles is shown in Figure 7. In general, the hammer blow per unit pile length increased with the increase in depth. However, it can be seen from Figure 7 that the number of hammer blows per unit pile length does not simply increase with depth. The geological conditions of each pile were different, and the corresponding Standard Penetration Test (SPT) value curve was also different. As can be seen from the figure, when the geological conditions are hard clay, dense sand, and weathered rock, the SPT value increases significantly, and the corresponding number of pile-driving hammer blows also increases significantly. The hammer weight of the PPT1 pile was 200 kN, the drop height was 0.7 m, the pile bottom elevation was −70 m, the hammer stop penetration S was 4.8 mm/blow, the total number of blows was 4728, the maximum compressive stress was 15.6 MPa, the maximum tensile stress was 3.8 MPa, and the compressive bearing capacity was 4726 kN. With this hammer type, the test pile PPT1 can be basically driven in place. The hammer weight of the PPT6 pile was 200 kN, the height of hammer fall was 1.2 m, the elevation of the pile bottom was −69.3 m, the stopping hammer penetration S was 8.1 mm/hammer blow, the total number of hammer blows was 5134, the maximum compressive stress was 41.8 MPa, the maximum tensile stress was 2.8 MPa, and the compressive bearing capacity was 6560 kN. The hammer weight of the PPT14 pile was 160 kN, the height of hammer fall was 0.6 m, the elevation of the pile bottom was −34.4 m, the stopping hammer penetration S was 3.5 mm/hammer blow, the total number of hammer blows was 1889, the maximum compressive stress was 14.4 MPa, the maximum tensile stress was 1.3 MPa, and the compressive bearing capacity was 4761 kN. It can be seen from the figure that the number of blows per meter corresponding to different depths is closely related to the properties of corresponding soil layers. In hard clay, dense sand, and weathered rock strata, the number of blows per unit pile length increases significantly.

3.3. Parameter Analysis of the Hiley Formula

Several parameters in Hiley's formula (such as hammer efficiency k, coefficient of restitution e, compression of pile cushion C_c, pile-soil elastic deformation ($C_p + C_q$)) were determined by assumptions, hence the final hammer penetration S obtained by Hiley's formula may differ significantly from the actual situation. The comparison between the measured and calculated values of the ultimate compressive bearing capacity of the single pile is shown in Table 5. The S_m in the table is the measured value, while S_p is the proposed set of values for stopping hammering. The S_p was calculated using Hiley's formula based on the design values of the bearing capacity and the assumed parameters. Most of the piles at the construction site did not meet the penetration requirements calculated according to the Hiley formula for stopping the hammer. As can be seen from Table 5, S_m is greater than S_p, especially since the difference between the two values of the PPT6 pile is large. The P_{um} in the table is the ultimate value of compressive bearing capacity measured by dynamic load tests, while P_{ud} is the design value. As can be seen from the table, although the final stopping hammering set S_m does not reach the requirement of S_p calculated according to Hiley's formula, the P_{um} values are all greater than the P_{ud} values and meet the design bearing capacity requirements. In order to meet the requirements of S_p calculated by the Hiley formula, excessive hammering would occur during pile driving on site, resulting in damage to the pile head, pile tip, and pile body.

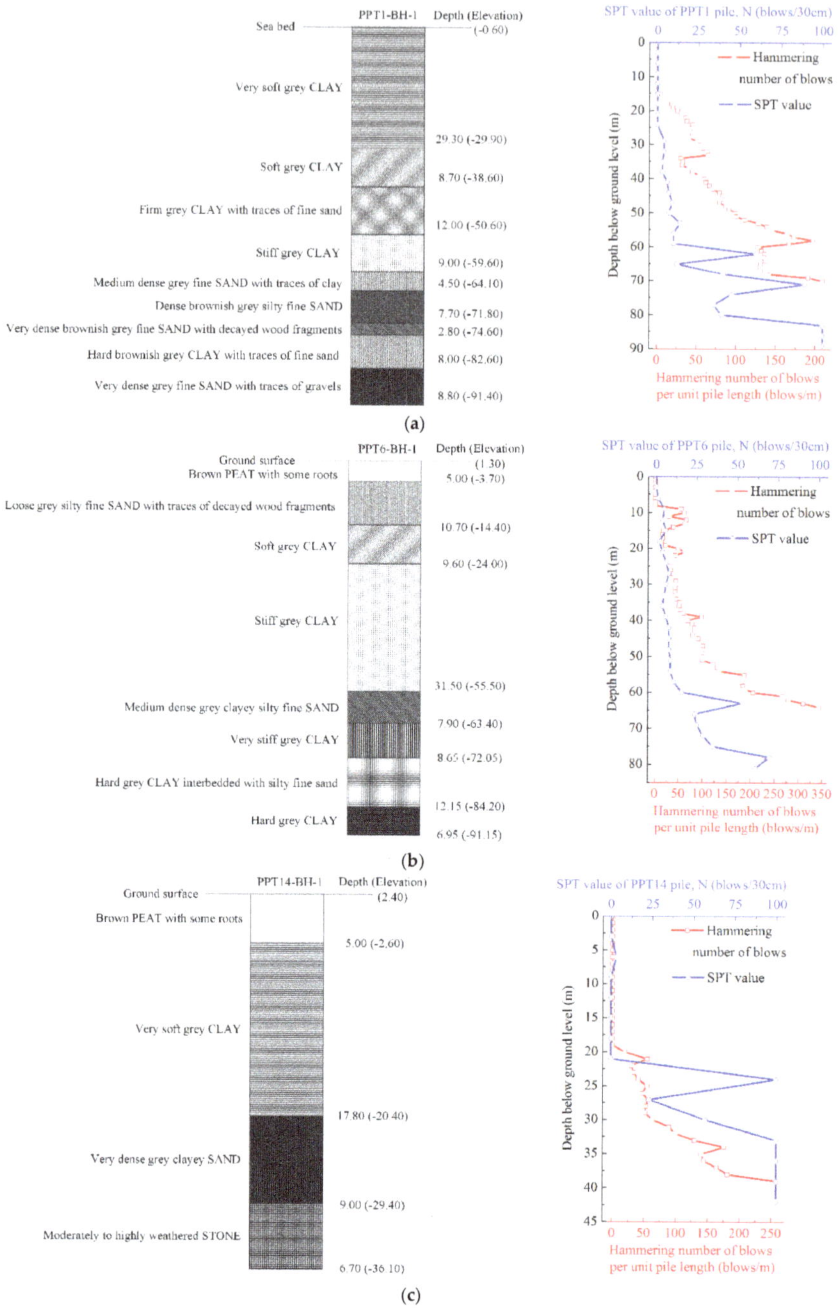

Figure 7. Strata parameter and monitoring of pile-driving blow count; (**a**) PPT1; (**b**) PPT6; (**c**) PPT14.

Table 5. Comparison between measured and calculated values of ultimate compressive bearing capacity of a single pile.

Pile No.	L (m)	H (m)	S_m (mm)	S_p (mm)	$C_p + C_q$ (mm)	C_c (mm)	P_{um} (kN)	P_{ud} (kN)	Percent Difference (%)
PPT6-1	78	1.2	81	21	14	3	6720	5200	29.23
PPT6-3	65	1.1	86	18	23	3	5980	5200	15.00
PPT6-5	65	1.1	114	18	16	3	6580	5200	26.54
PPT14-1	37	1.0	99	86	15	3	5190	4400	17.95
PPT14-2	39	0.6	24	21	13	3	4950	4400	12.50
PPT14-3	39	0.6	35	21	10	3	5230	4400	18.86

In order to analyze the differences in the values taken for the parameters of the Hiley formula, a comparison of the calculated and actual values of C_c, C_{p+q}, and e is presented in Table 6. According to the Hiley formula, C_c was taken as 3 mm in the calculation of this project. The C_c-C in the table is the positive arithmetic value, while C_c-B is the inverse arithmetic value introduced from the piling results. The Pd-C_c is the percentage difference between C_c-C and C_c-B. As can be seen from the table, the difference between the two values is relatively large, especially for the PPT14 pile, the C_c-B is surprisingly negative. The C_{p+q}-C in the table is the positive arithmetic value, while C_{p+q}-M is the dynamic load test measured value. The Pd-C_{p+q} is the percentage difference between C_{p+q}-C and C_{p+q}-M. The minimum value of Pd-C_{p+q} is 9.69% and the maximum value is 48.57%. According to the Hiley formula, e was taken as 0.5 in the calculation of this project. The e-C in the table is the positive arithmetic value, while e-B is the inverse arithmetic value introduced from the piling results. The Pd-e is the percentage difference between e-C and e-B. The minimum value of Pd-e is 25.60% and the maximum value is 72.20%. According to the Hiley formula parameter-taking method published by the British Institution of Civil Engineers, the values of C_c and C_q were determined by the pile material and the difficulty of pile driving. The value of C_p was taken not only related to these two factors but also related to the pile length. The values of C_c and C_q were small, and the ratio of C_c and C_q to all temporary compression was very low. The easier it was to sink the pile, the smaller the values of these two values were. The value of C_p was taken to be proportional to the pile length and accounted for a high ratio of all temporary compressions. Most of the PHC pipe piles in this project were extra-long piles, of which more than 93% were over 50 m in length. The pile length has a great influence on the C_p value, however, there are very few studies on the C_p value of extra-long PHC pipe piles.

Table 6. Comparison of calculated and actual values for C_c, C_{p+q}, and e.

Pile No.	C_c-C (mm)	C_c-B (mm)	Pd-C_c (%)	C_{p+q}-C (mm)	C_{p+q}-M (mm)	Pd-C_{p+q} (%)	e-C	e-B	Pd-e (%)
PPT6-1	3	5.23	74.33	20.80	14	48.57	0.5	0.759	51.80
PPT6-3	3	6.41	113.67	17.55	23	23.70	0.5	0.857	71.40
PPT6-5	3	0.84	72.00	17.55	16	9.69	0.5	0.796	59.20
PPT14-1	3	−3.15	205.00	10.55	15	29.67	0.5	0.628	25.60
PPT14-2	3	−10.67	455.67	11.05	13	15.00	0.5	0.668	33.60
PPT14-3	3	−7.15	338.33	11.05	10	10.50	0.5	0.861	72.20

The pile lengths and C_p values of all test piles in this project were counted and the relationship between the two was linearly fitted, as shown in Figure 8. The correlation coefficient R^2 is 0.7953, and the correlation between the two is good. As can be seen from Figure 8, for extra-long PHC pipe piles over 50 m, the relationship between C_p and pile length L is:

$$C_p = 0.36L - 11.37 \tag{6}$$

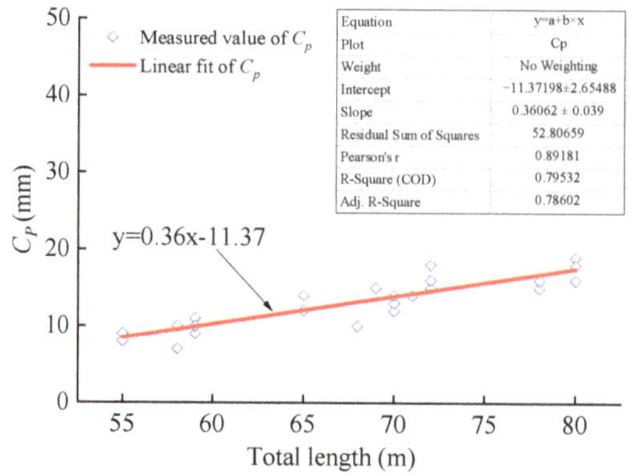

Figure 8. Fitting relationship between C_p and pile length.

The value of n is related to e and W_p/W, where e is the coefficient of restitution, and W_p/W is the ratio of pile weight to hammer weight. In this project, the minimum value of W_p/W of piles with a length of more than 50 m was 2.71, and the maximum value was 3.62. Within such a small range of variation, W_p/W had little influence on the value of n, corresponding to a minimum value of 0.41 and a maximum value of 0.45 for n. Therefore, a small range of W_p/W values has little effect on n values.

For piles with a longer length, the value of coefficient n in the Hiley formula is smaller. It means that the energy transferred from the hammer to the pile is very small. The reason is that the quality of the whole pile is considered. If the pile is relatively long, only the part of the pile body affected by the hammer should be considered during the process of bearing the hammer. However, it is very difficult to determine the length of the part of the pile body affected by the hammer. The coefficient of restitution and hammering efficiency have a very close correlation. The hammering efficiency data of the test piles in this project were counted and the relationship between them was fitted, as shown in Figure 9. The correlation coefficient R^2 is 0.9843, and the correlation between the two is good. As can be seen from Figure 9, for extra-long PHC pipe piles over 50 m, the relationship between n and e is:

$$n = 111.94e - 15.43 \tag{7}$$

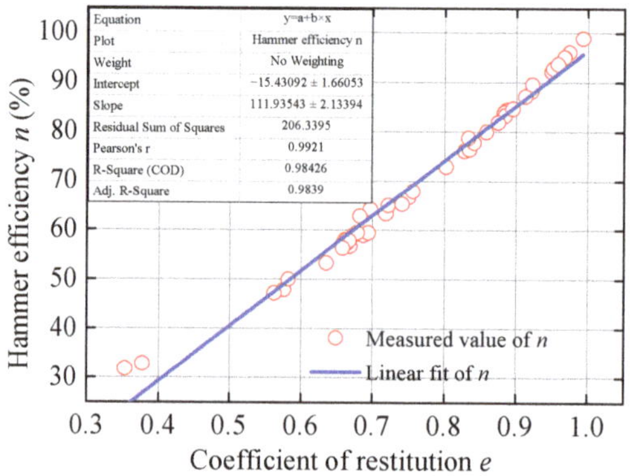

Figure 9. Fitting the relationship between hammer efficiency and the coefficient of restitution.

The value of the coefficient of restitution, e, was determined by the pile material and the piling conditions. According to the Hiley formula parameter value method published by the British Society of Civil Engineers, the value of e is 0.4 when the PHC pipe pile was driven by the single-acting hammer pile driver, and the value of e is 0.5 with a double-acting hammer. However, the coefficient of restitution of most piles in this project was greater than 0.5. The number of e greater than 0.5 accounts for 95.65% of all test piles. Obviously, this is quite different from the initial assumed value of e. With the development of piling technology, in addition to the steam hammer and diesel hammer, the hydraulic piling hammer has been widely used. The piling equipment used in this project was the hydraulic hammer. The hydraulic hammer does not emit any waste gas, has no noise, and has a high impact frequency. Each impact can achieve greater penetration. However, the value of e is not clear when the hydraulic hammer is used for the extra-long pile. Generally, in order to determine the penetration and verify whether the piling equipment, construction technology, and technical measures are suitable, the piling technology test must be carried out before the pile-driving construction. Therefore, the selection of the e value of super-long piles should be determined according to the test data of different pile lengths, different geological conditions, and different pile sinking techniques.

4. Conclusions

In this study, the extra-long PHC pipe pile was taken as the research object, and its driving resistance and pile-driving parameters were tested on the construction site. The following conclusions can be drawn.

- The hammer stop control criteria calculated according to the Hiley formula would lead to excessive hammering, with some piles having maximum tensile stresses exceeding the ultimate tensile strength of the concrete. Three types of damage occurred during construction: pile shaft breakage, weld tearing, and pile head breakage;
- The weight and drop height of the piling hammer selected for this project were appropriate, and the extra-long test piles can be hammered to the design depth. The number of blows per meter corresponding to different depths was closely related to the properties of corresponding soil layers. In hard clay, dense sand, and weathered rock strata, the number of blows per unit pile length increased significantly;
- In order to avoid pile damage caused by excessive hammering, the driving parameters of extra-long piles should be adjusted according to the high-strain dynamic test data. In this paper, the values of C_p and n were fitted based on the dynamic test data, which provided a more accurate reference for the selection of subsequent piling parameters of the project.

Author Contributions: Conceptualization, Y.X. and X.L.; methodology, Y.X. and X.L.; writing—review and editing, Y.X., X.L., J.Z., L.G. and Z.S.; project administration and funding acquisition, X.L. All authors have read and agreed to the published version of the manuscript.

Funding: This research was funded by the Science and Technology Research and Development Project of China State Construction Engineering Group Co., Ltd., grant number CSCEC-2020-Z-21.

Data Availability Statement: The data used to support the findings of this study are included in the article.

Conflicts of Interest: The authors declare no conflict of interest.

Nomenclature

P_u	ultimate bearing capacity of pile
k	mechanical efficiency of hammer
W, W_p	weight of hammer and pile
H	free drop height
e	coefficient of restitution
e-C	e value used for calculating S_p
e-B	backcalculation value of e based on test
S	penetration of pile per hammer blow

S_m	measured value of penetration
S_p	proposed value of penetration
C	summation of the temporary compression
C_c	compression of pile cushion
C_c-C	assumed value of C_c used for calculating S_p
C_c-B	backcalculation value of C_c based on test
C_p	compression of pile
C_q	compression of quake
C_{p+q}	sum of C_p and C_q
C_{p+q}-C	C_{p+q} value used for calculating S_p
C_{p+q}-M	measured value of C_{p+q} based on test
n	efficiency of the blow

References

1. Abadie, C.N.; Byrne, B.W.; Houlsby, G.T. Rigid pile response to cyclic lateral loading: Laboratory tests. *Geotechnique* **2019**, *69*, 863–876.
2. Flynn, K.N.; McCabe, B.A. Driven cast-in-situ pile capacity: Insights from dynamic and static load testing. *Can. Geotech. J.* **2021**, *58*, 1870–1883. [CrossRef]
3. Han, F.; Ganju, E.; Salgado, R.; Prezzi, M. Comparison of the load response of closed-ended and open-ended pipe piles driven in gravelly sand. *Acta Geotech.* **2019**, *14*, 1785–1803. [CrossRef]
4. Kou, H.L.; Chu, J.; Guo, W.; Zhang, M.Y. Pile load test of jacked open-ended prestressed high-strength concrete pipe pile in clay. *Proc. Inst. Civ. Eng. -Geotech. Eng.* **2018**, *171*, 243–251. [CrossRef]
5. Lanyi-Bennett, S.A.; Deng, L.J. Axial load testing of helical pile groups in glaciolacustrine clay. *Can. Geotech. J.* **2019**, *56*, 187–197. [CrossRef]
6. McAdam, R.A.; Byrne, B.W.; Houlsby, G.T.; Beuckelaers, W.; Burd, H.J.; Gavin, K.G.; Igoe, D.J.P.; Jardine, R.J.; Martin, C.M.; Wood, A.M.; et al. Monotonic laterally loaded pile testing in a dense marine sand at Dunkirk. *Geotechnique* **2020**, *70*, 986–998. [CrossRef]
7. Santrac, P.; Susic, N.; Bajic, Z. Determination of the Pile Stiffness Matrix Based on the Pile Load Test Results and the Effect of Pile Interaction. *Teh. Vjesn.-Tech. Gaz.* **2020**, *27*, 2016–2023.
8. Wang, S.; Lei, X.W.; Meng, Q.S.; Xu, J.L.; Wang, M.Z.; Guo, W. Model tests of single pile vertical cyclic loading in calcareous sand. *Mar. Georesources Geotechnol.* **2021**, *39*, 670–681. [CrossRef]
9. Cao, X.L.; Dai, G.L.; Gong, W.M.; Zhu, M.X.; Tang, J.A. Experimental study on the seismic behavior of new PHC piles. *Arab. J. Geosci.* **2020**, *13*. [CrossRef]
10. Huynh, V.H.; Nguyen, T.; Nguyen, D.P.; Nguyen, T.S.; Huynh, T.M.D.; Nguyen, T.C. A novel direct SPT method to accurately estimate ultimate axial bearing capacity of bored PHC nodular piles with 81 case studies in Vietnam. *Soils Found.* **2022**, *62*, 101163. [CrossRef]
11. Kou, H.L.; Chu, J.; Guo, W.; Zhang, M.Y. Field study of residual forces developed in pre-stressed high-strength concrete (PHC) pipe piles. *Can. Geotech. J.* **2016**, *53*, 696–707. [CrossRef]
12. Kou, H.L.; Guo, W.; Zhang, M.Y. Field study of set-up effect in open-ended PHC pipe piles. *Mar. Georesources Geotechnol.* **2017**, *35*, 208–215. [CrossRef]
13. Li, X.S.; Zhang, J.Q.; Xu, H.; Shi, Z.W.; Gao, Q.F. Static Load Test and Numerical Analysis of Influencing Factors of the Ultimate Bearing Capacity of PHC Pipe Piles in Multilayer Soil. *Sustainability* **2021**, *13*, 13166. [CrossRef]
14. Nguyen, T.; Ly, K.D.; Nguyen-Thoi, T.; Nguyen, B.P.; Doan, N.P. Prediction of axial load bearing capacity of PHC nodular pile using Bayesian regularization artificial neural network. *Soils Found.* **2022**, *62*, 101203. [CrossRef]
15. Oktiovan, Y.P.; Otaki, T.; Obara, T.; Kono, S.; Asai, Y.; Kobayashi, K.; Watanabe, H.; Mukai, D. Shear performance evaluation of PHC piles under different levels of axial load ratio. *Earthq. Eng. Struct. Dyn.* **2022**, *51*, 2091–2112. [CrossRef]
16. Meng, Q.J.; Shi, Z.F. Vibration Isolation of Plane Waves by Periodic Pipe Pile Barriers in Saturated Soil. *J. Aerosp. Eng.* **2019**, *32*, 11. [CrossRef]
17. Salgado, R.; Zhang, Y.B.; Abou-Jaoude, G.; Loukidis, D.; Bisht, V. Pile driving formulas based on pile wave equation analyses. *Comput. Geotech.* **2017**, *81*, 307–321. [CrossRef]
18. Wu, W.B.; El Naggar, M.H.; Abdlrahem, M.; Mei, G.X.; Wang, K.H. New interaction model for vertical dynamic response of pipe piles considering soil plug effect. *Can. Geotech. J.* **2017**, *54*, 987–1001. [CrossRef]
19. Zheng, C.J.; Ding, X.M.; Sun, Y.F. Vertical Vibration of a Pipe Pile in Viscoelastic Soil Considering the Three-Dimensional Wave Effect of Soil. *Int. J. Geomech.* **2016**, *16*, 10. [CrossRef]
20. Zheng, C.J.; Liu, H.L.; Ding, X.M.; Zhou, H. Vertical vibration of a large diameter pipe pile considering transverse inertia effect of pile. *J. Cent. South Univ.* **2016**, *23*, 891–897. [CrossRef]
21. Sun, Y.; Gao, Y.; Zhang, J.; Hu, J.; Huang, W. Feasibility analysis and field test on driving phc pile through deep dense sand. *J. Eng. Geol.* **2015**, *23*, 1079–1084.

22. Zhang, H.; He, R.; Chen, Q.; Xiao, J.; Chu, Y. Comparative evaluation of theoretical method and in situ tests method on vertical compressive bearing capacity of super-large-diameter monopiles for offshore wind turbines. *Ocean Eng.* **2022**, *40*, 104–115.
23. Liu, C.; Tang, M.; Hu, H.; Chen, H.; Liu, C.; Hou, Z. Comparative study on vertical compressive bearing behavior of PHC pipe piles based on high strain method and static load method. *Build. Struct.* **2022**, *52*, 136–140.
24. Zhang, L.; Chau, A.C.W.; Shek, M.P. Bearing capacity of long driven piles using dynamic methods. In Proceedings of the 12th Asian Regional Conference on Soil Mechanics and Geotechnical Engineering (12ARC), Singapore, 4–8 August 2003.
25. Arsyad, A.; Suyadi; Samang, L. Statistical Analysis of Dynamic Bearing Capacity of Driven Pile. In Proceedings of the 2nd International Conference on Civil, Offshore and Environmental Engineering (ICCOEE), Kuala Lumpur, Malaysia, 3–5 June 2014.
26. Tokhi, H.; Ren, G.; Xie, Y.M. Improving the Predictability of Hiley Pile Driving Formula in the Quality Control of the Pile Foundation. In Proceedings of the International Conference on Civil Engineering and Building Materials (CEBM), Kunming, China; 2011.
27. Xiao, Y.; Liu, X.; Zhou, J.; Song, L. Field Test Study on the Bearing Capacity of Extra-Long PHC Pipe Piles under Dynamic and Static Loads. *Sustainability* **2023**, *15*, 5161. [CrossRef]
28. Chellis, R.D. *Pile Foundations*; McGraw Hill: New York, NY, USA, 1961.
29. Chan, W.T.; Chow, Y.K.; Liu, L.F. Neural network: An alternative to pile driving formulas. *Comput. Geotech.* **1995**, *17*, 135–156. [CrossRef]
30. Goh, A.T.C. Back-propagation neural networks for modeling complex systems. *Artif. Intell. Eng.* **1995**, *9*, 143–151. [CrossRef]
31. El-Mossallamy, Y.M.; Tawfik, M.M.; Zayed, M.A. Application of ultimate limit state design for axially loaded single piles in Egyptian geotechnical practice. *HBRC J.* **2015**, *13*, 181–189. [CrossRef]
32. Sun, L.; Jia, T.; Yan, S.; Guo, W.; Ren, Y.; Lei, Z. Prediction of pile running during the driving process of large diameter pipe piles. *Ocean Eng.* **2016**, *128*, 48–57. [CrossRef]

Disclaimer/Publisher's Note: The statements, opinions and data contained in all publications are solely those of the individual author(s) and contributor(s) and not of MDPI and/or the editor(s). MDPI and/or the editor(s) disclaim responsibility for any injury to people or property resulting from any ideas, methods, instructions or products referred to in the content.

MDPI AG
Grosspeteranlage 5
4052 Basel
Switzerland
Tel.: +41 61 683 77 34

Buildings Editorial Office
E-mail: buildings@mdpi.com
www.mdpi.com/journal/buildings

Disclaimer/Publisher's Note: The title and front matter of this reprint are at the discretion of the Guest Editors. The publisher is not responsible for their content or any associated concerns. The statements, opinions and data contained in all individual articles are solely those of the individual Editors and contributors and not of MDPI. MDPI disclaims responsibility for any injury to people or property resulting from any ideas, methods, instructions or products referred to in the content.

www.ingramcontent.com/pod-product-compliance
Lightning Source LLC
Chambersburg PA
CBHW041938260926
51850CB00018B/265
9783725846672